LONDON MATHEMATICAL SOCIETY LECTURE NOTE SERIES

Managing Editor: Professor N.J. Hitchin, Mathematical Institute,
University of Oxford, 24–29 St. Giles, Oxford OX1 3LB, United Kingdom

The title below are available from booksellers, or from Cambridge University Press at www.cambridge.org

161 Lectures on block theory, BURKHARD KÜLSHAMMER
163 Topics in varieties of group representations, S.M. VOVSI
164 Quasi-symmetric designs, M.S. SHRIKANDE & S.S. SANE
166 Surveys in combinatorics, 1991, A.D. KEEDWELL (ed)
168 Representations of algebras, H. TACHIKAWA & S. BRENNER (eds)
169 Boolean function complexity, M.S. PATERSON (ed)
170 Manifolds with singularities and the Adams-Novikov spectral sequence, B. BOTVINNK
171 Squares, A.R. RAJWADE
172 Algebraic varieties, GEORGE R. KEMPF
173 Discrete groups and geometry, W.J. HARVEY & C. MACLACHLAN (eds)
174 Lectures on mechanics, J.E. MARSDEN
175 Adams memorial symposium on algebraic topology 1, N. RAY & G. WALKER (eds)
176 Adams memorial symposium on algebraic topology 2, N. RAY & G. WALKER (eds)
177 Applications of categories in computer science, M. FOURMAN, P. JOHNSTONE & A. PITTS (eds)
178 Lower K-and L-theory, A. RANICKI
179 Complex projective geometry, G. ELLINGSRUD et al
180 Lectures on ergodic theory and Pesin theory on compact manifolds, M. POLLICOTT
181 Geometric group theory I, G.A. NIBLO & M.A. ROLLER (eds)
182 Geometric Group Theory II, G.A. NIBLO & M.A. ROLLER (eds)
183 Shintani Zeta Functions, A. YUKIE
184 Arithmetical Functions, W. SCHWARZ & J. SPILKER
185 Representations of solvable groups. O. MANZ & T.R. WOLF
186 Complexity: knots, colotigs and counting, D.J.A. WELSH
187 Surveys in combinator. 1993 K. WALKER (ed)
188 Local'analysis for the odd order theorem, H. BENDER & G. GLAUBERMAN
189 Locally presentable and accessible categories, J. ADAMEK & J. ROSICKY
190 Polynomial inwariants of finite groups, D.J. BENSON
191 Finite geometry and combinatorics, F. DE CLERCK et al
192 Symplectic geometry, D. SALAMON (ed)
194 Independent random variables and rearrangment invariant spaces, M. BRAVERMAN
195 Arithmetic of blowup algebras, WOLMER VASCONCELOS
196 Microlocal analysis for differential operators, A. GRIGIS & J. SJÖSTRAND
197 Two-dimensional homotopy and combinatorial group theory, C. HOG-ANGELONI et al
198 The algebraic characterization of geometric 4-manifolds, J.A. HILLMAN
199 Invariant Potential theory in rhe unit ball of C^n, MANFRED STOLL
200 The Grothendieck theory of dessins d'enfant, L. SCHNEPS (ed)
201 Singularities, JEAN-PAUL BRASSELET (ed)
202 The technique of pseudodifferential operators, H.O. CORDES
203 Hochschild cohomology of von Neumann algebras, A. SINCLAIR & R. SMITH
204 Combinatorial and geometric group theory, A.J. DUNCAN, N.D. GILBERT & J. HOWEB (eds)
205 Ergodic theory and its connections with harmonic analysis, K. PETERSEN & I. SALAMA (eds)
207 Groups of Lie type and their geometries, W.M. KANTOR & L. DI MARTINO (eds)
208 Vector bundles in algebraic geometry, N.J. HITCHIN, P. NEWSTEAD & W.M. OXBURY (eds)
209 Arithmetic of diagonal hypersutrfaces over finite fields, Q. GOUVÉA & N. U
210 Hilbert C*-modules, E.C. LANCE
211 Groups 93 Galway / St Andrews I, CM. CAMPBELL et al (eds)
212 Groups 93 Galway / St Andrews II, C.M. CAMPBELL et al (eds)
214 Generalised Euler-Jacobi inversion formula and asymptotics beyond all orders, V. KOWALENKO el al
215 Number theory 1992–93, S. DAVID (ed)
216 Stochastic partial differential equations, A. ETHERIDGE (ed)
217 Quadratic form wirh applications to algebraic geometry and topology, A. PFISTER
218 Surveys in-combinatorics, 1995, PETER ROWLINSON (cd)
220 Algebraic set theory, A. JOYAL & I. MOERDIJK
221 Harmonic approximation, S.J. GARDINER
222 Advances in linear logic, J.-Y. GIRARD. Y. LAFONT & L. REGNIER (eds)
223 Analytic semigroups and semilinear initial boundary value problems, KAZUAKI TAIRA
224 Comrutabilitv, enumerability, unsolvability, S.B. COOPER, T.A. SLAMAN & S.S. WAINER (eds)
225 A mathematical introduction to string theory, S. ALBEVERIO et al
226 Novikov conjectures, index theorems and rigidity I. S. FERRY, A. RANICKI & 1. ROSENBERG (eds)
227 Novikov conjectures, index theorems and rigidity II, S. FERRY, A. RANICK1 & J. ROSENBERG (eds)
228 Ergodic theory of Z_d actions, M. POLLICOTT & K. SCHMIDT (eds)
229 Ergodicity for infinite dimensional systems, G. DA PRATO & J. ZABCZYK
230 Prolegomena to a middlebrow arithmetic of curves of genus 2, J.W.S. CASSELS & E.V. FLYNN
231 Semigmup theory and its applications. K.H. HOFMANN & M.W. MISLOVE (eds)
232 The descriptive set theory of Polish group actions, H. BECKER & A.S. KECHRIS
233 Finite fields and applications, S. COHEN & H. NIDERREITER (eds)
234 Introduction to subfactors, V. JONES & V.S. SUNDER
235 Number theory 1993–94. S. DAVID (ed)
236 The James forest, H. FETTER & B. GAMBOA DE BUEN

237 Sieve methods. exponential sums, and their applications in number theory, G.R.H. GREAVES *et al*
238 Representation theory and algebraic geometry, A. MARTSINKOVSKY & G. TODOROV (eds)
240 Stable groups, FRANK 0. WAGNER
241 Surveys in combinatorics, 1997. R.A. BAILEY (ed)
242 Geometric Galois actions I, L. SCHNEPS & P. LOCHAK (eds)
243 Geometric Galois actions II, L. SCHNEPS & P. LOCHAK (eds)
244 Model theory of groups and automrphism groups, D. EVANS (ed)
245 Geometry, combinatorial designs and related structures, J.W.P HIRSCHFELD *et al*
246 p-Automorphisms of Finite-groups, E. KHUKHRO
247 Analytic number theory, Y. MOTOHASHI (ed)
248 Tame topology and o-minimal structures, LOU VAN DEN DRIES
249 The atlas of finite gmups: ten years on, ROBERT CURTIS & ROBERT WILSON (eds)
250 Characters and blocks of finite groups. G. NAVARRO
251 Groner bases and applications, B. BUCHBERGER & E WINKLER (eds)
252 Geometry and cohomology in group theory, P. KROPHOLLER, G. NIBLO, R. STÖHR (eds)
253 The q-Schur algebra, S. DONKIN
254 Galois representations in arithmetic algebraic geometry, A.J. SCHOLL & R.L. TAYLOR (eds)
255 Symmetries and integrability of difference equations, P.A. CLARKSON & F.W. NIJHOFF (eds)
256 Aspects of Galois theory, HELMUT VOLKLBIN *et al*
257 An introduction to nancommutative differential geometty and its physical applications 2ed, J. MADORE
258 Sets and proofs, S.B. COOPER & J. TRUSS (eds)
259 Models and computability, S.B. COOPER & J. TRUSS (eds)
260 Groups St Andrews 1997 in Bath, I, CM. CAMPBELL *et al*
261 Groups St Andrews 1997 in Bath, II, C.M. CAMPBELL *et al*
262 Analysis and logic, C.W. HENSON, J. IOVINO, A.S. KECHRIS & E. ODELL
263 Singularity theoy, BILL BRUCE & DAVID MOND (eds)
264 New trends in algebraic geometry, K. HULEK, F. CATANESE, C. PETERS & M. REID (eds)
265 Elliptic curves in cryptography, I. BLAKE, G. SEROUSSI & N. SMART
267 Surveys in combinatorics, 1999, J.D. LAMB & D.A. PREECE (eds)
268 Spectral asymptatics in the semi-classical limit, M. DIMASSI & J. SJÖSTRAND
269 Ergodic theory and topological dynamics, M.B. BEKKA & M. MAYER
270 Analysis on Lie groups, N.T. VAROPOULOS & S. MUSTAPHA
271 Singular perturbations of differential operators, S. ALBEVERIO & P. KURASOV
272 Character theory for the odd order theorem, T. PETERFALVI
273 Spectral theory geometry, E.B. DAVIES & Y. SAFAROV (eds)
274 The Mandlebrot set, theme and variations, TAN LEI (ed)
275 Descriptive set theory and dynamical systems, M. FOREMAN *et al*
276 Singularities of plane curves, E. CASAS-ALVERO
277 Computatianal and geometric aspects of modern algebra, M.D. ATKINSON *et al*
278 Glbal attractors in abstract parabolic problems, J.W. CHOLEWA & T. DLOTKO
279 Topics in symbolic dynamics and applications, F. BLANCHARD, A. MAASS & A. NOGUEIRA (eds)
280 Chamctw and automorphism groups of compact Riemann surfaces, THOMAS BREUER
281 Explicit birational geometry of 3-folds, ALESSIO CORTI & MILES REID (eds)
282 Auslander-Buchweitz approximations of equivariant modules, M. HASHIMOTO
283 Nonlinear elasticity, Y. FU & R.W. OGDEN (eds)
284 Foundations of computational mathematics, R. DEVORE, A. ISERLES & E. SÜLI (eds)
285 Rational points on curves over finite fields. H. NIEDERREITER & C. XING
286 Clifford algebras and spinors 2ed, p. LOUNESTO
287 Topics on Riemann surfaces and Fuchsian groups, E. BUJALANCE, A.F. COSTA & E. MARTINEZ (eds)
288 Surveys in cambinatorics, 2001, J. HIRSCHFELD (ed)
289 Aspects of Sobolev-type inequalities. L. SALOFF-COSTE
290 Quantum groups and Lie theory, A. PRESSLEY (ed)
291 Tits buildings and the model theory of groups, K. TENT (ed)
292 A quantum groups primer, S. MAJID.
293 Second order partial diffetcntial equations in Hilbett spaces, G. DA PRATO & I. ZABCZYK
294 Imwduction to the theory of operator spaces, G. PISIER
295 Geometry and integrability. LIONEL MASON & YAVUZ NUTKU (eds)
296 Lectures on invariant theory, IGOR DOLGACHEV
297 The homotopy category of simply connected 4-manifolds, H.-J. BAUES
299 Kleinian groups and hyperbolic 3-manifolds, Y. KOMORI, V. MARKOVIC, & C. SERIES (eds)
300 Introduction to Möbius differential geometry, UDO HERTRICH-JEROMIN
301 Stable modules and the D(2)-problem, F.E.A. JOHNSON
302 Diicrete and continuous nonlinear Schrödinger systems, M.J. ABLOWITZ, B. PRINARI, & A.D. TRUBATCH
303 Number theory and algebraic geometry, MILES REID & ALEXEI SKOROBOOATOV (eds)
304 Groups St Andrews 2001 in Oxford Vol. I, COLIN CAMPBELL, EDMUND ROBERTSON
305 Groups St Andrews 2001 in Oxford Vol. 2. C.M. CAMPBELL, E.F. ROBERTSON & G.C. SMITH (eds)
307 Surveys in combinatorics 2003, C.D. WENSLEY (ed)
308 Topology, Geometry and Quantum Field Theory, U. TILLMAN (ed)
309 Corings and comodules, TOMASZ BRZEZINSKI & ROBERT WISBAUER (eds)
310 Topics in dynamics and ergodic theory, SERGEY BEZUGLYI & SERGIY KOLYADA (eds)
311 Groups, T.W. MÜLLER
312 Foundations of computational mathematics, Minneapolis 2002, FELIPE CUCKER *et al* (eds)
313 Transcendental Aspects of Algebraic Cycles, S. MÜLLER-STACH & C. PETERS (eds)

London Mathematical Society Lecture Note Series. 322

Recent Perspectives in Random Matrix Theory and Number Theory

Edited by

F. MEZZADRI
N. C. SNAITH
University of Bristol

CAMBRIDGE
UNIVERSITY PRESS

CAMBRIDGE UNIVERSITY PRESS
Cambridge, New York, Melbourne, Madrid, Cape Town, Singapore, São Paulo, Delhi

Cambridge University Press
The Edinburgh Building, Cambridge CB2 8RU, UK

Published in the United States of America by Cambridge University Press, New York

www.cambridge.org
Information on this title: www.cambridge.org/9780521620581

© Cambridge University Press 2005

First published 2005

A catalogue record for this publication is available from the British Library

ISBN 978-0-521-62058-1 paperback

Transferred to digital printing (with corrections) 2009

Contents

Introduction   vii
   F. Mezzadri and N.C. Snaith

Prime number theory and the Riemann zeta-function   1
   D. R. Heath-Brown

Introduction to the random matrix theory: Gaussian Unitary Ensemble and beyond   31
   Yan V. Fyodorov

Notes on pair correlation of zeros and prime numbers   79
   D.A. Goldston

Notes on eigenvalue distributions for the classical compact groups   111
   Brian Conrey

Compound nucleus resonances, random matrices, quantum chaos   147
   Oriol Bohigas

Basic analytic number theory   185
   David W. Farmer

Applications of mean value theorems to the theory of the Riemann zeta function   201
   S.M. Gonek

Families of L-functions and 1-level densities   225
   Brian Conrey

L-functions and the characteristic polynomials of random matrices   251
   J.P. Keating

Spacing distributions in random matrix ensembles   279
   Peter J. Forrester

Toeplitz determinants, Fisher-Hartwig symbols, and random matrices   309
   Estelle L. Basor

Mock-Gaussian behaviour   337
   C.P. Hughes

Some specimens of *L*-functions   **357**
   Philippe Michel

Computational methods and experiments in analytic number theory   **425**
   Michael Rubinstein

Introduction

F. Mezzadri and N. C. Snaith

This volume of proceedings stems from a school that was part of the programme *Random Matrix Approaches in Number Theory*, which ran at the Isaac Newton Institute for Mathematical Sciences, Cambridge, from 26 January until 16 July 2004. The purpose of these proceedings is twofold. Firstly, the impressive recent progress in analytic number theory brought about by the introduction of random matrix techniques has created a rapidly developing area of research. As a consequence there is not as yet a textbook on the subject. This volume is intended to fill this gap. There are, of course, well-established texts in both random matrix theory and analytic number theory, but very few of them treat in any length or detail these new applications of random matrix theory. Secondly, this new branch of mathematics is intrinsically multidisciplinary; teaching young researchers in random matrix theory, mathematical physics and number theory mathematical techniques that are not a natural part of their education is essential to introduce a new generation of scientists to this important and rapidly developing field. In writing their contributions to the proceedings, the lecturers kept in mind the diverse backgrounds of the audience to whom this volume is addressed.

The material in the volume includes the basic techniques of random matrix theory and number theory needed to understand the most important achievements in the subject; it also gives a comprehensive survey of recent results where random matrix theory has played a major role in advancing our understanding of open problems in number theory. We hope that the choice of topics will be useful to both the advanced graduate student and to the established researcher.

These proceedings contain a set of introductory lectures to analytic number theory and random matrix theory, written by Roger Heath-Brown and Yan Fyodorov respectively. The former includes a survey of elementary prime number theory and an introduction to the theory of the Riemann zeta function and other L-functions, while Fyodorov's lectures provide the reader with one of the main tools used in the theory of random matrices: the theory of orthogonal polynomials. This ubiquitous technique is then applied to the computation of the spectral correlation functions of eigenvalues of the Hermitian matrices which form the Gaussian Unitary Ensemble (GUE), as well as to computing the averages of moments and ratios of characteristic polynomials of these

Hermitian matrices. In contrast, fundamental techniques for calculating various eigenvalue statistics on ensembles of unitary matrices can be found in the "Notes on eigenvalue distributions for the classical compact groups" by Brian Conrey. These are the groups of matrices that are used in connection with L-functions, for example in the lectures of Hughes and Keating. The articles of Peter Forrester and Estelle Basor discuss more specific topics in random matrix theory. Forrester reviews in detail the theory of spacing distributions for various ensembles of matrices and emphasizes its connections with the theory of Painlevé equations and of Fredholm determinants; Basor's lectures introduce the reader to the theory of Toeplitz determinants, their asymptotic evaluations for both smooth and singular symbols and their connection to random matrix theory.

Dan Goldston reviews how random matrix theory and number theory came together unexpectedly when Montgomery, assuming the Riemann hypothesis, conjectured the two-point correlation function of the Riemann zeros, which Dyson recognized it as the two-point correlation function for eigenvalues of the random matrices in the CUE (or, equivalently, the GUE) ensemble. Looking toward applications to physics, Oriol Bohigas's article gives an historical survey of how random matrix theory was instrumental in the understanding of the statistical properties of spectra of complex nuclei and of individual quantum mechanical systems whose classical limit exhibits chaotic behaviour. After Montgomory's discovery overwhelming numerical evidence, largely produced by Andrew Odlyzko in the late 1980s, supported the hypothesis that the non-trivial zeros of the Riemann zeta function are locally correlated like eigenvalues of random matrices in the GUE ensemble. Later Hejhal (1994), and then Rudnick and Sarnak (1994,1996) proved similar results for the three and higher point correlations.

Several lectures are devoted to specific and more advanced topics in number theory. David Farmer introduces the reader to techniques in analytic number theory, discussing various ways to manipulate Dirichlet series, while Steve Gonek extends this to discuss mean-value theorems and their applications. Philippe Michel discusses the construction of many examples of L-functions, including those associated to elliptic curves and modular forms.

The remaining lectures highlight the connection between L-functions and random matrix theory. Brian Conrey's lectures "Families of L-functions and 1-level densities" concern the statistics of zeros of families of L-functions near the point where the line on which their Riemann hypothesis places their zeros crosses the real axis. Based on the example of the function field zeta functions, these statistics were proposed by Katz and Sarnak (1999) to be those of the eigenvalues of one of the classical compact groups, namely $U(N)$, $USp(2N)$ and $O(N)$. The lectures of Jon Keating reveal how the local statistical properties of the Riemann zeta function and other L-functions are inherently determined by the distribution of their zeros, thus high up the critical line $\zeta(s)$ can be modelled by the characteristic polynomial of random matrices belonging to

U(N). As a consequence of this property, techniques well developed in random matrix theory can lead to conjectures for quantities like moments and distributions of the values of L-functions, which have been open problems for almost eighty years. Chris Hughes discusses how the first few moments of the smooth counting functions of the eigenvalues of random matrices and of the zeros of L-functions are Gaussian while their distributions are not. Since much of the predictive power of random matrix theory is based on conjectures, numerical experiments play an important role in the theory; Michael Rubinstein's article introduces the reader to the most important techniques used in computational number theory and to conjectures and numerical experiments connecting number theory with random matrix theory.

We are particularly grateful to David Farmer and Brian Conrey for carefully reading many of the articles and to the staff of the Newton Institute for their invaluable assistance in making the school such a successful event. We also thankfully acknowledge financial contributions from the EU Network 'Mathematical Aspects of Quantum Chaos', the Institute of Physics Publishing, the Isaac Newton Institute for the Mathematical Sciences and the US National Science Foundation.

Francesco Mezzadri and Nina C. Snaith

July 2004

School of Mathematics,
University of Bristol,
Bristol BS8 1TW, UK

Prime Number Theory and the Riemann Zeta-Function

D.R. Heath-Brown

1 Primes

An integer $p \in \mathbb{N}$ is said to be "prime" if $p \neq 1$ and there is no integer n dividing p with $1 < n < p$. (This is not the algebraist's definition, but in our situation the two definitions are equivalent.)

The primes are multiplicative building blocks for $\mathbb{N}$, as the following crucial result describes.

Theorem 1. (The Fundamental Theorem of Arithmetic.) *Every* $n \in \mathbb{N}$ *can be written in exactly one way in the form*

$$n = p_1^{e_1} p_2^{e_2} \dots p_k^{e_k},$$

with $k \geq 0$, $e_1, \dots, e_k \geq 1$ *and primes* $p_1 < p_2 < \dots < p_k$.

For a proof, see Hardy and Wright [5, Theorem 2], for example. The situation for $\mathbb{N}$ contrasts with that for arithmetic in the set

$$\{m + n\sqrt{-5} : m, n \in \mathbb{Z}\},$$

where one has, for example,

$$6 = 2 \times 3 = (1 + \sqrt{-5}) \times (1 - \sqrt{-5}),$$

with $2, 3, 1 + \sqrt{-5}$ and $1 - \sqrt{-5}$ all being "primes".

A second fundamental result appears in the works of Euclid.

Theorem 2. *There are infinitely many primes.*

This is proved by contradiction. Assume there are only finitely many primes, $p_1, p_2, \ldots, p_n$, say. Consider the integer $N = 1 + p_1 p_2 \ldots p_n$. Then $N \geq 2$, so that N must have at least one prime factor p, say. But our list of primes was supposedly complete, so that p must be one of the primes p_i, say. Then p_i divides $N - 1$, by construction, while $p = p_i$ divides N by assumption. It follows that p divides $N - (N - 1) = 1$, which is impossible. This contradiction shows that there can be no finite list containing all the primes.

There have been many tables of primes produced over the years. They show that the detailed distribution is quite erratic, but if we define

$$\pi(x) = \#\{p \leq x : p \text{ prime}\},$$

then we find that $\pi(x)$ grows fairly steadily. Gauss conjectured that

$$\pi(x) \sim \mathrm{Li}(x),$$

where

$$\mathrm{Li}(x) = \int_2^x \frac{dt}{\log t},$$

that is to say that

$$\lim_{x \to \infty} \frac{\pi(x)}{\mathrm{Li}(x)} = 1.$$

The following figures bear this out.

$\pi(10^8)$	$=$	$5{,}776{,}455$	$\frac{\pi(x)}{\mathrm{Li}(x)} =$	$0.999869147\ldots,$
$\pi(10^{12})$	$=$	$37{,}607{,}912{,}018$	$\frac{\pi(x)}{\mathrm{Li}(x)} =$	$0.999989825\ldots,$
$\pi(10^{16})$	$=$	$279{,}238{,}341{,}033{,}925$	$\frac{\pi(x)}{\mathrm{Li}(x)} =$	$0.999999989\ldots.$

It is not hard to show that in fact

$$\mathrm{Li}(x) \sim \frac{x}{\log x},$$

but it turns out that $\mathrm{Li}(x)$ gives a better approximation to $\pi(x)$ than $x / \log x$ does. Gauss' conjecture was finally proved in 1896, by Hadamard and de la Vallée Poussin, working independently.

Theorem 3. (The Prime Number Theorem.) *We have*

$$\pi(x) \sim \frac{x}{\log x}$$

as $x \to \infty$.

One interesting interpretation of the Prime Number Theorem is that for a number n in the vicinity of x the "probability" that n is prime is asymptotically $1/\log x$, or equivalently, that the "probability" that n is prime is asymptotically $1/\log n$. Of course the event "n is prime" is deterministic — that is to say, the probability is 1 if n is prime, and 0 otherwise. None the less the probabilistic interpretation leads to a number of plausible heuristic arguments. As an example of this, consider, for a given large integer n, the probability that $n + 1, n + 2, \ldots, n + k$ are all composite. If k is at most n, say, then the probability that any one of these is composite is about $1 - 1/\log n$. Thus if the events were all independent, which they are not, the overall probability would be about

$$\left(1 - \frac{1}{\log n}\right)^k.$$

Taking $k = \mu(\log n)^2$ and approximating

$$\left(1 - \frac{1}{\log n}\right)^{\log n}$$

by e^{-1}, we would have that the probability that $n + 1, n + 2, \ldots, n + k$ are all composite, is around $n^{-\mu}$.

If E_n is the event that $n + 1, n + 2, \ldots, n + k$ are all composite, then the events E_n and E_{n+1} are clearly not independent. However we may hope that E_n and E_{n+k} are independent. If the events E_n were genuinely independent for different values of n then an application of the Borel-Cantelli lemma would tell us that E_n should happen infinitely often when $\mu < 1$, and finitely often for $\mu \geq 1$. With more care one can make this plausible even though E_n and $E_{n'}$ are correlated for nearby values n and n'. We are thus led to the following conjecture.

Conjecture 1. *If p' denotes the next prime after p then*

$$\limsup_{p \to \infty} \frac{p' - p}{(\log p)^2} = 1.$$

Numerical evidence for this is hard to produce, but what there is seems to be consistent with the conjecture.

In the reverse direction, our simple probabilistic interpretation of the Prime Number Theorem might suggest that the probability of having both n and $n+1$ prime should be around $(\log n)^{-2}$. This is clearly wrong, since one of n and $n+1$ is always even. However, a due allowance for such arithmetic effects leads one to the following.

Conjecture 2. *If*

$$c = 2\prod_{p > 2}\left(1 - \frac{1}{(p-1)^2}\right) = 1.3202\ldots,$$

the product being over primes, then

$$\#\{n \le x : n, n+2 \text{ both prime}\} \doteq c \int_2^x \frac{dt}{(\log t)^2}. \tag{1.1}$$

The numerical evidence for this is extremely convincing.

Thus the straightforward probabilistic interpretation of the Prime Number Theorem leads to a number of conjectures, which fit very well with the available numerical evidence. This probabilistic model is known as "Cramér's Model" and has been widely used for predicting the behaviour of primes.

One further example of this line of reasoning shows us however that the primes are more subtle than one might think. Consider the size of

$$\pi(N + H) - \pi(N) = \#\{p : N < p \le N + H\},$$

when H is small compared with N. The Prime Number Theorem leads one to expect that

$$\pi(N + H) - \pi(N) \doteq \int_N^{N+H} \frac{dt}{\log t} \sim \frac{H}{\log N}.$$

However the Prime Number Theorem only says that

$$\pi(x) = \int_2^x \frac{dt}{\log t} + o\left(\frac{x}{\log x}\right),$$

or equivalently that

$$\pi(x) = \int_2^x \frac{dt}{\log t} + f(x),$$

where

$$\frac{f(x)}{x/\log x} \to 0$$

as $x \to \infty$. Hence

$$\pi(N + H) - \pi(N) = \int_N^{N+H} \frac{dt}{\log t} + f(N + H) - f(N).$$

In order to assert that

$$\frac{f(N + H) - f(N)}{H/\log N} \to 0$$

as $N \to \infty$ we need $cN \le H \le N$ for some constant $c > 0$. None the less, considerably more subtle arguments show that

$$\pi(N + H) - \pi(N) \sim \frac{H}{\log N}$$

even when H is distinctly smaller than N.

A careful application of the Cramér Model suggests the following conjecture.

Conjecture 3. *Let $\kappa > 2$ be any constant. Then if $H = (\log N)^\kappa$ we should have*

$$\pi(N + H) - \pi(N) \sim \frac{H}{\log N}$$

as $N \to \infty$.

This is supported by the following result due to Selberg in 1943 [15].

Theorem 4. *Let $f(N)$ be any increasing function for which $f(N) \to \infty$ as $N \to \infty$. Assume the Riemann Hypothesis. Then there is a subset $\mathcal{E}$ of the integers $\mathbb{N}$, with*

$$\#\{n \in \mathcal{E} : n \leq N\} = o(N)$$

as $N \to \infty$, such that

$$\pi(n + f(n)\log^2 n) - \pi(n) \sim f(n)\log n$$

for all $n \notin \mathcal{E}$.

Conjecture 3 would say that one can take $\mathcal{E} = \emptyset$ if $f(N)$ is a positive power of $\log N$.

Since Cramér's Model leads inexorably to Conjecture 3, it came as quite a shock to prime number theorists when the conjecture was disproved by Maier [9] in 1985. Maier established the following result.

Theorem 5. *For any $\kappa > 1$ there is a constant $\delta_\kappa > 0$ such that*

$$\limsup_{N\to\infty} \frac{\pi(N + (\log N)^\kappa) - \pi(N)}{(\log N)^{\kappa-1}} \geq 1 + \delta_\kappa$$

and

$$\liminf_{N\to\infty} \frac{\pi(N + (\log N)^\kappa) - \pi(N)}{(\log N)^{\kappa-1}} \leq 1 - \delta_\kappa.$$

The values of N produced by Maier, where $\pi(N + (\log N)^\kappa) - \pi(N)$ is abnormally large, (or abnormally small), are very rare. None the less their existence shows that the Cramér Model breaks down. Broadly speaking one could summarize the reason for this failure by saying that arithmetic effects play a bigger rôle than previously supposed. As yet we have no good alternative to the Cramér model.

2 Open Questions About Primes, and Important Results

Here are a few of the well-known unsolved problems about the primes.

(1) Are there infinitely many "prime twins" $n, n+2$ both of which are prime? (Conjecture 2 gives a prediction for the rate at which the number of such pairs grows.)

(2) Is every even integer $n \geq 4$ the sum of two primes? (Goldbach's Conjecture.)

(3) Are there infinitely many primes of the form $p = n^2 + 1$?

(4) Are there infinitely many "Mersenne primes" of the form $p = 2^n - 1$?

(5) Are there arbitrarily long arithmetic progressions, all of whose terms are prime?

(6) Is there always a prime between any two successive squares?

However there have been some significant results proved too. Here are a selection.

(1) There are infinitely many primes of the form $a^2 + b^4$. (Friedlander and Iwaniec [4], 1998.)

(2) There are infinitely many primes p for which $p + 2$ is either prime or a product of two primes. (Chen [2], 1966.)

(3) There is a number n_0 such that any even number $n \geq n_0$ can be written as $n = p + p'$ with p prime and p' either prime or a product of two primes. (Chen [2], 1966.)

(4) There are infinitely many integers n such that $n^2 + 1$ is either prime or a product of two primes. (Iwaniec [8], 1978.)

(5) For any constant $c < \frac{243}{205} = 1.185\ldots$, there are infinitely many integers n such that $[n^c]$ is prime. Here $[x]$ denotes the integral part of x, that is to say the largest integer N satisfying $N \leq x$. (Rivat and Wu [14], 2001, after Piatetski-Shapiro, [11], 1953.)

(6) Apart from a finite number of exceptions, there is always a prime between any two consecutive cubes. (Ingham [6], 1937.)

(7) There is a number n_0 such that for every $n \geq n_0$ there is at least one prime in the interval $[n, n + n^{0.525}]$. (Baker, Harman and Pintz, [1], 2001.)

(8) There are infinitely many consecutive primes p', p such that $p' - p \leq (\log p)/4$. (Maier [10], 1988.)

(9) There is a positive constant c such that there are infinitely many consecutive primes p', p such that

$$p' - p \geq c \log p \frac{(\log \log p)(\log \log \log \log p)}{(\log \log \log p)^2}.$$

(Rankin [13], 1938.)

(10) For any positive integer q and any integer a in the range $0 \le a < q$, which is coprime to q, there are arbitrarily long strings of consecutive primes, all of which leave remainder a on division by q. (Shiu [16], 2000.)

By way of explanation we should say the following. The result (1) demonstrates that even though we cannot yet handle primes of the form $n^2 + 1$, we can say something about the relatively sparse polynomial sequence $a^2 + b^4$. The result in (5) can be viewed in the same context. One can think of $[n^c]$ as being a "polynomial of degree c" with $c > 1$. Numbers (2), (3) and (4) are approximations to, respectively, the prime twins problem, Goldbach's problem, and the problem of primes of the shape $n^2 + 1$. The theorems in (6) and (7) are approximations to the conjecture that there should be a prime between consecutive squares. Of these (7) is stronger, if less elegant. Maier's result (8) shows that the difference between consecutive primes is sometimes smaller than average by a factor $1/4$, the average spacing being $\log p$ by the Prime Number Theorem. (Of course the twin prime conjecture would be a much stronger result, with differences between consecutive primes sometimes being as small as 2.) Similarly, Rankin's result (9) demonstrates that the gaps between consecutive primes can sometimes be larger than average, by a factor which is almost $\log \log p$. Again this is some way from what we expect, since Conjecture 1 predict gaps as large as $(\log p)^2$. Finally, Shiu's result (10) is best understood by taking $q = 10^7$ and $a = 7,777,777$, say. Thus a prime leaves remainder a when divided by q, precisely when its decimal expansion ends in 7 consecutive 7's. Then (10) tells us that a table of primes will somewhere contain a million consecutive entries, each of which ends in the digits 7,777,777.

3   The Riemann Zeta-Function

In the theory of the zeta-function it is customary to use the variable $s = \sigma + it \in \mathbb{C}$. One then defines the complex exponential

$$n^{-s} := \exp(-s \log n), \quad \text{with} \quad \log n \in \mathbb{R}.$$

The Riemann Zeta-function is then

$$\zeta(s) := \sum_{n=1}^{\infty} n^{-s}, \quad \sigma > 1. \tag{3.1}$$

The sum is absolutely convergent for $\sigma > 1$, and for fixed $\delta > 0$ it is uniformly convergent for $\sigma \ge 1 + \delta$. It follows that $\zeta(s)$ is holomorphic for $\sigma > 1$. The function is connected to the primes as follows.

Theorem 6. (The Euler Product.) *If $\sigma > 1$ then we have*

$$\zeta(s) = \prod_{p}(1 - p^{-s})^{-1},$$

where p runs over all primes, and the product is absolutely convergent.

This result is, philosophically, at the heart of the theory. It relates a sum over all positive integers to a product over primes. Thus it relates the additive structure, in which successive positive integers are generated by adding 1, to the multiplicative structure. Moreover we shall see in the proof that the fact that the sum and the product are equal is exactly an expression of the Fundamental Theorem of Arithmetic.

To prove the result consider the finite product

$$\prod_{p \leq X} (1 - p^{-s})^{-1}.$$

Since $\sigma > 1$ we have $|p^{-s}| < p^{-1} < 1$, whence we can expand $(1 - p^{-s})^{-1}$ as an absolutely convergent series $1 + p^{-s} + p^{-2s} + p^{-3s} + \ldots$. We may multiply together a finite number of such series, and rearrange them, since we have absolute convergence. This yields

$$\prod_{p \leq X} (1 - p^{-s})^{-1} = \sum_{n=1}^{\infty} \frac{a_X(n)}{n^s},$$

where the coefficient $a_X(n)$ is the number of ways of writing n in the form

$$n = p_1^{e_1} p_2^{e_2} \ldots p_r^{e_r} \quad \text{with} \quad p_1 < p_2 < \ldots < p_r \leq X.$$

By the Fundamental Theorem of Arithmetic we have $a_X(n) = 0$ or 1, and if $n \leq X$ we will have $a_X(n) = 1$. It follows that

$$|\sum_{n=1}^{\infty} n^{-s} - \sum_{n=1}^{\infty} \frac{a_X(n)}{n^s}| \leq \sum_{n > X} |\frac{1}{n^s}| = \sum_{n > X} \frac{1}{n^{\sigma}}.$$

As $X \to \infty$ this final sum must tend to zero, since the infinite sum $\sum_{n=1}^{\infty} n^{-\sigma}$ converges. We therefore deduce that if $\sigma > 1$, then

$$\lim_{X \to \infty} \prod_{p \leq X} (1 - p^{-s})^{-1} = \sum_{n=1}^{\infty} \frac{1}{n^s},$$

as required. Of course the product is absolutely convergent, as one may see by taking $s = \sigma$.

One important deduction from the Euler product identity comes from taking logarithms and differentiating termwise. This can be justified by the local uniform convergence of the resulting series.

Corollary 1. *We have*

$$-\frac{\zeta'}{\zeta}(s) = \sum_{n=2}^{\infty} \frac{\Lambda(n)}{n^s}, \quad (\sigma > 1), \tag{3.2}$$

where

$$\Lambda(n) = \begin{cases} \log p, & n = p^e, \\ 0, & otherwise. \end{cases}$$

The function $\Lambda(n)$ is known as the von Mangoldt function.

4 The Analytic Continuation and Functional Equation of $\zeta(s)$

Our definition only gives a meaning to $\zeta(s)$ when $\sigma > 1$. We now seek to extend the definition to all $s \in \mathbb{C}$. The key tool is the Poisson Summation Formula .

Theorem 7. (The Poisson Summation Formula.) *Suppose that $f : \mathbb{R} \to \mathbb{R}$ is twice differentiable and that f, f' and f'' are all integrable over $\mathbb{R}$. Define the Fourier transform by*

$$\hat{f}(t) := \int_{-\infty}^{\infty} f(x) e^{-2\pi i t x} dx.$$

Then

$$\sum_{-\infty}^{\infty} f(n) = \sum_{-\infty}^{\infty} \hat{f}(n),$$

both sides converging absolutely.

There are weaker conditions under which this holds, but the above more than suffices for our application. The reader should note that there are a number of conventions in use for defining the Fourier transform, but the one used here is the most appropriate for number theoretic purposes.

The proof (see Rademacher [12, page 71], for example) uses harmonic analysis on $\mathbb{R}^+$. Thus it depends only on the additive structure and not on the multiplicative structure.

If we apply the theorem to $f(x) = \exp\{-x^2\pi v\}$, which certainly fulfils the conditions, we have

$$
\begin{aligned}
\hat{f}(n) &= \int_{-\infty}^{\infty} e^{-x^2 \pi v} e^{-2\pi i n x} dx \\
&= \int_{-\infty}^{\infty} e^{-\pi v (x + i n / v)^2} e^{-\pi n^2 / v} dx \\
&= e^{-\pi n^2 / v} \int_{-\infty}^{\infty} e^{-\pi v y^2} dy \\
&= \frac{1}{\sqrt{v}} e^{-\pi n^2 / v},
\end{aligned}
$$

providing that v is real and positive. Thus if we define

$$\theta(v) := \sum_{-\infty}^{\infty} \exp(-\pi n^2 v),$$

then the Poisson Summation Formula leads to the transformation formula

$$\theta(v) = \frac{1}{\sqrt{v}} \theta(1/v).$$

The function $\theta(v)$ is a *theta-function,* and is an example of a *modular form.* It is the fact that $\theta(v)$ not only satisfies the above transformation formula when v goes to $1/v$ but is also periodic, that makes $\theta(v)$ a modular form.

The "Langlands Philosophy" says that all reasonable generalizations of the Riemann Zeta-function are related to modular forms, in a suitably generalized sense.

We are now ready to consider $\zeta(s)$, but first we introduce the function

$$\psi(v) = \sum_{n=1}^{\infty} e^{-n^2\pi v}, \tag{4.1}$$

so that $\psi(v) = (\theta(v) - 1)/2$ and

$$2\psi(v) + 1 = \frac{1}{\sqrt{v}}\{2\psi(\frac{1}{v}) + 1\}. \tag{4.2}$$

We proceed to compute that, if $\sigma > 1$, then

$$
\begin{aligned}
\int_0^\infty x^{s/2-1}\psi(x)dx &= \sum_{n=1}^\infty \int_0^\infty x^{s/2-1}e^{-n^2\pi x}dx \\
&= \sum_{n=1}^\infty \frac{1}{(n^2\pi)^{s/2}} \int_0^\infty y^{s/2-1}e^{-y}dy \\
&= \sum_{n=1}^\infty \frac{1}{(n^2\pi)^{s/2}}\Gamma(\frac{s}{2}) \\
&= \zeta(s)\pi^{-s/2}\Gamma(\frac{s}{2}),
\end{aligned}
$$

on substituting $y = n^2\pi x$. The interchange of summation and integration is justified by the absolute convergence of the resulting sum.

We now split the range of integration in the original integral, and apply the transformation formula (4.2). For $\sigma > 1$ we obtain the expression

$$
\begin{aligned}
\zeta(s)\pi^{-s/2}\Gamma(\frac{s}{2}) &= \int_1^\infty x^{s/2-1}\psi(x)dx + \int_0^1 x^{s/2-1}\psi(x)dx \\
&= \int_1^\infty x^{s/2-1}\psi(x)dx + \int_0^1 x^{s/2-1}\{\frac{1}{\sqrt{x}}\psi(\frac{1}{x}) + \frac{1}{2\sqrt{x}} - \frac{1}{2}\}dx \\
&= \int_1^\infty x^{s/2-1}\psi(x)dx + \int_0^1 x^{s/2-3/2}\psi(\frac{1}{x})dx + \frac{1}{s-1} - \frac{1}{s} \\
&= \int_1^\infty x^{s/2-1}\psi(x)dx + \int_1^\infty y^{(1-s)/2-1}\psi(y)dy - \frac{1}{s(1-s)},
\end{aligned}
$$

where we have substituted y for $1/x$ in the final integral.

We therefore conclude that

$$\zeta(s)\pi^{-s/2}\Gamma(\frac{s}{2}) = \int_1^\infty \{x^{s/2-1} + x^{(1-s)/2-1}\}\psi(x)dx - \frac{1}{s(1-s)}, \tag{4.3}$$

whenever $\sigma > 1$. However the right-hand side is meaningful for all values $s \in \mathbb{C} - \{0, 1\}$, since the integral converges by virtue of the exponential decay of $\psi(x)$. We may therefore use the above expression to define $\zeta(s)$ for all $s \in \mathbb{C} - \{0, 1\}$, on noting that the factor $\pi^{-s/2}\Gamma(s/2)$ never vanishes. Indeed, since $\Gamma(s/2)^{-1}$ has a zero at $s = 0$ we see that the resulting expression for $\zeta(s)$ is regular at $s = 0$. Finally we observe that the right-hand side of (4.3) is invariant on substituting s for $1 - s$. We are therefore led to the the following conclusion.

Theorem 8. (Analytic Continuation and Functional Equation.) *The function $\zeta(s)$ has an analytic continuation to $\mathbb{C}$, and is regular apart from a simple pole at $s = 1$, with residue 1. Moreover*

$$\pi^{-s/2}\Gamma(\frac{s}{2})\zeta(s) = \pi^{-(1-s)/2}\Gamma(\frac{1-s}{2})\zeta(1-s).$$

Furthermore, if $a \leq \sigma \leq b$ and $|t| \geq 1$, then $\pi^{-s/2}\Gamma(\frac{s}{2})\zeta(s)$ is bounded in terms of a and b.

To prove the last statement in the theorem we merely observe that

$$|\pi^{-s/2}\Gamma(\frac{s}{2})\zeta(s)| \leq 1 + \int_1^\infty (x^{b/2-1} + x^{(1-a)/2-1})\psi(x)dx.$$

5   Zeros of $\zeta(s)$

It is convenient to define

$$\xi(s) = \frac{1}{2}s(s-1)\pi^{-s/2}\Gamma(\frac{s}{2})\zeta(s) = (s-1)\pi^{-s/2}\Gamma(1+\frac{s}{2})\zeta(s), \qquad (5.1)$$

so that $\xi(s)$ is entire. The functional equation then takes the form $\xi(s) = \xi(1-s)$. It is clear from (3.2) that $\zeta(s)$ can have no zeros for $\sigma > 1$, since the series converges. Since $1/\Gamma(z)$ is entire, the function $\Gamma(s/2)$ is non-vanishing, so that $\xi(s)$ also has no zeros in $\sigma > 1$. Thus, by the functional equation, the zeros of $\xi(s)$ are confined to the "critical strip" $0 \leq \sigma \leq 1$. Moreover any zero of $\zeta(s)$ must either be a zero of $\xi(s)$, or a pole of $\Gamma(s/2)$. We then see that the zeros of $\zeta(s)$ lie in the critical strip, with the exception of the "trivial zeros" at $s = -2, -4, -6, \ldots$ corresponding to poles of $\Gamma(s/2)$.

We may also observe that if ρ is a zero of $\xi(s)$ then, by the functional equation, so is $1 - \rho$. Moreover, since $\overline{\xi(s)} = \xi(\bar{s})$, we deduce that $\bar{\rho}$ and $1 - \bar{\rho}$ are also zeros. Thus the zeros are symmetrically arranged about the real axis, and also about the "critical line" given by $\sigma = 1/2$. With this picture in mind we mention the following important conjectures.

Conjecture 4. (The Riemann Hypothesis.) *We have $\sigma = 1/2$ for all zeros of $\xi(s)$.*

Conjecture 5. *All zeros of $\xi(s)$ are simple.*

In the absence of a proof of Conjecture 5 we adopt the convention that in any sum or product over zeros, we shall count them according to multiplicity.

6   The Product Formula

There is a useful product formula for $\xi(s)$, due to Hadamard. In general we have the following result, for which see Davenport [3, Chapter 11] for example.

Theorem 9. *Let $f(z)$ be an entire function with $f(0) \neq 0$, and suppose that there are constants $A > 0$ and $\theta < 2$ such that $f(z) = O(\exp(A|z|^\theta))$ for all complex z. Then there are constants α and β such that*

$$f(z) = e^{\alpha + \beta z} \prod_{n=1}^{\infty} \{(1 - \frac{z}{z_n})e^{z/z_n}\},$$

where z_n runs over the zeros of $f(z)$ counted with multiplicity. The infinite sum $\sum_{n=1}^{\infty} |z_n|^{-2}$ converges, so that the product above is absolutely and uniformly convergent in any compact set which includes none of the zeros.

We can apply this to $\xi(s)$, since it is apparent from Theorem 8, together with the definition (5.1) that

$$\xi(0) = \xi(1) = \frac{1}{2}\pi^{-1/2}\Gamma(\frac{1}{2})\mathrm{Res}\{\zeta(s); s = 1\} = \frac{1}{2}.$$

For $\sigma \geq 2$ one has $\zeta(s) = O(1)$ directly from the series (3.1), while Stirling's approximation yields $\Gamma(s/2) = O(\exp(|s| \log |s|))$. It follows that $\xi(s) = O(\exp(|s| \log |s|))$ whenever $\sigma \geq 2$. Moreover, when $\frac{1}{2} \leq \sigma \leq 2$ one sees from Theorem 8 that $\xi(s)$ is bounded. Thus, using the functional equation, we can deduce that $\xi(s) = O(\exp(|s| \log |s|))$ for all s with $|s| \geq 2$.

We may therefore deduce from Theorem 9 that

$$\xi(s) = e^{\alpha + \beta s} \prod_{\rho} \{(1 - \frac{s}{\rho})e^{s/\rho}\},$$

where ρ runs over the zeros of $\xi(s)$. Thus, with appropriate branches of the logarithms, we have

$$\log \xi(s) = \alpha + \beta s + \sum_{\rho} \{\log(1 - \frac{s}{\rho}) + \frac{s}{\rho}\}.$$

We can then differentiate termwise to deduce that

$$\frac{\xi'}{\xi}(s) = \beta + \sum_{\rho} \{\frac{1}{s - \rho} + \frac{1}{\rho}\},$$

the termwise differentiation being justified by the local uniform convergence of the resulting sum. We therefore deduce that

$$\frac{\zeta'}{\zeta}(s) = \beta - \frac{1}{s-1} + \frac{1}{2}\log \pi - \frac{1}{2}\frac{\Gamma'}{\Gamma}(\frac{s}{2}+1) + \sum_{\rho}\{\frac{1}{s-\rho} + \frac{1}{\rho}\}, \qquad (6.1)$$

where, as ever, ρ runs over the zeros of ξ counted according to multiplicity. In fact, on taking $s \to 1$, one can show that

$$\beta = -\frac{1}{2}\gamma - 1 - \frac{1}{2}\log 4\pi,$$

where γ is Euler's constant. However we shall make no use of this fact.

7   The Functions $N(T)$ and $S(T)$

We shall now investigate the frequency of the zeros ρ. We define

$$N(T) = \#\{\rho = \beta + i\gamma : 0 \le \beta \le 1, 0 \le \gamma \le T\}.$$

The notation $\beta = \Re(\rho)$, $\gamma = \Im(\rho)$ is standard. In fact one can easily show that $\psi(x) < (2\sqrt{x})^{-1}$, whence (4.3) suffices to prove that $\zeta(s) < 0$ for real $s \in (0,1)$. Thus we have $\gamma > 0$ for any zero counted by $N(T)$.

The first result we shall prove is the following.

Theorem 10. *If T is not the ordinate of a zero, then*

$$N(T) = \frac{T}{2\pi} \log \frac{T}{2\pi} - \frac{T}{2\pi} + \frac{7}{8} + S(T) + O(1/T),$$

where

$$S(T) = \frac{1}{\pi} \arg \zeta(\frac{1}{2} + iT),$$

is defined by continuous variation along the line segments from 2 to $2 + iT$ to $\frac{1}{2} + iT$.

We shall evaluate $N(T)$ using the Principle of the Argument, which shows that

$$N(T) = \frac{1}{2\pi} \Delta_R \arg \xi(s),$$

providing that T is not the ordinate of any zero. Here R is the rectangular path joining 2, $2 + iT$, $-1 + iT$, and -1. To calculate $\Delta_R \arg \xi(s)$ one starts with any branch of $\arg \xi(s)$ and allows it to vary continuously around the path. Then $\Delta_R \arg \xi(s)$ is the increase in $\arg \xi(s)$ along the path. Our assumption about T ensures that $\xi(s)$ does not vanish on R.

Now $\xi(s) = \xi(1 - s)$ and $\xi(1 - s) = \overline{\xi(1 - \bar{s})}$, whence $\xi(\frac{1}{2} + a + ib)$ is conjugate to $\xi(\frac{1}{2} - a + ib)$. (In particular this shows that $\xi(\frac{1}{2} + it)$ is always real.) It follows that

$$\Delta_R \xi(s) = 2\Delta_P \xi(s),$$

where P is the path $\frac{1}{2} \to 2 \to 2 + iT \to \frac{1}{2} + iT$. On the first line segment $\xi(s)$ is real and strictly positive, so that the contribution to $\Delta_P \xi(s)$ is zero. Let L be the remaining path $2 \to 2 + iT \to \frac{1}{2} + iT$. Then

$$\Delta_L \xi(s) = \Delta_L \{\arg(s - 1)\pi^{-s/2}\Gamma(\frac{s}{2} + 1)\} + \Delta_L \arg \zeta(s).$$

Now on L the function $s - 1$ goes from 1 to $-\frac{1}{2} + iT$, whence

$$\Delta_L \arg(s - 1) = \arg(-\frac{1}{2} + iT) = \frac{\pi}{2} + O(T^{-1}).$$

We also have

$$\arg \pi^{-s/2} = \Im \log \pi^{-s/2} = \Im(-\frac{s}{2} \log \pi),$$

so that $\arg \pi^{-s/2}$ goes from 0 to $-(T \log \pi)/2$ and

$$\Delta_L \arg \pi^{-s/2} = -\frac{T}{2} \log \pi.$$

Finally, Stirling's formula yields

$$\log \Gamma(z) = (z - \frac{1}{2}) \log z - z + \frac{1}{2} \log(2\pi) + O(|z|^{-1}), \quad (|\arg(z)| \le \pi - \delta), \quad (7.1)$$

whence

$$\begin{aligned}
\Delta_L \arg \Gamma(\frac{s}{2} + 1) &= \Im \log \Gamma(\frac{\frac{1}{2} + iT}{2} + 1) \\
&= \Im\{(\frac{3}{4} + i\frac{T}{2}) \log(\frac{5}{4} + i\frac{T}{2}) - (\frac{5}{4} + i\frac{T}{2}) + \frac{1}{2}\log(2\pi)\} \\
&\qquad\qquad\qquad\qquad\qquad +O(1/T) \\
&= \frac{T}{2} \log \frac{T}{2} - \frac{T}{2} + \frac{3\pi}{8} + O(1/T),
\end{aligned}$$

since

$$\log(\frac{5}{4} + i\frac{T}{2}) = \log \frac{T}{2} + i\frac{\pi}{2} + O(1/T).$$

These results suffice for Theorem 10

We now need to know about $S(T)$. Here we show the following.

Theorem 11. *We have $S(T) = O(\log T)$.*

Corollary 2. (The Riemann – von Mangoldt Formula). *We have*

$$N(T) = \frac{T}{2\pi} \log \frac{T}{2\pi} - \frac{T}{2\pi} + O(\log T).$$

We start the proof by taking $s = 2 + iT$ in (3.2) and noting that

$$|\frac{\zeta'}{\zeta}(s)| \le \sum_{n=2}^{\infty} \frac{\Lambda(n)}{n^2} = O(1).$$

Thus the partial fraction decomposition (6.1) yields

$$\sum_\rho \{\frac{1}{2 + iT - \rho} + \frac{1}{\rho}\} = \frac{1}{2}\frac{\Gamma'}{\Gamma}(2 + \frac{iT}{2}) + O(1).$$

We may differentiate (7.1), using Cauchy's formula for the first derivative, to produce

$$\frac{\Gamma'}{\Gamma}(z) = \log z + O(1), \quad (|\arg(z)| \le \pi - \delta), \tag{7.2}$$

and then deduce that

$$\sum_\rho \{\frac{1}{2 + iT - \rho} + \frac{1}{\rho}\} = O(\log(2 + T)). \tag{7.3}$$

We have only assumed here that $T \geq 0$, not that $T \geq 2$. In order to get the correct order estimate when $0 \leq T \leq 2$ we have therefore written $O(\log(2+T))$, which is $O(1)$ for $0 \leq T \leq 2$.

Setting $\rho = \beta + i\gamma$ we now have

$$\Re \frac{1}{2+iT-\rho} = \frac{2-\beta}{(2-\beta)^2+(T-\gamma)^2} \geq \frac{1}{4+(T-\gamma)^2}$$

and

$$\Re \frac{1}{\rho} = \frac{\beta}{\beta^2+\gamma^2} \geq 0,$$

since $0 \leq \beta \leq 1$. We therefore produce the useful estimate

$$\sum_\rho \frac{1}{4+(T-\gamma)^2} = O(\log(2+T)), \tag{7.4}$$

which implies in particular that

$$\#\{\rho : T-1 \leq \gamma \leq T+1\} = O(\log(2+T)). \tag{7.5}$$

We now apply (6.1) with $s = \sigma + iT$ and $0 \leq \sigma \leq 2$, and subtract (7.3) from it to produce

$$\frac{\zeta'}{\zeta}(\sigma+iT) = -\frac{1}{\sigma+iT-1} + \sum_\rho \{\frac{1}{\sigma+iT-\rho} - \frac{1}{2+iT-\rho}\} + O(\log(2+T)).$$

Terms with $|\gamma - T| > 1$ have

$$|\frac{1}{\sigma+iT-\rho} - \frac{1}{2+iT-\rho}| = |\frac{2-\sigma}{(\sigma+iT-\rho)(2+iT-\rho)}|$$
$$\leq \frac{2}{|\gamma-T|.|\gamma-T|}$$
$$\leq \frac{2}{\frac{1}{5}\{4+(T-\gamma)^2\}}.$$

Thus (7.4) implies that

$$\sum_{\rho: |\gamma-T|>1} \{\frac{1}{\sigma+iT-\rho} - \frac{1}{2+iT-\rho}\} = O(\log(2+T)),$$

and hence that

$$\frac{\zeta'}{\zeta}(\sigma+iT) = -\frac{1}{\sigma+iT-1} + \sum_{\rho: |\gamma-T|\leq 1} \{\frac{1}{\sigma+iT-\rho} - \frac{1}{2+iT-\rho}\}$$
$$+ O(\log(2+T)).$$

However we also have

$$\left|\frac{1}{2+iT-\rho}\right| \le \frac{1}{2-\beta} \le 1,$$

whence (7.5) produces

$$\sum_{\rho:\,|\gamma-T|\le 1} \frac{1}{2+iT-\rho} = O(\log(2+T)).$$

We therefore deduce the following estimate.

Lemma 1. *For $0 \le \sigma \le 2$ and $T \ge 0$ we have*

$$\frac{\zeta'}{\zeta}(\sigma+iT) = -\frac{1}{\sigma+iT-1} + \sum_{\rho:\,|\gamma-T|\le 1} \frac{1}{\sigma+iT-\rho} + O(\log(2+T)).$$

We are now ready to complete our estimation of $S(T)$. Taking $T \ge 2$, we have

$$\arg\zeta(\tfrac{1}{2}+iT) = \Im\log\zeta(\tfrac{1}{2}+iT) = \Im\int_2^{1/2+iT} \frac{\zeta'}{\zeta}(s)ds,$$

the path of integration consisting of the line segments from 2 to $2+iT$ and from $2+iT$ to $1/2+iT$. Along the first of these we use the formula (3.2), which yields

$$\int_2^{2+iT} \frac{\zeta'}{\zeta}(s)ds = \left[\sum_{n=2}^{\infty} \frac{\Lambda(n)}{n^s\log n}\right]_2^{2+iT} = O(1).$$

For the remaining range we use Lemma 1, which produces

$$\begin{aligned}
\Im\int_{2+iT}^{1/2+iT} \frac{\zeta'}{\zeta}(s)ds &= \sum_{\rho:\,|\gamma-T|\le 1}\Im\int_{2+iT}^{1/2+iT}\frac{ds}{s-\rho} + O(\log T)\\
&= \sum_{\rho:\,|\gamma-T|\le 1}\Im\{\log(\tfrac{1}{2}+iT-\rho)-\log(2+iT-\rho)\}\\
&\qquad\qquad +O(\log T)\\
&= \sum_{\rho:\,|\gamma-T|\le 1}\{\arg(\tfrac{1}{2}+iT-\rho)-\arg(2+iT-\rho)\}\\
&\qquad\qquad +O(\log T)\\
&= \sum_{\rho:\,|\gamma-T|\le 1} O(1) + O(\log T)\\
&= O(\log T),
\end{aligned}$$

by (7.5). This suffices for the proof of Theorem 11.

8 The Non-Vanishing of $\zeta(s)$ on $\sigma = 1$

So far we know only that the non-trivial zeros of $\zeta(s)$ lie in the critical strip $0 \le \sigma \le 1$. Qualitatively the only further information we have is that there are no zeros on the boundary of this strip.

Theorem 12. *(Hadamard and de la Vallée Poussin, independently, 1896.) We have $\zeta(1 + it) \ne 0$, for all real t.*

This result was the key to the proof of the Prime Number Theorem. Quantitatively one can say a little more.

Theorem 13. *(De la Vallée Poussin.) There is a positive absolute constant c such that for any $T \ge 2$ there are no zeros of $\zeta(s)$ in the region*

$$\sigma \ge 1 - \frac{c}{\log T}, \quad |t| \le T.$$

In fact, with much more work, one can replace the function $c/\log T$ by one that tends to zero slightly more slowly, but that will not concern us here. The proof of Theorem 13 uses the following simple fact.

Lemma 2. *For any real θ we have*

$$3 + 4\cos\theta + \cos 2\theta \ge 0.$$

This is obvious, since

$$3 + 4\cos\theta + \cos 2\theta = 2\{1 + \cos\theta\}^2.$$

We now use the identity (3.2) to show that

$$-3\frac{\zeta'}{\zeta}(\sigma) - 4\Re\frac{\zeta'}{\zeta}(\sigma + it) - \Re\frac{\zeta'}{\zeta}(\sigma + 2it)$$

$$= \sum_{n=2}^{\infty} \frac{\Lambda(n)}{n^\sigma}\{3 + 4\cos(t\log n) + \cos(2t\log n)\}$$

$$\ge 0,$$

for $\sigma > 1$. When $1 < \sigma \le 2$ we have

$$-\frac{\zeta'}{\zeta}(\sigma) = \frac{1}{\sigma - 1} + O(1),$$

from the Laurent expansion around the pole at $s = 1$. For the remaining two terms we use Lemma 1, to deduce that

$$\frac{3}{\sigma - 1} + O(1) - 4\Re \sum_{\rho:\, |\gamma - t| \le 1} \frac{1}{\sigma + it - \rho} - \Re \sum_{\rho:\, |\gamma - 2t| \le 1} \frac{1}{\sigma + 2it - \rho} + O(\log T)$$

$$\geq 0$$

for $1 < \sigma \leq 2$, $T \geq 2$, and $|t| \leq T$. Suppose we have a zero $\rho_0 = \beta_0 + i\gamma_0$, say, with $0 \leq \gamma_0 \leq T$. Set $t = \gamma_0$. We then observe that for any zero we have

$$\Re \frac{1}{\sigma + it - \rho} = \frac{\sigma - \beta}{(\sigma - \beta)^2 + (t - \gamma)^2} \geq 0,$$

since $\sigma > 1 \geq \beta$, and similarly

$$\Re \frac{1}{\sigma + 2it - \rho} \geq 0.$$

We can therefore drop all terms from the two sums above, with the exception of the term corresponding to $\rho = \rho_0$, to deduce that

$$\frac{4}{\sigma - \beta_0} \leq \frac{3}{\sigma - 1} + O(\log T).$$

Suppose that the constant implied by the $O(\ldots)$ notation is c_0. This is just a numerical value that one could calculate with a little effort. Then

$$\frac{4}{\sigma - \beta_0} \leq \frac{3}{\sigma - 1} + c_0 \log T$$

whenever $1 < \sigma \leq 2$. If $\beta_0 = 1$ we get an immediate contradiction by choosing $\sigma = 1 + (2c_0 \log T)^{-1}$. If $\beta_0 < 3/4$ the result of Theorem 13 is immediate. For the remaining range of β_0 we choose $\sigma = 1 + 4(1 - \beta_0)$, which will show that

$$\frac{4}{5(1 - \beta_0)} \leq \frac{3}{4(1 - \beta_0)} + c_0 \log T.$$

Thus

$$\frac{1}{20(1 - \beta_0)} \leq c_0 \log T,$$

and hence

$$1 - \beta_0 \geq \frac{1}{20 c_0 \log T}.$$

This completes the proof of Theorem 13.

The reader should observe that the key feature of the inequality given in Lemma 2 is that the coefficients are non-negative, and that the coefficient of $\cos \theta$ is strictly greater than the constant term. In particular, the inequality

$$1 + \cos \theta \geq 0$$

just fails to work.

Theorem 13 has a useful corollary.

Corollary 3. *Let c be as in Theorem 13, and let $T \geq 2$. Then if*

$$1 - \frac{c}{2 \log T} \leq \sigma \leq 2$$

and $|t| \leq T$, we have

$$\frac{\zeta'}{\zeta}(\sigma + it) = -\frac{1}{\sigma + it - 1} + O(\log^2 T).$$

For the proof we use Lemma 1. The sum over zeros has $O(\log T)$ terms, by (7.5), and each term is $O(\log T)$, since

$$\sigma - \beta \geq \frac{c}{2 \log T},$$

by Theorem 13.

9 Proof of the Prime Number Theorem

Since our argument is based on the formula (3.2), it is natural to work with $\Lambda(n)$. We define

$$\psi(x) = \sum_{n \leq x} \Lambda(n) = \sum_{p^k \leq x} \log p. \tag{9.1}$$

This is not the same function as that defined in (4.1)! Our sum $\psi(x)$ is related to $\pi(x)$ in the following lemma.

Lemma 3. *For $x \geq 2$ we have*

$$\pi(x) = \frac{\psi(x)}{\log x} + \int_2^x \frac{\psi(t)}{t \log^2 t} dt + O(x^{1/2}).$$

For the proof we begin by setting

$$\theta(x) = \sum_{p \leq x} \log p.$$

Then

$$
\begin{aligned}
\int_2^x \frac{\theta(t)}{t \log^2 t} dt &= \int_2^x \sum_{p \leq t} \frac{\log p}{t \log^2 t} dt \\
&= \sum_{p \leq x} \int_p^x \frac{\log p}{t \log^2 t} dt \\
&= \sum_{p \leq x} \left[-\frac{\log p}{\log t} \right]_p^x \\
&= \pi(x) - \frac{\theta(x)}{\log x},
\end{aligned}
$$

so that

$$\pi(x) = \frac{\theta(x)}{\log x} + \int_2^x \frac{\theta(t)}{t \log^2 t} dt. \tag{9.2}$$

However it is clear that terms in (9.1) with $k \geq 2$ have $p \leq x^{1/2}$, and there are at most $x^{1/2}$ such p. Moreover $k \leq \log x / \log p$, whence the total contribution from terms with $k \geq 2$ is $O(x^{1/2} \log x)$. Thus

$$\psi(x) = \theta(x) + O(x^{1/2} \log x).$$

If we substitute this into (9.2) the required result follows.

We will use contour integration to relate $\psi(x)$ to $\zeta'(s)/\zeta(s)$. This will be done via the following result.

Lemma 4. *Let $y > 0$, $c > 1$ and $T \geq 1$. Define*

$$I(y,T) = \frac{1}{2\pi i} \int_{c-iT}^{c+iT} \frac{y^s}{s} ds.$$

Then

$$I(y,T) = \left\{ \begin{array}{ll} 0, & 0 < y < 1 \\ 1, & y > 1 \end{array} \right\} + O(\frac{y^c}{T |\log y|}).$$

When $0 < y < 1$ we replace the path of integration by the line segments $c - iT \to N - iT \to N + iT \to c + iT$, and let $N \to \infty$. Then

$$\int_{N-iT}^{N+iT} \frac{y^s}{s} ds \to 0,$$

while

$$\int_{c-iT}^{N-iT} \frac{y^s}{s} ds = O(\int_c^N \frac{y^\sigma}{T} d\sigma) = O(\frac{y^c}{T |\log y|}),$$

and similarly for the integral from $N + iT$ to $c + iT$. It follows that

$$I(y,T) = O(\frac{y^c}{T |\log y|})$$

for $0 < y < 1$. The case $y > 1$ can be treated analogously, using the path $c - iT \to -N - iT \to -N + iT \to c + iT$. However in this case we pass a pole at $s = 0$, with residue 1, and this produces the corresponding main term for $I(y,T)$.

We can now give our formula for $\psi(x)$.

Theorem 14. *For $x - \frac{1}{2} \in \mathbb{N}$, $\alpha = 1 + 1/\log x$ and $T \geq 1$ we have*

$$\psi(x) = \frac{1}{2\pi i} \int_{\alpha-iT}^{\alpha+iT} \{-\frac{\zeta'}{\zeta}(s)\} \frac{x^s}{s} ds + O(\frac{x \log^2 x}{T}).$$

For the proof we integrate termwise to get

$$\frac{1}{2\pi i} \int_{\alpha - iT}^{\alpha + iT} \{-\frac{\zeta'}{\zeta}(s)\} \frac{x^s}{s} ds = \sum_{n=2}^{\infty} \Lambda(n) I(\frac{x}{n}, T)$$

$$= \sum_{n \leq x} \Lambda(n) + O(\sum_{n=2}^{\infty} \Lambda(n)(\frac{x}{n})^{\alpha} \frac{1}{T|\log x/n|}).$$

Since we are taking $x - \frac{1}{2} \in \mathbb{N}$ the case $x/n = 1$ does not occur. In the error sum those terms with $n \leq x/2$ or $n \geq 2x$ have $|\log x/n| \geq \log 2$. Such terms therefore contribute

$$O(\sum_{n=2}^{\infty} \Lambda(n) \frac{x^{\alpha}}{Tn^{\alpha}}) = O(\frac{x^{\alpha}}{T} |\frac{\zeta'}{\zeta}(\alpha)|)$$

$$= O(\frac{x^{1+1/\log x}}{T} \frac{1}{\alpha - 1})$$

$$= O(\frac{x \log x}{T}).$$

When $x/2 < n < 2x$ we have

$$|\log x/n| \geq \frac{1}{2} \frac{|x - n|}{x}$$

and

$$\Lambda(n)(\frac{x}{n})^{\alpha} = O(\log x).$$

These terms therefore contribute

$$\sum_{x/2 < n < 2x} O(\frac{x \log x}{T|x - n|}) = O(\frac{x \log^2 x}{T})$$

on bearing in mind that $x - \frac{1}{2} \in \mathbb{N}$. The theorem now follows.

We are now ready to prove the following major result.

Theorem 15. *There is a positive constant c_0 such that*

$$\psi(x) = x + O(x \exp\{-c_0 \sqrt{\log x}\}) \tag{9.3}$$

for all $x \geq 2$. Moreover we have

$$\pi(x) = \mathrm{Li}(x) + O(x \exp\{-c_0 \sqrt{\log x}\})$$

for all $x \geq 2$.

The error terms here can be improved slightly, but with considerably more work.

It clearly suffices to consider the case in which $x - \frac{1}{2} \in \mathbb{N}$. To prove the result we set

$$\mu = 1 - \frac{c}{2\log T}, \quad T \geq 2,$$

as in Lemma 3, and replace the line of integration in Theorem 14 by the path $\alpha - iT \to \mu - iT \to \mu + iT \to \alpha + iT$. The integrand has a pole at $s = 1$ with residue x, arising from the pole of $\zeta(s)$, but no other singularities, by virtue of Theorem 13. On the new path of integration Lemma 3 shows that

$$\frac{\zeta'}{\zeta}(s) = O(\log^2 T).$$

We therefore deduce that

$$\psi(x) = x + O(\frac{x\log^2 x}{T}) + O(\int_\mu^\alpha \frac{\log^2 T}{T}x^\sigma d\sigma) + O(\int_{-T}^{T} \frac{\log^2 T}{|\mu + it|}x^\mu dt),$$

where the first error integral corresponds to the line segments $\alpha - iT \to \mu - iT$ and $\mu + iT \to \alpha + iT$, and the second error integral to the segment $\mu - iT \to \mu + iT$. These integrals are readily estimated to yield

$$\psi(x) = x + O(\frac{x\log^2 x}{T}) + O(\frac{\log^2 T}{T}x^\alpha) + O(x^\mu \log^3 T).$$

Of course $x^\alpha = O(x)$ here. Thus if $T \leq x$ we merely get

$$\psi(x) = x + O(x\log^3 x\{\frac{1}{T} + x^{\mu-1}\}).$$

We now choose

$$T = \exp\{\sqrt{\log x}\},$$

whence

$$\psi(x) = x + O(x(\log x)^3 \exp\{-\min(1, \frac{c}{2})\sqrt{\log x}\}).$$

We may therefore choose any positive constant $c_0 < \min(1, \frac{c}{2})$ in Theorem 15. This establishes (9.3). To prove the remaining assertion, it suffices to insert (9.3) into Lemma 3.

Finally we should stress that the success of this argument depends on being able to take $\mu < 1$, since there is an error term which is essentially of order x^μ. Thus it is crucial that we should at least know that $\zeta(1 + it) \neq 0$.

If we assume the Riemann Hypothesis, then we may take any $\mu > \frac{1}{2}$ in the above analysis. This leads to the following estimates.

Theorem 16. *On the Riemann Hypothesis we have*

$$\psi(x) = x + O(x^\theta)$$

and

$$\pi(x) = \mathrm{Li}(x) + O(x^\theta)$$

for any $\theta > \frac{1}{2}$ and all $x \geq 2$.

One cannot reduce the exponent below $1/2$, since there is a genuine contribution to $\pi(x)$ arising from the zeros of $\zeta(s)$.

10 Explicit Formulae

In this section we shall argue somewhat informally, and present results without proof.

If $f : (0, \infty) \to \mathbb{C}$ we define the Mellin transform of f to be the function

$$F(s) := \int_0^\infty f(x) x^{s-1} dx.$$

By a suitable change of variables one sees that this is essentially a form of Fourier transform. Indeed all the properties of Mellin transforms can readily be translated from standard results about Fourier transforms. In particular, under suitable conditions one has an inversion formula

$$f(x) = \frac{1}{2\pi i} \int_{\sigma - i\infty}^{\sigma + i\infty} F(s) x^{-s} ds.$$

Arguing purely formally one then has

$$\sum_{n=2}^\infty \Lambda(n) f(n) = \sum_{n=2}^\infty \Lambda(n) \frac{1}{2\pi i} \int_{2-i\infty}^{2+i\infty} F(s) n^{-s} ds$$

$$= \frac{1}{2\pi i} \int_{2-i\infty}^{2+i\infty} \{-\frac{\zeta'}{\zeta}(s)\} F(s) ds.$$

If one now moves the line of integration to $\Re(s) = -N$ one passes poles at $s = 1$ and at $s = \rho$ for every non-trivial zero ρ, as well as at the trivial zeros $-2n$. Under suitable conditions the integral along $\Re(s) = -N$ will tend to 0 as $N \to \infty$. This argument leads to the following result.

Theorem 17. *Suppose that $f \in C^2(0, \infty)$ and that $\mathrm{supp}(f) \subseteq [1, X]$ for some X. Then*

$$\sum_{n=2}^\infty \Lambda(n) f(n) = F(1) - \sum_\rho F(\rho) - \sum_{n=1}^\infty F(-2n).$$

One can prove such results subject to weaker conditions on f. If x is given, and

$$f(t) = \begin{cases} 1, & t \le x \\ 0, & t > x, \end{cases}$$

then the conditions above are certainly not satisfied, but we have the following related result.

Theorem 18. (The Explicit Formula.) *Let $x \ge T \ge 2$. Then*

$$\psi(x) = x - \sum_{\rho : |\gamma| \le T} \frac{x^\rho}{\rho} + O(\frac{x \log^2 x}{T}).$$

For a proof of this see Davenport [3, Chapter 17], for example. There are variants of this result containing a sum over all zeros, and with no error term, but the above is usually more useful.

The explicit formula shows exactly how the zeros influence the behaviour of $\psi(x)$, and hence of $\pi(x)$. The connection between zeros and primes is even more clearly shown by the following result of Landau.

Theorem 19. *For fixed positive real x define $\Lambda(x) = 0$ if $x \notin \mathbb{N}$ and $\Lambda(x) = \Lambda(n)$ if $x = n \in \mathbb{N}$. Then*

$$\Lambda(x) = -\frac{2\pi}{T} \sum_{\rho:\, 0<\gamma\leq T} x^\rho + O_x(\frac{\log T}{T}),$$

where $O_x(\ldots)$ indicates that the implied constant may depend on x.

This result shows that the zeros precisely determine the primes. Thus, for example, one can reformulate the conjecture (1.1) as a statement about the zeros of the zeta-function. All the unevenness of the primes, for example the behaviour described by Theorem 5, is encoded in the zeros of the zeta-function. It therefore seems reasonable to expect that the zeros themselves should have corresponding unevenness.

11   Dirichlet Characters

We now turn to the simplest type of generalization of the Riemann Zeta-function, namely the Dirichlet L-functions. In the remainder of these notes we shall omit most of the proofs, being content merely to describe what can be proved.

A straightforward example of a Dirichlet L-function is provided by the infinite series

$$1 - \frac{1}{3^s} + \frac{1}{5^s} - \frac{1}{7^s} + \frac{1}{9^s} - \frac{1}{11^s} + \ldots. \tag{11.1}$$

We first need to describe the coefficients which arise.

Definition . *Let $q \in \mathbb{N}$. A "(Dirichlet) character χ to modulus q" is a function $\chi : \mathbb{Z} \to \mathbb{C}$ such that*

(i) $\chi(mn) = \chi(m)\chi(n)$ for all $m, n \in \mathbb{Z}$;

(ii) $\chi(n)$ has period q;

(iii) $\chi(n) = 0$ whenever $(n, q) \neq 1$; and

(iv) $\chi(1) = 1$.

Part (iv) of the definition is necessary merely to rule out the possibility that χ is identically zero.

As an example we can take the function

$$\chi(n) = \begin{cases} 1, & n \equiv 1 \ (\text{mod } 4), \\ -1, & n \equiv 3 \ (\text{mod } 4), \\ 0, & n \equiv 0 \ (\text{mod } 2). \end{cases} \tag{11.2}$$

This is a character modulo 4, and is the one generating the series (11.1). A second example is the function

$$\chi_0(n) := \begin{cases} 1, & (n,q) = 1, \\ 0, & (n,q) \neq 1. \end{cases}$$

This produces a character for every modulus q, known as the *principal character* modulo q.

A number of key facts are gathered together in the following theorem.

Theorem 20. *(i) We have $|\chi(n)| = 1$ for every n coprime to q.*

(ii) If χ_1 and χ_2 are two characters to modulus q, then so is $\chi_1\chi_2$, where we define $\chi_1\chi_2(n) = \chi_1(n)\chi_2(n)$.

(iii) There are exactly $\varphi(q)$ different characters to modulus q.

(iv) If $n \not\equiv 1 (\text{mod } q)$ then there is at least one character χ modulo q for which $\chi(n) \neq 1$.

In part (iii) the function $\varphi(q)$ is the number of positive integers $n \leq q$ for which n and q are coprime.

To prove part (i) we note that the sequence $n^k \ (\text{mod } q)$ must eventually repeat when k runs through $\mathbb{N}$. Thus there exist $k < j$ with $\chi(n^k) = \chi(n^j)$, and hence $\chi(n)^k = \chi(n)^j$. Since n is coprime to q we have $\chi(n) \neq 0$, so that $\chi(n)^{j-k} = 1$.

Part (ii) of the theorem is obvious, but parts (iii) and (iv) are harder, and we refer the reader to Davenport [3, Chapter 4] for the details. As an example of (iii) we note that $\varphi(4) = 2$, and we have already found two characters modulo 4. There are no others.

One further fact may elucidate the situation. Consider a general finite abelian group G. In our case we will have $G = (\mathbb{Z}/q\mathbb{Z})^\times$. Thus G will consist of those residue classes $n \mod q$ for which $(n,q) = 1$, with the multiplication operation. Define $\widehat{G}$ to be the group of homomorphisms $\theta : G \to \mathbb{C}^\times$, where the group action is given by $(\theta_1\theta_2)(g) := \theta_1(g)\theta_2(g)$. In our case these homomorphisms are, in effect, the characters. Then the groups G and $\widehat{G}$ are isomorphic, and part (iii) above is an immediate consequence. The details can be found in Ireland and Rosen [7, pages 253 and 254], for example. Indeed there is a duality between G and $\widehat{G}$. The isomorphism between them is not "natural", but there is a natural isomorphism

$$G \simeq \widehat{\widehat{G}}.$$

There are two orthogonality relations satisfied by the characters to a given modulus q. The first of these is the following.

Theorem 21. *If a and q are coprime then*

$$S := \sum_{\chi \pmod{q}} \chi(n)\overline{\chi(a)} = \begin{cases} \varphi(q), & n \equiv a \pmod{q}, \\ 0, & n \not\equiv a \pmod{q}. \end{cases}$$

When $n \equiv a \pmod{q}$ this is immediate since then $\chi(n)\overline{\chi(a)} = 1$ for all χ. In the remaining case, choose an element b with $ab \equiv 1 \pmod{q}$. By (iv) of Theorem 20 there exists a character χ_1 such that $\chi_1(nb) \neq 1$. Then

$$\chi_1(nb)S = \sum_{\chi \pmod{q}} \chi_1(n)\chi(n)\chi_1(b)\overline{\chi(a)}.$$

However

$$\chi_1(b)\chi_1(a) = \chi_1(ab) = \chi_1(1) = 1,$$

whence $\chi_1(b) = \overline{\chi_1(a)}$. We therefore deduce that

$$\begin{aligned} \chi_1(nb)S &= \sum_{\chi \pmod{q}} \chi_1(n)\chi(n)\overline{\chi_1(a)}\overline{\chi(a)} \\ &= \sum_{\chi \pmod{q}} \chi_1\chi(n)\overline{\chi_1\chi(a)}. \end{aligned}$$

As χ runs over the complete set of characters to modulus q the product $\chi_1\chi$ does as well, since $\chi_1\chi = \chi_1\chi'$ implies $\chi = \chi'$. Thus

$$\sum_{\chi \pmod{q}} \chi_1\chi(n)\overline{\chi_1\chi(a)} = S$$

and hence $\chi_1(nb)S = S$. Since $\chi_1(nb) \neq 1$ we deduce that $S = 0$, as required.

The second orthogonality relation is the following.

Theorem 22. *If $\chi \neq \chi_0$ then $\sum_{n=1}^{q} \chi(n) = 0$.*

The proof is analogous to the previous result, and is based on the obvious fact that if $\chi \neq \chi_0$ then there is some integer n coprime to q such that $\chi(n) \neq 1$. The details are left as an exercise for the reader.

If q has a factor r and χ is a character modulo r we can define the character ψ modulo q which is "induced by" χ. This is done by setting

$$\psi(n) = \begin{cases} \chi(n), & (n, q) = 1, \\ 0, & (n, q) \neq 1. \end{cases}$$

For example, we may take χ to be the character modulo 4 given by (11.2). Then if $q = 12$ we induce a character ψ modulo 12, as in the following table.

	1	2	3	4	5	6	7	8	9	10	11	12
χ	1	0	-1	0	1	0	-1	0	1	0	-1	0
ψ	1	0	0	0	1	0	-1	0	0	0	-1	0

A character $\chi \pmod{q}$ which *cannot* be produced this way from some divisor $r < q$ is said to be "primitive". The principal character is induced by the character $\chi_d \pmod 1$, that is to say by the character which is identically 1. If q is prime, then all the characters except for the principal character are primitive. In general there will be both primitive and imprimitive characters to each modulus. Imprimitive characters are a real nuisance!!

12 Dirichlet L-functions

For any character χ to modulus q we will define the corresponding Dirichlet L-function by setting

$$L(s,\chi) = \sum_{n=1}^{\infty} \frac{\chi(n)}{n^s}, \quad (\sigma > 1).$$

We content ourselves here with describing the key features of these functions, and refer the reader to Davenport [3], for example, for details.

The sum is absolutely convergent for $\sigma > 1$ and is locally uniformly convergent, so that $L(s,\chi)$ is holomorphic in this region. If χ is the principal character modulo q then the series fails to converge when $\sigma \le 1$. However for non-principal χ the series is conditionally convergent when $\sigma > 0$, and the series defines a holomorphic function in this larger region.

There is an Euler product identity

$$L(s,\chi) = \prod_{p}(1 - \chi(p)p^{-s})^{-1}, \quad (\sigma > 1).$$

This can be proved in the same way as for $\zeta(s)$ using the multiplicativity of the function χ.

Suppose that χ is primitive, and that $\chi(-1) = 1$. If we apply the Poisson summation formula to

$$f(x) = e^{-(a+qx)^2 \pi v/q},$$

multiply the result by $\chi(a)$, and sum for $1 \le a \le q$, we find that

$$\theta(v,\chi) = \frac{\tau(\chi)}{\sqrt{q}} \frac{1}{\sqrt{v}} \theta(\frac{1}{v}, \overline{\chi}),$$

where

$$\theta(v,\chi) := \sum_{n=-\infty}^{\infty} \chi(n)e^{-n^2 \pi v/q}.$$

is a generalisation of the theta-function, and

$$\tau(\chi) := \sum_{a=1}^{q} \chi(a) e^{2\pi i a/q}$$

is the *Gauss sum.*

When χ is primitive and $\chi(-1) = -1$ the function $\theta(v, \chi)$ vanishes identically. Instead we use

$$\theta_1(v, \chi) := \sum_{n=-\infty}^{\infty} n\chi(n) e^{-n^2 \pi v/q},$$

for which one finds the analogous transformation formula

$$\theta_1(v, \chi) = \frac{i\tau(\chi)}{\sqrt{q}} \frac{1}{v^{3/2}} \theta_1(\frac{1}{v}, \overline{\chi}).$$

These two transformation formulae then lead to the analytic continuation and functional equation for $L(s, \chi)$. The conclusion is that, if χ is primitive then $L(s, \chi)$ has an analytic continuation to the whole complex plane, and is regular everywhere, except when χ is identically 1, (in which case $L(s, \chi)$ is just the Riemann Zeta-function $\zeta(s)$). Moreover, still assuming that χ is primitive, with modulus q, we set

$$\xi(s, \chi) = (\frac{q}{\pi})^{(s+a)/2} \Gamma(\frac{s+a}{2}) L(s, \chi),$$

where

$$a = a(\chi) := \begin{cases} 0, & \chi(-1) = 1, \\ 1, & \chi(-1) = -1. \end{cases}$$

Then

$$\xi(1 - s, \overline{\chi}) = \frac{i^a q^{1/2}}{\tau(\chi)} \xi(s, \chi).$$

Notice in particular that, unless the values taken by χ are all real, this functional equation relates $L(s, \chi)$ not to the same function at $1 - s$ but to the conjugate L-function, with character $\overline{\chi}$.

It follows from the Euler product and the functional equation that there are no zeros of $\xi(s, \chi)$ outside the critical strip. The zeros will be symmetrically distributed about the critical line $\sigma = 1/2$, but unless χ is real they will not necessarily be symmetric about the real line. Hence in general it is appropriate to define

$$N(T, \chi) := \#\{\rho : \xi(\rho, \chi) = 0 \, |\gamma| \leq T\},$$

counting zeros both above and below the real axis. We then have

$$\frac{1}{2} N(T, \chi) = \frac{T}{2\pi} \log \frac{qT}{2\pi} - \frac{T}{2\pi} + O(\log qT)$$

for $T \geq 2$, which can be seen as an analogue of the Riemann – von Mangoldt formula. This shows in particular that the interval $[T, T + 1]$ contains around

$$\frac{1}{2\pi} \log \frac{qT}{2\pi}$$

zeros, on average.

The work on regions without zeros can be generalized, but there are serious problems with possible zeros on the real axis. Thus one can show that there is a constant $c > 0$, which is independent of q, such that if $T \geq 2$ then $L(s, \chi)$ has no zeros in the region

$$\sigma \geq 1 - \frac{c}{\log qT}, \quad 0 < |t| \leq T.$$

If χ is not a real-valued character then we can extend this result to the case $t = 0$, but is a significant open problem to deal with the case in which χ is real. However in many other important aspects techniques used for the Riemann Zeta-function can be successfully generalized to handle Dirichlet L-functions.

References

[1] R.C. Baker, G. Harman, and J. Pintz, The difference between consecutive primes. II., *Proc. London Math. Soc. (3)*, 83 (2001), 532-562.

[2] J.-R. Chen, on the representation of a large even integer as a sum of a prime and a product of at most two primes, *Kexue Tongbao*, 17 (1966), 385-386.

[3] H. Davenport, *Multiplicative number theory,* Graduate Texts in Mathematics, 74. (Springer-Verlag, New York-Berlin, 1980).

[4] J.B. Friedlander and H. Iwaniec, The polynomial $X^2 + Y^4$ captures its primes, *Annals of Math. (2)*, 148 (1998), 945-1040.

[5] G.H. Hardy and E.M. Wright, *An introduction to the theory of numbers,* (Oxford University Press, New York, 1979).

[6] A.E. Ingham, On the difference between consecutive primes, *Quart. J. Math. Oxford,* 8 (1937), 255-266.

[7] K. Ireland and M. Rosen, *A classical introduction to modern number theory,* Graduate Texts in Math., 84, (Springer, Heidelberg-New York, 1990).

[8] H. Iwaniec, Almost-primes represented by quadratic polynomials, *Invent. Math.,* 47 (1978), 171-188.

[9] H. Maier, Primes in short intervals, *Michigan Math. J.,* 32 (1985), 221-225.

[10] H. Maier, Small differences between prime numbers, *Michigan Math. J.*, 35 (1988), 323-344.

[11] I.I. Piatetski-Shapiro, On the distribution of prime numbers in sequences of the form $[f(n)]$, *Mat. Sbornik N.S.*, 33(75) (1953), 559-566.

[12] H. Rademacher, *Topics in analytic number theory*, Grundlehren math. Wiss., 169, (Springer, New York-Heidelberg, 1973).

[13] R.A. Rankin, The difference between consecutive prime numbers, *J. London Math. Soc.*, 13 (1938), 242-247.

[14] J. Rivat and J. Wu, Prime numbers of the form $[n^c]$, *Glasg. Math. J.*, 43 (2001), 237-254.

[15] A. Selberg, On the normal density of primes in small intervals, and the difference between consecutive primes, *Arch. Math. Naturvid.*, 47, (1943), 87-105.

[16] D.K.L. Shiu, Strings of congruent primes, *J. London Math. Soc. (2)*, 61 (2000), 359-373.

Mathematical Institute,
24-29, St. Giles',
Oxford OX1 3LB

rhb@maths.ox.ac.uk

Introduction to the Random Matrix Theory: Gaussian Unitary Ensemble and Beyond

Yan V. Fyodorov

Abstract

These lectures provide an informal introduction into the notions and tools used to analyze statistical properties of eigenvalues of large random Hermitian matrices. After developing the general machinery of orthogonal polynomial method, we study in most detail Gaussian Unitary Ensemble (GUE) as a paradigmatic example. In particular, we discuss Plancherel-Rotach asymptotics of Hermite polynomials in various regimes and employ it in spectral analysis of the GUE. In the last part of the course we discuss general relations between orthogonal polynomials and characteristic polynomials of random matrices which is an active area of current research.

1   Preface

Gaussian Ensembles of random Hermitian or real symmetric matrices always played a prominent role in the development and applications of Random Matrix Theory. Gaussian Ensembles are uniquely singled out by the fact that they belong both to the family of invariant ensembles, and to the family of ensembles with independent, identically distributed (i.i.d) entries. In general, mathematical methods used to treat those two families are very different.

In fact, all random matrix techniques and ideas can be most clearly and consistently introduced using Gaussian case as a paradigmatic example. In the present set of lectures we mainly concentrate on consequences of the invariance of the corresponding probability density function, leaving aside methods of exploiting statistical independence of matrix entries. Under these circumstances the method of orthogonal polynomials is the most adequate one, and for the Gaussian case the relevant polynomials are Hermite polynomials. Being mostly interested in the limit of large matrix sizes we will spend a considerable amount of time investigating various asymptotic regimes of Hermite polynomials, since the latter are main building blocks of various correlation functions of interest. In the last part of our lecture course we will discuss why statistics of characteristic polynomials of random Hermitian matrices turns out to be interesting and informative to investigate, and will make a contact with recent results in the domain.

The presentation is quite informal in the sense that I will not try to prove various statements in full rigor or generality. I rather attempt outlining the main concepts, ideas and techniques preferring a good illuminating example to a general proof. A much more rigorous and detailed exposition can be found in the cited literature. I will also frequently employ the symbol $\propto$. In the present set of lectures it always means that the expression following $\propto$ contains a multiplicative constant which is of secondary importance for our goals and can be restored when necessary.

2 Introduction

In these lectures we use the symbol ^{T} to denote matrix or vector transposition and the asterisk * to denote Hermitian conjugation. In the present section the bar $\overline{z}$ denotes complex conjugation.

Let us start with a square complex matrix $\hat{Z}$ of dimensions $N \times N$, with complex entries $z_{ij} = x_{ij} + iy_{ij}$, $1 \leq i,j \leq N$. Every such matrix can be conveniently looked at as a point in a $2N^2$-dimensional Euclidean space with real Cartesian coordinates x_{ij}, y_{ij}, and the length element in this space is defined in a standard way as:

$$(ds)^2 = \mathrm{Tr}\left(d\hat{Z}d\hat{Z}^*\right) = \sum_{ij} dz_{ij}\overline{dz_{ij}} = \sum_{ij}\left[(dx)^2_{ij} + (dy)^2_{ij}\right]. \qquad (2.1)$$

As is well-known (see e.g.[1]) any surface embedded in an Euclidean space inherits a natural Riemannian metric from the underlying Euclidean structure. Namely, let the coordinates in a $n-$dimensional Euclidean space be $(x_1, \ldots, x_n)$, and let a $k-$dimensional surface embedded in this space be parameterized in terms of coordinates $(q_1, \ldots, q_k)$, $k \leq n$ as $x_i = x_i(q_1, \ldots, q_k)$, $i =$

$1, \ldots n$. Then the Riemannian metric $g_{ml} = g_{lm}$ *on the surface* is defined from the Euclidean length element according to

$$(ds)^2 = \sum_{i=1}^{n} (dx_i)^2 = \sum_{i=1}^{n} \left(\sum_{m=1}^{k} \frac{\partial x_i}{\partial q_m} dq_m \right)^2 = \sum_{m,l=1}^{k} g_{mn} dq_m dq_l. \qquad (2.2)$$

Moreover, such a Riemannian metric induces the corresponding integration measure on the surface, with the volume element given by

$$d\mu = \sqrt{|g|} dq_1 \ldots dq_k, \quad g = \det (g_{ml})_{l,m=1}^{k}. \qquad (2.3)$$

For $k = n$ these are just the familiar formulae for the lengths and volume associated with change of coordinates in an Euclidean space. For example, for $n = 2$ we can pass from Cartesian coordinates $-\infty < x, y < \infty$ to polar coordinates $r > 0$, $0 \leq \theta < 2\pi$ by $x = r \cos \theta$, $y = r \sin \theta$, so that $dx = dr \cos \theta - r \sin \theta d\theta$, $dy = dr \sin \theta + r \cos \theta d\theta$, and the Riemannian metric is defined by $(ds)^2 = (dx)^2 + (dy)^2 = (dr)^2 + r^2 (d\theta)^2$. We find that $g_{11} = 1$, $g_{12} = g_{21} = 0$, $g_{22} = r^2$, and the volume element of the integration measure in the new coordinates is $d\mu = r dr d\theta$; as it should be. As the simplest example of a "surface" with $k < n = 2$ embedded in such a two-dimensional space we consider a circle $r = R = const$. We immediately see that the length element $(ds)^2$ restricted to this "surface" is $(ds)^2 = R^2 (d\theta)^2$, so that $g_{11} = R^2$, and the integration measure induced on the surface is correspondingly $d\mu = R d\theta$. The "surface" integration then gives the total "volume" of the embedded surface (i.e. circle length $2\pi R$).

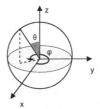

Figure 1: The spherical coordinates for a two dimensional sphere in the three-dimensional Euclidean space.

Similarly, we can consider a two-dimensional ($k = 2$) sphere $R^2 = x^2 + y^2 + z^2$ embedded in a three-dimensional Euclidean space ($n = 3$) with coordinates x, y, z and length element $(ds)^2 = (dx)^2 + (dy)^2 + (dz)^2$. A natural parameterization of the points on the sphere is possible in terms of the spherical coordinates ϕ, θ (see Fig. 1)

$$x = R \sin \theta \cos \phi, \ y = R \sin \theta \sin \phi, \ z = R \cos \theta; \quad 0 \leq \theta \leq \pi, \ 0 \leq \phi < 2\pi,$$

which results in $(ds)^2 = R^2(d\theta)^2 + R^2 \sin^2\theta(d\phi)^2$. Hence the matrix elements of the metric are $g_{11} = R^2$, $g_{12} = g_{21} = 0$, $g_{22} = R^2 \sin^2\theta$, and the corresponding "volume element" on the sphere is the familiar elementary area $d\mu = R^2 \sin\theta d\theta d\phi$.

As a less trivial example to be used later on consider a 2−dimensional manifold formed by 2×2 unitary matrices $\hat{U}$ embedded in the 8 dimensional Euclidean space of $Gl(2; C)$ matrices. Every such matrix can be represented as the product of a matrix $\hat{U}_c$ from the coset space $U(2)/U(1) \times U(1)$ parameterized by $k = 2$ real coordinates $0 \le \phi < 2\pi, 0 \le \theta \le \pi/2$, and a diagonal unitary matrix U_d, that is $\hat{U} = \hat{U}_d\hat{U}_c$, where

$$\hat{U}_c = \begin{pmatrix} \cos\theta & -\sin\theta e^{-i\phi} \\ \sin\theta e^{i\phi} & \cos\theta \end{pmatrix}, \quad \hat{U}_d = \begin{pmatrix} e^{-i\phi_1} & 0 \\ 0 & e^{i\phi_2} \end{pmatrix}. \quad (2.4)$$

Then the differential $d\hat{U}$ of the matrix $\hat{U} = \hat{U}_d\hat{U}_c$ has the following form:

$$d\hat{U} = \begin{pmatrix} -[d\theta\sin\theta + i\cos\theta d\phi_1]e^{-i\phi_1} & e^{-i(\phi_1+\phi)}[-d\theta\cos\theta + i(d\phi_1 + d\phi)\sin\theta] \\ e^{i(\phi+\phi_2)}[d\theta\cos\theta + i(d\phi + d\phi_2)\sin\theta] & [-d\theta\sin\theta + id\phi_2\cos\theta]e^{i\phi_2} \end{pmatrix}, \quad (2.5)$$

which yields the length element and the induced Riemannian metric:

$$\begin{aligned} (ds)^2 &= \mathrm{Tr}\left(d\hat{U}d\hat{U}^*\right) \quad (2.6) \\ &= 2(d\theta)^2 + (d\phi_1)^2 + (d\phi_2)^2 + 2\sin^2\theta(d\phi)^2 \\ &\quad + 2\sin^2\theta(d\phi\, d\phi_1 + d\phi\, d\phi_2). \end{aligned}$$

We see that the nonzero entries of the Riemannian metric tensor g_{mn} in this case are $g_{11} = 2$, $g_{22} = g_{33} = 1$, $g_{44} = 2\sin^2\theta$, $g_{24} = g_{42} = g_{34} = g_{43} = \sin^2\theta$, so that the determinant $\det[g_{mn}] = 4\sin^2\theta\cos^2\theta$. Finally, the induced integration measure on the group $U(2)$ is given by

$$d\mu(\hat{U}) = 2\sin\theta\cos\theta\, d\theta\, d\phi\, d\phi_1\, d\phi_2. \quad (2.7)$$

It is immediately clear that the above expression is invariant, by construction, with respect to multiplications $\hat{U} \to \hat{V}\hat{U}$, for any fixed unitary matrix V from the same group. Therefore, Eq.(2.7) is just the Haar measure on the group.

We will make use of these ideas several times in our lectures. Let us now concentrate on the N^2−dimensional subspace of Hermitian matrices in the $2N^2-$ dimensional space of all complex matrices of a given size N. The Hermiticity condition $\hat{H} = \hat{H}^* \equiv \overline{\hat{H}^T}$ amounts to imposing the following restrictions on the coordinates: $x_{ij} = x_{ji}$, $y_{ij} = -y_{ji}$. Such a restriction from the space of general complex matrices results in the length and volume element on the subspace of Hermitian matrices:

$$(ds)^2 = \mathrm{Tr}\left(d\hat{H}d\hat{H}^*\right) = \sum_i (dx_{ii})^2 + 2\sum_{i<j}\left[(dx_{ij})^2 + (dy_{ij})^2\right] \quad (2.8)$$

$$d\mu(\hat{H}) = 2^{\frac{N(N-1)}{2}}\prod_i dx_{ii}\prod_{i<j} dx_{ij}dy_{ij}. \quad (2.9)$$

Obviously, the length element $(ds)^2 = \mathrm{Tr} d\hat{H} d\hat{H}^*$ is invariant with respect to an automorphism (a mapping of the space of Hermitian matrices to itself) by a similarity transformation $\hat{H} \to U^{-1}\hat{H}\hat{U}$, where $\hat{U} \in U(N)$ is any given unitary $N \times N$ matrix: $\hat{U}^* = \hat{U}^{-1}$. Therefore the corresponding integration measure $d\mu(\hat{H})$ is also invariant with respect to all such "rotations of the basis".

The above-given measure $d\mu(\hat{H})$ written in the coordinates x_{ii}, $x_{i<j}$, $y_{i<j}$ is frequently referred to as the "flat measure". Let us discuss now another, very important coordinate system in the space of Hermitian matrices which will be in the heart of all subsequent discussions. As is well-known, every Hermitian matrix $\hat{H}$ can be represented as

$$\hat{H} = \hat{U}\hat{\Lambda}\hat{U}^{-1}, \quad \hat{\Lambda} = \mathrm{diag}(\lambda_1, \ldots, \lambda_N), \quad \hat{U}^*\hat{U} = \hat{I}, \qquad (2.10)$$

where real $-\infty < \lambda_k < \infty$, $k = 1, \ldots, N$ are eigenvalues of the Hermitian matrix, and rows of the unitary matrix $\hat{U}$ are corresponding eigenvectors. Generically, we can consider all eigenvalues to be simple (non-degenerate). More precisely, the set of matrices $\hat{H}$ with non-degenerate eigenvalues is dense and open in the N^2-dimensional space of all Hermitian matrices, and has full measure (see [3], p.94 for a formal proof). The correspondence $\hat{H} \to \left(\hat{U} \in U(N), \hat{\Lambda}\right)$ is, however, not one-to-one, namely $\hat{U}_1\hat{\Lambda}\hat{U}_1^{-1} = \hat{U}_2\hat{\Lambda}\hat{U}_2^{-1}$ if $\hat{U}_1^{-1}\hat{U}_2 = \mathrm{diag}\left(e^{i\phi_1}, \ldots, e^{i\phi_N}\right)$ for any choice of the phases $\phi_1, \ldots, \phi_N$. To make the correspondence one-to-one we therefore have to restrict the unitary matrices to the coset space $U(N)/U(1) \otimes \ldots \otimes U(1)$, and also to order the eigenvalues, e.g. requiring $\lambda_1 < \lambda_2 < \ldots < \lambda_N$. Our next task is to write the integration measure $d\mu(\hat{H})$ in terms of eigenvalues $\hat{\Lambda}$ and matrices $\hat{U}$. To this end, we differentiate the spectral decomposition $\hat{H} = \hat{U}\hat{\Lambda}\hat{U}^*$, and further exploit: $d\left(\hat{U}^*\hat{U}\right) = d\hat{U}^*\hat{U} + \hat{U}^*d\hat{U} = 0$. This leads to

$$d\hat{H} = \hat{U}\left[d\hat{\Lambda} + \hat{U}^*d\hat{U}\hat{\Lambda} - \hat{\Lambda}\hat{U}^*d\hat{U}\right]\hat{U}^*. \qquad (2.11)$$

Substituting this expression into the length element $(ds)^2$, see Eq.(2.8), and using the short-hand notation $\delta\hat{U}$ for the matrix $\hat{U}^*d\hat{U}$ satisfying anti-Hermiticity condition $\delta\hat{U}^* = -\delta\hat{U}$, we arrive at:

$$(ds)^2 = \mathrm{Tr}\left[\left(d\hat{\Lambda}\right)^2 + 2d\hat{\Lambda}\left(\delta\hat{U}\hat{\Lambda} - \hat{\Lambda}\delta\hat{U}\right)\right. \qquad (2.12)$$
$$\left. + \left(\delta\hat{U}\hat{\Lambda}\right)^2 + \left(\hat{\Lambda}\delta\hat{U}\right)^2 - 2\delta\hat{U}\hat{\Lambda}^2\delta\hat{U}\right].$$

Taking into account that $\hat{\Lambda}$ is purely diagonal, and therefore the diagonal entries of the commutator $\left(\delta\hat{U}\hat{\Lambda} - \hat{\Lambda}\delta\hat{U}\right)$ are zero, we see that the second term in Eq.(2.12) vanishes. On the other hand, the third and subsequent terms when added up are equal to

$$2\mathrm{Tr}\left[\delta\hat{U}\hat{\Lambda}\delta\hat{U}\hat{\Lambda} - \delta\hat{U}^2\hat{\Lambda}^2\right] = 2\sum_{ij}\left[\delta U_{ij}\lambda_j\delta U_{ji}\lambda_i - \lambda_i^2\delta U_{ij}\delta U_{ji}\right]$$

$$= -\sum_{ij}(\lambda_i - \lambda_j)^2\,\delta U_{ji}\delta U_{ij}$$

which together with the first term yields the final expression for the length element in the "spectral" coordinates

$$(ds)^2 = \sum_i (d\lambda_i)^2 + \sum_{i<j}(\lambda_i - \lambda_j)^2\,\overline{\delta U_{ij}}\delta U_{ij} \tag{2.13}$$

where we exploited the anti-Hermiticity condition $-\delta U_{ji} = \overline{\delta U_{ij}}$. Introducing the real and imaginary parts $\delta U_{ij} = \delta p_{ij} + i\delta q_{ij}$ as independent coordinates we can calculate the corresponding integration measure $d\mu(\hat{H})$ according to the rule in Eq.(2.3), to see that it is given by

$$d\mu(\hat{H}) = \prod_{i<j}(\lambda_i - \lambda_j)^2\prod_i d\lambda_i \times d\mathcal{M}(\hat{U}). \tag{2.14}$$

The last factor $d\mathcal{M}(U)$ stands for the part of the measure depending only on the U−variables. A more detailed consideration shows that, in fact, $d\mathcal{M}(\hat{U}) \equiv d\mu(\hat{U})$, which means that it is given (up to a constant factor) by the invariant Haar measure on the unitary group $U(N)$. This fact is however of secondary importance for the goals of the present lecture.

Having an integration measure at our disposal, we can introduce a probability density function (p.d.f.) $\mathcal{P}(\hat{H})$ on the space of Hermitian matrices, so that $\mathcal{P}(\hat{H})d\mu(\hat{H})$ is the probability that a matrix $\hat{H}$ belongs to the volume element $d\mu(\hat{H})$. Then it seems natural to require for such a probability to be invariant with respect to all the above automorphisms, i.e. $\mathcal{P}(\hat{H}) = \mathcal{P}\left(\hat{U}^*\hat{H}\hat{U}\right)$. It is easy to understand that this "postulate of invariance" results in $\mathcal{P}$ being a function of N first traces $\mathrm{Tr}\hat{H}^n$, $n = 1,\ldots,N$ (the knowledge of first N traces fixes the coefficients of the characteristic polynomial of $\hat{H}$ uniquely, and hence the eigenvalues. Therefore traces of higher order can always be expressed in terms of the lower ones). Of particular interest is the case

$$\mathcal{P}(\hat{H}) = C\exp -Tr\,Q(\hat{H}), \quad Q(x) = a_{2j}x^{2j} + \ldots + a_0, \tag{2.15}$$

where $2j \le N$, the parameters a_{2l} and C are real constants, and $a_{2j} > 0$. Observe that if we take

$$Q(x) = ax^2 + bx + c, \tag{2.16}$$

then $e^{-Tr\,Q(\hat{H})}$ takes the form of the product

$$e^{-a\left[\sum_i x_{ii}^2 + 2\sum_{i<j}(x_{ij}^2 + y_{ij}^2)\right]}e^{-b\sum_i x_{ii}}e^{-cN} \tag{2.17}$$

$$= e^{-cN}\prod_{i=1}^N\left(e^{-ax_{ii}^2 - bx_{ii}}\right)\prod_{i<j}e^{-2ax_{ij}^2}\prod_{i<j}e^{-2ay_{ij}^2}.$$

We therefore see that the probability distribution of the matrix $\hat{H}$ can be represented as a product of factors, each factor being a suitable Gaussian distribution depending only on one variable in the set of all coordinates x_{ii}, $x_{i<j}$, $y_{i<j}$. Since the same factorization is valid also for the integration measure $d\mu(\hat{H})$, see Eq.(2.9), we conclude that all these N^2 variables are statistically independent and Gaussian-distributed.

A much less obvious statement is that if we impose *simultaneously* two requirements:

- The probability density function $\mathcal{P}(\hat{H})$ is invariant with respect to all conjugations $\hat{H} \to \hat{H}' = U^{-1}\hat{H}\hat{U}$ by unitary matrices $\hat{U}$, that is $\mathcal{P}(\hat{H}') = \mathcal{P}(\hat{H})$; and

- the N^2 variables x_{ii}, $x_{i<j}$, $y_{i<j}$ are statistically independent, i.e.

$$\mathcal{P}(\hat{H}) = \prod_{i=1}^{N} f_i(x_{ii}) \prod_{i<j}^{N} f_{ij}^{(1)}(x_{ij}) f_{ij}^{(2)}(y_{ij}), \qquad (2.18)$$

then the function $\mathcal{P}(\hat{H})$ is *necessarily* of the form $\mathcal{P}(\hat{H}) = Ce^{-\left(a\mathrm{Tr}\hat{H}^2 + b\mathrm{Tr}\hat{H} + cN\right)}$, for some constants $a > 0$, b, c. The proof for any N can be found in [2], and here we just illustrate its main ideas for the simplest, yet nontrivial case $N = 2$. We require invariance of the distribution with respect to the conjugation of $\hat{H}$ by $\hat{U} \in U(2)$, and first consider a particular choice of the unitary matrix $\hat{U} = \begin{pmatrix} 1 & -\theta \\ \theta & 1 \end{pmatrix}$ corresponding to $\phi = \phi_1 = \phi_2 = 0$, and small values $\theta \ll 1$ in Eq.(2.4). In this approximation the condition $\hat{H}' = U^{-1}\hat{H}\hat{U}$ amounts to

$$\begin{pmatrix} x'_{11} & x'_{12} + iy'_{12} \\ x'_{12} - iy'_{12} & x'_{22} \end{pmatrix} \qquad (2.19)$$
$$= \begin{pmatrix} x_{11} + 2\theta x_{12} & x_{12} + iy_{12} + \theta\left(x_{22} - x_{11}\right) \\ x_{12} - iy_{12} + \theta\left(x_{22} - x_{11}\right) & x_{22} - 2\theta x_{12} \end{pmatrix},$$

where we kept only terms linear in θ. With the same precision we expand the factors in Eq.(2.18):

$$f_1(x'_1) = f_1(x_1)\left[1 + 2\theta x_{12}\frac{1}{f_1}\frac{df_1}{dx_{11}}\right], \quad f_2(x'_{22}) = f_2(x_{22})\left[1 - 2\theta x_{12}\frac{1}{f_2}\frac{df_2}{dx_{22}}\right]$$

$$f_{12}^{(1)}(x'_{12}) = f_{12}^{(1)}(x_{12})\left[1 + \theta(x_{22} - x_{11})\frac{1}{f_{12}^{(1)}}\frac{df_{12}^{(1)}}{dx_{12}}\right], \quad f_{12}^{(2)}(y'_{21}) = f_{12}^{(2)}(y_{12}).$$

The requirements of statistical independence and invariance amount to the product of the left-hand sides of the above expressions to be equal to the product of the right-hand sides, for any θ. This is possible only if:

$$2x_{12}\left[\frac{d\ln f_1}{dx_{11}} - \frac{d\ln f_2}{dx_{22}}\right] + (x_{22} - x_{11})\frac{d\ln f_{12}^{(1)}}{dx_{12}} = 0, \qquad (2.20)$$

which can be further rewritten as

$$\frac{1}{(x_{22} - x_{11})} \left[\frac{d \ln f_1}{dx_{11}} - \frac{d \ln f_2}{dx_{22}} \right] = const = \frac{1}{2x_{12}} \frac{d \ln f_{12}^{(1)}}{dx_{12}}, \qquad (2.21)$$

where we used that the two sides in the equation above depend on essentially different sets of variables. Denoting $const_1 = -2a$, we see immediately that

$$f_{12}^{(1)}(x_{12}) \propto e^{-2ax_{12}^2},$$

and further notice that

$$\frac{d \ln f_1}{dx_{11}} + 2ax_{11} = const_2 = \frac{d \ln f_2}{dx_{22}} + 2ax_{22}$$

by the same reasoning. Denoting $const_2 = -b$, we find:

$$f_1(x_{11}) \propto e^{-ax_{11}^2 - bx_{11}}, \quad f_2(x_{22}) \propto e^{-ax_{22}^2 - bx_{22}}, \qquad (2.22)$$

and thus we are able to reproduce the first two factors in Eq.(2.17). To reproduce the remaining factors we consider the conjugation by the unitary matrix $\hat{U}_d = \begin{pmatrix} 1 - i\alpha & 0 \\ 0 & 1 + i\alpha \end{pmatrix}$, which corresponds to the choice $\theta = 0$, $\phi_1 = \phi_2 = -\alpha = $ in Eq.(2.4), and again we keep only terms linear in the small parameter $\alpha \ll 1$. Within such a precision the transformation leaves the diagonal entries x_{11}, x_{22} unchanged, whereas the real and imaginary parts of the off-diagonal entries are transformed as

$$x'_{12} = x_{12} - 2\alpha y_{12}, \quad y'_{12} = y_{12} + 2\alpha x_{12}.$$

In this case the invariance of the p.d.f. $\mathcal{P}(\hat{H})$ together with the statistical independence of the entries amount, after straightforward manipulations, to the condition

$$\frac{1}{x_{12}} \frac{d \ln f_{12}^{(1)}}{dx_{12}} = \frac{1}{y_{12}} \frac{d \ln f_{12}^{(2)}}{dy_{12}}$$

which together with the previously found $f_{12}^{(1)}(x_{12})$ yields

$$f_{12}^{(2)}(y_{12}) \propto e^{-2ay_{12}^2},$$

completing the proof of Eq.(2.17).

The Gaussian form of the probability density function, Eq.(2.17), can also be found as a result of rather different lines of thought. For example, one may invoke an information theory approach *a la* Shanon-Khinchin and define the amount of information $\mathcal{I}[\mathcal{P}(\hat{H})]$ associated with any probability density function $\mathcal{P}(\hat{H})$ by

$$\mathcal{I}[\mathcal{P}(\hat{H})] = - \int d\mu(\hat{H}) \, \mathcal{P}(\hat{H}) \ln \mathcal{P}(\hat{H}) \qquad (2.23)$$

This is a natural extension of the corresponding definition $\mathcal{I}[p_1, \ldots, p_m] = -\sum_{l=1}^{m} p_m \ln p_m$ for discrete events $1, \ldots, m$.

Now one can argue that in order to have matrices $\hat{H}$ as random as possible one has to find the p.d.f. minimizing the information associated with it for a certain class of $\mathcal{P}(H)$ satisfying some conditions. The conditions usually have a form of constraints ensuring that the probability density function has desirable properties. Let us, for example, impose the only requirement that the ensemble average for the two lowest traces $\mathrm{Tr}\hat{H}, \mathrm{Tr}\hat{H}^2$ must be equal to certain prescribed values, say $E\left[\mathrm{Tr}\hat{H}\right] = b$ and $E\left[\mathrm{Tr}\hat{H}^2\right] = a > 0$, where the $E[\ldots]$ stand for the expectation value with respect to the p.d.f. $\mathcal{P}(H)$. Incorporating these constraints into the minimization procedure in a form of Lagrange multipliers ν_1, ν_2, we seek to minimize the functional

$$\mathcal{I}[\mathcal{P}(\hat{H})] = -\int d\mu(\hat{H})\,\mathcal{P}(\hat{H})\left\{\ln\mathcal{P}(\hat{H}) - \nu_1\mathrm{Tr}\hat{H} - \nu_2\mathrm{Tr}\hat{H}^2\right\}. \qquad (2.24)$$

The variation of such a functional with respect to $\delta\mathcal{P}(\mathcal{H})$ results in

$$\delta\mathcal{I}[\mathcal{P}(\hat{H})] = -\int d\mu(\hat{H})\,\delta\mathcal{P}(\hat{H})\left\{1 + \ln\mathcal{P}(\hat{H}) - \nu_1\mathrm{Tr}\hat{H} - \nu_2\mathrm{Tr}\hat{H}^2\right\} = 0 \qquad (2.25)$$

possible only if

$$\mathcal{P}(\hat{H}) \propto \exp\{\nu_1\mathrm{Tr}\hat{H} + \nu_2\mathrm{Tr}\hat{H}^2\}$$

again giving the Gaussian form of the p.d.f. The values of the Lagrange multipliers are then uniquely fixed by constants a, b, and the normalization condition on the probability density function. For more detailed discussion, and for further reference see [2], p.68.

Finally, let us discuss yet another construction allowing one to arrive at the Gaussian Ensembles exploiting the idea of Brownian motion. To start with, consider a system whose state at time t is described by one real variable x, evolving in time according to the simplest linear differential equation $\frac{d}{dt}x = -x$ describing a simple exponential relaxation $x(t) = x_0 e^{-t}$ towards the stable equilibrium $x = 0$. Suppose now that the system is subject to a random additive Gaussian white noise $\xi(t)$ function of intensity D [1], so that the corresponding equation acquires the form

$$\frac{d}{dt}x = -x + \xi(t), \qquad E_\xi\left[\xi(t_1)\xi(t_1)\right] = D\delta(t_1 - t_2), \qquad (2.26)$$

---

[1]The following informal but instructive definition of the white noise process may be helpful for those not very familiar with theory of stochastic processes. For any $0 < t < 2\pi$ and integer $k \geq 1$ define the random function $\xi_k(t) = \sqrt{2/\pi}\sum_{n=0}^{k} a_n \cos nt$, where real coefficients a_n are all independent, Gaussian distributed with zero mean $E[a_n] = 0$ and variances $E[a_0^2] = D/2$ and $E[a_n^2] = D$ for $1 \leq n \leq k$. Then one can, in a certain sense, consider white noise as the limit of $\xi_k(t)$ for $k \to \infty$. In particular, the Dirac $\delta(t - t')$ is approximated by the limiting value of $\frac{\sin[(k+1/2)(t-t')]}{2\pi\sin(t-t')/2}$.

where $E_\xi[\ldots]$ stands for the expectation value with respect to the random noise. The main characteristic property of a Gaussian white noise process is the following identity:

$$E_\xi \left[\exp\left\{ \int_a^b \xi(t)v(t)dt \right\} \right] = \exp\left\{ \frac{D}{2} \int_a^b v^2(t)dt \right\} \qquad (2.27)$$

valid for any (smooth enough) test function $v(t)$. This is just a direct generalization of the standard Gaussian integral identity:

$$\int_{-\infty}^\infty \frac{dq}{\sqrt{2\pi a}}\, e^{-\frac{1}{2a}q^2 + qb} = e^{\frac{ab^2}{2}}, \qquad (2.28)$$

valid for $\mathrm{Re}\, a > 0$, and any (also complex) parameter b.

For any given realization of the Gaussian random process $\xi(t)$ the solution of the stochastic differential equation Eq.(2.26) is obviously given by

$$x(t) = e^{-t}\left[x_0 + \int_0^t e^\tau \xi(\tau)d\tau \right]. \qquad (2.29)$$

This is a random function, and our main goal is to find the probability density function $\mathcal{P}(t, x)$ for the variable $x(t)$ to take value x at any given moment in time t, if we know surely that $x(0) = x_0$. This p.d.f. can be easily found from the characteristic function

$$\mathcal{F}(t, q) = E_\xi\left[e^{-iqx(t)} \right] = \exp\left\{ -iqx_0 e^{-t} - \frac{Dq^2}{4}(1 - e^{-2t}) \right\} \qquad (2.30)$$

obtained by using Eqs. (2.27) and (2.29). The p.d.f. is immediately recovered by employing the inverse Fourier transform:

$$
\begin{aligned}
\mathcal{P}(t, x) &= \int_{-\infty}^\infty \frac{dq}{2\pi} e^{iqx} E_\xi\left[e^{-iqx(t)} \right] \qquad (2.31) \\
&= \frac{1}{\sqrt{\pi D(1 - e^{-2t})}} \exp\left\{ -\frac{(x - x_0 e^{-t})^2}{D(1 - e^{-2t})} \right\}.
\end{aligned}
$$

The formula Eq.(2.31) is called the Ornstein-Uhlenbeck (OU) probability density function, and the function $x(t)$ satisfying the equation Eq.(2.26) is known as the O-U process. In fact, such a process describes an interplay between the random "kicks" forcing the system to perform a kind of Brownian motion and the relaxation towards $x = 0$. It is easy to see that when time grows the OU p.d.f. "forgets" about the initial state and tends to a *stationary* (i.e. time-independent) universal Gaussian distribution:

$$\mathcal{P}(t \to \infty, x) = \frac{1}{\sqrt{\pi D}} \exp\left\{ -\frac{x^2}{D} \right\}. \qquad (2.32)$$

Coming back to our main topic, let us consider N^2 independent OU processes: N of them denoted as

$$\frac{d}{dt}x_i = -x_i + \xi_i(t), \quad 1 \leq i \leq N \tag{2.33}$$

and the rest $N(N-1)$ given by

$$\frac{d}{dt}x_{ij} = -x_{ij} + \xi_{ij}^{(1)}(t), \quad \frac{d}{dt}y_{ij} = -y_{ij} + \xi_{ij}^{(2)}(t), \tag{2.34}$$

where the indices satisfy $1 \leq i < j \leq N$. Stochastic processes $\xi(t)$ in the above equations are taken to be all mutually independent Gaussian white noises characterized by the correlation functions:

$$\begin{aligned}
E_\xi\left[\xi_{i_1}(t_1)\xi_{i_2}(t_2)\right] &= 2D\delta_{i_1,i_2}\delta(t_1-t_2), \quad E_\xi\left[\xi_{ij}^{\sigma_1}(t_1)\xi_{kl}^{\sigma_2}(t_2)\right] \\
&= D\delta_{\sigma_1,\sigma_2}\delta_{i,k}\delta_{j,l}\delta(t_1-t_2).
\end{aligned} \tag{2.35}$$

As initial values $x_i(0), x_{ij}(0), y_{ij}(0)$ for each OU process we choose diagonal and off-diagonal entries $H_{ii}^{(0)}$, $i = 1, \ldots, N$ and $\mathrm{Re}H_{i<j}^{(0)}$, $\mathrm{Im}H_{i<j}^{(0)}$ of a fixed $N \times N$ Hermitian matrix $\hat{H}^{(0)}$. Let us now consider the Hermitian matrix $\hat{H}(t)$ whose entries are $H_{ii}(t) = x_i(t)$, $H_{i<j}(t) = x_{i<j}(t) + iy_{i<j}(t)$ for any $t \geq 0$. It is immediately clear that the joint p.d.f. $\mathcal{P}\left(t, \hat{H}\right)$ of the entries of such a matrix $\hat{H}(t)$ will be given for any $t \geq 0$ by the OU-type formula:

$$\mathcal{P}(t,\hat{H}) \propto Const \times \frac{1}{\sqrt{(1-e^{-2t})^{N^2}}} \exp\left\{-\frac{1}{D(1-e^{-2t})}\mathrm{Tr}\left(\hat{H} - \hat{H}_0 e^{-t}\right)^2\right\}. \tag{2.36}$$

In the limit $t \to \infty$ this p.d.f. converges to a stationary, t-independent expression

$$\mathcal{P}(t,\hat{H}) \propto C\,e^{-\frac{1}{D}\mathrm{Tr}\hat{H}^2} \tag{2.37}$$

independent of the initial matrix $\hat{H}_0$. We see therefore that the familiar Gaussian ensemble in the space of Hermitian matrices arises as the result of the stochastic relaxation from any initial condition, in particular, from any diagonal matrix with uncorrelated entries. In the next step one may try to deduce the stochastic dynamics of the *eigenvalues* of the corresponding matrices. Those eigenvalues obviously evolve from completely uncorrelated to highly correlated patterns. This very interesting set of question goes beyond our present goals and we refer to [2] for an introduction into the problem.

After specifying the probability density function $\mathcal{P}(H)$ the main question of interest is to characterize the statistical properties of the sequence of eigenvalues $\lambda_1, \ldots, \lambda_N$ of $\hat{H}$. A convenient way of doing this is to start with the joint p.d.f. of all these eigenvalues. Because of the "rotational invariance" assumption the function $\mathcal{P}(\hat{H})$ depends in fact only on the eigenvalues, for example for the "symmetric" Gaussian case $b = 0$ we have $\mathcal{P}(\hat{H}) \propto e^{-a\sum_{i=1}^N \lambda_i^2}$.

Moreover, we have seen that the integration measure $d\mu(\hat{H})$ when expressed in terms of eigenvalues and eigenvectors effectively factorizes, see Eq.(2.14). Collecting all these facts we arrive at the conclusion, that the relevant joint p.d.f of all eigenvalues can be always written, up to a normalization constant, as

$$\mathcal{P}(\lambda_1, \ldots, \lambda_N)\, d\lambda_1 \ldots d\lambda_N \propto e^{-\sum_{i=1}^{N} Q(\lambda_i)} \prod_{i<j} (\lambda_i - \lambda_j)^2 \; d\lambda_1 \ldots d\lambda_N \qquad (2.38)$$

for a general, non-gaussian weight $e^{-\mathrm{Tr}Q(\hat{H})}$. We immediately see that the presence of the "Jacobian factor" $\prod_{i<j} (\lambda_i - \lambda_j)^2$ is responsible of the fact that the eigenvalues are correlated in a non-trivial way. In what follows we are going to disregard the fact that eigenvalues λ_i were initially put in increasing order. More precisely, for any *symmetric* function f of N real variables $\lambda_1, \ldots, \lambda_N$ the expected value will be calculated as

$$\int_{\mathbf{R}^N} f(\lambda_1, \ldots, \lambda_N) \mathcal{P}(\lambda_1, \ldots, \lambda_N) d\lambda_1 \ldots d\lambda_N.$$

Indeed, with p.d.f. being symmetric with respect to permutations of the eigenvalue set, disregarding the ordering amounts to a simple multiplicative combinatorial factor $n!$ in the normalization constant.

Our main goal is to extract the information about these eigenvalue correlations in the limit of large size N. From this point of view it is pertinent to discuss a few quantitative measures frequently used to characterize correlations in sequences of real numbers.

3   Characterization of Spectral Sequences

Let $-\infty < \lambda_1, \lambda_2, \ldots, \lambda_N < \infty$ be the positions of N points on the real axis, characterized by the joint probability density function (JPDF)

$$\mathcal{P}(\lambda_1, \lambda_2, \ldots, \lambda_N)\, d\lambda_1 \ldots d\lambda_N$$

of having, *regardless of labelling*, one point in the interval $[\lambda_1, \lambda_1 + d\lambda_1]$, another in the interval $[\lambda_2, \lambda_2 + d\lambda_2]$,..., another in $[\lambda_N, \lambda_N + d\lambda_N]$. Since in this section we deal exclusively with real variables, the bar will stand for the expectation value with respect to such a JPDF.

The statistical properties of the sequence $\{\lambda_i\}$ are conveniently characterized by the set of $n-$point *correlation functions*, defined as

$$\mathcal{R}_n(\lambda_1, \lambda_2, \ldots, \lambda_n) = \frac{N!}{(N-n)!} \int \mathcal{P}(\lambda_1, \lambda_2, \ldots, \lambda_N)\, d\lambda_{n+1} \ldots d\lambda_N. \qquad (3.1)$$

It is obvious from this definition that the lower correlation functions can be obtained from the higher-order ones:

$$\mathcal{R}_n(\lambda_1, \lambda_2, \ldots, \lambda_n) = \frac{1}{(N-n)} \int \mathcal{R}_{n+1}(\lambda_1, \lambda_2, \ldots, \lambda_{n+1})\, d\lambda_{n+1}. \qquad (3.2)$$

To provide a more clear interpretation of these correlation functions we relate them to the statistics of the number N_B of points of the sequence $\{\lambda_i\}$ within any set B of the real axis (e.g an interval $[a,b]$). Let $\chi_B(x)$ be the characteristic function of the set B, equal to unity if $x \in B$ and zero otherwise. Introduce the exact density function $\rho_N(\lambda)$ of the points $\{\lambda_i\}$ around the point λ on the real axis. It can be conveniently written using the Dirac's $\delta-$function as $\rho_N(\lambda) = \sum_{i=1}^{N} \delta(\lambda - \lambda_i)$. Then $N_B = \int \chi_B(\lambda)\rho_N(\lambda)d\lambda$.

On the other hand, consider

$$\int_B \mathcal{R}_1(\lambda_1)d\lambda_1 = \int \chi_B(\lambda_1)\mathcal{R}_1(\lambda_1)d\lambda_1 \tag{3.3}$$

$$= N \int \chi_B(\lambda_1)\mathcal{P}(\lambda_1, \lambda_2, \ldots, \lambda_N) \, d\lambda_1 \ldots d\lambda_N$$

$$= \int \sum_{i=1}^{N} \chi_B(\lambda_i)\mathcal{P}(\lambda_1, \lambda_2, \ldots, \lambda_N) \, d\lambda_1 \ldots d\lambda_N$$

and therefore

$$\int_B \mathcal{R}_1(\lambda_1)d\lambda_1 = \overline{N_B} = \text{expectation of the number of points in B.} \tag{3.4}$$

Similarly, consider

$$\int \chi_B(\lambda_1)\chi_B(\lambda_2)\mathcal{R}_2(\lambda_1, \lambda_2)d\lambda_1 d\lambda_2 \tag{3.5}$$

$$= N(N-1) \int \chi_B(\lambda_1)\chi_B(\lambda_2)\mathcal{P}(\lambda_1, \lambda_2, \ldots, \lambda_N) \, d\lambda_1 \ldots d\lambda_N$$

$$= \int \sum_{i \neq j}^{N} \chi_B(\lambda_i)\chi_B(\lambda_j)\mathcal{P}(\lambda_1, \lambda_2, \ldots, \lambda_N) \, d\lambda_1 \ldots d\lambda_N,$$

which can be interpreted as

$$\int_{B \times B} \mathcal{R}_2(\lambda_1, \lambda_2)d\lambda_1 d\lambda_2 = \text{expectation of the number of pairs of points in B} \tag{3.6}$$

where if, say, λ_1 and λ_2 are in B, then the pair $\{1,2\}$ and $\{2,1\}$ are both counted.

To relate the two-point correlation function to the variance of N_B we notice that in view of Eq.(3.4) the one-point correlation function $\mathcal{R}_1(\lambda)$ coincides with the mean density $\overline{\rho}_N(\lambda)$ of the points $\{\lambda_i\}$ around the point λ on the real axis. Similarly, write the mean square $\overline{N_B^2}$

$$\overline{N_B^2} = \int \chi_B(\lambda)\chi_B(\lambda')\overline{\rho_N(\lambda)\rho_N(\lambda')} \, d\lambda d\lambda' \tag{3.7}$$

and notice that

$$\overline{\rho_N(\lambda)\rho_N(\lambda')} = \overline{\sum_{ij} \delta(\lambda - \lambda_i)\delta(\lambda' - \lambda_j)} \tag{3.8}$$

$$= \delta(\lambda - \lambda')\overline{\sum_i \delta(\lambda - \lambda_i)} + \overline{\sum_{i \neq j} \delta(\lambda - \lambda_i)\delta(\lambda' - \lambda_j)}$$

$$= \delta(\lambda - \lambda')\mathcal{R}_1(\lambda) + \mathcal{R}_2(\lambda, \lambda').$$

In this way we arrive at the relation:

$$\overline{N_B^2} = \overline{N_B} + \int_{B \times B} \mathcal{R}_2(\lambda, \lambda') \, d\lambda d\lambda'. \tag{3.9}$$

In fact, it is natural to introduce the so-called "number variance" statistics $\Sigma_2(B) = \overline{N_B^2} - \left[\overline{N_B}\right]^2$ describing the variance of the number of points of the sequence inside the set B. Obviously,

$$\Sigma_2(B) = \overline{N_B} + \int_{B \times B} [\mathcal{R}_2(\lambda, \lambda') - \mathcal{R}_1(\lambda)\mathcal{R}_1(\lambda')] \, d\lambda d\lambda' \tag{3.10}$$

$$\equiv \overline{N_B} - \int_{B \times B} Y_2(\lambda, \lambda') \, d\lambda d\lambda'$$

where we introduced the so-called *cluster function* $Y_2(\lambda, \lambda') = \mathcal{R}_1(\lambda)\mathcal{R}_1(\lambda') - \mathcal{R}_2(\lambda, \lambda')$ frequently used in applications.

Finally, in principle the knowledge of all $n-$point correlation functions provides one with the possibility of calculating an important characteristic of the spectrum known as the "hole probability" $A(L)$. This quantity is defined as the probability for a random matrix to have *no* eigenvalues in the interval $(-L/2, L/2)$ [2]. Define $\chi_L(\lambda)$ to be the characteristic function of this interval. Obviously,

$$A(L) = \int \ldots \int \mathcal{P}(\lambda_1, \ldots, \lambda_N) \prod_{k=1}^{N} (1 - \chi_L(\lambda_k)) \, d\lambda_1 \ldots d\lambda_N \tag{3.11}$$

$$= \sum_{j=0}^{N} (-1)^j \int \ldots \int \mathcal{P}(\lambda_1, \ldots, \lambda_N) h_j \{\chi_L(\lambda_1), \ldots, \chi_L(\lambda_N)\} \, d\lambda_1 \ldots d\lambda_N,$$

where $h_j\{x_1, \ldots, x_N\}$ is the $j - th$ symmetric function:

$$h_0\{x_1, \ldots, x_N\} = 1, \quad h_1\{x_1, \ldots, x_N\} = \sum_{i=1}^{N} x_i,$$

$$h_2\{x_1, \ldots, x_N\} = \sum_{i<j}^{N} x_i x_j, \quad \ldots, \quad h_N\{x_1, \ldots, x_N\} = x_1 x_2 \ldots x_N.$$

---

[2] Sometimes one uses instead the interval $[-L, L]$ to define $A(L)$, see e.g. [3].

Now, for $1 \leq j \leq N$

$$\int \ldots \int \prod_{k=1}^{j} \chi_L(\lambda_k) \, \mathcal{P}(\lambda_1, \ldots, \lambda_N) d\lambda_1 \ldots d\lambda_N$$

$$= \frac{(N-j)!}{N!} \int \ldots \int \prod_{k=1}^{j} \chi_L(\lambda_k) \mathcal{R}_j(\lambda_1, \ldots, \lambda_j) \, d\lambda_1 \ldots d\lambda_j$$

$$= \frac{(N-j)!}{N!} \int_{|x_1|<L/2} \ldots \int_{|x_j|<L/2} \mathcal{R}_j(\lambda_1, \ldots, \lambda_j) \, d\lambda_1 \ldots d\lambda_j. \quad (3.12)$$

As h_j contains $\binom{N}{j}$ terms and as $\mathcal{R}_j(\lambda_1, \ldots, \lambda_j)$ is invariant under permutations of the arguments, it follows that

$$\int \ldots \int \mathcal{P}(\lambda_1, \ldots, \lambda_N) h_j \left\{ \chi_L(\lambda_1) \ldots \chi_L(\lambda_N) \right\} \, d\lambda_1 \ldots d\lambda_j$$

$$= \frac{(N-j)!}{N!} \binom{N}{j} \int_{|x_1|<L/2} \ldots \int_{|x_j|<L/2} \mathcal{R}_j(\lambda_1, \ldots, \lambda_j) \, d\lambda_1 \ldots d\lambda_j$$

$$= \frac{1}{j!} \int_{|x_1|<L/2} \ldots \int_{|x_j|<L/2} \mathcal{R}_j(\lambda_1, \ldots, \lambda_j) \, d\lambda_1 \ldots d\lambda_j. \quad (3.13)$$

Thus, we arrive at the following relation between the hole probability and the n−point correlation functions:

$$A(L) = \sum_{j=0}^{N} \frac{(-1)^j}{j!} \int_{-L/2}^{L/2} \ldots \int_{-L/2}^{L/2} \mathcal{R}_j(\lambda_1, \ldots, \lambda_j) \, d\lambda_1 \ldots d\lambda_j. \quad (3.14)$$

One of the main goals of this set of lectures is to develop a method allowing to evaluate all the n−point correlation functions of the eigenvalues for any JPDF corresponding to unitary invariant ensembles of the form Eq.(2.15). After that we will concentrate on a particular case of Gaussian weight and will investigate the limiting behaviour of the kernel function $K_n(\lambda, \lambda')$ as $N \to \infty$. But even before doing this it is useful to keep in mind for reference purposes the results corresponding to completely uncorrelated (a.k.a. Poissonian) spectra. Those are described by a sequence of real points $\lambda_1, \ldots, \lambda_N$, characterized by the fully factorized JPDF:

$$\mathcal{P}(\lambda_1, \lambda_2, \ldots, \lambda_N) = p(\lambda_1) \ldots p(\lambda_N). \quad (3.15)$$

The normalization condition requires $\int_{-\infty}^{\infty} p(\lambda) \, d\lambda = 1$, and we further assume $p(\lambda)$ to be a smooth enough integrable function. Obviously, for this case

$$\mathcal{R}_n(\lambda_1, \lambda_2, \ldots, \lambda_n) = \frac{N!}{(N-n)!} p(\lambda_1) \ldots p(\lambda_n). \quad (3.16)$$

In particular, $\mathcal{R}_n(\lambda) = Np(\lambda)$ which is just the mean density $\overline{\rho(\lambda)}$ of points $\{\lambda_i\}$ around the point λ on the real axis, and $\mathcal{R}_n(\lambda_1, \lambda_2) = N(N-1)p(\lambda_1)p(\lambda_2)$, etc.. From this we easily find for the number of levels in the domain B and for its mean square:

$$\overline{N_B} = N \int_B p(\lambda) \, d\lambda, \quad \overline{N_B^2} = \overline{N_B}(N-1)/N + \left[\overline{N_B}\right]^2 \tag{3.17}$$

and for the hole probability

$$A(L) = \sum_{j=0}^{N} \frac{(-1)^j}{j!} \frac{N!}{(N-j)!} \left[\int_{-L/2}^{L/2} p(\lambda) \, d\lambda\right]^j \tag{3.18}$$

$$= \left[1 - \int_{-L/2}^{L/2} p(\lambda) \, d\lambda\right]^N .$$

Finally, let us specify B to be the interval $[-L/2, L/2]$ around the origin, and being interested mainly in large $N \gg 1$ consider the length L comparable with the mean spacing between neighbouring points in the sequence $\{\lambda_i\}$ close to the origin, given by $\Delta \equiv \left[\overline{\rho(0)}\right]^{-1} = 1/[Np(0)]$. In other words $s = L/\Delta = LNp(0)$ stays finite when $N \to \infty$. On the other hand, for large enough N the function $p(\lambda)$ can be considered practically constant through the interval of the length $L = O(1/N)$, and therefore the mean number of points of the sequence $\{\lambda_i\}$ inside the interval $[-L/2, L/2]$ will be asymptotically given by $\overline{N(s)} \approx N L p(0) = s$. Similarly, using Eq.(3.17) one can easily calculate the "number variance" $\Sigma_2(s) = \overline{N_{[-\frac{L}{2}, \frac{L}{2}]}^2} - \left[\overline{N_{[-\frac{L}{2}, \frac{L}{2}]}}\right]^2 = (N-1) \int_{-L/2}^{L/2} p(\lambda) \, d\lambda \approx s$. In the same approximation the hole probability, Eq.(3.18), tends asymptotically to $A(s) \approx e^{-s}$. Later on we shall compare these results with the corresponding behaviour emerging from the random matrix calculations.

4 The method of orthogonal polynomials

In the heart of the method developed mainly by Dyson, Mehta and Gaudin lies an "integrating-out" Lemma [2]. In presenting this material I follow very closely [3], pp.103-105.

- Let $J_n = J_n(\mathbf{x}) = (J_{ij})_{1 \le i,j \le n}$ be an $n \times n$ matrix whose entries depend on a real vector $\mathbf{x} = (x_1, x_2, \ldots, x_n)$ and have the form $J_{ij} = f(x_i, x_j)$, where f is some (in general, complex-valued) function satisfying for some measure $d\mu(x)$ the "reproducing kernel" property:

$$\int f(x, y) f(y, z) \, d\mu(y) = f(x, z). \tag{4.1}$$

  Then

$$\int \det J_n(\mathbf{x}) \, d\mu(x_n) = [q - (n-1)] \det J_{n-1} \tag{4.2}$$

  where $q = \int f(x, x) \, d\mu(x)$, and the matrix $J_{n-1} = (J_{ij})_{1 \le i,j \le n-1}$ have the same functional form as J_n with $\mathbf{x}$ replaced by $(x_1, x_2, \ldots, x_{n-1})$.

Before giving the idea of the proof for an arbitrary n it is instructive to have a look on the simplest case $n = 2$, when

$$J_2 = \begin{pmatrix} f(x_1, x_1) & f(x_1, x_2) \\ f(x_2, x_1) & f(x_2, x_2) \end{pmatrix},$$

hence $\quad \det J_n = f(x_1, x_1)f(x_2, x_2) - f(x_1, x_2)f(x_2, x_1).$

Integrating the latter expression over x_2, and using the "reproducing kernel" property, we immediately see that the result is indeed just $(q - 1)f(x, x) = (q - 1) \det J_1$ in full agreement with the statement of the Lemma.

For general n one should follow essentially the same strategy and expand the determinant as a sum over $n!$ permutations $P_n(\sigma) = (\sigma_1, \ldots \sigma_n)$ of the index set $1, \ldots, n$ as

$$\int \det J_n(\mathbf{x}) \, d\mu(x_n) = \sum_{P_n} (-1)^{P_n} \int f(x_1, x_{\sigma_1}) \ldots f(x_n, x_{\sigma_n}) \, d\mu(x_n), \quad (4.3)$$

where $(-1)^{P_n}$ stands for the sign of permutations. Now, we classify the terms in the sum according to the actual value of the index $\sigma_n = k$, $k = 1, 2, \ldots, n$. Consider first the case $\sigma_n = n$, when effectively only the last factor $f(x_n, x_n)$ in the product is integrated yielding d upon the integration. Summing up over the remaining $(n - 1)!$ permutations $P_{n-1}(\sigma)$ of the index set $(1, 2, \ldots, n - 1)$ we see that:

$$\sum_{P_{n-1}} (-1)^{P_n} \int f(x_1, x_{\sigma_1}) \ldots f(x_n, x_n) \, d\mu(x_n)$$

$$= q \sum_{P_{n-1}} (-1)^{P_{n-1}} f(x_1, x_{\sigma_1}) \ldots f(x_{n-1}, x_{\sigma_{n-1}}),$$

which is evidently equal to $q \det J_{n-1}$. Now consider $(n - 1)!$ terms with $\sigma_n = k < n$, when we have $\sigma_j = n$ for some $j < n$. For every such term we have by the "reproducing property"

$$\int f(x_1, x_{\sigma_1}) \ldots f(x_j, x_n) \ldots f(x_n, x_k) \, d\mu(x_n)$$

$$= f(x_1, x_{\sigma_1}) \ldots f(x_j, x_k) \ldots f(x_{n-1}, x_{\sigma_{n-1}}).$$

Therefore

$$\int \det J_n(\mathbf{x}) \, d\mu(x_n) = \quad (4.4)$$

$$q \det J_{n-1} + \sum_{k=1}^{n-1} \sum_{P_n : \sigma_n = k)} (-1)^{P_n} f(x_1, x_{\sigma_1}) \ldots f(x_j, x_k) \ldots f(x_{n-1}, x_{\sigma_{n-1}}).$$

It is evident that the structure and the number of terms is as required, and the remaining task is to show that the summation over all possible $(n - 1)!$

permutation of the index set for fixed k yields always $-\det J_{n-1}$, see [3]. Then the whole expression is indeed equal to $[q - (n - 1)] \det J_{n-1}$ as required.

Our next step is to apply this Lemma to calculating the $n-$point correlation functions of the eigenvalues $\lambda_1, \ldots, \lambda_n$ starting from the JPDF, Eq.(2.38).

For this we notice that

$$\prod_{i<j}^{N}(\lambda_i - \lambda_j) = (-1)^{\frac{N(N-1)}{2}} \det \begin{pmatrix} 1 & \cdots & 1 \\ \lambda_1 & \cdots & \lambda_N \\ \cdot & \cdot & \cdot \\ \cdot & \cdot & \cdot \\ \cdot & \cdot & \cdot \\ \lambda_1^{N-1} & \cdots & \lambda_N^{N-1} \end{pmatrix} \equiv \Delta_N(\lambda_1, \ldots, \lambda_N), \quad (4.5)$$

where the determinant in the right-hand side is the famous van der Monde determinant. Since the determinant cannot change upon linearly combining its rows, the entries λ_i^k in $(k+1)-th$ row of the van der Monde determinant can be replaced, up to a constant factor $a_0 a_1 \ldots a_{N-1}$, by a polynomial of degree k of the form: $\pi_k(\lambda_i) = a_k \lambda_i^k +$ any polynomial in λ_i of degree less than k, with any choice of the coefficients a_l, $l = 0, \ldots, k$. Therefore:

$$\prod_{i<j}^{N}(\lambda_i - \lambda_j) = \frac{(-1)^{\frac{N(N-1)}{2}}}{a_0 a_1 \ldots a_{N-1}} \det \begin{pmatrix} \pi_0(\lambda_1) & \cdots & \pi_0(\lambda_N) \\ \pi_1(\lambda_1) & \cdots & \pi_1(\lambda_N) \\ \cdot & \cdot & \cdot \\ \cdot & \cdot & \cdot \\ \cdot & \cdot & \cdot \\ \pi_{N-1}(\lambda_1) & \cdots & \pi_{N-1}(\lambda_1) \end{pmatrix}$$

$$\equiv \frac{(-1)^{\frac{N(N-1)}{2}}}{a_0 a_1 \ldots a_{N-1}} \det \left(\pi_{i-1}(\lambda_j) \right)_{1 \le i,j \le N}. \quad (4.6)$$

Multiplying every entry in j_{th} column in the above determinant with the factor $e^{-\frac{1}{2}Q(\lambda_j)}$ we see that the JPDF can be conveniently written, up to a multiplicative constant, as

$$\mathcal{P}(\lambda_1, \ldots, \lambda_N) \propto \left[\det \left(e^{-\frac{1}{2}Q(\lambda_j)} \pi_{i-1}(\lambda_j) \right)_{1 \le i,j \le N} \right]^2. \quad (4.7)$$

If we let $\hat{A}$ be the matrix with the entries $A_{ij} = (\phi_{i-1}(x_j))_{1 \le i,j \le N}$, then

$$[\det A]^2 = \det \hat{A}^T \hat{A} = \det \left(\sum_{j=1}^{n} A_{ji} A_{jk} \right). \quad (4.8)$$

This implies the following form of the JPDF:

$$\mathcal{P}(\lambda_1, \ldots, \lambda_N) \propto \det \left(\sum_{j=1}^{N} \phi_{j-1}(\lambda_i) \phi_{j-1}(\lambda_k) \right)_{1 \le i,k \le N}$$

$$\equiv \det \left(K_N(\lambda_i, \lambda_k) \right)_{1 \le i,k \le N} \quad (4.9)$$

where we introduced the notation:

$$K_N(\lambda, \lambda') = \sum_{j=0}^{N-1} \phi_j(\lambda)\phi_j(\lambda') \tag{4.10}$$

usually called "kernel" in the literature. In our particular case

$$\phi_{i-1}(\lambda) = e^{-\frac{1}{2}Q(\lambda)}\pi_{i-1}(\lambda) \tag{4.11}$$

so that the kernel is given explicitly by

$$K_N(\lambda, \lambda') = e^{-\frac{1}{2}(Q(\lambda)+Q(\lambda'))} \sum_{j=0}^{N-1} \pi_j(\lambda)\pi_j(\lambda'). \tag{4.12}$$

Now it is easy to see that if we take the polynomials $\pi_i(x)$ such that they form an *orthonormal system* with respect to the weight $e^{-Q(x)}$, the corresponding kernel will be a "reproducing" one with respect to the measure $d\mu(x) \equiv dx$, in the sense of the "integrating-out" Lemma. Indeed, suppose that $\pi_i(x)$ satisfy the orthonormality conditions:

$$\int e^{-Q(x)}\pi_i(x)\pi_j(x)\,dx = \delta_{ij}, \tag{4.13}$$

for any indices $i \geq 1$, $j \geq 1$. Then we obviously have

$$\int K_N(x,y)K_N(y,z)dy$$

$$= \sum_{j=0}^{N-1}\sum_{k=0}^{N-1} e^{-\frac{1}{2}(Q(x)+Q(z))}\pi_j(x)\pi_k(z)\int \pi_j(y)\pi_k(y)e^{-Q(y)}dy$$

$$= \sum_{j=0}^{N-1} e^{-\frac{1}{2}(Q(x)+Q(z))}\pi_j(x)\pi_j(z) = K_N(x,z) \tag{4.14}$$

exactly as required by the reproducing property. Moreover, in this case obviously

$$q_N = \int K_N(x,x)dx = \sum_{j=0}^{N-1}\int e^{-Q(x)}\pi_j(x)\pi_j(x)\,dx = N,$$

and therefore the relation (4.2) amounts to

$$\int \det\left(K_N(x_i,x_j)\right)_{1\leq i,j\leq N}\,dx_N = \det\left(K_N(x_i,x_j)\right)_{1\leq i,j\leq N-1}. \tag{4.15}$$

Continuing this process one step further we see

$$\int\int \det\left(K_N(x_i,x_j)\right)_{1\leq i,j\leq N}\,dx_{N-1}dx_N$$

$$= \int \det\left(K_N(x_i,x_j)\right)_{1\leq i,j\leq N-1}\,dx_{N-1}$$

$$= [N-(N-2)]\det\left(K_N(x_i,x_j)\right)_{1\leq i,j\leq N-2} \tag{4.16}$$

and continuing by induction

$$\int \ldots \int \det \left(K_N(x_i, x_j) \right)_{1 \leq i, j \leq N} \, dx_{k+1} \ldots dx_N$$
$$= (N-k)! \det \left(K_N(x_i, x_j) \right)_{1 \leq i, j \leq k} \qquad (4.17)$$

for $k = 1, 2, \ldots$, and the result is $N!$ for $k = 0$. Remembering the expression of the JPDF, Eq.(4.9), in terms of the kernel $K_N(x_i, x_j)$ we see that, in fact, the theory developed provided simultaneously the explicit formulae for all $n-$point correlation functions of the eigenvalues $\mathcal{R}_n(\lambda_1, \ldots, \lambda_n)$, introduced by us earlier, Eq.(3.3):

$$\mathcal{R}_n(\lambda_1, \ldots, \lambda_n) = \det \left(K_N(\lambda_i, \lambda_j) \right)_{1 \leq i, j \leq n} \qquad (4.18)$$

expressed, in view of the relations Eq.(4.12) effectively in terms of the orthogonal polynomials $\pi_k(\lambda)$. In particular, remembering the relation between the mean eigenvalue density and the one-point function derived by us earlier, we have:

$$\overline{\rho_N(\lambda)} = K_N(\lambda, \lambda) = \sum_{j=1}^{N-1} e^{-Q(\lambda)} \pi_{j-1}(\lambda) \pi_{j-1}(\lambda). \qquad (4.19)$$

The latter result allows to represent the "connected" (or "cluster") part of the two-point correlation function introduced by us in Eq.(3.10) in the form:

$$Y_2(\lambda_1, \lambda_2) = \overline{\rho_N(\lambda_1)} \; \overline{\rho_N(\lambda_2)} - \mathcal{R}_2(\lambda_1, \lambda_2) = \left[K_N(\lambda_1, \lambda_2) \right]^2. \qquad (4.20)$$

Finally, combining the relation Eq.(3.14) between the hole probability $A(L)$ and the n-point correlation functions, and on the other hand the expression of the latter in terms of the kernel $K_N(\lambda, \lambda')$, see Eq.(4.18), we arrive at

$$A(L) = \qquad\qquad\qquad\qquad\qquad\qquad\qquad\qquad\qquad\qquad (4.21)$$

$$\sum_{j=0}^{N} \frac{(-1)^j}{j!} \int_{-L/2}^{L/2} \ldots \int_{-L/2}^{L/2} \det \begin{pmatrix} K_N(\lambda_1, \lambda_1) & \ldots & K_N(\lambda_1, \lambda_j) \\ \cdot & \cdot & \cdot \\ \cdot & \cdot & \cdot \\ \cdot & \cdot & \cdot \\ K_N(\lambda_j, \lambda_1) & \ldots & K_N(\lambda_j, \lambda_j) \end{pmatrix} d\lambda_1 \ldots d\lambda_j.$$

In fact, the last expression can be written in a very compact form by noticing that it is just a Fredholm determinant $\det(\mathcal{I} - \mathcal{K}_N)$, where $\mathcal{K}_N$ is a (finite rank) integral operator with the kernel $K_N(\lambda, \lambda') = \sum_{i=0}^{N-1} \phi_i(\lambda) \phi_i(\lambda')$ acting on square-integrable functions on the interval $\lambda \in (-L/2, L/2)$.

5 Properties of Hermite polynomials

5.1 Orthogonality, Recurrence Relations and Integral Representation

Consider the set of polynomials $h_k(x)$ defined as [3]

$$h_k(x) = (-1)^k e^{N\frac{x^2}{2}} \frac{d^k}{dx^k}\left(e^{-N\frac{x^2}{2}}\right) = N^k x^k + \cdots, \qquad (5.1)$$

and consider, for $k \geq l$

$$
\begin{aligned}
\int_{-\infty}^{\infty} e^{-N\frac{x^2}{2}} h_l(x) h_k(x) dx &= (-1)^k \int_{-\infty}^{\infty} dx\, h_l(x) \frac{d^k}{dx^k}\left(e^{-N\frac{x^2}{2}}\right) \qquad (5.2) \\
&= (-1)^{k+1} \int_{-\infty}^{\infty} dx\, h_l'(x) \frac{d^{k-1}}{dx^{k-1}}\left(e^{-N\frac{x^2}{2}}\right) \\
&= \cdots \\
&= (-1)^{2k} \int_{-\infty}^{\infty} dx\, e^{-N\frac{x^2}{2}} \frac{d^k}{dx^k} h_l(x).
\end{aligned}
$$

Obviously, for $k > l$ we have $\frac{d^k}{dx^k} h_l(x) = 0$, whereas for $k = l$ we have $\frac{d^k}{dx^k} h_k(x) = k! N^k$. In this way we verified the orthogonality relations and the normalization conditions

$$\int_{-\infty}^{\infty} e^{-N\frac{x^2}{2}} \tilde{h}_l(x) \tilde{h}_k(x) dx = \delta_{kl} \qquad (5.3)$$

for normalized polynomials

$$\tilde{h}_k(x) =: \frac{1}{\left[k! N^k \sqrt{\frac{2\pi}{N}}\right]^{1/2}} h_k(x). \qquad (5.4)$$

In the theory of orthogonal polynomials an important role is played by recurrence relations:

$$
\begin{aligned}
h_{k+1}(x) &= (-1)^{k+1} e^{N\frac{x^2}{2}} \frac{d^k}{dx^k}\left(\frac{d}{dx} e^{-N\frac{x^2}{2}}\right) \qquad (5.5) \\
&= (-1)^{k+2} N e^{N\frac{x^2}{2}} \frac{d^k}{dx^k}\left(x e^{-N\frac{x^2}{2}}\right) \\
&= (-1)^{k+2} N e^{N\frac{x^2}{2}} \left[\binom{0}{k} x \frac{d^k}{dx^k}\left(e^{-N\frac{x^2}{2}}\right) + \binom{1}{k} \frac{d^{k-1}}{dx^{k-1}}\left(e^{-N\frac{x^2}{2}}\right)\right] \\
&= N\left[x\, h_k(x) - k\, h_{k-1}(x)\right],
\end{aligned}
$$

---

[3] The standard reference to the Hermite polynomials uses the definition

$$H_k(x) = (-1)^k e^{x^2} \frac{d^k}{dx^k}\left(e^{-x^2}\right) = 2^k x^k + \cdots,$$

Such a choice ensures $H_k(x)$ to be orthogonal with respect to the weight e^{-x^2}. Our choice is motivated by random matrix applications, and is related to the standard one as $h_k(x) = H_k\left(\sqrt{\frac{N}{2}} x\right)$.

where we exploited the Leibniz formula for the $k-$th derivative of a product. After normalization we therefore have

$$\left[\frac{k+1}{N}\right]^{1/2}\tilde{h}_{k+1}(x) = x\,\tilde{h}_k(x) - \left[\frac{k}{N}\right]^{1/2}\tilde{h}_{k-1}(x). \tag{5.6}$$

Let us multiply this relation with $\tilde{h}_k(y)$, and then replace x by y. In this way we arrive at two relations:

$$\left[\frac{k+1}{N}\right]^{1/2}\tilde{h}_{k+1}(x)\tilde{h}_k(y) = x\,\tilde{h}_k(x)\tilde{h}_k(y) - \left[\frac{k}{N}\right]^{1/2}\tilde{h}_{k-1}(x)\tilde{h}_k(y), \tag{5.7}$$

$$\left[\frac{k+1}{N}\right]^{1/2}\tilde{h}_{k+1}(y)\tilde{h}_k(x) = y\,\tilde{h}_k(x)\tilde{h}_k(y) - \left[\frac{k}{N}\right]^{1/2}\tilde{h}_{k-1}(y)\tilde{h}_k(x). \tag{5.8}$$

The difference between the upper and the lower line can be written for any $k = 1, 2, \ldots$ as

$$(x-y)\tilde{h}_k(x)\tilde{h}_k(y) = A_{k+1} - A_k$$

$$, \quad A_k = \left[\frac{k}{N}\right]^{1/2}\{\tilde{h}_{k-1}(y)\tilde{h}_k(x) - \tilde{h}_{k-1}(x)\tilde{h}_k(y)\}.$$

Summing up these expressions over k:

$$(x-y)\sum_{k=1}^{n-1}\tilde{h}_k(x)\tilde{h}_k(y) = (A_2 + \ldots + A_n) - (A_1 + \ldots + A_{n-1}) = A_n - A_1$$

and remembering that $A_1 = \sqrt{\frac{N}{2\pi}}(x-y) = (x-y)\tilde{h}_0(x)\tilde{h}_0(y)$ we arrive at a very important relation:

$$\sum_{k=0}^{n-1}\tilde{h}_k(x)\tilde{h}_k(y) = \sqrt{\frac{n}{N}}\frac{\tilde{h}_{n-1}(y)\tilde{h}_n(x) - \tilde{h}_{n-1}(x)\tilde{h}_n(y)}{x-y}, \tag{5.9}$$

or, for the original (not-normalized) polynomials:

$$\sum_{k=0}^{n-1}\frac{1}{k!N^k}h_k(x)h_k(y) = \frac{1}{(n-1)!N^n}\frac{h_{n-1}(y)h_n(x) - h_{n-1}(x)h_n(y)}{x-y}, \tag{5.10}$$

which are known as the *Christoffel-Darboux formulae*. Finally, taking the limit $x \to y$ in the above expression we see that

$$\sum_{k=0}^{n-1}\frac{1}{k!N^k}h_k^2(x) = \frac{1}{(n-1)!N^n}\left[h_{n-1}(x)h_n'(x) - h_{n-1}'(x)h_n(x)\right]. \tag{5.11}$$

Most of the properties and relations discussed above for Hermite polynomials have their analogues for general class of orthogonal polynomials. Now we are going to discuss another very useful property which is however shared only by few families of *classical* orthogonal polynomials: Hermite, Laguerre, Legendre, Gegenbauer and Jacoby. All these polynomials have one of few *integral representations* which are frequently exploited when analyzing their properties. For the case of Hermite polynomials we can most easily arrive to the corresponding representation by using the familiar Gaussian integral identity, cf. Eq.(2.28):

$$e^{-N\frac{x^2}{2}} = \sqrt{\frac{N}{2\pi}} \int_{-\infty}^{\infty} dq\, e^{-\frac{N}{2}q^2 + ixqN}. \tag{5.12}$$

Substituting such an identity to the original definition, Eq.(5.1), we immediately see that

$$h_k(x) = (-iN)^k \sqrt{\frac{N}{2\pi}}\, e^{N\frac{x^2}{2}} \int_{-\infty}^{\infty} dq\, q^k\, e^{-\frac{N}{2}q^2 + ixqN}, \tag{5.13}$$

which is the required integral representation, to be mainly used later on when addressing the large-N asymptotics of the Hermite polynomials. Meanwhile, let us note that differentiating the above formula with respect to x one arrives at the useful relation $\frac{d}{dx}h_k(x) = Nxh_k(x) - h_{k+1}(x) = Nk\,h_{k-1}(x)$. This can be further used to simplify the formula Eq.(5.11) bringing it to the form

$$\sum_{k=0}^{n-1} \frac{1}{k!N^k} h_k^2(x) = \frac{1}{(n-2)!N^{n-1}} \left[h_n^2(x) - h_{n-1}(x)h_{n+1}(x) \right]. \tag{5.14}$$

5.2 Saddle-point method and Plancherel-Rotach asymptotics of Hermite polynomials

In our definition, the Hermite polynomials $h_k(x)$ depend on two parameters: explicitly on the order index $k = 0, 1, \ldots$ and implicitly on the parameter N due to the fact that the weight function $e^{-N\frac{x^2}{2}}$ contains this parameter. Invoking the random matrix background for the use of orthogonal polynomials, we associate the parameter N with the size of the underlying random matrix. From this point of view, the limit $N \gg 1$ arises naturally as we are interested in investigating the spectral characteristics of large matrices. A more detailed consideration reveals that, from the random matrix point of view, the most interesting task is to extract the asymptotic behaviour of the Hermite polynomials with index k large and comparable with N, i.e. $k = N + n$, where the parameter n is considered to be of the order of unity. Such behaviour is known as Plancherel-Rotach asymptotics.

To understand this fact it is enough to invoke the relation (4.19) expressing the mean eigenvalue density in terms of the set of orthogonal polynomials:

$$\overline{\rho_N(\lambda)} = K_N(\lambda, \lambda) = e^{-\frac{N}{2}\lambda^2} \sum_{j=0}^{N-1} \tilde{h}_j^2(\lambda), \qquad (5.15)$$

$$= e^{-\frac{N}{2}\lambda^2} \frac{1}{(N-2)!N^{N-1}} \left[h_N^2(\lambda) - h_{N-1}(\lambda)h_{N+1}(\lambda) \right], \qquad (5.16)$$

where we used the expressions pertinent to the Gaussian weight: $Q(\lambda) \equiv \frac{N}{2}\lambda^2$, $\pi_k(\lambda) \equiv \tilde{h}_k(\lambda)$, and further exploited the variant of the Christoffel-Darboux formula, Eq.(5.14). It is therefore evident that the limiting shape of the mean eigenvalue density for large random matrices taken from the Gaussian Unitary Ensemble is indeed controlled by the Plancherel-Rotach asymptotics of the Hermite polynomials. In fact, similar considerations exploiting the original Christoffel-Darboux formula, Eq.(5.9), show that our main object of interest -the kernel $K_N(\lambda, \lambda')$ - can be expressed as

$$K_N(\lambda, \lambda') = e^{-\frac{N}{4}(\lambda^2 + \lambda'^2)} \frac{\tilde{h}_{N-1}(\lambda)\tilde{h}_N(\lambda') - \tilde{h}_{N-1}(\lambda)\tilde{h}_N(\lambda')}{\lambda - \lambda'} \qquad (5.17)$$

and therefore all the higher correlation functions are controlled by the Plancherel-Rotach asymptotics as well.

For extracting the required asymptotics we are going to use the integral representation for the Hermite polynomials. We start with rewriting the expression Eq.(5.13) as

$$\begin{aligned} h_{N+n}(x) &= (-iN)^{N+n} \sqrt{\frac{N}{2\pi}} \int_{-\infty}^{\infty} dq\, q^{N+n}\, e^{-\frac{N}{2}(q-ix)^2} \\ &= (-iN)^{N+n} \sqrt{\frac{N}{2\pi}} \left[I_{N+n}(x) + (-1)^{N+n} I_{N+n}(-x) \right], \quad (5.18) \end{aligned}$$

where

$$I_{N+n}(x) = \int_0^{\infty} dq\, q^n\, e^{Nf(q)}, \quad f(q) = \ln q - \frac{1}{2}(q - ix)^2. \qquad (5.19)$$

The latter form is suggestive of exploiting the so-called *saddle-point* method (also known as the method of *steepest descent* or method of *stationary phase*) of asymptotic evaluation of integrals of the form

$$\int_\Gamma \phi(z) e^{NF(z)} dz, \qquad (5.20)$$

where the integration goes along a contour Γ in the complex plane, $F(z)$ is an analytic function of z in some domain containing the contour of integration, and N is a large parameter. The main idea of the method can be informally outlined as follows. Suppose that the contour Γ is such that: (i) the value of $\mathrm{Re}F$ has its *maximum* at a point $z_0 \in \Gamma$, and decreases fast enough when

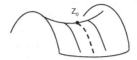

Figure 2: Schematic structure of a harmonic function in the vicinity of a stationary point z_0.

we go along Γ away from z_0, and (ii) the value of $\mathrm{Im}F$ stays constant along Γ (to avoid fast oscillations of the integrand). Then we can expect the main contribution for $N \gg 1$ to come from a small vicinity of $z_0 = x_0 + iy_0$.

Since the function $\mathrm{Re}F$ is a harmonic function of $x = \mathrm{Re}z$, $y = \mathrm{Im}z$, it can have only *saddle points* (see Fig. 2) found from the condition of stationarity $F'(z_0) = 0$. Let us suppose that there exists only one such saddle point $z = z_0$, close to which we can expand $F(z) \approx F(z_0) + C(z - z_0)^2$, where $C = \frac{1}{2}F''(z_0)$. Consider the level curves $[\mathrm{Re}F](x, y) = [\mathrm{Re}F](x_0, y_0)$, which are known either to go to infinity, or end up at a boundary of the domain of analyticity. In the vicinity of the chosen saddle-point the equation for the level curves is $\mathrm{Re}[F(z) - F(z_0)] = 0$, hence

$$\mathrm{Re}[|C|e^{i\theta}(z - z_0)^2] = \left[(x - x_0)^2(y - y_0)^2\right]\cos\theta - 2(x - x_0)(y - y_0)\sin(\theta) = 0,$$

which describes two orthogonal straight lines passing through the saddle-point

$$y = y_0 + \tan\left(\frac{\pi}{4} - \frac{\theta}{2}\right)(x - x_0), \quad y = y_0 - \tan\left(\frac{\pi}{4} + \frac{\theta}{2}\right)(x - x_0)$$

partitioning the x, y plane into four sectors: two "positive" ones: $\mathrm{Re}F(z) >$

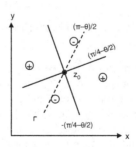

Figure 3: Partitioning of the $x - y$ plane in a vicinity of the stationary point z_0 into four sectors by "level curves" (solid lines). Dashed line shows the bi-sector of the negative sectors: the direction of the steepest descent contour.

$\mathrm{Re}F(z_0)$, and two "negative" ones $\mathrm{Re}F(z) < \mathrm{Re}F(z_0)$, see Fig. 3. If the "edge points" of the integration contour Γ (denoted z_1 and z_2) both belong to the *same* sector, and $\mathrm{Re}F(z_1) \neq \mathrm{Re}F(z_2)$, one always can deform the contour in such a way that $\mathrm{Re}F(z)$ is monotonically increasing along the contour. Then obviously the main contribution to the integral comes from the vicinity of the endpoint (of the largest value of $\mathrm{Re}F(z)$). Essentially the same situation happens when z_1 belongs to a negative (positive) sector, and z_2 is in a positive (resp., negative) sector. And only if the two endpoints belong to two *different negative* sectors, we can deform the contour in such a way, that $\mathrm{Re}F(z)$ has its maximum along the contour at $z = z_0$, and decays away from this point. Moreover, it is easy to understand that the fastest decay away from z_0 will occur along the *bi-sector* of the negative sectors, i.e. along the line $y - y_0 = \tan\frac{\pi-\theta}{2}(x - x_0)$. Approximating the integration contour in the vicinity of z_0 as this bi-sector, i.e. by $z = z_0 + (x - x_0)\frac{e^{-i\frac{\pi-\theta}{2}}}{\sin(\theta/2)}$, we get the leading term of the large-N asymptotics for the original integral by extending the limits of integration in the variable $\tilde{x} = x - x_0$ from $-\infty$ to ∞:

$$\int_\Gamma \phi(z)e^{NF(z)}dz \approx \phi(z_0)e^{NF(z_0)}\frac{e^{-i\frac{\pi-\theta}{2}}}{\sin(\theta/2)}\int_{-\infty}^{\infty}d\tilde{x}e^{-N|C|\tilde{x}^2\sin^2\theta/2}$$

$$= \phi(z_0)\sqrt{\frac{2\pi}{N|F''(z_0)|}}\exp\{NF(z_0) + \frac{i}{2}(\pi - Arg[F''(z_0)/2)])\}. \quad (5.21)$$

It is not difficult to make our informal consideration rigorous, and to calculate systematic corrections to the leading-order result, as well as to consider the case of several isolated saddle-points, the case of a saddle-point coinciding with an end of the contour, etc., see [7] for more detail.

After this long exposition of the method we proceed by applying it to our integral, Eq.(5.19). The saddle-point equation and its solution in that case amount to:

$$F'(q) = \frac{1}{q} - q + ix = 0, \quad q = q_\pm = \frac{1}{2}\left(ix \pm \sqrt{4 - x^2}\right).$$

It is immediately clear that we have essentially three different cases: a) $|x| < 2$ (b) $|x| > 2$ and (c) $|x| = 2$.

1. $|x| < 2$. In this case we can introduce $x = 2\cos\phi$, $0 < \phi < \pi$, so that $q_\pm = i\cos\phi \pm \sin\phi$, or $q_+ = e^{-i(\phi-\pi/2)}$, $q_- = e^{i(\phi+\pi/2)}$. It is easy to understand that we are interested only in q_+ (see Fig.4) and to calculate that $\mathrm{Re}f(q_+) = \frac{1}{2}\cos(2\phi)$. On the other hand $\mathrm{Re}f(q) \to -\infty$ when either $q \to \infty$ or $q \to 0$, so that both endpoints belong to negative sectors. To understand whether they belong to the same or different sectors, we consider the values of $\mathrm{Re}f(q) = \ln R - \frac{1}{2}(R^2 - x^2)$ along the real axis, $q = R$-real. As a function of the variable R this expression has its maximal value $\mathrm{Re}f(q = 1) = -\frac{1}{2} + 2\cos^2\phi$ at $q = R = 1$.

Noting that $\mathrm{Re}f(q = 1) - \mathrm{Re}f(q_+) = \cos^2\phi > 0$, we conclude that the point $q = 1$ belongs to a positive sector, and therefore the existence of

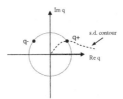

Figure 4: Structure of the saddle-points $q_\pm$ and the relevant steepest descent contour for $|x| < 2$.

this positive sector makes the endpoints $q = 0$ and $q = \infty$ belonging to two *different* negative sectors, as required by the saddle-point method. Calculating

$$f''(q_+) = -\left(1 + \frac{1}{q_+^2}\right) = 2i \sin \phi e^{i\phi}$$

we see that $|C| = \sin \phi$, $\theta = \phi + \pi/2$, and further

$$f''(q_+) = \frac{1}{2} \cos(2\phi) + i\left[\frac{1}{2}\sin(2\phi) - \phi + \pi/2\right].$$

Now we have all the ingredients to enter in Eq.(5.21), and can find the leading order contribution to $I_{N+n}(x)$. Further using $I_{N+n}(-x) = \overline{I_{N+n}(x)}$, valid for real x, we obtain the required Plancherel-Rotach asymptotics of the Hermite polynomial:

$$h_{N+n}(x) \approx N^{N+n} \sqrt{\frac{2}{\sin \phi}} e^{\frac{N}{2} \cos 2\phi} \tag{5.22}$$

$$\times \cos\left\{(n + 1/2)\phi - \pi/4 + N\left(\phi - \frac{1}{2}\sin 2\phi\right)\right\},$$

where $x = 2\cos\phi$, $0 < \phi < \pi$, $n \ll N$.

Now we consider the opposite case:

2. $|x| > 2$. It is enough to consider explicitly the case $x > 2$ and parameterize $x = 2\cosh\phi$, $0 < \phi < \infty$. The saddle points in this case are purely imaginary:

$$q_\pm = \frac{i}{2}(2\cosh\phi \pm 2\sinh\phi) = ie^{\pm\phi}. \tag{5.23}$$

One possible contour of the constant phase passing through both points is just the imaginary axis $q = iy$, where $\mathrm{Im}f(q) = \pi/2$ and $\mathrm{Re}f(q) = \ln y + \frac{1}{2}(y - x)^2$. Simple consideration gives that $y_- = e^{-\phi}$ corresponds to the maximum, and $y_+ = e^\phi$ to the minimum of $\mathrm{Re}f(q)$ along such a contour. It is also clear that for $q = iy_+$ the expression $\mathrm{Re}f(q)$ has a local maximum along the path going through this point in the direction

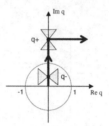

Figure 5: The saddle-points $q_\pm$, the corresponding positive sectors (shaded), and the relevant steepest descent contour (bold) for $|x| > 2$.

transverse to the imaginary axis. The "topography" of $\mathrm{Re}f(q)$ in the vicinity of the two saddle-points is sketched in Fig. 5

This discussion suggests a possibility to deform the path of integration Γ to be a contour of constant phase $\mathrm{Im}f(q)$ consisting of two pieces - $\Gamma_1 = \{q = iy, 0 \le y \le y_+\}$ and Γ_2 starting from $q = iy_+$ perpendicular to the imaginary axis and then going towards $q = \infty$. Correspondingly,

$$I_{N+n}(x > 2) = \int_0^\infty dq\, q^n\, e^{Nf(q)} = \int_0^{iy_+} dq\, q^n\, e^{Nf(q)} + \int_{\Gamma_2} dq\, q^n\, e^{Nf(q)}.$$
$$(5.24)$$

The second integral is dominated by the vicinity of the saddle-point $q = iy_+$, and its evaluation by the saddle-point technique gives:

$$\int_{\Gamma_2} dq\, q^n\, e^{Nf(q)} \approx \frac{1}{2}\sqrt{\frac{\pi e^\phi}{N\sinh\phi}} i^{n+N} e^{n\phi + N\left(\phi + \frac{1}{2}e^{-2\phi}\right)},$$

where the factor $\frac{1}{2}$ arises due to the saddle-point being simultaneously the end-point of the contour. As to the first integral, it is dominated by the vicinity of iy_-, and can also be evaluated by the saddle-point method. However, it is easy to verify that when calculating $h_{N+n}(x) \propto \left[I_{N+n}(x) + (-1)^{N+n} I_{N+n}(-x)\right]$ the corresponding contribution is cancelled out. As a result, we recover the asymptotic behaviour of Hermite polynomials for $x > 2$ to be given by:

$$h_{N+n}(x = 2\cosh(2\phi) > 2) = \frac{N^{n+N} e^{-\frac{N}{2}}}{\sqrt{\sinh\phi}} e^{\left(n+\frac{1}{2}\right)\phi - \frac{N}{2}(\sinh(2\phi) - 2\phi)}. \quad (5.25)$$

Now we come to the only remaining possibility,

3. $|x| = 2$. It is again enough to consider only the case $x = 2$ explicitly. In fact, this is quite a special case, since for $x \to 2$ two saddle-points $q_\pm$ degenerate into one: $q_+ \to q_- \to i$. Under such exceptional circumstances the standard saddle-point method obviously fails. Indeed,

the method assumed that different saddle-points do not interfere, which means the distance $|q_+ - q_-| = \sqrt{|4 - x^2|}$ is much larger than the typical widths $W \sim \frac{1}{\sqrt{N|f''(q_\pm)|}}$ of the regions around individual saddle-points which yield the main contribution to the integrand. Simple calculation gives $|f''(q_\pm)| = |1 + q_\pm^{-2}| = \sqrt{|4 - x^2|}$, and the criterion of two separate saddle-points amounts to $|x - 2| \gg N^{-2/3}$. We therefore see that in the vicinity of $x = 2$ such that $|x - 2| \sim N^{-2/3}$ additional care must be taken when extracting the leading order behaviour of the corresponding integral $I_{N+n}(x = 2)$ as $N \to \infty$.

To perform the corresponding calculation, we introduce a new scaling variable $\xi = N^{2/3}(2 - x)$, and consider ξ to be fixed and finite when $N \to \infty$. We also envisage from the discussion above that the main contribution to the integral comes from the domain around the saddle-point $q_{sp} = i$ of the widths $|q - i| \sim \sqrt{|2 - x|} \sim N^{-1/3}$. The integral we are interested in is given by

$$J_N(\xi) = \int_{-\infty}^{\infty} dq \, q^{N+n} \, e^{-\frac{N}{2}(q-ix)^2} \tag{5.26}$$

$$= N^{-1/3} \int_{-\infty}^{\infty} dt \left(i + \frac{t}{N^{1/3}}\right)^n e^{N\left[\ln\left(i + \frac{t}{N^{1/3}}\right) - \frac{1}{2}\left\{i + \frac{t}{N^{1/3}} - i\left(2 - \frac{\xi}{N^{2/3}}\right)\right\}^2\right]}$$

where we shifted the contour of integration from the real axis to the line $q = i + \frac{t}{N^{1/3}}$, $-\infty < t < \infty$ to ensure that it passes through the expected saddle-point $q_{sp} = i$, and also scaled the integration variable appropriately. Now we can consider ξ, t-finite when $N \gg 1$, and expand the integrand accordingly. A simple computation yields:

$$J_{N \gg 1}(\xi) \approx N^{-1/3} i^{N+n} \, e^{N/2 - N^{1/3}\xi} \int_{-\infty}^{\infty} dt \, e^{-i\xi t + i\frac{t^3}{3}}. \tag{5.27}$$

Up to a constant factor the integral appearing in this expression is, in fact, a representation of a special function known as Airy function $Ai(\xi)$:

$$Ai(\xi) = \frac{1}{\pi} \int_0^{\infty} dt \, \cos\left(\xi t + \frac{t^3}{3}\right) \tag{5.28}$$

which is a solution of the second-order linear differential equation $\frac{d^2}{d\xi^2} F(\xi) - \xi F(\xi) = 0$. A typical behaviour of such a solution is shown in Fig. 6.

All this results in the asymptotic behaviour of the Hermite polynomials in the so-called "scaling vicinity" of the point $x = 2$:

$$h_{N+n}\left(x = 2 - \frac{\xi}{N^{2/3}}\right) \approx \frac{N^{1/6}}{\sqrt{2\pi}} N^{N+n} \, e^{N/2 - N^{1/3}\xi} \, Ai(-\xi). \tag{5.29}$$

Such scaling vicinity of $x = 2$ is what gives room for a transitional regime between the oscillating asymptotics of the Hermite polynomials

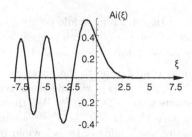

Figure 6: The Airy function $Ai(\xi)$.

for $|x| < 2$, see Eq.(5.22), and the exponential decay typical for $|x| > 2$ as described in Eq.(5.25). Formula (5.29) indeed matches Eq.(5.22) as $\xi \to \infty$ and Eq.(5.25) as $\xi \to -\infty$. This statement is most easily verified by invoking the known asymptotics of the Airy function:

$$Ai(-\xi) \approx \begin{cases} \xi^{-1/4}\pi^{-1/2}\cos\left(-\frac{2}{3}\xi^{3/2} + \frac{\pi}{4}\right), & \xi \to \infty \\ \frac{1}{\pi^{1/2}|\xi|^{1/4}}\, e^{-\frac{2}{3}|\xi|^{3/2}}, & \xi \to -\infty \end{cases} \qquad (5.30)$$

and identifying $\phi = |\xi|^{1/2}N^{-1/3} \ll 1$ in the corresponding expressions.

Now we are going to apply the derived formulae for extracting the large-N behaviour of the mean eigenvalue density and the kernel as described in Eqs.(5.15) and (5.17), respectively. In fact, it is more conventional in the random matrix literature to use the mean density to be normalized to unity, rather than to N. Such a density will have a well-defined large-N limit which we will denote as $\rho_\infty(\lambda)$.

6 Scaling regimes for GUE

6.1 Bulk scaling: Wigner semicircle and Dyson kernel.

The first case to be considered is the spectral parameter $|\lambda| < 2$ when we can parameterize $\lambda = 2\cos\phi$, and exploit the Plancherel-Rotach expression (5.22) for the Hermite polynomials. Furthermore, denoting $\alpha = \frac{1}{2}\phi - \frac{\pi}{4} + N\left(\phi - \frac{1}{2}\sin 2\phi\right)$, and using the identity $\cos^2\alpha - \cos(\alpha + \phi)\cos(\alpha - \phi) = \sin^2\phi$ we find that $h_N^2(\lambda) - h_{N-1}(\lambda)h_{N+1}(\lambda) \approx 2N^{2N}\sin\phi e^{N\cos(2\phi)}$. Furthermore, using for large N the Stirling formula: $(N-1)! \approx \sqrt{\frac{2\pi}{N}}N^N e^{-N}$ and remembering that $\sin\phi = \frac{1}{2}\sqrt{4 - \lambda^2}$ we arrive, after collecting all factors, to the famous Wigner semicircular law for the mean (normalized) spectral density:

$$\lim_{N\to\infty}\left[\frac{1}{N}\overline{\rho(\lambda)}\right] = \rho_\infty(\lambda) = \frac{1}{2\pi}\sqrt{4 - \lambda^2}, \quad |\lambda| < 2. \qquad (6.1)$$

We see that in the limit of large N all N eigenvalues of GUE matrices are concentrated in the interval $[-2, 2]$, and the typical separation of two neighbouring eigenvalues close to an "internal" point $\lambda \in (-2, 2)$ is $\Delta = \frac{1}{N\rho_\infty(\lambda)} = O(N^{-1})$, see Fig. 7. That is why the case $\lambda \in (-2, 2)$ is frequently referred to as the "bulk of the spectrum" regime.

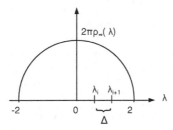

Figure 7: Wigner semicircular density. Sketch shows a typical spacing between neighbouring levels in the bulk of the spectrum.

Let us now follow the same strategy for obtaining, under the same conditions, the limiting expression for the kernel $K(\lambda, \lambda')$ using for this goal formula (5.17). We have:

$$h_N(\lambda)h_{N-1}(\lambda') - h_N(\lambda')h_{N-1}(\lambda) \qquad (6.2)$$

$$\approx 2N^{2N}\frac{1}{\sqrt{\sin\phi\sin\phi'}}e^{\frac{N}{2}(\cos(2\phi)+\cos(2\phi'))}\left[\cos\alpha_1^+\cos\alpha_2^- - \cos\alpha_1^-\cos\alpha_2^+\right]$$

where $\alpha_1^\pm = \pm\frac{1}{2}\phi - \frac{\pi}{4} + N\left(\phi - \frac{1}{2}\sin 2\phi\right)$, $\alpha_2^\pm = \pm\frac{1}{2}\phi' - \frac{\pi}{4} + N\left(\phi' - \frac{1}{2}\sin 2\phi'\right)$. The next step is to introduce $\psi = (\phi + \phi')/2$ and $\Omega = (\phi - \phi')/2$, and to consider the parameter Ω to be of the order of $O(N^{-1})$ when taking the limit. This choice ensures that the distance $\lambda - \lambda' = 2[\cos\phi - \cos\phi'] \approx 4\Omega\sin\psi \approx 4\Omega\pi\rho_\infty(\lambda)$ is of the order of the mean eigenvalue separation Δ - the typical scale for the correlations between the eigenvalues in the *bulk* of the spectrum- and thus must be reflected in the structure of the kernel. To this end we denote $\Omega = w/N$, and keep in the expressions for $\alpha_{1,2}^\pm$ terms up to the order $O(1)$, i.e. writing $\alpha_{1,2}^\pm = N\beta + \left[\pm\frac{1}{2}\psi - \frac{\pi}{4} \pm 2w\sin^2\psi\right]$, where $\beta = \left(\psi - \frac{1}{2}\sin 2\psi\right)$. With the same precision:

$$\cos\alpha_1^+\cos\alpha_2^- - \cos\alpha_1^-\cos\alpha_2^+ \approx \sin\psi\sin\left(4w\sin^2\psi\right)$$
$$\approx \sin\psi\sin\left[\pi\rho_\infty(\lambda)N(\lambda_1 - \lambda_2)\right]$$

and $\cos 2\phi_1 + \cos 2\phi_2 \approx 2\left(\frac{\lambda^2}{2} - 1\right)$ substituting all these factors back into Eq.(6.2), we get

$$h_N(\lambda)h_{N-1}(\lambda') - h_N(\lambda')h_{N-1}(\lambda) \approx 2N^{2N}e^{2N\left(\frac{\lambda^2}{2}-1\right)}\sin\left[\pi\rho_\infty(\lambda)N(\lambda - \lambda')\right] \quad (6.3)$$

Now, taking into account the normalization factors in $\tilde{h}_N(\lambda)$ and $\tilde{h}_{N-1}(\lambda)$, see Eq.(5.4), using again the Stirling formula and invoking Eq.(4.19) we arrive at the following asymptotic expression for the kernel, Eq.(5.17):

$$\lim_{N\to\infty}\left[\frac{K_N(\lambda,\lambda')}{K_N(\lambda,\lambda)}\right] = K_\infty\left[N\rho_\infty(\lambda)(\lambda - \lambda')\right], \quad K_\infty(r) = \frac{\sin\pi r}{\pi r} \tag{6.4}$$

where $K_\infty(r)$ is the famous *Dyson scaling* form for the kernel. The formula is valid as long as both λ and λ' are within the range $(-2, 2)$, and $\lambda - \lambda' = O(N^{-1})$. Such choice of the parameters is frequently referred to as the "bulk scaling" limit.

Having at our disposal the limiting form of both mean eigenvalue density and the two-point kernel we can analyse such important statistical characteristics of the spectra as e.g. the "number variance", see Eq.(3.10), for an interval of the length L comparable with the mean spacing close to the origin $\Delta = [N\rho_\infty(0)]^{-1}$. Under such a condition we can legitimately employ the scaling form Eq.(6.4) of the kernel when substituting it into formula (4.20) for the cluster function $Y_2(\lambda, \lambda')$. In this way we arrive at

$$\Sigma_2(L) = N\int_{-L/2}^{L/2}\rho_\infty(\lambda)d\lambda - N^2\int_{-L/2}^{L/2}d\lambda\int_{-L/2}^{L/2}d\lambda'\rho_\infty(\lambda)\rho_\infty(\lambda')$$
$$\times K_\infty^2\left[N\rho_\infty(\lambda)(\lambda-\lambda')\right]$$
$$= s - \int_{-s/2}^{s/2}du\int_{-s/2}^{s/2}du'K_\infty^2\left[(u-u')\right]. \tag{6.5}$$

Here we used the fact that with the same precision we can put $\rho_\infty(\lambda) \approx \rho_\infty(\lambda') \approx \rho_\infty(0)$ in the above expression, and introduced the natural scaling variables: $u = \lambda/\Delta$, $u = \lambda'/\Delta$ as well as the scaled length of the interval $s = L/\Delta$ (cf. a similar procedure for Poissonian sequences after Eq.(3.14)). To simplify this expression further we introduce $u_+ = (u + u')/2$, $r = u - u'$ as integration variables, and use that, in fact $K_\infty(r) \equiv K_\infty(|r|)$. The number variance takes the final form:

$$\Sigma_2(s) = s - \int_{-s}^{s}dr\int_{-\frac{s}{2}+\frac{|r|}{2}}^{\frac{s}{2}-\frac{|r|}{2}}du_+K_\infty^2(|r|) = s - 2\int_0^s dr\,(s-r)\left[\frac{\sin\pi r}{\pi r}\right]^2. \tag{6.6}$$

In fact, we are mainly interested in the large-s behaviour of this expression. To extract it, we use the identity: $2\int_0^\infty dr\left[\frac{\sin\pi r}{\pi r}\right]^2 = \frac{2}{\pi}\int_0^\infty dx\left[\frac{\sin x}{x}\right]^2 = 1$, and rewrite the above expression as

$$\Sigma_2(s) = \frac{2s}{\pi}\int_{\pi s}^\infty dx\frac{\sin^2 x}{x^2} + \frac{1}{\pi^2}\int_0^{2\pi s}\frac{1-\cos x}{x}dx. \tag{6.7}$$

The second integral obviously grows logarithmically with s and dominates at large s. A more accurate evaluation gives the asymptotic formula:

$$\Sigma_2(s \gg 1) = \frac{1}{\pi^2}\left[\ln 2\pi s + \gamma + 1\right] + O(1/s). \tag{6.8}$$

where $\gamma = 0.5772...$ is Euler's constant. This is much slower than the linear growth $\Sigma_2(s \gg 1) = s$ typical for uncorrelated (Poissonian) sequence, see Fig. 8. The explanation of the slow growth is that the sequence of eigenvalues is, in fact, quite ordered, with quite regular spacings of the order of Δ, and therefore the number of points in the interval does not fluctuate as much as it does for uncorrelated sequence.

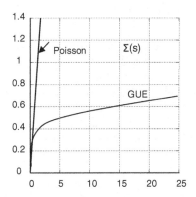

Figure 8: The logarithmic growth of the number variance $\Sigma(s)$ for large GUE matrices versus linear growth for uncorrelated (Poissonian) spectrum.

As to another important and frequently used statistical characteristic of spectral sequences - the "hole probability"- its calculation amounts to investigating the asymptotics of the Fredholm determinant of the kernel $K_{N \to \infty}$, see Eq.(4.21). This is a very difficult mathematical problem, and the most elegant solution uses an advanced mathematical technique known as the Riemann-Hilbert method[3]. Let us just quote the result:

$$A(s \gg 1) \propto \frac{1}{s^{1/4}} e^{-\frac{\pi^2}{8} s^2}. \qquad (6.9)$$

This Gaussian decay should be again contrasted with a much slower exponential decay typical for uncorrelated sequences as indeed in full correspondence with a "quasiregular" structure of the random matrix spectrum.

6.2   Edge scaling regime and Airy kernel

As we already know, in the vicinity of the "spectral edge" $x = 2$ (and its counterpart $x = -2$) the Plancherel-Rotach asymptotics of the Hermite polynomials changes, and is basically given by the Airy function, see Eq.(5.29). This certainly results in essential modifications of the large$-N$ behaviour of the mean eigenvalue density and of the two-point kernel as long as $|\lambda - 2| \sim N^{-2/3}$. To extract the explicit formulae for this so-called "edge scaling" limit one may try the same strategy as in the bulk. However, one immediately discovers that simple substitution of Eq.(5.29) into formula (5.16) for the mean density yields zero. A possible way out may be to calculate the next-to-leading order corrections to the asymptotics of $h_N(x)$, but we will rather follow a slightly different (and more direct) route and consider the integral representation for the main combination of interest:

$$
\begin{aligned}
\mathcal{D}_N(\lambda) &= h_N^2(\lambda) - h_{N-1}(\lambda)h_{N+1}(\lambda) \qquad\qquad\qquad (6.10) \\
&= \frac{(-1)^N}{2\pi} N^{2N} \int_{-\infty}^{\infty} dq_1 \int_{-\infty}^{\infty} dq_2 \frac{q_1 - q_2}{q_1} e^{N[f(q_1)+f(q_2)]}
\end{aligned}
$$

where we exploited Eq.(5.18,5.19), and defined, as before, $f(q) = \ln q - \frac{1}{2}(q - i\lambda)^2$. To evaluate this integral in the edge scaling limit, we follow a familiar procedure: introduce the scaling variable $\xi = N^{2/3}(\lambda - 2)$, shift the contours of integration from the real axis to the lines $q_{1,2} = i + \frac{t_{1,2}}{N^{1/3}}$, $-\infty < t_{1,2} < \infty$, consider ξ, $t_{1,2}$ to be fixed and finite when $N \to \infty$, and expand the integrand accordingly around the saddle-points $t_{1,2} = 0$. Simple calculation yields, in complete analogy with Eq.(5.27), the expression:

$$
\mathcal{D}_N(\xi) \approx \frac{N^{2N}}{2\pi} N^{-1/3} e^{N+2N^{1/3}\xi} \qquad\qquad\qquad\qquad (6.11)
$$

$$
\times \left\{ \int_\Gamma dt_1\, e^{i\xi t_1 + i\frac{t_1^3}{3}} \int_\Gamma dt_2\, t_2^2\, e^{i\xi t_2 + i\frac{t_2^3}{3}} - \int_\Gamma dt_1\, t_1\, e^{i\xi t_1 + i\frac{t_1^3}{3}} \int_\Gamma dt_2\, t_2\, e^{i\xi t_2 + i\frac{t_2^3}{3}} \right\}.
$$

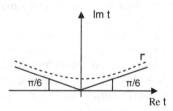

Figure 9: The contour of integration Γ in the definition of the Airy function.

The only essential difference from Eq.(5.27) which deserves mentioning is the choice of the integration contour Γ which ensures the existence of all the integrals involved. Obviously, one can not simply take $\Gamma = (-\infty, \infty)$, but a more detailed investigation shows that the correct contour must be chosen in such a way as to be asymptotically tangent to the line $\mathrm{Arg}(t) = 5\pi/6$ for $\mathrm{Re}\, t \to -\infty$, and asymptotically tangent to $\mathrm{Arg}(t) = \pi/6$ for $\mathrm{Re}\, t \to \infty$, see Fig.9. It is then evident, that

$$
Ai(\xi) = \frac{1}{\pi} \int_\Gamma dt\, e^{i\xi t + i\frac{t^3}{3}}, \quad -iAi'(\xi) = \frac{1}{\pi} \int_\Gamma dt\, t\, e^{i\xi t + i\frac{t^3}{3}}, \quad -Ai''(\xi)
$$

$$
= \frac{1}{\pi} \int_\Gamma dt\, t^2\, e^{i\xi t + i\frac{t^3}{3}}
$$

and collecting all factors we find the expression of the mean eigenvalue density close to the "spectral edge":

$$\overline{\rho}(\lambda = 2 + \xi \, N^{-2/3}) \propto \rho_e(\xi) = Ai'(\xi)^2 - Ai''(\xi)Ai(\xi). \qquad (6.12)$$

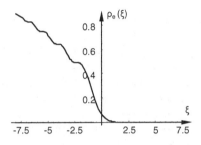

Figure 10: Behaviour of the spectral density close to the spectral edge.

For $\xi < 0$ the function shows noticeable oscillations , see Fig.10, with neighbouring maxima separated by distance of the order of $\lambda_i - \lambda_{i-1} \propto \Delta_{edge} =\sim N^{-1/3}$ and reflecting typical positions of individual eigenvalues close to the "spectral edge". In contrast, for $\xi > 0$ the mean density decays extremely fast, reflecting the typical absence of the eigenvalues beyond the spectral edge.

A very similar calculation shows that under the same conditions the kernel $K_N(\lambda, \lambda')$ assumes the form:

$$K(\xi_1, \xi_2) = \frac{Ai(\xi_1)Ai'(\xi_2) - Ai(\xi_2)Ai'(\xi_1)}{\xi_1 - \xi_2} \qquad (6.13)$$

known as the *Airy kernel*, see [8].

7   Orthogonal polynomials versus characteristic polynomials

Our efforts in studying Hermite polynomials in detail were amply rewarded by the provided possibility to arrive at the bulk and edge scaling forms for the matrix kernel in the corresponding large-N limits. It is those forms which turn out to be *universal*, which means independent of the particular detail of the random matrix probability distribution, provided size of the corresponding matrices is large enough. This is why one can hope that the Dyson kernel would be relevant to many applications, including properties of the Riemann ζ-function. An important issue for many years was to prove the universality for unitary-invariant ensembles which was finally achieved, first in [10].

In fact, quite a few basic properties of the Hermite polynomials are shared also by any other set of orthogonal polynomials $\pi_k(x)$. Among those worth of particular mentioning is the Christoffel-Darboux formula for the combination entering the two-point kernel Eq.(4.12), (cf. Eq.(5.10)):

$$\sum_{k=0}^{n-1} \pi_k(x)\pi_k(y) = b_n \frac{\pi_{n-1}(y)\pi_n(x) - \pi_{n-1}(x)\pi_n(y)}{x - y},$$ (7.1)

where b_n are some constants. So the problem of the universality of the kernel (and hence, of the n-point correlation functions) amounts to finding the appropriate large-N scaling limit for the right-hand side of Eq.(7.1) (in the "bulk" of the spectrum, or close to the spectral "edge").

The main dissatisfaction is that explicit formulas for orthogonal polynomials (most important, an integral representation similar to Eq.(5.13)) are not available for general weight functions $dw(\lambda) = e^{-Q(\lambda)}d\lambda$. For this reason we have to devise alternative tools of constructing the orthogonal polynomials and extracting their asymptotics. Any detailed discussion of the relevant technique goes far beyond the modest goals of the present set of lectures. Nevertheless, some hints towards the essence of the powerful methods employed for that goal will be given after a digression.

Namely, I find it instructive to discuss first a question which seems to be quite unrelated,- the statistical properties of the characteristic polynomials

$$Z_N(\mu) = \det\left(\mu \mathbf{1}_N - \hat{H}\right) = \prod_{i=1}^{N}(\mu - \lambda_i)$$ (7.2)

for any Hermitian matrix ensemble with invariant JPDF $\mathcal{P}(\hat{H}) \propto \exp\{-N\mathrm{Tr}Q(\hat{H})\}$. Such objects are very interesting on their own for many reasons. Moments of characteristic polynomials for various types of random matrices were much studied recently, in particular due to an attractive possibility to use them, in a very natural way, for characterizing "universal" features of the Riemann ζ-function along the critical line, see the pioneering paper[11] and the lectures by Jon Keating in this volume. The same moments also have various interesting combinatorial interpretations, see e.g. [12, 13], and are important in applications to physics, as I will elucidate later on.

On the other hand, addressing those moments will allow us to arrive at the most natural way of constructing polynomials orthogonal with respect to an arbitrary weight $dw(\lambda) = e^{-Q(\lambda)}d\lambda$. To understand this, we start with considering the lowest moment, which is just the expectation value of the characteristic polynomial:

$$E\left[Z_N(\mu)\right] = \int_{-\infty}^{\infty} dw(\lambda_1)\ldots\int_{-\infty}^{\infty} dw(\lambda_N) \prod_{i<j}^{N}(\lambda_i - \lambda_j)^2 \prod_{i=1}^{N}(\mu - \lambda_i)$$ (7.3)

We first notice that

$$\prod_{i<j}^{N}(\lambda_i - \lambda_j)\prod_{i=1}^{N}(\mu - \lambda_i) \propto \det\begin{pmatrix} 1 & \cdots & 1 & 1 \\ \lambda_1 & \cdots & \lambda_N & \mu \\ \cdot & \cdot & \cdot & \cdot \\ \cdot & \cdot & \cdot & \cdot \\ \cdot & \cdot & \cdot & \cdot \\ \lambda_1^{N-1} & \cdots & \lambda_N^{N-1} & \mu^{N-1} \\ \lambda_1^{N} & \cdots & \lambda_N^{N} & \mu^{N} \end{pmatrix} \tag{7.4}$$

Indeed, the right-hand side is obviously a polynomial of degree N in the variable μ, with roots at $\mu = \mu_1, \mu_2, \ldots, \mu_N$. Therefore it must be of the form $C \times \prod_{i=1}^{N}(\mu - \lambda_i)$, with prefactor C being a function of $\lambda_1, \ldots, \lambda_N$. The value of such a prefactor can be easily established by comparing both sides as $\mu \to \infty$: the left-hand side behaves as $C \mu^N$, whereas expanding the determinant with respect to the last column and using the expression for the van der Monde determinant, Eq.(4.5), we see that the right-hand side grows as $\mu^N \prod_{i<j}^{N}(\lambda_i - \lambda_j)$.

Exploiting Eq.(7.4) allows us to rewrite the expectation value for the characteristic polynomial as

$$E\left[Z_N(\mu)\right] \propto \int_{-\infty}^{\infty}\prod_{i=1}^{N} dw(\lambda_i) \tag{7.5}$$

$$\times \det\begin{pmatrix} 1 & \cdots & 1 \\ \lambda_1 & \cdots & \lambda_N \\ \cdot & \cdot & \cdot \\ \cdot & \cdot & \cdot \\ \lambda_1^{N-1} & \cdots & \lambda_N^{N-1} \end{pmatrix} \det\begin{pmatrix} 1 & \cdots & 1 & 1 \\ \lambda_1 & \cdots & \lambda_N & \mu \\ \cdot & \cdot & \cdot & \cdot \\ \cdot & \cdot & \cdot & \cdot \\ \lambda_1^{N} & \cdots & \lambda_N^{N} & \mu^{N} \end{pmatrix},$$

which can be further written down as the standard sum over all permutations $P_\sigma = (\sigma_1, \ldots, \sigma_N)$ of the index set $(1, 2, \ldots, N)$:

$$E\left[Z_N(\mu)\right] \propto \sum_{P_\sigma}(-1)^{|P_\sigma|}\int_{-\infty}^{\infty}\prod_{i=1}^{N} dw(\lambda_i)\lambda_{\sigma_1}^{0}\ldots\lambda_{\sigma_N}^{N-1} \tag{7.6}$$

$$\times \det\begin{pmatrix} 1 & \cdots & 1 & 1 \\ \lambda_1 & \cdots & \lambda_N & \mu \\ \cdot & \cdot & \cdot & \cdot \\ \cdot & \cdot & \cdot & \cdot \\ \lambda_1^{N} & \cdots & \lambda_N^{N} & \mu^{N} \end{pmatrix},$$

where $|P_\sigma| = 0(1)$ for even(odd) permutations. The symmetry of the remaining determinant with respect to permutation of its columns ensures that every term in the sum above yields exactly the same contribution, and it is enough to consider only the first term with $P_\sigma = (1, 2, \ldots, N)$, and multiply the result with

$N!$. For such a choice, the product of factors $\lambda_1^0 \ldots \lambda_N^{N-1}$ can be "absorbed" in the determinant by multiplying the j-th column of the latter with the factor λ_j^{j-1}, for all $j = 1, \ldots, N$. This gives

$$E\left[Z_N(\mu)\right] \propto \int_{-\infty}^{\infty} \prod_{i=1}^{N} dw(\lambda_i) \det \begin{pmatrix} 1 & \lambda_2 & \ldots & \lambda_N^{N-1} & 1 \\ \lambda_1 & \lambda_2^2 & \ldots & \lambda_N^N & \mu \\ \cdot & \cdot & \cdot & \cdot & \\ \cdot & \cdot & \cdot & \cdot & \\ \cdot & \cdot & \cdot & \cdot & \\ \lambda_1^{N-1} & \lambda_2^N & \ldots & \lambda_N^{2N-2} & \mu^{N-1} \\ \lambda_1^N & \lambda_2^{N+1} & \ldots & \lambda_N^{2N-1} & \mu^N \end{pmatrix}. \quad (7.7)$$

The integral in the right-hand side is obviously a polynomial of degree N in μ, which we denote $D_N(\mu)$ and write in the final form as

$$D_N(\mu) = \det \begin{pmatrix} \int_{-\infty}^{\infty} dw(\lambda) & \int_{-\infty}^{\infty} dw(\lambda)\lambda & \ldots & \int_{-\infty}^{\infty} dw(\lambda)\lambda^{N-1} & 1 \\ \int_{-\infty}^{\infty} dw(\lambda)\lambda & \int_{-\infty}^{\infty} dw(\lambda)\lambda^2 & \ldots & \int_{-\infty}^{\infty} dw(\lambda)\lambda^N & \mu \\ \vdots & \vdots & \vdots & \vdots & \\ \int_{-\infty}^{\infty} dw(\lambda)\lambda^{N-1} & \int_{-\infty}^{\infty} dw(\lambda)\lambda^N & \ldots & \int_{-\infty}^{\infty} dw(\lambda)\lambda^{2N-2} & \mu^{N-1} \\ \int_{-\infty}^{\infty} dw(\lambda)\lambda^N & \int_{-\infty}^{\infty} dw(\lambda)\lambda^{N+1} & \ldots & \int_{-\infty}^{\infty} dw(\lambda)\lambda^{2N-1} & \mu^N \end{pmatrix}.$$
$$(7.8)$$

The last form makes evident the following property. Multiply the right-hand side with $dw(\mu)\mu^p$ and integrate over μ. By linearity, the factor and the integration can be "absorbed" in the last column of the determinant. For $p = 0, 1, \ldots, N - 1$ this last column will be identical to one of preceding columns, making the whole determinant vanishing, so that

$$\int_{-\infty}^{\infty} dw(\mu)\mu^p D_N(\mu) = 0, \quad p = 0, 1, \ldots, N - 1. \quad (7.9)$$

Moreover, it is easy to satisfy oneself that the polynomial $D_N(\mu)$ can be written as $D_N(\mu) = D_{N-1}\mu^N + \ldots$, where the leading coefficient $D_{N-1} = \det \left(\int_{-\infty}^{\infty} dw(\lambda)\lambda^{i+j} \right)_{i,j=0}^{N}$ is necessarily positive: $D_{n-1} > 0$. The last fact immediately follows from the positivity of the quadratic form:

$$G(x_1, \ldots, x_N) = \int_{-\infty}^{\infty} dw(\lambda) \left(\sum_{i=1}^{N} x_i \lambda^i \right)^2 = \sum_{i,j}^{N} x_i x_j \int_{-\infty}^{\infty} dw(\lambda)\lambda^{i+j}$$

Finally, notice that

$$\int_{-\infty}^{\infty} dw(\mu) D_N^2(\mu) = \int_{-\infty}^{\infty} dw(\mu) D_N(\mu) \left[D_{N-1}\mu^N + lower\ powers \right] \quad (7.10)$$

$$= D_{N-1} \int_{-\infty}^{\infty} dw(\mu) D_N(\mu)\mu^N = D_{N-1}D_N$$

where we first exploited Eq.(7.9) and at the last stage Eq.(7.8). Combining all these facts together we thus proved that the polynomials $\pi_N(\lambda) = \frac{1}{\sqrt{D_{N-1}D_N}}D(\lambda)$ form the orthogonal (and normalized to unity) set with respect to the given measure $dw(\lambda)$. Moreover, our discussion makes it immediately clear that the expectation value of the *characteristic* polynomial $Z_N(\mu)$ for any given random matrix ensemble is nothing else, but just the corresponding *monic* orthogonal polynomial:

$$E\left[Z_N(\mu)\right] = \pi_N^{(m)}(\mu), \tag{7.11}$$

whose leading coefficient is unity. Leaving aside the modern random matrix interpretation the combination of the right hand sides of the formulas Eq.(7.11) and Eq.(7.3) goes back, according to [6], to Heine-Borel work of 1878, and as such is completely classical.

The random matrix interpretation is however quite instructive, since it suggests to consider also higher moments of the characteristic polynomials, and even more general objects like the correlation functions

$$C_k(\mu_1,\mu_2,\ldots,\mu_k) = E\left[Z_N(\mu_1)Z_N(\mu_2)\ldots Z_N(\mu_k)\right]. \tag{7.12}$$

Let us start with considering

$$C_2(\mu_1,\mu_2) = \int_{-\infty}^{\infty} dw(\lambda_1)\ldots\int_{-\infty}^{\infty} dw(\lambda_N)\prod_{i<j}^{N}(\lambda_i-\lambda_j)^2\prod_{i=1}^{N}(\mu_1-\lambda_i)\prod_{i=1}^{N}(\mu_2-\lambda_i). \tag{7.13}$$

Using the notation $\Delta_N(\lambda_1,\ldots,\lambda_N)$ for the van der Monde determinant, see Eq.(4.5), we further notice that

$$\Delta_{N+2}(\lambda_1,\ldots,\lambda_N,\mu_1,\mu_2) = \Delta_N(\lambda_1,\ldots,\lambda_N)\times(\mu_1-\mu_2)\prod_{i=1}^{N}(\mu_1-\lambda_i)\prod_{i=1}^{N}(\mu_2-\lambda_i),$$

which allows us to rewrite the correlation function as

$$C_2(\mu_1,\mu_2) = \frac{1}{(\mu_1-\mu_2)}\int_{-\infty}^{\infty}\prod_{i=1}^{N} dw(\lambda_i)\,\Delta_N(\lambda_1,\ldots,\lambda_N)\Delta_{N+2}(\lambda_1,\ldots,\lambda_N,\mu_1,\mu_2).$$

Now we replace each entry λ_i^j in both van der Monde determinant factors with the orthogonal polynomial $\pi_j(\lambda_i)$ (cf. eq.(4.6)), and further expand the first factor as a sum over permutations: $\Delta_N(\lambda_1,\ldots,\lambda_N) \propto \sum_P(-1)^{|P|}\pi_0(\lambda_{\sigma_1})\ldots\pi_{N-1}(\lambda_{\sigma_N})$. Further using permutational symmetry of the second determinant, we again see that every term yields after integration the same contribution. Up to a proportionality factor we can therefore rewrite the correlation function as

$$C_2(\mu_1, \mu_2) = \frac{1}{(\mu_1 - \mu_2)} \int_{-\infty}^{\infty} \prod_{i=1}^{N} dw(\lambda_i)\, \pi_0(\lambda_1) \ldots \pi_{N-1}(\lambda_N) \qquad (7.14)$$

$$\times\ \det \begin{pmatrix} \pi_0(\lambda_1) & \pi_0(\lambda_2) & \cdots & \pi_0(\mu_1) & \pi_0(\mu_2) \\ \pi_1(\lambda_1) & \pi_1(\lambda_2) & \cdots & \pi_1(\mu_1) & \pi_1(\mu_2) \\ \cdot & \cdot & \cdot & \cdot & \\ \cdot & \cdot & \cdot & \cdot & \\ \cdot & \cdot & \cdot & \cdot & \\ \pi_N(\lambda_1) & \pi_N(\lambda_2) & \cdots & \pi_N(\mu_1) & \pi_N(\mu_2) \\ \pi_{N+1}(\lambda_1) & \pi_{N+1}(\lambda_2) & \cdots & \pi_{N+1}(\mu_1) & \pi_{N+1}(\mu_2) \end{pmatrix}.$$

At the next step we absorb the factors $\pi_0(\lambda_1), \ldots, \pi_{N-1}(\lambda_N)$ inside the determinant by multiplying the first column with $\pi_0(\lambda_1)$,..., the $N-th$ column with $\pi_{N-1}(\lambda_N)$, and leaving the last two columns intact. By linearity, we can also absorb the product of the integrals inside the determinant by integrating the first column over λ_1,..., and $N-th$ column over λ_N. Due to the orthogonality, the first N columns of the resulting determinant after integration contain zero components off-diagonal, whereas the entries on the main diagonal are equal to the normalization constants $c_k = \int_{-\infty}^{\infty} dw(\lambda)\, \pi_k^2(\lambda)$, $k = 0, \ldots, N$. Therefore, the resulting determinant is easy to calculate and, up to a multiplicative constant we arrive to the following simple formula:

$$C_2(\mu_1, \mu_2) \propto \frac{1}{(\mu_1 - \mu_2)} \det \begin{pmatrix} \pi_N(\mu_1) & \pi_N(\mu_2) \\ \pi_{N+1}(\mu_1) & \pi_{N+1}(\mu_2) \end{pmatrix}. \qquad (7.15)$$

In particular, for the second moment of the characteristic polynomial we have the expression

$$E[Z^2(\mu)] = \lim_{\mu_1 \to \mu_2 = \mu} C_2(\mu_1, \mu_2) \propto \det \begin{pmatrix} \pi_N(\mu) & \pi_N'(\mu) \\ \pi_{N+1}(\mu) & \pi_{N+1}'(\mu) \end{pmatrix}. \qquad (7.16)$$

This procedure can be very straightforwardly extended to higher order correlation functions[14, 16], and higher order moments[15] of the characteristic polynomials. The general structure is always the same, and is given in the form of a determinant whose entries are orthogonal polynomials of increasing order.

One more observation deserving mentioning here is that the structure of the two-point correlation function of characteristic polynomials is identical to that of the Christoffel-Darboux, which is the main building block of the kernel function, Eq.(4.12). Moreover, comparing the above formula (7.16) for the gaussian case with expressions (5.15,5.14), one notices a great degree in similarity between the structure of mean eigenvalue density and that for the second moment of the characteristic polynomial. All these similarities are not accidental, and there exists a general relation between the two types of quantities as I proceed to demonstrate on the simplest example. For this we recall that the mean eigenvalue density $\rho_N(\lambda)$ is just the one-point correlation function, see Eq.(3.4), and according to Eq.(3.1) and Eq.(2.38) can be written as

$$\mathcal{R}_1(\lambda) = N \int \mathcal{P}(\lambda, \lambda_2, \ldots, \lambda_N) \, d\lambda_2 \ldots \lambda_N \tag{7.17}$$

$$\propto e^{-Q(\lambda)} \int d\lambda_2 \ldots \lambda_N e^{-\sum_{i=2}^{N} Q(\lambda_i)} \prod_{i=2}^{N} (\lambda - \lambda_i)^2 \prod_{2 \leq i < j \leq N} (\lambda_i - \lambda_j)^2.$$

It is immediately evident after simple renumbering $(\lambda_2, \ldots, \lambda_N) \rightarrow (\lambda_1, \ldots, \lambda_{N-1})$ that the integral in the second line allows a clear interpretation as the second moment of the characteristic polynomial $E[Z_{N-1}^2(\lambda)]$ of a random matrix H_{N-1} distributed according to the same joint probability density function $\mathcal{P}(H_{N-1}) \, d\hat{H}_{N-1}$, but of reduced size $N - 1$, see Eq.(7.3) for comparison. We therefore have a general relation between the mean eigenvalue density and the second moment of the characteristic polynomial of the reduced-size matrix:

$$\rho_N(\lambda) \propto e^{-Q(\lambda)} \overline{\left[\det\left(\lambda \mathbf{1}_N - \hat{H}_{N-1}\right)\right]^2} \tag{7.18}$$

which explains the observed similarity. This type of relations, and their natural generalizations to higher-order correlation functions hold for general invariant ensembles and were found helpful in several applications; e.g. for the so-called "chiral" ensembles (notion of such ensembles is shortly discussed in the very end of these notes) in [18], for non-Hermitian matrices with complex eigenvalues see examples and further references in [19]); for real symmetric matrices see the recent paper[20].

Now let us discuss another important class of correlation functions involving characteristic polynomials, - namely one combining both positive and negative moments, the simplest example being the expectation value of the ratio:

$$\mathcal{K}_N(\mu, \nu) = E\left[\frac{Z_N(\mu)}{Z_N(\nu)}\right]. \tag{7.19}$$

For such an object to be well-defined it is necessary to regularize the characteristic polynomial in the denominator $Z_N(\nu) = \det\left(\nu \mathbf{1}_N - \hat{H}\right)$ by considering the complex-valued spectral parameter ν such that $\mathrm{Im}\,\nu \neq 0$. Further generalizations include more than one polynomial in numerator and/or denominator.

Such objects turned out to be indispensable tools in applications of random matrices to physical problems. In fact, in all applications a very fundamental role is played by the resolvent matrix $(\mu \mathbf{1}_N - \hat{H})^{-1}$, and statistics of its entries is of great interest. In particular, the familiar eigenvalue density $\rho(\nu)$ can be extracted from the trace of the resolvent as

$$\rho(\nu) = \frac{1}{\pi} \lim_{\mathrm{Im}\mu \to 0^-} \mathrm{Im}\,\mathrm{Tr}\frac{1}{\mu \mathbf{1}_N - \hat{H}}. \tag{7.20}$$

It is easy to understand that one can get access to such an object, and more general correlation functions of the traces of the resolvent by using the identity:

$$\mathrm{Tr}\frac{1}{\mu\mathbf{1}_N - \hat{H}} = -\frac{\partial}{\partial\nu}\frac{Z_N(\mu)}{Z_N(\nu)}\big|_{\mu=\nu}. \tag{7.21}$$

We conclude that the products of ratios of characteristic polynomials can be used to extract the multipoint correlation function of spectral densities (see an example below). Moreover, distributions of some other interesting quantities as, e.g. individual entries of the resolvent, or statistics of eigenvalues as functions of some parameter can be characterized in terms of general correlation functions of ratios, see [21] for more details and examples. Thus, that type of the correlation function is even more informative than one containing products of only positive moments of the characteristic polynomials.

In fact, it turns out that there exists a general relation between the two types of the correlation functions, which is discussed in full generality in recent papers [17, 22, 23, 24]. Here we would like to illustrate such a relation on the simplest example, Eq.(7.19). To this end let us use the following identity:

$$[Z_N(\nu)]^{-1} = \frac{1}{\prod_{i=1}^N(\nu - \lambda_i)} = \sum_{k=1}^N\frac{1}{\nu - \lambda_k}\prod_{i\neq k}^N\frac{1}{\lambda_i - \lambda_k} \tag{7.22}$$

and integrate the ratio of the two characteristic polynomials over the joint probability density of all the eigenvalues. When performing integrations, each of N terms in the sum in Eq.(7.22) produces identical contributions, so that we can take one term with $k = 1$ and multiply the result by N. Representing $\Delta^2(\lambda_1, \ldots, \lambda_N) = \prod_{2\leq i}(\lambda_1 - \lambda_i)^2\prod_{2\leq i<j}(\lambda_i - \lambda_j)^2$, and observing some cancellations, we have

$$\mathcal{K}_N(\mu, \nu) \propto \int dw(\lambda_1)\frac{\mu - \lambda_1}{\nu - \lambda_1}\int dw(\lambda_2)\ldots dw(\lambda_N) \tag{7.23}$$

$$\times \prod_{2\leq i<j}^N(\lambda_i - \lambda_j)^2\prod_{i=2}^N(\lambda_1 - \lambda_i)(\mu - \lambda_i)$$

$$\propto \int dw(\lambda_1)\frac{\mu - \lambda_1}{\nu - \lambda_1}\times\overline{\det\left(\lambda_1 - \hat{H}_{N-1}\right)\det\left(\mu - \hat{H}_{N-1}\right)}.$$

The average value of the products of two characteristic polynomials found by us in Eq.(7.15) can now be inserted into the integral entering Eq.(7.24), and the resulting expression can be again written in the form of a 2×2 determinant:

$$\mathcal{K}_N(\mu, \nu) \propto \det\begin{pmatrix} \pi_{N-1}(\mu) & f_{N-1}(\nu) \\ \pi_N(\mu) & f_N(\nu) \end{pmatrix} \tag{7.24}$$

where $f_N(\nu)$ stands for the so-called Cauchy transform of the orthogonal polynomial

$$f_N(\nu) = \frac{1}{2\pi i}\int_{-\infty}^{\infty}\frac{dw(\lambda)}{\nu - \lambda}\pi_N(\lambda) \tag{7.25}$$

The emerged functions $f_N(\nu)$ is a rather new feature in Random Matrix Theory. It is instructive to have a closer look at their properties for the simplest case of the Gaussian Ensemble, $Q(\lambda) = N\lambda^2/2$. It turns out that for such a case the functions $f_N(\nu)$ are, in fact, related to the so-called generalized Hermite functions $\mathcal{H}_N$ which are second -non-polynomial- solutions of the same differential equation which is satisfied by Hermite polynomials themselves. The functions also have a convenient integral representations, which can be obtained in the most straightforward way by substituting the identity

$$\frac{1}{\nu - \lambda} \propto \int_0^\infty dt\, e^{it\,\mathrm{sgn}[\mathrm{Im}\nu](\nu - \lambda)}$$

into the definition (7.25), replacing the Hermite polynomial with its integral representation, Eq.(5.13), exchanging the order of integrations and performing the λ–integral explicitly. Such a procedure results in

$$f_{N+n}(\nu) \propto \int_0^\infty dt\, t^{N+n} e^{-N\left(\frac{t^2}{2} - it\,\mathrm{sgn}[\mathrm{Im}\nu]\nu\right)}.$$

Note, that this is precisely the integral (5.19) whose large-N asymptotics for real ν we studied in the course of our saddle-point analysis. The results can be immediately extended to complex ν, and in the "bulk scaling" limit we arrive to the following asymptotics of the correlation function (7.24) close to the origin

$$\lim_{N\to\infty} \mathcal{K}_N(\mu, \nu) = \mathcal{K}_\infty\left[N\rho_\infty(0)(\mu - \nu)\right], \qquad (7.26)$$

$$\mathcal{K}_\infty(r) \propto \begin{cases} e^{-i\pi r} & \text{if } \mathrm{Im}(\nu) > 0 \\ e^{i\pi r} & \text{if } \mathrm{Im}(\nu) > 0 \end{cases}.$$

In a similar, although more elaborate way one can calculate an arbitrary correlation function containing ratios and products of characteristic polynomials [17, 23, 24]. The detailed analysis shows that the kernel $S(\mu, \nu) = \mathcal{K}(\mu, \nu)/(\mu - \nu)$ and its scaling form $S_\infty(r) \propto \frac{\mathcal{K}_\infty(r)}{r}$ play the role of a building block for more general correlation functions involving ratios, in the same way as the Dyson kernel (6.4) plays similar role for the n-point correlation functions of eigenvalue densities. This is a new type of "kernel function" with structure different from the standard random matrix kernel Eq.(4.10). The third type of such kernels - made from functions $f_N(\nu)$ alone - arises when considering only negative moments of the characteristic polynomials.

To give an instructive example of the form emerging consider

$$\mathcal{K}_N(\mu_1, \mu_2, \nu_1, \nu_2) = E\left[\frac{Z_N(\mu_1)}{Z_N(\nu_1)}\frac{Z_N(\mu_2)}{Z_N(\nu_2)}\right] \qquad (7.27)$$

$$= \frac{(\mu_1 - \nu_1)(\mu_1 - \nu_2)(\mu_2 - \nu_1)(\mu_2 - \nu_2)}{(\mu_1 - \mu_2)(\nu_1 - \nu_2)} \det\begin{pmatrix} S(\mu_1, \nu_1) & S(\mu_1, \nu_2) \\ S(\mu_2, \nu_1) & S(\mu_2, \nu_2) \end{pmatrix}.$$

Assuming $\mathrm{Im}\,\nu_1 > 0$, $\mathrm{Im}\,\nu_2 < 0$, both infinitesimal, we find in the bulk scaling limit such that both $N\rho(0)\mu_{1,2} = \zeta_{1,2}$ and $N\rho(0)\nu_{1,2} = \kappa_{1,2}$ are finite the following expression (see e.g. [22], or [21])

$$\lim_{N\to\infty} \mathcal{K}_N(\mu_1,\mu_2,\nu_1,\nu_2) = \mathcal{K}_\infty(\zeta_1,\zeta_2,\kappa_1,\kappa_2) \qquad (7.28)$$

$$= \frac{e^{i\pi(\zeta_1-\zeta_2)}}{\zeta_1-\zeta_2} \left[e^{i\pi(\kappa_1-\kappa_2)} \frac{(\kappa_1-\zeta_1)(\kappa_2-\zeta_2)}{\kappa_1-\kappa_2} - e^{-i\pi(\kappa_1-\kappa_2)} \frac{(\kappa_1-\zeta_2)(\kappa_2-\zeta_1)}{\kappa_1-\kappa_2} \right].$$

This formula can be further utilized for many goals. For example, it is a useful exercise to understand how the scaling limit of the two-point cluster function (4.20) can be extracted from such an expression (hint: the cluster function is related to the correlation function of eigenvalue densities by Eq.(3.7); exploit the relations (7.20),(7.21)).

All these developments, important and interesting on their own, indirectly prepared the ground for discussing the mathematical framework for a proof of universality in the large-N limit. As was already mentioned, the main obstacle was the absence of any sensible integral representation for general orthogonal polynomials and their Cauchy transforms. The method which circumvents this obstacle in the most elegant fashion is based on the possibility to define *both* orthogonal polynomials *and* their Cauchy transforms in a way proposed by Fokas, Its and Kitaev, see references in [3], as elements of a (matrix valued) solution of the following (Riemann-Hilbert) problem. The latter can be introduced as follows. Let the contour Σ be the real axis orientated from the left to the right. The upper half of the complex plane with respect to the contour will be called the positive one and the lower half - the negative one. Fix an integer $n \geq 0$ and the measure $w(z) = e^{-Q(z)}$ and define the Riemann-Hilbert problem as that of finding a 2×2 matrix valued function $Y = Y^{(n)}(z)$ satisfying the following conditions:

- $Y^{(n)}(z)$ – analytic in $\mathbb{C} \setminus \Sigma$

- $Y_+^{(n)}(z) = Y_-^{(n)}(z) \begin{pmatrix} 1 & w(z) \\ 0 & 1 \end{pmatrix}$, $z \in \Sigma$

- $Y^{(n)}(z) \mapsto (I + \mathcal{O}(z^{-1})) \begin{pmatrix} z^n & 0 \\ 0 & z^{-n} \end{pmatrix}$ as $z \mapsto \infty$

Here $Y_\pm^{(n)}(z)$ denotes the limit of $Y^{(n)}(z')$ as $z' \mapsto z \in \Sigma$ from the positive/negative side of the complex plane. It may be proved (see [3]) that the solution of such a problem is unique and is given by

$$Y^{(n)}(z) = \begin{pmatrix} \pi_n(z) & f_n(z) \\ \gamma_{n-1}\pi_{n-1}(z) & \gamma_{n-1}f_{n-1}(z) \end{pmatrix}, \quad \mathrm{Im}\,z \neq 0 \qquad (7.29)$$

where the constants γ_n are simply related to the normalization of the corresponding polynomials: $\gamma_n = -2\pi i [\int_{-\infty}^{\infty} dw \pi_n^2]^{-1}$.

On comparing formulae (7.24) and (7.29) we observe that the structure of the correlation function $\mathcal{K}_N(\mu, \nu)$ is very intimately related to the above Riemann-Hilbert problem. In fact, for $\mu = \nu = z$ the matrices involved are identical (even the constant γ_{n-1} in Eq.(7.24) emerges when we replace $\propto$ with exact equality sign). Actually, all three types of kernels can be expressed in terms of the solution of the Riemann-Hilbert problem. The original works [3, 25] dealt only with the standard kernel built from polynomials alone. From that point of view the presence of Cauchy transforms in the Riemann-Hilbert problem might seem to be quite mysterious, and even superfluous. Now, after revealing the role and the meaning of more general kernels the picture can be considered complete, and the presence of the Cauchy transforms has its logical justification.

The relation to the Riemann-Hilbert problem is the starting point for a very efficient method of extracting the large$-N$ asymptotics for essentially any potential function $Q(x)$ entering the probability distribution measure. The corresponding machinery is known as the variant of the steepest descent/stationary phase method introduced for Riemann-Hilbert problems by Deift and Zhou. It is discussed at length in the book by Deift[3] which can be recommended to the interested reader for further details. In this way the universality was verified for all three types of kernels pertinent to the random matrix theory not only for bulk of the spectrum[22], but also for the spectral edges . In our considerations of the Gaussian Unitary Ensemble we already encountered the edge scaling regime where the spectral properties were parameterized by the Airy functions $Ai(x)$. Dealing with ratios of characteristic polynomials in such a regime requires second solution of the Airy equation denoted by $Bi(x)$, see [28].

We finish our exposition by claiming that there exist other interesting classes of matrix ensembles which attracted a considerable attention recently, see the paper[26] for more detail on the classification of random matrices by underlying symmetries. In the present framework we only mention one of them - the so-called *chiral* GUE. The corresponding $2N \times 2N$ matrices are of the form $\hat{H}_{ch} = \begin{pmatrix} \mathbf{0}_N & \hat{J} \\ \hat{J}^\dagger & \mathbf{0}_N \end{pmatrix}$, where $\hat{J}$ of a general complex matrix. They were introduced to provide a background for calculating the universal part of the microscopic level density for the Euclidian QCD Dirac operator, see [27] and references therein, and also have relevance for applications to condensed matter physics. The eigenvalues of such matrices appear in pairs $\pm\lambda_k$, $k = 1, ..., N$. It is easy to understand that the origin $\lambda = 0$ plays a specific role in such matrices, and close to this point eigenvalue correlations are rather different from those of the GUE, and described by the so-called Bessel kernels[29]. An alternative way of looking essentially at the same problem is to consider the random matrices of Wishart type $\hat{W} = \hat{J}^\dagger \hat{J}$, where the role of the special point is again played by the origin (in such context the origin is frequently referred to as the "hard spectral edge", since no eigenvalues are possible beyond that point. This should be contrasted with the Airy regime close to the semicircle edge, the latter being sometimes referred to as the "soft edge" of the spectrum.). The corresponding problems for products and ratios of characteristic polynomials were treated in full rigor by Riemann-Hilbert technique by Vanlessen[30], and in a less formal way in [28].

7.1 Acknowledgement

My own understanding in the topics covered was shaped in the course of numerous discussions and valuable contacts with many colleagues. I am particularly grateful to Gernot Akemann, Jon Keating, Boris Khoruzhenko and Eugene Strahov for fruitful collaborations on various facets of the Random Matrix theory which I enjoyed in recent years. I am indebted to Nina Snaith and Francesco Mezzadri for their invitation to give this lecture course, and for careful editing of the notes. It allowed me to spend a few wonderful weeks in the most pleasant and stimulating atmosphere of the Newton Institute, Cambridge whose hospitality and financial support I acknowledge with thanks. Finally, it is my pleasure to thank Guler Ergun for her assistance with preparing this manuscript for publication.

References

[1] B.A. Dubrovin, S.P. Novikov and A.T. Fomenko, *Modern Geometry: Methods and applications* (Springer, NY, 1984).

[2] M.L. Mehta, *Random matrices and the statistical theory of energy levels*, 2nd ed. (Academic, NY, 1991).

[3] P. Deift, *Orthogonal Polynomials and Random Matrices: a Riemann-Hilbert approach*, Courant Inst. Lecture Notes (AMS, Rhode Island, 2000).

[4] F. Haake, *Quantum Signatures of Chaos*, 2nd ed. (Springer, Berlin, 1999).

[5] O. Bohigas in: Les Houches Summer School, Session LII *Chaos and Quantum Physics*, ed. by M.-J. Giannoni et al. (Amsterdam, North-Holland, 1991).

[6] G. Szegö, *Orthogonal Polynomials*, 4th ed. American Mathematical Society, (Colloquium Publications, 23, Providence, 1975).

[7] R. Wong, *Asymptotic Approximations of integrals*, (Academic Press, New York, 1989).

[8] C.A. Tracy and H. Widom, Level spacing distributions and the Airy kernels, *Commun. Math. Phys.* **159**, 151-174 (1994).

[9] P.J. Forrester, The spectrum edge of random matrix ensembles, *Nucl. Phys. B[FS]* **402**, 709-728 (1993).

[10] L. Pastur and M. Scherbina, Universality of the local eigenvalue statistics for a class of unitary-invariant random matrix ensembles, *J. Stat. Phys.* **86**, 109 (1997).

[11] J.P. Keating and N.C. Snaith, Random Matrix Theory and L-functions at $s = 1/2$, *Commun. Math. Phys.* **214**, 91-110 (2000).

[12] E. Strahov, Moments of characteristic polynomials enumerate two-rowed lexicographic arrays, *Elect. Journ. Combinatorics* **10**, (1) R24 (2004).

[13] P. Diaconis and A. Gamburd, Random Matrices, Magic squares and Matching Polynomials, *Elect. Journ. Combinatorics* **11** (2004).

[14] E. Brezin and S. Hikami, Characteristic polynomials of random matrices, *Commun. Math. Phys.* **214**, 111 (2000).

[15] P. Forrester and N. Witte, Application of the $\tau-$ function theory of Painleve equations to random matrices, *Commun. Math. Phys.* **219**, 357-398 (2001).

[16] M.L. Mehta and J.-M. Normand, Moments of the characteristic polynomial in the three ensembles of random matrices, *J. Phys. A: Math.Gen.* **34**, 4627-4639 (2001).

[17] Y.V. Fyodorov and E. Strahov, An exact formula for general spectral correlation function of random Hermitian matrices *J. Phys. A: Math.Gen.* **36**, 3203-3213 (2003).

[18] G. Akemann and E. Kanzieper Spectra of Massive and massless QCD Dirac Operators: a Novel Link, *Phys. Rev. Lett.* **85**, 1174-1177 (2000).

[19] Y.V. Fyodorov and H.-J. Sommers, Random Matrices close to Hermitian or Unitary: overview of methods and results, *J. Phys. A: Math. Gen.* **36**, 3303-3348 (2003).

[20] Y.V. Fyodorov, Complexity of Random Energy landscapes, Glass Transition and Absolute value of Spectral Determinant of Random Matrices, *Phys. Rev. Lett.* **92**: art. no. 240601 (2004); Erratum: *ibid* **93**: art. no. 149901 (E) (2004).

[21] A.V. Andreev and B.D. Simons, Correlator of the Spectral Determinants in Quantum Chaos, *Phys. Rev. Lett.* **75**, 2304-2307 (1995).

[22] E. Strahov and Y.V. Fyodorov, Universal results for Correlations of characteristic polynomials: Riemann-Hilbert approach matrices *Commun. Math. Phys.* **219**, 343-382 (2003).

[23] J. Baik, P. Deift and E. Strahov, Products and ratios of characteristic polynomials of random Hermitian matrices, *J. Math. Phys.* **44**, 3657-3670 (2003).

[24] A. Borodin, and E. Strahov , Averages of Characteristic Polynomials in Random Matrix Theory, *Comm. Pure and Applied Math.* **59**, Issue 2, 161-253 (2006).

[25] P. Bleher and A. Its, Semiclassical asymptotics of orthogonal polynomials, Riemann-Hilbert problem, and universality in the matrix model, *Ann. Mathematics* **150**, 185-266 (1999).

[26] M.R. Zirnbauer Symmetry classes in Random Matrix Theory, e-preprint ArXiv:math-ph/0404058.

[27] J.J. Verbaarschot and T. Wettig Random Matrix Theory and Chiral Symmetry in QCD, *Annu. Rev. NUcl. Part. Sci.* **50**, 343-410 (2000).

[28] G. Akemann and Y.V. Fyodorov, Universal random matrix correlations of ratios of characteristic polynomials at the spectral edges, *Nucl. Phys.* B **664**, 457-476 (2003).

[29] C.A. Tracy and H. Widom, Level spacing distributions and the Bessel kernels, *Commun. Math. Phys.* **161**, 289-309 (1994).

[30] M. Vanlessen, Universal behaviour for averages of characteristic polynomials at the origin of the spectrum, *Commun. Math. Phys.* **253** No. 3, 535-560 (2005).

Department of Mathematical Sciences
Brunel University
Uxbridge UB8 3PH
UK

Notes on Pair Correlation of Zeros and Prime Numbers

D. A. Goldston *

These notes are based on my four lectures given at the Newton Institute in April 2004 during the Recent Perspectives in Random Matrix Theory and Number Theory Workshop. Their purpose is to introduce the reader to the analytic number theory necessary to understand Montgomery's work on the pair correlation of the zeros of the Riemann zeta-function and subsequent work on how this relates to prime numbers. A very brief introduction to Selberg's work on the moments of $S(T)$ is also given.

1   Introduction and Some Personal History

In 1973 Montgomery's paper [28], "The Pair Correlation of Zeros of the Zeta Function," appeared in the AMS series of Proceedings of Symposia in Pure Mathematics, and a new field of study was born — slowly. I first came across this paper in 1977, and was probably the only person at Berkeley to read it. Most zeta-function people (as some of us refer to ourselves) recognized the importance of this work and the new phenomena discovered, but it was not clear what to do next. At first, the main interest was in using Montgomery's conjectures to refine the classical results on primes obtained assuming the Riemann Hypothesis. Gallagher and Mueller [12] wrote an important paper on this in

---

*The author was supported by an NSF FRG grant

1978, followed by further results from Heath-Brown [21]. In 1981 I wrote my
Ph.D. thesis on this topic. A few years later Montgomery and I [18] obtained
an equivalence between the pair correlation conjecture and primes. However,
this work attracted little attention — probably because the results were obtained using Montgomery's conjectures. Then, in the early 1980's everything
changed: Odlyzko [31] computed statistics on the zeros and convinced even
the most skeptical that after almost a century of intensive study a totally new,
unsuspected, and fundamental property of the zeta-function had been discovered. The field has since had a flood of activity, with the generalization of
Montgomery's work to higher correlations by Hejhal [23] and Rudnick-Sarnak
[32], the interpretation of these results in terms of mathematical physics by
Berry and Bogomolny-Keating [1, 2], the function field case of Katz-Sarnak
[26], the random matrix model for moments of the zeta-function of Keating
and Snaith culminating in [4], and a profusion of new work.

In these notes, I will discuss Montgomery's results and their relations to
primes. As a unifying tool, I will use Montgomery's explicit formula [28] to
prove a number of later results that were originally obtained by other methods.
This approach was first made use of in part of my Ph.D. thesis, and was based
on a suggestion of Montgomery in a letter. At that time Heath-Brown had
just finished his paper which covered the same ground, and I saw no need to
publish this material beyond the summary that appeared in [15]. My goal, in
line with the emphasis of the workshop on reaching out to beginners in the
field, is to provide some of the main ideas used without technicalities and at the
same time supply simple details which would be accepted without comment by
experts. I have intentionally left out many things to keep these notes focused.
The last section on Selberg's theory of $S(T)$ and $\log \zeta(s)$ is somewhat different
from the previous ones, and I have decided to state only the main results and
present a few of the ideas that are used.

I would like to thank Andrew Ledoan for the many improvements he suggested for these notes.

2   Basic Facts and Notation

Following Riemann, we use the complex variable $s = \sigma + it$. The Riemann
zeta-function $\zeta(s)$ is defined, for $\sigma > 1$, by either the Dirichlet series or the
Euler product

$$\zeta(s) = \sum_{n=1}^{\infty} \frac{1}{n^s} = \prod_{p} \left(1 - \frac{1}{p^s}\right)^{-1}. \qquad (2.1)$$

Here p will always denote a prime, so the product is over all the prime numbers.
To extract information about primes from the Euler product, we compute
the logarithmic derivative of the zeta-function and use the power series for

$- \log(1 - z)$, to obtain, for $\sigma > 1$,

$$\frac{\zeta'}{\zeta}(s) := \frac{\zeta'(s)}{\zeta(s)} = \frac{d}{ds} \log \zeta(s) = \frac{d}{ds} \Big(\sum_{m=1}^{\infty} \sum_p \frac{1}{mp^{ms}} \Big) = - \sum_{n=1}^{\infty} \frac{\Lambda(n)}{n^s}, \quad (2.2)$$

where the von Mangoldt function $\Lambda(n)$ is given by

$$\Lambda(n) = \begin{cases} \log p, & \text{if } n = p^m, \ p \text{ prime}, \ m \geq 1, \\ 0, & \text{otherwise.} \end{cases} \quad (2.3)$$

The Chebyshev function $\psi(x)$ is the counting function for $\Lambda(n)$ given by

$$\psi(x) = \sum_{n \leq x} \Lambda(n). \quad (2.4)$$

Because of the simple relationship with the zeta-function, it is preferable to use $\Lambda(n)$ in place of the indicator function for the primes, and $\psi(x)$ in place of the counting function $\pi(x)$ for the number of primes up to x. If needed, one can usually recover $\pi(x)$ from $\psi(x)$ by simple arguments. The Prime Number Theorem (PNT) states that as $x \to \infty$

$$\psi(x) \sim x, \quad \text{or} \quad \pi(x) \sim \frac{x}{\log x}. \quad (2.5)$$

The PNT with the error term obtained by de la Vallée Poussin in 1899 is, for a small constant c,

$$\psi(x) = x + O\left(xe^{-c\sqrt{\log x}}\right), \quad (2.6)$$

which on returning to $\pi(x)$ gives (c may differ from equation to equation)

$$\pi(x) = \text{li}(x) + O\left(xe^{-c\sqrt{\log x}}\right), \quad (2.7)$$

where the logarithmic integral

$$\text{li}(x) = \int_2^x \frac{du}{\log u}$$

is the actual main term in the theorem. For the error term above, we have for any constant $A > 0$

$$e^{-c\sqrt{\log x}} \ll \frac{1}{(\log x)^A}. \quad (2.8)$$

Here, the Vinogradov notation[1] $\ll$ is equivalent to "big oh" of the right-hand side. This estimate is freely used when the PNT is invoked.

---

[1] Editors' comment: See the Appendix of the lectures of D.W. Farmer, page 185, for a discussion of the $\ll$ notation.

We frequently need the Dirichlet series for $\zeta(s)^{-1}$, which from the Euler product is

$$\frac{1}{\zeta(s)} = \sum_{n=1}^{\infty} \frac{\mu(n)}{n^s} \tag{2.9}$$

for $\sigma > 1$, where the Möbius function is defined by $\mu(1) = 1$ and

$$\mu(n) = \begin{cases} (-1)^m, & \text{if } n = p_1 p_2 \cdots p_m, \ p_i\text{'s distinct}, \\ 0, & \text{if } p^2 | n, \text{ some } p. \end{cases} \tag{2.10}$$

The zeta-function has a simple pole with residue 1 at $s = 1$, trivial zeros at $s = -2n$, $n = 1, 2, 3, \ldots$. and complex zeros

$$\rho = \beta + i\gamma, \quad 0 < \beta < 1. \tag{2.11}$$

The inequality $\beta < 1$ is the key result needed in the analytic proofs of the PNT. The zeros are positioned symmetrically with the real line and the "$\frac{1}{2}$-line" $\frac{1}{2} + it$, so that ρ, $\bar{\rho}$, $1 - \rho$, and $1 - \bar{\rho}$ are all zeros. The Riemann Hypothesis (RH) is the conjecture that $\beta = \frac{1}{2}$, and thus $\rho = \frac{1}{2} + i\gamma$. For example, the first 6 zeros in the upper half of the critical strip are

$$\begin{aligned} &\frac{1}{2} + i14.13472\ldots, \quad \frac{1}{2} + i21.02203\ldots, \quad \frac{1}{2} + i25.01085\ldots, \\ &\frac{1}{2} + i30.42487\ldots, \quad \frac{1}{2} + i32.93506\ldots, \quad \frac{1}{2} + i37.58617\ldots. \end{aligned} \tag{2.12}$$

To count the number of complex zeros in a given region, we define

$$n(T) = |\{\gamma : 0 < \gamma \le T\}|, \qquad N(T) = \frac{n(T+0) + n(T-0)}{2}, \tag{2.13}$$

where $|A|$ denotes the number of elements of the set A. Note that $N(T)$ counts any zeros $\gamma = T$ with weight one-half, which arises naturally in the theory; therefore we always use $N(T)$ in preference to $n(T)$. The Riemann-von Mangoldt formula for $N(T)$, obtained by applying the argument principle to ζ and using the functional equation (see [8], [25], [39]), is

$$N(T) = \frac{T}{2\pi} \log \frac{T}{2\pi e} + \frac{7}{8} + R(T) + S(T), \tag{2.14}$$

where

$$R(T) \ll \frac{1}{T} \tag{2.15}$$

and

$$S(T) = \frac{1}{\pi} \arg \zeta\left(\frac{1}{2} + iT\right) \ll \log T. \tag{2.16}$$

In fact, $R(T)$ is continuous, differentiable, and can be expanded into a series in inverse powers of T. We see that (2.14) provides a remarkably precise formula for the number of zeros up to height T, with the finer details of the vertical distribution of zeros wrapped up in the study of $S(T)$. In particular, we have

$$N(T) \sim \frac{T}{2\pi} \log T. \tag{2.17}$$

Another consequence of (2.14) – (2.16) which we make frequent use of is the sharp estimate

$$N(T+1) - N(T) = \sum_{T < \gamma \leq T+1} 1 \ll \log T. \tag{2.18}$$

3 Explicit Formulas

To study the relationship between the zeros of the zeta-function and primes you need to be able to work with explicit formulas. There are many such formulas but the best known is the Riemann-von Mangoldt explicit formula, which states that, for $x > 1$,

$$\psi_0(x) = x - \sum_\rho \frac{x^\rho}{\rho} - \log 2\pi - \frac{1}{2} \log \left(1 - \frac{1}{x^2} \right), \tag{3.1}$$

where $\psi_0(x) = \frac{1}{2}(\psi(x+0) + \psi(x-0))$. By (2.14), the sum is not absolutely convergent and the terms are added with ρ and $\bar{\rho}$ grouped together. The explicit formula also contains this information, since on taking $x \to 1^+$ and letting $x = e^u$ we see that

$$\sum_\rho \frac{e^{\rho u}}{\rho} = \frac{1}{2} \log \frac{1}{u} + O(1) \quad \text{as} \quad u \to 0^+. \tag{3.2}$$

For applications we usually use the truncated version of (3.1),

$$\psi(x) = x - \sum_{|\gamma| \leq T} \frac{x^\rho}{\rho} + O \left(\frac{x}{T} (\log xT)^2 \right) + O \left((\log x) \min \left(1, \frac{x}{T\|x\|} \right) \right), \tag{3.3}$$

where $\|x\|$ denotes the distance of x to the closest integer, and the last term reflects the jumps of $\psi(x)$ at the primes and prime powers. As an example of an application of (3.3), assuming RH we have

$$\frac{x^\rho}{\rho} \ll \frac{x^{\frac{1}{2}}}{|\gamma|},$$

and so by (2.12) and (2.18), with $[x]$ denoting the integer part of x,

$$\sum_{|\gamma| \leq T} \frac{1}{|\gamma|} = 2 \sum_{1 < \gamma \leq T} \frac{1}{\gamma} \leq \sum_{n=1}^{[T]+1} \frac{1}{n} \sum_{n < \gamma \leq n+1} 1 \ll \sum_{n \leq 2T} \frac{\log 2n}{n} \ll (\log T)^2.$$

Thus, taking $T = x$ in (3.3) we have

$$\psi(x) = x + O\left(x^{\frac{1}{2}}(\log x)^2\right).\tag{3.4}$$

It can also be proved that this estimate implies RH, and therefore is equivalent to the RH.[2] Equation (3.4) is due to von Koch in 1901 and has never been improved.

We next apply (3.4) to the problem of large gaps between primes. Let p_n denote the n-th prime number. The highest power of a prime $\leq x$ is the largest k for which $2^k \leq x$, so $k = [\log_2 x]$. By the PNT,

$$\begin{aligned}
\psi(x) &= \sum_{p \leq x} \log p + \sum_{2 \leq m \leq \log_2 x} \sum_{p^m \leq x} \log p \\
&= \sum_{p \leq x} \log p + \sum_{p \leq \sqrt{x}} \log p + O\left(\pi\left(x^{\frac{1}{3}}\right)(\log x)^2\right) \\
&= \sum_{p \leq x} \log p + O\left(x^{\frac{1}{2}}\right).
\end{aligned}$$

Thus (3.4) continues to hold when we only sum over primes. For $1 \leq h \leq x$, we have by (3.4) and differencing that

$$\sum_{x < p \leq x+h} \log p = h + O\left(x^{\frac{1}{2}}(\log x)^2\right).$$

On taking $h = Cx^{\frac{1}{2}}(\log x)^2$, with the constant C being larger than the implicit absolute constant in the error term, we conclude that the sum on the left is positive and $\gg h$, and thus the interval $(x, x+h]$ must contain $\gg \frac{h}{\log x}$ primes. If p_n is the first prime in $(x, x+h)$, then

$$p_{n+1} - p_n < h \ll p_n^{\frac{1}{2}}(\log p_n)^2.\tag{3.5}$$

An explicit formula that also exhibits the close connection between zeros and primes is the Landau formula, which states that (for x fixed) as $T \to \infty$

$$\sum_{0 < \gamma \leq T} x^\rho = -\frac{T\Lambda(x)}{2\pi} + O(\log T).\tag{3.6}$$

Here we define $\Lambda(x)$ to be zero for real non-integer x. Formally this is obtained by differentiating (3.1) with respect to x. The exponential sum over the zeros encodes the information on which integers are primes or prime powers. Equation (3.6) is not particularly useful, but Fujii [10] and independently Gonek [19] have developed uniform versions which can be used in applications.

---

[2] Even the seemingly weaker estimate $\psi(x) = x + O\left(x^{\frac{1}{2}+\epsilon}\right)$ for any $\epsilon > 0$ is equivalent to RH.

An explicit formula of at least historic interest is the Cramér explicit formula, which states that for $\text{Im}(z) > 0$

$$\sum_{\gamma>0} e^{\rho z} = \frac{e^z}{2\pi i} \sum_{n=2}^{\infty} \frac{\Lambda(n)}{n} \left(\frac{1}{z - \log n} + \frac{1}{\log n} \right)$$

$$+ \frac{1}{2\pi i} \sum_{n=2}^{\infty} \frac{\Lambda(n)}{n} \left(\frac{1}{z + \log n} - \frac{1}{\log n} \right)$$

$$+ \left(\frac{1}{4} + \frac{\gamma + \log 2\pi}{2\pi i} \right) \left(1 + \frac{1}{z} \right) + \frac{1}{2\pi i} \frac{\Gamma'}{\Gamma} \left(\frac{z}{2\pi i} \right) + \frac{1}{2} e^z \qquad (3.7)$$

$$- \frac{z}{2\pi i} \int_0^1 e^{sz} \log |\zeta(z)| \, ds - \frac{1}{2\pi i z} \int_0^{\infty} \frac{t}{e^t - 1} \frac{dt}{t+z}.$$

On taking $z = -\log \tau + iy$, $0 < y \leq 1$, and letting $\tau \to \infty$ Cramér [7] proved that

$$-2\pi \text{Re} \sum_{\gamma>0} e^{\rho(-\log \tau + iy)} = \sum_{n=2}^{\infty} \frac{\Lambda(n)}{n} \left(\frac{y}{\left(\log \frac{\tau}{n}\right)^2 + y^2} \right) - \pi + O\left(\frac{1}{\log \tau} \right). \quad (3.8)$$

He used (3.8) and related formulas in a series of papers starting in 1920 to prove results on primes. One such result is that on RH

$$p_{n+1} - p_n \ll p_n^{\frac{1}{2}} \log p_n. \qquad (3.9)$$

This only saves a logarithm over the trivial use of (3.4) in (3.5) but is the best result known on RH. We will later assume a much stronger hypothesis and only improve (3.9) by a half-power of a logarithm. On the other hand, Cramér conjectured [7] that the gaps between consecutive primes are always much smaller than this size. Recent work indicates that Cramér's original conjecture may be slightly too strong, but all evidence still suggests

$$p_{n+1} - p_n \ll (\log p_n)^2. \qquad (3.10)$$

At one time I had a fondness for Cramér's formula and made use of it in my thesis, but I later decided that nothing was to be gained by its use except complicated arguments. The proof of (3.9), for instance, can now be done from a smoothed version of (3.1) in just a few lines. However, there have been a number of recent papers on the structure of Cramér's formula (see [24]).

Most of these explicit formulas are based on evaluating the contour integral

$$\mathcal{I} = \frac{1}{2\pi i} \int_{c-i\infty}^{c+i\infty} -\frac{\zeta'}{\zeta}(s) K_z(s) \, ds, \qquad (3.11)$$

where the kernel $K_z(s)$ is a meromorphic function. Frequently $K_z(s) = K(s + z)$ or $K_z(s) = K(zs)$. If $c > 1$ the Dirichlet series for $\frac{\zeta'}{\zeta}(s)$ converges absolutely and

$$\mathcal{I} = \sum_{n=2}^{\infty} \Lambda(n) \hat{K}_z(n), \quad \hat{K}_z(n) = \frac{1}{2\pi i} \int_{c-i\infty}^{c+i\infty} K_z(s) n^{-s} \, ds. \qquad (3.12)$$

One then obtains an explicit formula by moving the contour to the left, thus encountering poles at $s = 1$ and at the zeros of $\zeta(s)$, as well as any poles of $K_z(s)$.

Another explicit formula frequently used is the Weil explicit formula, which contains a general weight function and has the advantage of allowing the relationships between terms to be explicitly exhibited. Our approach here, however, is to use explicit formulas only as tools for studying zeros and primes. Therefore we will take the opposite path and stay specific. The formula we will base our work on is due to Montgomery [28].

Proposition 1. *Assume the Riemann Hypothesis. For $x \geq 1$,*

$$2x^{\frac{1}{2}-it} \sum_{\gamma} \frac{x^{i\gamma}}{1+(t-\gamma)^2} = -\sum_{n=1}^{\infty} \frac{\Lambda(n)a_n(x)}{n^{it}} + \frac{2x^{1-it}}{\left(\frac{1}{2}+it\right)\left(\frac{3}{2}-it\right)}$$
$$+ x^{-\frac{1}{2}} \left(\log(|t|+2) + O(1)\right) + O\left(\frac{x^{-2}}{|t|+2}\right), \tag{3.13}$$

where

$$a_n(x) = \min\left(\left(\frac{n}{x}\right)^{\frac{1}{2}}, \left(\frac{x}{n}\right)^{\frac{3}{2}}\right). \tag{3.14}$$

Proof. This proposition is proved by using an explicit formula from Landau's 1909 Handbuch [27], which states that (unconditionally) for $x > 1$, $x \neq p^m$,

$$\sum_{n \leq x} \frac{\Lambda(n)}{n^s} = -\frac{\zeta'}{\zeta}(s) + \frac{x^{1-s}}{1-s} - \sum_{\rho} \frac{x^{\rho-s}}{\rho-s} + \sum_{n=1}^{\infty} \frac{x^{-2n-s}}{2n+s} \tag{3.15}$$

provided $s \neq 1$, $s \neq \rho$, $s \neq -2n$.[3] Rewriting (3.15), we have

$$\sum_{\rho} \frac{x^{\rho}}{\rho-s} = -x^s \left(\frac{\zeta'}{\zeta}(s) + \sum_{n \leq x} \frac{\Lambda(n)}{n^s}\right) + \frac{x}{1-s} + \sum_{n=1}^{\infty} \frac{x^{-2n}}{2n+s}. \tag{3.16}$$

This equation holds independently of RH, but assuming RH we have $\rho = \frac{1}{2}+i\gamma$. Letting $s = \frac{3}{2} + it$ and using (2.2), the above equation simplifies to read

$$-x^{\frac{1}{2}} \sum_{\gamma} \frac{x^{i\gamma}}{1+i(t-\gamma)} = x^{it} \sum_{n>x} \frac{\Lambda(n)a_n(x)}{n^{it}} + \frac{x}{-\frac{1}{2}-it} + \sum_{n=1}^{\infty} \frac{x^{-2n}}{2n+\frac{3}{2}+it}.$$

On the other hand, if $s = -\frac{1}{2} + it$ in (3.16) we have

$$x^{\frac{1}{2}} \sum_{\gamma} \frac{x^{i\gamma}}{1-i(t-\gamma)} = -x^{it} \sum_{n \leq x} \frac{\Lambda(n)a_n(x)}{n^{it}} - x^{-\frac{1}{2}+it}\frac{\zeta'}{\zeta}\left(-\frac{1}{2}+it\right)$$
$$+ \frac{x}{\frac{3}{2}-it} + \sum_{n=1}^{\infty} \frac{x^{-2n}}{2n-\frac{1}{2}+it}.$$

---

[3] If $s = 0$ we get (3.1). Landau used (3.15) to prove Riemann's original explicit formula for $\pi(x)$.

Subtracting the latter from the former and using

$$\frac{\zeta'}{\zeta}\left(-\frac{1}{2}+it\right) = -\log(|t|+2) + O(1),$$

which follows easily from the functional equation, we obtain the proposition. By continuity the values $x = 1, p^m$ no longer need to be excluded.

The role of RH in Proposition 1 is notational. Recently, a new notation has emerged which is very convenient. We write the complex zeros of the zeta function as $\rho = \frac{1}{2} + i\gamma$, $\gamma \in \mathbb{C}$, so that γ is complex when that zero is off the $\frac{1}{2}$-line. Thus, the RH becomes the statement that γ is real. With this notation we see that the proof is unchanged, and Proposition 1 holds unconditionally. We will not make any further use of this notation since the size of the terms in our sums over zeros become important and the RH is often needed.

4   Montgomery's theorem

We first examine Montgomery's explicit formula heuristically and see what each term means. The weight in the sum over zeros concentrates the sum to zeros in a short bounded interval around t, and therefore behaves similarly to

$$\sum_{t<\gamma\leq t+1} x^{i\gamma}.$$

By (2.18), if this sum is substantially smaller than $\log t$ then we will have detected cancelation from $x^{i\gamma}$. If $x = 1$ or is close to 1 no cancelation can occur, and this is reflected by the term $x^{-1/2}\log(|t|+2)$ in (3.13). The sum over primes is concentrated around x, and therefore behaves similarly to

$$\sum_{\frac{1}{2}x<n\leq 2x} \frac{\Lambda(n)}{n^{it}}.$$

The expected value of the original sum over primes is obtained by the PNT and equals the remaining term

$$\frac{2x^{1-it}}{\left(\frac{1}{2}+it\right)\left(\frac{3}{2}-it\right)}.$$

How does one extract information from (3.13)? Montgomery was interested in studying the distribution of the differences of pairs of zeros, and for this it is clear one needs to square the absolute value of the sum over zeros. It would be nice to be able to obtain this distribution in an interval of length one around t, but the pointwise dependence on t in the Dirichlet sum over primes is intractable. To circumvent this problem, we also integrate with respect to t to obtain our distribution in a longer range. To this end we consider

$$\int_0^T \left|\sum_\gamma \frac{x^{i\gamma}}{1+(t-\gamma)^2}\right|^2 dt.$$

Since the weight in the sum will be small when $|t - \gamma|$ is large, which is the case over most of the integration range unless $0 < \gamma \leq T$, we may restrict the sum to this range with a small error. With the sum restricted to the zeros $0 < \gamma \leq T$, we may extend the integration range to $(-\infty, \infty)$ with a small error. Using (2.18), Montgomery showed

$$\int_0^T \left| \sum_\gamma \frac{x^{i\gamma}}{1 + (t - \gamma)^2} \right|^2 dt = \int_{-\infty}^\infty \left| \sum_{0 < \gamma \leq T} \frac{x^{i\gamma}}{1 + (t - \gamma)^2} \right|^2 dt + O\left((\log T)^3\right).$$

Multiplying out the integral on the right-hand side, we find

$$\int_{-\infty}^\infty \left| \sum_{0 < \gamma \leq T} \frac{x^{i\gamma}}{1 + (t - \gamma)^2} \right|^2 dt = \frac{\pi}{2} \sum_{0 < \gamma, \gamma' \leq T} x^{i(\gamma - \gamma')} w(\gamma - \gamma'),$$

where the weight

$$w(u) = \frac{4}{4 + u^2} \tag{4.1}$$

is obtained on evaluating the integral either by residues, convolution, or otherwise. We thus define for $x > 0$

$$F(x, T) = \sum_{0 < \gamma, \gamma' \leq T} x^{i(\gamma - \gamma')} w(\gamma - \gamma') = \frac{2}{\pi} \int_{-\infty}^\infty \left| \sum_{0 < \gamma \leq T} \frac{x^{i\gamma}}{1 + (t - \gamma)^2} \right|^2 dt. \tag{4.2}$$

Then

$$F(x, T) \geq 0, \qquad F(x, T) = F\left(\frac{1}{x}, T\right), \tag{4.3}$$

and

$$F(x, T) = \frac{2}{\pi} \int_0^T \left| \sum_\gamma \frac{x^{i\gamma}}{1 + (t - \gamma)^2} \right|^2 dt + O\left((\log T)^3\right). \tag{4.4}$$

The next step is to use Proposition 1 to evaluate $F(x, T)$. Denoting (3.13) by

$$L(x, t) = R(x, t),$$

we have just shown that

$$\int_0^T |L(x, t)|^2 dt = 2\pi x F(x, T) + O(x(\log T)^3). \tag{4.5}$$

For $R(x, T)$, we compute the mean-square of each term. For the Dirichlet series, we use a standard mean value theorem of Montgomery and Vaughan [29], which states that

$$\int_0^T \left| \sum_{n=1}^\infty \frac{a_n}{n^{it}} \right|^2 dt = \sum_{n=1}^\infty |a_n|^2 (T + O(n)). \tag{4.6}$$

Hence

$$\int_0^T \left| \sum_{n=1}^{\infty} \frac{\Lambda(n)a_n(x)}{n^{it}} \right|^2 dt = \sum_{n=1}^{\infty} |\Lambda(n)a_n(x)|^2 (T + O(n))$$

$$= xT(\log x + O(1)) + O\left(x^2 \log x\right),$$

by Stieltjes integration and the PNT (with remainder). The remaining terms are elementary:

$$\int_0^T \left| \frac{2x^{1-it}}{(\frac{1}{2}+it)(\frac{3}{2}-it)} \right|^2 dt \ll x^2,$$

$$\int_0^T \left| x^{-\frac{1}{2}} \left(\log(|t|+2) + O(1) \right) \right|^2 dt = \frac{T}{x} \left((\log T)^2 + O(\log T) \right),$$

and

$$\int_0^T \left| \frac{x^{-2}}{|t|+2} \right|^2 dt \ll x^{-4}.$$

We thus have two main terms, the Dirichlet series term for $(\log T)^{3/2} \leq x \leq o(T)$ and the term $\log(|t|+2)$ which dominates for $1 \leq x \leq (\log T)^{3/4}$. In the intermediate range all terms are $o(xT \log T)$. By the Cauchy-Schwarz inequality, the largest term among these provides the main term in an asymptotic formula. Therefore,

$$\int_0^T |R(x,t)|^2 dt = xT(\log x + o(\log T)) + O(x^2 \log x) + \frac{T}{x}(\log T)^2(1 + o(1)),$$

and we conclude that

$$F(x,T) = \frac{T}{2\pi} \log x + o(T \log T) + O(x \log x) + \frac{T}{2\pi x^2}(\log T)^2(1 + o(1)). \quad (4.7)$$

Following Montgomery, we set

$$x = T^\alpha \quad (4.8)$$

and normalize by defining

$$F(\alpha) = F(\alpha, T) = \left(\frac{T}{2\pi} \log T \right)^{-1} \sum_{0 < \gamma, \gamma' \leq T} T^{i\alpha(\gamma-\gamma')} w(\gamma - \gamma'). \quad (4.9)$$

Thus we have arrived at Montgomery's theorem.

Theorem 1. *Assume the Riemann Hypothesis. Then $F(\alpha)$ is real, even, and non-negative. Further, uniformly for $0 \leq \alpha \leq 1 - \epsilon$, we have*

$$F(\alpha) = \alpha + o(1) + (1 + o(1))T^{-2\alpha} \log T. \quad (4.10)$$

The error term $O(x \log x)$ in (4.7) can be improved to $O(x)$ by using a sieve bound for prime twins [18], which shows the theorem holds for

$$0 \leq \alpha \leq 1. \quad (4.11)$$

A detailed analysis of the above proof has recently been done by Tsz Ho Chan, with all second order terms obtained.

5   Application to simple zeros and small gaps between zeros

The function $F(\alpha)$ is useful for evaluating sums over differences of zeros. Let $r(u) \in L^1$, and define the Fourier transform by

$$\hat{r}(\alpha) = \int_{-\infty}^{\infty} r(u)e(\alpha u)\,du, \qquad e(u) = e^{2\pi i u}. \qquad (5.1)$$

If $\hat{r}(\alpha) \in L^1$, we have almost everywhere

$$r(u) = \int_{-\infty}^{\infty} \hat{r}(\alpha)e(-u\alpha)\,d\alpha. \qquad (5.2)$$

On multiplying (4.9) by $\hat{r}(\alpha)$ and integrating, we obtain

$$\sum_{0<\gamma,\gamma'\leq T} r\left((\gamma-\gamma')\frac{\log T}{2\pi}\right)w(\gamma-\gamma') = \frac{T}{2\pi}\log T\int_{-\infty}^{\infty}\hat{r}(\alpha)F(\alpha)\,d\alpha. \qquad (5.3)$$

Using Theorem 1, we can evaluate the right-hand side provided $\hat{r}(\alpha)$ has support in $[-1, 1]$. Thus, we can evaluate sums over differences of zeros on the class of functions whose Fourier transforms are supported in $[-1, 1]$. Using the Fourier pair

$$k(u) = \left(\frac{\sin \pi \lambda u}{\pi \lambda u}\right)^2, \qquad \hat{k}(\alpha) = \frac{1}{\lambda}\max\left(1 - \frac{|\alpha|}{\lambda}, 0\right) \qquad (\lambda > 0) \qquad (5.4)$$

we have for $0 < \lambda \leq 1$

$$\begin{aligned}
\sum_{0<\gamma,\gamma'\leq T} &\left(\frac{\sin(\frac{\lambda}{2}(\gamma-\gamma')\log T)}{\frac{\lambda}{2}(\gamma-\gamma')\log T}\right)^2 w(\gamma-\gamma') \\
&= \left(\frac{1}{\lambda}\int_{-\lambda}^{\lambda}\left(1-\frac{|\alpha|}{\lambda}\right)F(\alpha)\,d\alpha\right)\frac{T}{2\pi}\log T \\
&\sim \left(\frac{2}{\lambda}\int_0^{\lambda}\left(1-\frac{\alpha}{\lambda}\right)\left(\alpha + T^{-2\alpha}\log T\right)d\alpha\right)\frac{T}{2\pi}\log T \\
&\sim \left(\frac{1}{\lambda}+\frac{\lambda}{3}\right)\frac{T}{2\pi}\log T.
\end{aligned} \qquad (5.5)$$

This result has an important application to simple zeros of $\zeta(s)$.

Theorem 2. *Assume the Riemann Hypothesis. At least two thirds of the zeros of the Riemann zeta-function are simple in the sense that as $T \to \infty$*

$$N_s(T) := \sum_{\substack{0<\gamma\leq T \\ \rho \text{ simple}}} 1 \geq \left(\frac{2}{3} - o(1)\right)N(T). \qquad (5.6)$$

Proof. The sum in (5.5) over pairs of zeros counts distinct zeros weighted by their multiplicity. Thus a double zero gets counted 4 times, a triple zero 9 times, etc. Denoting the multiplicity of ρ by m_ρ, we have

$$\sum_{0<\gamma\leq T} m_\rho = \sum_{\substack{0<\gamma,\gamma'\leq T \\ \gamma=\gamma'}} 1$$

$$\leq \sum_{0<\gamma,\gamma'\leq T} \left(\frac{\sin\left(\frac{\lambda}{2}(\gamma-\gamma')\log T\right)}{\frac{\lambda}{2}(\gamma-\gamma')\log T} \right)^2 w(\gamma-\gamma')$$

$$\leq (1+o(1))\left(\frac{1}{\lambda}+\frac{\lambda}{3}\right)\frac{T}{2\pi}\log T.$$

Choosing $\lambda=1$, we have

$$\sum_{0<\gamma\leq T} m_\rho \leq \left(\frac{4}{3}+\epsilon\right)\frac{T}{2\pi}\log T. \tag{5.7}$$

But

$$\sum_{\substack{0<\gamma\leq T \\ \rho \text{ simple}}} 1 \geq \sum_{0<\gamma\leq T} (2-m_\rho),$$

and applying (2.17) completes the proof.

It is possible to make very small improvements in the value $\frac{2}{3}$ in Theorem 2. It would be a major advance to be able to prove that almost all the zeros are simple, even on RH. Conrey, Ghosh, and Gonek [6] have proved using a different method that assuming RH and the Generalized Lindelöf Hypothesis,

$$N_s(T) \geq \left(\frac{19}{27}-\epsilon\right)N(T).$$

Montgomery also proved that there are gaps between zeros closer than the average. He used the transform pair (5.4) with their roles reversed to obtain

$$\liminf_{n\to\infty} (\gamma_{n+1}-\gamma_n)\frac{\log\gamma_n}{2\pi} \leq 0.669\ldots.$$

Consider the Fourier pair

$$h(u) = \left(\frac{\sin\pi u}{\pi u}\right)^2\left(\frac{1}{1-u^2}\right), \tag{5.8}$$

$$\hat{h}(\alpha) = \max\left(1-|u|+\frac{\sin 2\pi|u|}{2\pi},0\right),$$

where $h(u)$ is the Selberg minorant of the characteristic function of the interval $[-1,1]$ in the class of functions with Fourier transforms with support in $[-1,1]$. We prove

Theorem 3. *Assume the Riemann Hypothesis. We have*

$$\liminf_{n\to\infty} (\gamma_{n+1} - \gamma_n) \frac{\log \gamma_n}{2\pi} \leq 0.6072\ldots . \tag{5.9}$$

Proof. Take $r(u) = h\left(\frac{u}{\lambda}\right)$. Then $r(u)$ is a minorant of the characteristic function of the interval $[-\lambda, \lambda]$. Thus

$$\sum_{0<\gamma\leq T} m_\rho + 2 \sum_{0<\gamma-\gamma'\leq \frac{2\pi\lambda}{\log T}} 1 \geq \sum_{0<\gamma,\gamma'\leq T} h\left((\gamma-\gamma')\frac{\log T}{2\pi\lambda}\right) w(\gamma-\gamma')$$

$$= \left(\frac{T}{2\pi}\log T\right) \int_{-\frac{1}{\lambda}}^{\frac{1}{\lambda}} \lambda\hat{h}(\lambda\alpha)F(\alpha)\,d\alpha.$$

Assume $\lambda < 1$. Since the integrand is positive we obtain a lower bound by decreasing the integration range to $[-1, 1]$. We can assume

$$\sum_{0<\gamma\leq T} m_\rho \sim \frac{T}{2\pi}\log T,$$

since otherwise we would have infinitely many multiple zeros and the theorem holds for this reason. Thus

$$\sum_{0<\gamma-\gamma'\leq \frac{2\pi\lambda}{\log T}} 1 \geq \left(\frac{1}{2} - \epsilon\right) \frac{T}{2\pi}\log T \left(\lambda - 1 + 2\lambda \int_0^1 \alpha h(\lambda\alpha)\,d\alpha\right).$$

By an easy numerical calculation, we find that the right-hand side is positive for $\lambda > 0.6072\ldots$, which proves the result.

By a different method (on RH), Montgomery and Odlyzko [30] improved on this result and obtained the upper bound 0.5179. Conrey, Ghosh, and Gonek [5] later replaced this by 0.5172.

6   Montgomery's Conjectures

What if $\alpha > 1$? It is not difficult to see from the proof of Montgomery's theorem that for $x \geq T$

$$F(x,T) = \frac{1}{2\pi x} \int_0^T \left| \sum_{n=1}^\infty \frac{\Lambda(n)a_n(x)}{n^{it}} - \frac{2x^{1-it}}{\left(\frac{1}{2}+it\right)\left(\frac{3}{2}-it\right)} \right|^2 dt + o(T\log T). \tag{6.1}$$

We saw that the diagonal terms in the sum contribute $\frac{T}{2\pi}\log x$, while the expected value term contributes cx. On the other hand, we have the trivial bound

$$F(x,T) \leq F(0,T) \sim \frac{T}{2\pi}\log^2 T, \tag{6.2}$$

where the last relation follows from Theorem 1 (or unconditionally from (2.14)). Thus $F(x,T)$ never gets as large as x for $x \gg T(\log T)^2$, and therefore the off-diagonal terms in the sum over primes must almost perfectly cancel the expected value term.

Montgomery proceeded by multiplying out the integrand in (6.1) and integrating term by term. For the off-diagonal terms, one needs to assume the Hardy-Littlewood k-tuple conjecture [20] for 2-tuples (or prime pairs) with a strong error term. This conjecture states that for $0 < k \le N$

$$\sum_{n \le N} \Lambda(n)\Lambda(n+k) = \mathfrak{S}(k)N + O\left(N^{\frac{1}{2}+\epsilon}\right), \qquad (6.3)$$

where

$$\mathfrak{S}(k) = \begin{cases} 2C_2 \displaystyle\prod_{\substack{p|k \\ p>2}} \left(\dfrac{p-1}{p-2}\right), & \text{if } k \text{ is even, } n \ne 0; \\[2em] 0, & \text{if } k \text{ is odd;} \end{cases} \qquad (6.4)$$

and

$$C_2 = \prod_{p>2} \left(1 - \frac{1}{(p-1)^2}\right). \qquad (6.5)$$

Montgomery stated that this conjecture "would allow us to carry out our program" for $x \le T \le x^{2-\epsilon}$ and obtain

$$F(x,T) \sim \frac{T}{2\pi} \log T.$$

Further, there is no reason to expect any change in behavior for bounded $\alpha \ge 2$. On this basis Montgomery made the following conjecture.

Strong Pair Correlation Conjecture (SPC). For any fixed bounded M^4,

$$F(\alpha) = 1 + o(1), \qquad \text{for } 1 \le \alpha \le M. \qquad (6.6)$$

A question left unanswered by (6.6) is the rate at which the function $M = M(T)$ tends to infinity.

With regard to Montgomery's heuristics for making SPC, the argument that (6.3) implies SPC in the range $1 \le \alpha \le 2 - \epsilon$ was carried out by Bolanz in a 1987 Diplomarbeit (in 131 pages).[5] At the cost of slightly weaker but acceptable error terms, one can greatly simplify Bolanz's proof by smoothing (6.1) (see [16]). In section 9, we will see that one can go further by never multiplying out the integrand in (6.1).

---

[4]Editors' comment: The form factor, $F(\alpha)$, is also discussed in Section 8 of the lectures of S.M. Gonek, page 201.

[5]This thesis only proves the result in the range $x \le T \le x^{\frac{3}{2}-\epsilon}$, but Bolanz extended the result to the wider range (written communication).

With SPC and Theorem 1 we can now evaluate almost any sum over differences of zeros. In particular, Montgomery was lead to make the following now famous conjecture.[6]

Pair Correlation Conjecture (PCC). For any fixed $\beta > 0$,

$$N(T, \beta) := \left(\frac{T}{2\pi} \log T\right)^{-1} \sum_{\substack{0 < \gamma, \gamma' \leq T \\ 0 < \gamma' - \gamma \leq \frac{2\pi\beta}{\log T}}} 1 \sim \int_0^\beta 1 - \left(\frac{\sin \pi u}{\pi u}\right)^2 du. \qquad (6.7)$$

The density here for the number of pairs of zeros within β of the average spacing between zeros is where the connection with random matrix theory first occurred. [7]

One can now replace Theorems 2 and 3 with completely satisfactory results. From the PCC we immediately see that the following conjecture is true.

Small Gaps Conjecture (SGC). We have

$$\liminf_{n \to \infty} (\gamma_{n+1} - \gamma_n) \frac{\log \gamma_n}{2\pi} = 0. \qquad (6.8)$$

We also have

Simple Zeros Conjecture (SZC). We have

$$N^*(T) := \left(\frac{T}{2\pi} \log T\right)^{-1} \sum_{0 < \gamma \leq T} m_\rho \sim 1. \qquad (6.9)$$

Technically this is a conjecture on the average multiplicity which implies almost all the zeros are simple, but there is no need to make this distinction here. Another related conjecture that follows immediately from the PCC is that

$$N(T, \beta) = o(1), \qquad \text{as } \beta \to 0^+; \qquad (6.10)$$

this conjecture and SZC together are sometimes refereed to as the Essential Simplicity Conjecture (ESC). Of course, the PCC itself implies a stronger repulsion between zeros: as $\beta \to 0^+$,

$$N(T, \beta) \ll \beta^3. \qquad (6.11)$$

We now prove the following result.

Theorem 4. *Assume the Riemann Hypothesis. SPC implies PCC and SZC.*

---

[6]The SPC conjecture doesn't explicitly say anything about pair correlation, and was often not distinguished from the PCC. It is also sometimes called Montgomery's $F(\alpha)$ conjecture.

[7]Editors' comment: Note that the integrand in (6.7) is the two-point correlation function from random matrix theory as described by Y.V. Fyodorov (page 31) in equations (3.1), (4.18) and (6.4). See also the discussion of pair correlation of random matrix eigenvalues in the lectures by J.B. Conrey, page 111, Section 8.1.

First, we need a simple consequence of Theorem 1 to handle the range when $\alpha \geq M$.

Lemma 1. *Assume the Riemann Hypothesis. We have uniformly for any B, possibly depending on T,*

$$\int_B^{B+1} F(\alpha)\, d\alpha \leq 3. \tag{6.12}$$

Proof. With $B = C - \frac{1}{2}$, we have

$$\int_{C-\frac{1}{2}}^{C+\frac{1}{2}} F(\alpha)\, d\alpha \leq 2 \int_{C-1}^{C+1} \left(1 - |\alpha - C|\right) F(\alpha)\, d\alpha$$

$$= \frac{2}{\frac{T}{2\pi} \log T} \sum_{0<\gamma,\gamma'\leq T} T^{iC(\gamma-\gamma')} \left(\frac{\sin\left(\frac{1}{2}(\gamma-\gamma')\log T\right)}{\frac{1}{2}(\gamma-\gamma')\log T}\right)^2 w(\gamma-\gamma')$$

$$\leq \frac{2}{\frac{T}{2\pi} \log T} \sum_{0<\gamma,\gamma'\leq T} \left(\frac{\sin\left(\frac{1}{2}(\gamma-\gamma')\log T\right)}{\frac{1}{2}(\gamma-\gamma')\log T}\right)^2 w(\gamma-\gamma')$$

$$\leq \frac{8}{3} + \epsilon,$$

by (5.5).

Proof of Theorem 4. For SZC, we repeat the calculation in (5.5) but now assume $\lambda \geq 1$ and use SPC for that range to find

$$\sum_{0<\gamma,\gamma'\leq T} \left(\frac{\sin\left(\frac{\lambda}{2}(\gamma-\gamma')\log T\right)}{\frac{\lambda}{2}(\gamma-\gamma')\log T}\right)^2 w(\gamma-\gamma') \sim \left(1 + \frac{1}{3\lambda^2}\right) \frac{T}{2\pi}\log T. \tag{6.13}$$

The result now follows on letting $\lambda \to \infty$.

To prove the PCC, we use the Fejer kernel from (5.4) and apply (5.3) to get

$$\int_{-\infty}^{\infty} F(\alpha) \left(\frac{\sin(\pi\beta\alpha)}{\pi\beta\alpha}\right)^2 d\alpha$$

$$= \left(\frac{T}{2\pi}\log T\right)^{-1} \sum_{\substack{0<\gamma,\gamma'\leq T \\ |\gamma'-\gamma|\leq \frac{2\pi\beta}{\log T}}} \frac{1}{\beta}\left(1 - \left|\frac{(\gamma-\gamma')\log T}{2\pi\beta}\right|\right) w(\gamma-\gamma')$$

$$= \frac{1}{\beta}N^*(T) + \frac{2}{\beta^2}\int_0^{\beta} N(T,u)\, du + O\left(\frac{\beta(1+\beta)}{(\log T)^2}\right), \tag{6.14}$$

where the error term comes from removing the factor $w(\gamma-\gamma')$. By SZC, $N^*(T) \sim 1$. We now evaluate the left-hand side using Theorem 1 in the range

$|\alpha| \leq 1$, SPC in $1 < |\alpha| \leq M$, and Lemma 1 in $|\alpha| > M$. On letting $M \to \infty$, we have

$$\int_0^\beta N(T, u)\, du \sim \frac{\beta^2}{2} + \int_0^\beta (u - \beta) \left(\frac{\sin \pi u}{\pi u} \right)^2 du. \qquad (6.15)$$

Since

$$\frac{1}{h} \int_{\beta-h}^\beta N(T, u)\, du \leq N(T, \beta) \leq \frac{1}{h} \int_\beta^{\beta+h} N(T, u)\, du,$$

we obtain the PCC on differencing (6.15).

7 Gallagher and Mueller's Work on Pair Correlation

A few years after Montgomery's work, Gallagher and Mueller [12] proved a number of interesting results on pair correlation. Their starting point is the counting function $N(T, \beta)$ in (6.7), but rather than assuming it satisfies the PCC they assumed

$$N(T, \beta) \sim \int_0^\beta 1 - \mu(\alpha)\, d\alpha, \qquad (7.1)$$

uniformly for $0 \leq \beta_0 \leq \beta \leq \beta_1 < \infty$, as $T \to \infty$, where μ is a real, even, continuous, L^1 function. Thus they assumed an asymptotic density function for pair correlation, where $\mu(\alpha)$ measures the deviation from a uniform distribution corresponding to a totally random distribution of zeros. They then proved the following result.

Theorem 5. *With $N^*(T)$ given in (6.9), we have*

$$\int_{-\infty}^\infty \mu(\alpha)\, d\alpha \sim N^*(T). \qquad (7.2)$$

In particular, PCC implies SZC.

From this we see that

$$\int_{-\infty}^\infty \mu(\alpha)\, d\alpha \geq 1,$$

which shows that if the zeros of the zeta-function have an asymptotic pair correlation density, then the zeros must repulse each other somewhat. Further evidence of this was later obtained by Gallagher [11].

That the PCC implies SZC follows from

$$\int_{-\infty}^\infty \left(\frac{\sin \pi \alpha}{\pi \alpha} \right)^2 d\alpha = 1.$$

The notable feature here is that this result holds unconditionally. One can obtain this result on RH by first using Theorem 1 to prove (5.5), and then evaluating the off-diagonal terms in the sum over zeros by partial summation with $N(T,\beta)$ to determine the diagonal terms $N^*(T)$. Gallagher and Mueller replaced Theorem 1 by a result of Fujii [9] (also obtained by Selberg) on the function $S(T)$ from (2.16) . Let

$$R(T,h) = \int_0^T (S(t+h) - S(t))^2 \, dt. \tag{7.3}$$

Fujii proved that

$$R(T,h) \ll T \log(2 + h \log T) \quad \text{if } \frac{1}{\log T} \ll h \ll 1, \tag{7.4}$$

and

$$R(T,h) \sim \frac{T}{\pi^2} \log(h \log T) \quad \text{if } h \log T \to \infty, \ h \ll 1. \tag{7.5}$$

Proof of Theorem 5. We have

$$\int_0^T (N(t+h) - N(t))^2 dt = \int_0^T \left(\sum_{t < \gamma \le t+h} 1 \right)^2 dt$$

$$\sim \left(hN^*(T) + \frac{4\pi}{\log T} \int_0^{\frac{h \log T}{2\pi}} N(T,u) \, du \right) \left(\frac{T}{2\pi} \log T \right).$$

(This is (6.14) from a different perspective.) By (2.14), the left-hand side is also

$$\sim T \left(h \frac{\log T}{2\pi} \right)^2 + R(T,h).$$

On substituting $N(T,u)$ from (7.1) and letting $h \log T \to \infty$ and $h \to 0$ so that (7.5) applies, the theorem follows.

Gallagher and Mueller proved that for $h = \frac{2\pi\beta}{\log T}$

$$R(T,h) \sim T \int_{-\infty}^{\infty} \min(|\alpha|, \beta)\mu(\alpha) \, d\alpha, \tag{7.6}$$

a result essentially equivalent to PCC.

Gallagher and Mueller also studied some consequences of (7.1) for primes. In particular they proved that the error in the PNT can be improved on assuming (7.1) and RH to

$$\psi(x) = x + o\left(x^{\frac{1}{2}} (\log x)^2 \right), \tag{7.7}$$

and obtained an asymptotic formula for a weighted second moment for primes in short intervals first studied by Selberg [34]. Their proof is quite complicated, since the approach in using (7.1) requires partial summation to evaluate sums over differences of zeros, introducing many complications to handle the "edges" of the summation. An interesting consequence is a form of Theorem 1 obtained for $\mu(\alpha)$. Assuming RH and also SZC, Gallagher and Mueller proved

$$\hat{\mu}(\alpha) = 1 - |\alpha|, \qquad |\alpha| \leq 1. \tag{7.8}$$

The PCC density agrees with this, and has $\hat{\mu}(\alpha) = 0$ elsewhere.

Related to this, there is an alternative form of the PCC which has been found useful when generalizing to higher correlations. Starting from (5.3), and supposing $\hat{r}(\alpha)$ has support in $[-1, 1]$, we have by Theorem 1 on RH that

$$\int_{-\infty}^{\infty} \hat{r}(\alpha) F(\alpha)\, d\alpha = r(0) + \int_{-\infty}^{\infty} \hat{r}(\alpha)\, (F(\alpha) - 1)\, d\alpha$$

$$\sim r(0) + \hat{r}(0) - \int_{-1}^{1} (1 - |\alpha|)\hat{r}(\alpha)\, d\alpha \tag{7.9}$$

$$= r(0) + \int_{-\infty}^{\infty} r(\alpha) \left(1 - \left(\frac{\sin \pi\alpha}{\pi\alpha} \right)^2 \right) d\alpha,$$

by Plancherel's formula. The second line is still true if $F(\alpha) \sim 1$ for $|\alpha| \geq 1$ even if $\hat{r}(\alpha)$ does not have support in $[-1, 1]$. It is no accident that the PCC density occurs in the integrand. The conclusion is that assuming RH, Theorem 1 implies that

$$\sum_{0 < \gamma, \gamma' \leq T} r\left((\gamma - \gamma') \frac{\log T}{2\pi} \right) w(\gamma - \gamma') \sim r(0) + \int_{-\infty}^{\infty} r(\alpha) \left(1 - \left(\frac{\sin \pi\alpha}{\pi\alpha} \right)^2 \right) d\alpha,$$
$$\tag{7.10}$$

for all r with $\hat{r}$ having support in $[-1, 1]$. Moreover, PCC is equivalent to the conjecture that (7.10) holds for all test functions r in some dense subset of L^1. Here, the factor $w(\gamma - \gamma')$ may be removed, if desired.

8 Heath-Brown's Results on Primes

In [21] Heath-Brown proved a number of results on primes using Montgomery's $F(\alpha)$ function. By (6.2), the trivial bound for $F(\alpha)$ is

$$F(\alpha) \leq (1 + o(1)) \log T. \tag{8.1}$$

Heath-Brown showed that any improvement in the order of magnitude of this bound would have important implications for primes. First, he proved that the improvement in the error in the PNT (7.7) also holds if one assumes RH and $F(\alpha) = o(\log T)$ uniformly for $1 \leq \alpha \leq M$, for any bounded M. It should be pointed out that further improvements in the error depend not only on the size of $F(\alpha)$ but also on the growth of $M(T)$.

Heath-Brown next proved a number of results on gaps between primes, which take their strongest form if we assume

$$F(\alpha) \ll 1, \tag{8.2}$$

for various ranges of α. With regard to Cramér's bound (3.9), assuming RH and that (8.2) holds for α in any small interval around $\alpha = 2$, he proved that

$$p_{n+1} - p_n \ll \sqrt{p_n \log p_n}. \tag{8.3}$$

Assuming $F(\alpha) \sim 1$ in this range one can improve (8.3) on RH to little oh [22][8]. (This also follows from a result in the next section.) Next, assuming (8.2) for $1 \leq \alpha \leq 2 + \epsilon$ and RH,

$$\sum_{\substack{p_n \leq x \\ p_{n+1} - p_n \geq H}} (p_{n+1} - p_n) \ll \frac{x \log x}{H}. \tag{8.4}$$

This becomes non-trivial as soon as $\frac{H}{\log x} \to \infty$. On integrating with respect to H, we obtain

$$\sum_{p_n \leq x} (p_{n+1} - p_n)^2 \ll x(\log x)^2. \tag{8.5}$$

Previously Selberg [34], improving on earlier work of Cramér [7], obtained these results on RH alone with an extra $\log x$ in each bound. Finally, Heath-Brown proved, assuming RH and that $F(\alpha) \sim 1$ in any small interval around $\alpha = 1$,

$$\liminf_{n \to \infty} \left(\frac{p_{n+1} - p_n}{\log p_n} \right) = 0, \tag{8.6}$$

so that there exist primes much closer together than the average spacing between primes. This result can be made to depend on the size of the error term in the asymptotic formula for $F(\alpha)$ for α in a neighborhood of 1. If the error term is a logarithm smaller than the main term, then one actually gets that there are infinitely often primes a bounded distance apart.

In the next section, we shall prove these results by following a method that is structurally different but fundamentally the same as Heath-Brown's arguments. A very useful idea of Heath-Brown is the following bound for the sum over zeros (3.6).

Theorem 6. *For $T \geq 2$,*

$$\left| \sum_{0 \leq \gamma \leq T} x^{i\gamma} \right| \ll \sqrt{T \max_{t \leq T} F(x,t)}. \tag{8.7}$$

Note that this becomes non-trivial as soon as our bound for F is non-trivial. *Proof.* We have

$$F(x,T) = \int_{-\infty}^{\infty} \left| \sum_{0 < \gamma \leq T} x^{i\gamma} e^{i\gamma u} \right|^2 e^{-2|u|} \, du. \tag{8.8}$$

---

[8] The unusual order of the listed authors was due to a typo in the manuscript.

This is related to (4.2) by Plancherel's theorem but may be verified directly. By Gallagher's inequality [8],

$$|f(0)| \ll \int_{-1}^{1} |f(u)|\, du + \int_{-1}^{1} |f'(u)|\, du$$

for $f \in C^1$. Then with

$$f(u) = \left| \sum_{0 < \gamma \le T} x^{i\gamma} e^{i\gamma u} \right|^2,$$

we obtain

$$\left| \sum_{0 \le \gamma \le T} x^{i\gamma} \right|^2 \ll \int_{-1}^{1} \left| \sum_{0 < \gamma \le T} x^{i\gamma} e^{i\gamma u} \right|^2 du$$

$$+ \int_{-1}^{1} \left| \sum_{0 < \gamma \le T} x^{i\gamma} e^{i\gamma u} \right| \left| \frac{\partial}{\partial u} \sum_{0 < \gamma \le T} x^{i\gamma} e^{i\gamma u} \right| du.$$

In the first integral on the right we insert the weight $e^{-2|u|}$ and extend the limits of integration to $(-\infty, \infty)$ to see that, by (8.8), this is bounded by $F(x, T)$. To complete the proof, the second integral is handled similarly following an application of the Cauchy-Schwarz inequality and partial summation.

9　Equivalence between SPC and Primes

In [18] Montgomery and I proved the following equivalence between the SPC and the second moment for primes in short intervals.

Theorem 7. *Assume the Riemann Hypothesis. If* $0 < B_1 \le B_2 \le 1$, *then*

$$I(x, \delta) := \int_{1}^{X} \left(\psi\left((1+\delta)x\right) - \psi(x) - \delta x \right)^2 dx \sim \frac{1}{2} \delta X^2 \log \frac{1}{\delta} \qquad (9.1)$$

holds uniformly for $X^{-B_2} \le \delta \le X^{-B_1}$ *provided*

$$F(x, T) \sim \frac{T}{2\pi} \log T \qquad (9.2)$$

holds for
$$X^{B_1} (\log x)^{-3} \le T \le X^{B_2} (\log x)^3.$$

Conversely, if $1 \le A_1 \le A_2 < \infty$, *then* (9.2) *holds uniformly for* $T^{A_1} \le X \le T^{A_2}$ *provided that* (9.1) *holds uniformly for*

$$X^{-\frac{1}{A_1}} (\log x)^{-3} \le \delta \le X^{-\frac{1}{A_2}} (\log x)^3.$$

In particular, one can prove on RH that SPC is equivalent to

$$\int_1^X (\psi(x+h) - \psi(x) - h)^2 \, dx \sim hX \log \frac{X}{h} \tag{9.3}$$

for $1 \le h \le X^{1-\epsilon}$,[9] where an argument of Saffari and Vaughan [33] is used to move from primes in the interval $(x, x + \delta x]$ to the fixed interval $(x, x + h]$. Results (8.3) – (8.6) are consequences of Theorem 7. Further, it is a straightforward exercise to show that the twin prime conjecture in the form (6.3) implies (9.1) and (9.3) in the ranges $X^{-1} \le \delta \le X^{-\frac{1}{2}-\epsilon}$ and $1 \le h \le X^{\frac{1}{2}-\epsilon}$ respectively, and consequently we again obtain that (6.3) implies SPC in the range $1 \le \alpha \le 2 - \epsilon$. Of course, the second moment for primes in short intervals (9.1) or (9.3) is a considerably weaker hypothesis and gives the full range for SPC.

Proof of Theorem 7. We follow initially the analysis in [15]. Let us consider again Montgomery's explicit formula (3.13) but now aim towards obtaining a sum over primes in a short interval. This is usually done by differencing values of x but Montgomery showed me the following elegant approach. Let κ, δ, and T be related by

$$e^\kappa = 1 + \delta = 1 + \frac{1}{T} \tag{9.4}$$

so that $\delta = \frac{1}{T}$, and define

$$G_\kappa(t) = \left(\frac{\sin \frac{\kappa}{2} t}{\frac{\kappa}{2} t} \right) \left(\sum_{n=1}^\infty \frac{\Lambda(n) a_n(x)}{n^{it}} - \frac{2x^{1-it}}{(\frac{1}{2} + it)(\frac{3}{2} - it)} \right). \tag{9.5}$$

The Fourier transform of $G_\kappa(t)$ is

$$\hat{G}_\kappa(y) = \frac{2\pi}{\kappa} \sum_{\left| y + \frac{\log n}{2\pi} \right| < \frac{\delta}{4\pi}} \Lambda(n) a_n(x) - \frac{2\pi x}{\delta} \int_{xe^{-\kappa/2}}^{xe^{\kappa/2}} a_v \left(e^{-2\pi y} \right) \frac{dv}{v}$$

which has the desired (but weighted) sum over primes in a short interval. By Parseval's identity, we have

$$\int_{-\infty}^\infty |G_\kappa(t)|^2 \, dt = \int_{-\infty}^\infty |\hat{G}_\kappa(y)|^2 \, dy. \tag{9.6}$$

Using (3.13) to express $G_\kappa(t)$ in terms of a sum over zeros with the remaining terms estimated as error terms and simplifying we find, assuming RH,

$$\int_0^\infty \left(\sum_{y < n \le y + \frac{y}{T}} \Lambda(n) a_n(x) - x \int_{xe^{-\kappa}}^x a_v(y) \frac{dv}{v} \right)^2 \frac{dy}{y}$$

$$= \frac{4x\kappa^2}{\pi} \int_0^\infty \left(\frac{\sin \frac{\kappa}{2} t}{\frac{\kappa}{2} t} \right)^2 \left| \sum_\gamma \frac{x^{i\gamma}}{1 + (t - \gamma)^2} \right|^2 dt + O\left(\frac{(\log T)^2}{T} \right). \tag{9.7}$$

---

[9] If (9.3) holds for this range of h it implies RH.

We abbreviate this equation as

$$L(x,T) = R(x,T). \tag{9.8}$$

To prove Heath-Brown's results (8.3) and (8.4) from the last section, it is easy to see that, taking $T = \frac{3x}{H}$,

$$L(x,T) \gg \frac{x}{T^2} \sum_{\substack{\frac{x}{2} < p_n \le x \\ p_{n+1}-p_n \ge H}} (p_{n+1} - p_n)$$

and, assuming (8.2),

$$R(x,T) \ll \frac{x}{T} \log T.$$

Equation (8.4) follows from this, and (8.3) follows by taking only the last term in (8.4).

For the proof of Theorem 7 we would like to remove the weight $a_n(x)$ in $L(x,T)$ and thus obtain an expression involving $I(X,\delta)$. In view of (4.4), since $\kappa \sim \frac{1}{T}$, $R(x,T)$ can be related to $F(x,T)$ through Abelian and Tauberian theorems. If one assumes an asymptotic formula for $I(X,\delta)$ one then obtains an asymptotic formula for $L(x,\delta)$ which gives an asymptotic formula for $R(x,\delta)$ and then a Tauberian theorem gives an asymptotic formula for $F(x,T)$. The converse direction works similarly using an Abelian theorem. All the details may be found in [18] except how $L(x,T)$ is related to $I(X,\delta)$, since the proof there proceeds from (2.14) rather than Proposition 1. It took me a long time to figure out how to remove the weight $a_n(y)$ even though it is actually obvious. If $\frac{y}{T}$ is small, then for $y < n \le y + \frac{y}{T}$ it is reasonable to replace n by y and thus replace $a_n(x)$ with $a_y(x)$ in (9.7) with a small error. Thus the weight is removed, and one finds that

$$L(x,T) = 4x^3 \int_x^\infty I(y,\delta)\frac{dy}{y^5} + O\left(\frac{x^2(\log T)^2}{T^3}\right). \tag{9.9}$$

Since the integrand is non-negative, if we have an asymptotic formula for $L(x,T)$ then a simple differencing argument will give an asymptotic formula for $I(x,\delta)$. The converse is immediate. Here the error term is smaller than the main term when $T \le x \le T^{2-\epsilon}$. To obtain the full range, rather than replacing $a_n(x)$ by $a_y(x)$, we use Stieltjes integration and the PNT with the RH error (3.4) to evaluate the sum over primes, and together with the Cauchy Schwarz inequality we find that the error term in (9.9) can be replaced by

$$O\left(\frac{x(\log x)^4}{T^{\frac{3}{2}}}\right).$$

This suffices for the full range.

10 Selberg's theory of $S(T)$

For more than 50 years, Selberg has been working on the distribution of values of $\log \zeta(s)$ and related functions. In the early 1940's and he made major contributions on $S(T)$ [35, 36]. Further results for Dirichlet L-functions were obtained in [37]. Selberg has continued to work on these problems, and while he has lectured on his results, his next published paper on this subject [38] only appeared in 1992. In this already famous paper Selberg introduced the properties of a general class of Dirichlet series, now referred to as the "Selberg class". Selberg showed that his theory, originally devised for the Riemann zeta-function, carries over to the Selberg class with remarkably few changes. To learn more about this subject, I recommend first reading Selberg's 1992 paper. Second, Kai-man Tsang (Selberg's only Ph.D. student) wrote a thesis [40] in 1984 which contains full details of the proofs for some of Selberg's more recent work on $\log \zeta(s)$. Also, the two papers of A. Ghosh [13, 14] refine some of Selberg's work from the 1940's.

As examples, we state two of Selberg's results proved in Tsang's thesis. Selberg has developed methods for evaluating

$$\int_0^T F(\log \zeta(\sigma + it)) \, dt,$$

for functions $F(z)$ such as $\operatorname{sgn}(\operatorname{Re}(z))$, $\operatorname{sgn}(\operatorname{Im}(z))$, $|\operatorname{Re}(z)|$, and $|\operatorname{Im}(z)|$. Let $\chi_{\alpha,\beta}(u)$ be 1 if $\alpha \leq u \leq \beta$ and zero otherwise. Then for $\alpha < \beta$

$$\int_0^T \chi_{\alpha,\beta} \left(\sqrt{\frac{\pi}{\log \log T}} S(t) \right) dt = T \int_\alpha^\beta e^{-\pi u^2} \, du + O\left(\frac{T \log \log \log T}{\sqrt{\log \log T}} \right). \tag{10.1}$$

We have similar results for the real and imaginary parts of $\log \zeta(\sigma + it)$.

For the second result, let $Z(T)$ denote the number of sign changes of $S(t)$ in $[0, T]$. Selberg proved $Z(T) \gg T(\log T)^{\frac{1}{3} - \epsilon}$ on RH in [35], and unconditionally (and with an improvement on the ϵ) in [36]. Ghosh [13] improved this to $Z(T) \gg T(\log T)^{1-\epsilon}$.[10] Tsang's thesis contains the following remarkable improvements on these results. For some $c > 0$,

$$Z(T) \gg T \log T e^{-c(\log \log \log T)^2} \tag{10.2}$$

and

$$Z(T) \ll T \log T \frac{\log \log \log T}{\sqrt{\log \log T}}. \tag{10.3}$$

If the analysis of an error term could be improved then one would obtain

$$Z(T) \sim T \frac{\log T}{\sqrt{\pi \log \log T}}. \tag{10.4}$$

---

[10] Also obtained earlier but unpublished by Selberg

I will now describe some key ideas that went into Selberg's work on $S(t)$. The very remarkable result that Selberg proved in 1946 is that all the even moments of $S(t)$ can be computed unconditionally [36]. He proved this on intervals $(T, T + H]$, where $T^a \le H \le T$ and $a > \frac{1}{2}$, but for simplicity we will consider the interval $[0, T]$.

Theorem 8. *For $k \ge 1$, we have*

$$\int_0^T \left| S(t) + \frac{1}{\pi} \sum_{p \le T^{\frac{1}{k}}} \frac{\sin(t \log p)}{\sqrt{p}} \right|^{2k} dt \ll_k T \qquad (10.5)$$

and

$$\int_0^T |S(t)|^{2k} \, dt = \frac{(2k)!}{k!(2\pi)^{2k}} T (\log \log T)^k + O_k \left(T (\log \log T)^{k - \frac{1}{2}} \right). \qquad (10.6)$$

This last relation is the $2k$th moment of a Gaussian. Earlier Selberg [35] proved (10.5) assuming the RH, and also (10.6) on RH but with an error term $O_k(T(\log \log T)^{k-1})$. These results were a great advance over previous work, which had failed to even obtain an asymptotic formula for the second moment. From (10.5) we see that $S(t)$ can be approximated well in L^{2k} norm by the imaginary part of a short Dirichlet series. This series is short enough so that its L^{2k} norm is determined by diagonal terms, and has the Gaussian property in (10.6). Thus $S(t)$ has this property too.

The proof of (10.5) and (10.6) is based on an approximate formula for $S(t)$, which has its origin in Selberg's earlier paper [34] on primes in short intervals. There, he proved on RH that for $\sigma = \frac{1}{2} + \frac{\beta}{\log T}$ and $\beta \gg 1$,

$$\int_0^T \left| \frac{\zeta'}{\zeta}(\sigma + it) \right|^2 \, dt \ll_\beta T(\log T)^2. \qquad (10.7)$$

Selberg's work was ahead of its time, since we now know that replacing the bound in (10.7) by an asymptotic formula is equivalent to the PCC [17].

Selberg first found an approximate formula for $\frac{\zeta'}{\zeta}(s)$. This is not straightforward. For $\sigma > 1$, we have the Dirichlet series representation (2.2) for $\frac{\zeta'}{\zeta}(s)$. As we bring s into the critical strip the Dirichlet series fails to converge. It is a familiar fact that an appropriate partial sum of a Dirichlet series will still provide a good approximation for the analytic continuation of the series. However, on or near the critical line we expect the poles from the zeros $\zeta(s)$ to dominate, as reflected in the partial-fraction formula, for $s \ne \rho$, $t \ge 2$,

$$\frac{\zeta'}{\zeta}(s) = \sum_\rho \left(\frac{1}{s - \rho} + \frac{1}{\rho} \right) + O(\log t). \qquad (10.8)$$

Since

$$S(t) = \frac{1}{\pi} \arg \zeta \left(\frac{1}{2} + it \right) = -\frac{1}{\pi} \int_{\frac{1}{2}}^\infty \mathrm{Im} \left(\frac{\zeta'}{\zeta}(\sigma + it) \right) dt, \qquad (10.9)$$

it is (maybe) plausible that the Dirichlet series part of $S(t)$ will usually dominate. A candidate for an approximate formula is (3.15) which we can rewrite as, for $x > 1$, $s \neq 1$, $s \neq \rho$, $s \neq -2k$,

$$
\frac{\zeta'}{\zeta}(s) = -\sum_{n \leq x} \frac{\Lambda(n)}{n^s} + \frac{x^{1-s}}{1-s} - \sum_{\rho} \frac{x^{\rho-s}}{\rho-s} + \sum_{n=1}^{\infty} \frac{x^{-2n-s}}{2n+s}. \tag{10.10}
$$

In hindsight (10.10) looks even better, because

$$
-\frac{1}{\pi} \int_{\frac{1}{2}}^{\infty} \operatorname{Im} \left(\sum_{n \leq x} \frac{\Lambda(n)}{n^{\sigma+it}} \right) d\sigma = -\frac{1}{\pi} \sum_{n \leq x} \frac{\Lambda(n) \sin(t \log n)}{\sqrt{n} \log n}
$$

which gives exactly the approximation in (10.5) from the terms where n is prime. (The prime powers will contribute an error term.) The problem here is that the sum over zeros does not converge absolutely, and consequently (10.10) has never been used successfully for this problem. Earlier work had smoothed this formula (or rather over-smoothed it), so that the correct approximation was lost. Selberg had the innovative idea that one only needs to smooth slightly in order to obtain absolute convergence in the sum over zeros.

Let

$$
\Lambda_x(n) = \begin{cases} \Lambda(n), & \text{for } 1 \leq n \leq x, \\ \Lambda(n)\frac{\log \frac{x^2}{n}}{\log n}, & \text{for } x \leq n \leq x^2. \end{cases} \tag{10.11}
$$

Then, for $x > 1$, $s \neq 1$, $s \neq \rho$, $s \neq -2k$,

$$
\frac{\zeta'}{\zeta}(s) = -\sum_{n \leq x^2} \frac{\Lambda_x(n)}{n^s} + \frac{x^{2(1-s)} - x^{1-s}}{(1-s)^2 \log x} + \frac{1}{\log x} \sum_{\rho} \frac{x^{\rho-s} - x^{2(\rho-s)}}{(\rho-s)^2}
$$

$$
+ \frac{1}{\log x} \sum_{n=1}^{\infty} \frac{x^{-2n-s} - x^{-2(2n+s)}}{(2n+s)^2}. \tag{10.12}
$$

This formula is much easier to prove than (10.10). Selberg next argues as follows. Assume RH, and suppose $4 \leq x \leq t^2$. Let

$$
\sigma_1 = \frac{1}{2} + \frac{1}{\log x}, \tag{10.13}
$$

which is at the transition from the region where the Dirichlet series dominates to the region where the zeros dominate. From (10.12), we see that for $\sigma \geq \sigma_1$ and some complex number ω with $|\omega| \leq 1$

$$
\frac{\zeta'}{\zeta}(\sigma + it) = -\sum_{n \leq x^2} \frac{\Lambda_x(n)}{n^{\sigma+it}} + O\left(x^{\frac{1}{2}-\sigma}\right) \tag{10.14}
$$

$$
+ 2\omega x^{\frac{1}{2}-\sigma} \sum_{\gamma} \frac{\sigma_1 - \frac{1}{2}}{\left(\sigma_1 - \frac{1}{2}\right)^2 + (t-\gamma)^2}.
$$

We also have, on taking the real part of (10.8),

$$\text{Re}\frac{\zeta'}{\zeta}(\sigma+it) = \sum_{\gamma} \frac{\sigma-\frac{1}{2}}{\left(\sigma-\frac{1}{2}\right)^2 + (t-\gamma)^2} + O(\log t),$$

Thus, taking the real part of (10.14) with $\sigma = \sigma_1$ gives for some $-1 \leq \omega' \leq 1$,

$$\sum_{\gamma} \frac{\sigma_1-\frac{1}{2}}{\left(\sigma_1-\frac{1}{2}\right)^2 + (t-\gamma)^2} + O(\log t) = -\text{Re}\sum_{n\leq x^2} \frac{\Lambda_x(n)}{n^{\sigma_1+it}}$$

$$+ O(1) + \frac{2\omega'}{e}\sum_{\gamma} \frac{\sigma_1-\frac{1}{2}}{\left(\sigma_1-\frac{1}{2}\right)^2 + (t-\gamma)^2}.$$

Since $1 - \frac{2\omega'}{e} > 1 - \frac{2}{e} > \frac{1}{4}$, we conclude that

$$\sum_{\gamma} \frac{\sigma_1-\frac{1}{2}}{\left(\sigma_1-\frac{1}{2}\right)^2 + (t-\gamma)^2} = O\left(\left|\sum_{n\leq x^2} \frac{\Lambda_x(n)}{n^{\sigma_1+it}}\right|\right) + O(\log t).$$

Substituting this back into (10.14), we obtain

$$\frac{\zeta'}{\zeta}(\sigma+it) = -\sum_{n\leq x^2} \frac{\Lambda_x(n)}{n^{\sigma+it}} + O\left(x^{\frac{1}{2}-\sigma}\left|\sum_{n\leq x^2} \frac{\Lambda_x(n)}{n^{\sigma_1+it}}\right|\right) + O\left(x^{\frac{1}{2}-\sigma}\log t\right). \quad (10.15)$$

Selberg next substitutes (10.15) into (10.9) for the integration range $\sigma_1 \leq \sigma < \infty$. For $\frac{1}{2} < \sigma \leq \sigma_1$, he uses (10.8) and (10.15) to show this range only contributes to the error terms. The conclusion is the following theorem, which is the primary tool for obtaining Theorem 8 assuming the RH.

Theorem 9. *Assume the Riemann Hypothesis. For $t \geq 2$, $4 \leq x \leq t^2$, and σ_1 given in (10.13), we have*

$$S(t) = -\sum_{n<x^2} \frac{\Lambda_x(n)}{n^{\sigma_1}} \frac{\sin(t\log n)}{\log n} + O\left(\frac{1}{\log x}\left|\sum_{n\leq x^2} \frac{\Lambda_x(n)}{n^{\sigma_1+it}}\right|\right) + O\left(\frac{\log t}{\log x}\right).$$
$$(10.16)$$

How do you remove the RH from the above analysis? I think it takes great insight to even suspect that this can be done. Selberg makes a much more subtle choice for σ_1. He defines

$$\sigma_{x,t} = \frac{1}{2} + 2\max_{\rho\in\mathcal{A}}\left(\beta - \frac{1}{2}, \frac{2}{\log x}\right), \quad (10.17)$$

where

$$\mathcal{A} = \left\{\rho : |t-\gamma| \leq \frac{x^{3|\beta-\frac{1}{2}|}}{\log x}\right\}. \quad (10.18)$$

Thus, we move towards or away from the critical line depending on how far off the line nearby zeros lie. There is also an issue of convergence, and the explicit formula (10.12) needs to be replaced by a similar formula where the sum over zeros has a factor of $(s - \rho)^3$ in the denominator. Ultimately the contribution from zeros off the $\frac{1}{2}$-line is bounded by a density estimate proved in [36].

References

[1] E. B. Bogomolny and J. P. Keating, *Random matrix theory and the Riemann zeros I: three- and four-point correlations*, Nonlinearity **8** (1995), 1115–1131.

[2] E. B. Bogomolny and J. P. Keating, *Random matrix theory and the Riemann zeros II: n-point correlations*, Nonlinearity **9** (1996), 911–935.

[3] Joachim Bolanz, *Uber Die Montgomery'she Paarvermutung*, Diplomarbeit 1987, 131 pages.

[4] J. B. Conrey, D. W. Farmer, J. P. Keating, M. O. Rubinstein, N. C. Snaith, *Integral moments of L-functions*, Proc. Lond. Math. Soc. (1) **91** (2005), 33–104.

[5] J. B. Conrey, A. Ghosh, and S. M. Gonek, *A note on gaps between zeros of the zeta-function*, Bull. London Math. Soc. **16** (1984), 421–424.

[6] J. B. Conrey, A. Ghosh, and S. M. Gonek, *Simple zeros of the Riemann zeta-function*, Proc. London Math. Soc. (3) **76** (1998), 497–522.

[7] H. Cramér, *On the order of magnitude of the difference between consecutive prime numbers*, Acta Arithmetica, **2** (1936), 23–46.

[8] Harold Davenport, *Multiplicative number theory. Revised and with a preface by Hugh L. Montgomery. 3rd ed.* Graduate Texts in Mathematics, **74** New York, NY: Springer, 177 pp.

[9] A. Fujii, *On the zeros of Dirichlet L-functions, I*, Trans. A. M. S. , **196** (1974), 225–235.

[10] Akio Fujii, *On a theorem of Landau*, Proc. Japan Acad., Ser. A 65, No.2 (1989), 51–54.

[11] P. X. Gallagher, *Pair correlation of zeros of the zeta function*, J. Reine Angew. Math. **362** (1985), 72–86.

[12] P. X. Gallagher and J. Mueller, *Primes and zeros in short intervals*, J. Reine Angew. Math. **303/304** (1978), 205–220.

[13] A. Ghosh, *On Riemann's zeta function—sign changes of S(T)*, Recent progress in analytic number theory, Vol. 1 (Durham, 1979), Academic Press, London-New York, 1981, 25–46.

[14] A. Ghosh, *On the Riemann zeta function—mean value theorems and the distribution of |S(T)|*, J. Number Theory **17** (1983), no. 1, 93–102.

[15] D. A. Goldston, *Prime numbers and the pair correlation of zeros of the zeta-function*, Proc. of the Texas Conference on Number Theory 1982, Univ. of Texas Press.

[16] D. A. Goldston and S. M. Gonek, *Mean value theorems for long Dirichlet polynomials and tails of Dirichlet series*, Acta Arith. **84** (1998), 155–192.

[17] D. A. Goldston, S. M. Gonek, and H. L. Montgomery, *Mean values of the logarithmic derivative of the Riemann zeta-function with applications to primes in short intervals*, J. Reine Angew. Math. **537** (2001), 105-126.

[18] D. A. Goldston and H. L. Montgomery *Pair correlation of zeros and primes in short intervals*, Analytic Number Theory and Diophantine Problems, Birkhaüser, Boston, Mass. 1987, 183–203.

[19] S. M. Gonek, *An explicit formula of Landau and its applications to the theory of the zeta-function*, Knopp, Marvin (ed.) et al., A tribute to Emil Grosswald: number theory and related analysis. Providence, RI: American Mathematical Society. Contemp. Math. 143, 395-413 (1993).

[20] G. H. Hardy and J. E. Littlewood, *Some problems of 'Partitio Numerorum': III On the expression of a number as a sum of primes*, Acta Math. **44** (1923), 1–70.

[21] D. R. Heath-Brown *Gaps between primes, and the pair correlation of zeros of the zeta-function* Acta Arith. **41** (1982), 85–99.

[22] D. R. Heath-Brown and D. A. Goldston, *A note on the difference between consecutive primes*, Math. Ann. **266** (1984), 317–320.

[23] Dennis A. Hejhal, *On the triple correlation of zeros of the zeta function*, Internat. Math. Res. Notices **293ff** Issue 7, (1994), 10pp (electronic).

[24] Georg Illies, *Cramér functions and Guinand equations*, Acta Arith. **105** no. 2, (2002), 103–118.

[25] A. E. Ingham, *The distribution of prime numbers*, Cambridge Tracts in Mathematics and Mathematical Physics, 30; Cambridge Mathematical Library, Cambridge University Press (1990), 114 pp .

[26] N. Katz and P. Sarnak, *Random matrices, Frobenius eigenvalues, and monodromy*, Colloquium Publications. American Mathematical Society (AMS). **45**Providence, RI, (1999) 419 pp .

[27] E. Landau, *Handbuch der Lehre von der Verteilung der Primzchlen,* Teubner, Leipzig (1909). Reprinted by Chelsea Publishing Co., New York, (1953).

[28] H. L. Montgomery *The pair correlation of zeros of the zeta function* Proc. Sympos. Pure Math. **24** AMS, Providence, R. I., 1973, 181–193.

[29] Hugh L. Montgomery, *The analytic principle of the large sieve,* Bull. Am. Math. Soc. **84** (1978), 547-567.

[30] H. L. Montgomery and A. Odlyzko, *Gaps between zeros of the zeta-function,* Topics in Classical Number Theory, Vol I, II, Budapest, 1981, Colloquia Math. Soc. János Bolyai, **34**, North-Holland, Amsterdam-New York, 1984, 1079–1106.

[31] A. M. Odlyzko, *On the distribution of spacings between zeros of the zeta function,* Math. Comp. **48** (1987), 273–308.

[32] Zeév Rudnick and Peter Sarnak, *Zeros of principal L-functions and random matrix theory.* A celebration of John F. Nash, Jr , Duke Math. J. **81** Issue 2, (1996), 269–322.

[33] Saffari, B. and Vaughan, R. C., *On the fractional parts of x/n and related sequences, II.* Ann. Inst. Fourier (Grenoble) **27** (1977), no. 2, 1–30.

[34] A. Selberg, *On the normal density of primes in small intervals,and the difference between consecutive primes,* Arch. Math. Naturvid. **47** No. 6, (1943), 87-105.

[35] A. Selberg, *On the remainder in the formula for $N(T)$, the number of zeros of $\zeta(s)$ in the strip $0 < t < T$,* Avh. Norske Vid. Akad. Oslo. I. No.1, (1944) 27 pp.

[36] A. Selberg, *Contributions to the theory of the Riemann zeta-function,* Archiv for Mathematik og Naturvidenskab B. **48** (1946), No. 5, 89–155.

[37] A. Selberg, *Contributions to the theory of Dirichlet's L-functions,* Skrifter utgitt av Det Norske Videnskaps-Akademi i Oslo. I. Math.-Naturv. Klasse (1946), No. 3, 1–62.

[38] A. Selberg, *Old and new conjectures and results about a class of Dirichlet series,* Bombieri, E. (ed.) et al., Proceedings of the Amalfi conference on analytic number theory, held at Maiori, Amalfi, Italy, Sept. 25–29, 1989. Salerno: Universitá di Salerno, (1992) 367-385 .

[39] E. C. Titchmarsh, *The theory of the Riemann zeta-function,* Second edition. Edited and with a preface by D. R. Heath-Brown. The Clarendon Press, Oxford University Press, New York, 1986.

[40] Kai-man Tsang, *The distribution of the values of the Riemann zeta-function,* Thesis, Princeton University, October 1984, 179pp.

Department of Mathematics
San Jose State University
San Jose, CA 95192, USA

e-mail: goldston@math.sjsu.edu

Notes on eigenvalue distributions for the classical compact groups

Brian Conrey

Contents

1 Some notation   112

2 Introduction   114

3 Definitions and Haar measures   115

   3.1 Unitary 115

   3.2 Orthogonal and Symplectic 116

4 Vandermonde determinants and orthogonal polynomials   117

   4.1 An alternate formula for the Haar measure on $U(N)$ 118

   4.2 Alternate formulas for orthogonal and symplectic Haar measures 119

5 Andréief's identity   **123**

   5.1 Proof of Andréief's identity   123

   5.2 Verification that the Haar measures have total mass 1   124

6 Gaudin's Lemma   **125**

   6.1 Calculation for orthogonal and symplectic cases   128

7 n-level density   **129**

   7.1 Unitary   129

   7.2 Normalized eigenangles and large N limits   130

   7.3 Orthogonal and symplectic   131

8 Correlations   **132**

   8.1 Pair correlation for $U(N)$ 132

   8.2 n-correlation for $U(N)$ 133

   8.3 n-correlation for orthogonal and symplectic   134

9 Neighbor spacings   **135**

   9.1 Nearest neighbor for $U(N)$ 135

   9.2 Gram's identity   137

   9.3 Intervals with precisely n eigenvalues   140

   9.4 jth lowest eigenvalue   142

   9.5 Relations between the eigenvalues   143

10 Conclusion   **144**

1   Some notation

Before we get started we record some notation that will be used in these notes. This section is merely to serve as a convenient reference; the notions are defined at the appropriate places in the notes.

- The sine ratios are:

$$S(x) = \frac{\sin \pi x}{\pi x}$$

$$S_N(x) = \frac{\sin Nx/2}{\sin x/2}$$

$$S_N^*(x) = \frac{1}{2\pi} S_N(x)$$

Note: We are using the Katz and Sarnak [KaSa] definition (5.4.2) of S_N; Mehta's S_N (see 11.1.6 of [Meh]) is our S_N^*. It is useful to have both functions available.

- $G(N)$ stands for one of the groups $U(N)$, $USp(2N)$, $SO(2N)$, $SO(2N+1)$ and G by itself stands for one of the symmetry types U (unitary), Sp (symplectic), O (orthogonal) even, O (orthogonal) odd

- The kernel functions are

$$K_{U(N)}(x, y) = S_N^*(y - x)$$

$$K_{SO(2N)}(x, y) = S_{2N-1}^*(y - x) + S_{2N-1}^*(y + x)$$

$$K_{USp(2N)}(x, y) = S_{2N+1}^*(y - x) - S_{2N+1}^*(y + x)$$

$$K_{SO(2N+1)}(x, y) = S_{2N}^*(y - x) - S_{2N}^*(y + x).$$

- The scaled limit of these kernel functions are

$$K_U(x, y) = S(y - x)$$

$$K_{Sp}(x, y) = S(y - x) - S(y + x)$$

$$K_{O,even}(x, y) = S(y - x) + S(y + x)$$

$$K_{O,odd}(x, y) = S(y - x) - S(y + x).$$

This means that $\lim_{N \to \infty} \frac{2\pi}{N} K_{U(N)} \left(\frac{2\pi}{N} x, \frac{2\pi}{N} y \right) = K_U(x, y)$ and also that $\lim_{N \to \infty} \frac{\pi}{N} K_{G(N)} \left(\frac{\pi}{N} x, \frac{\pi}{N} y \right) = K_G(x, y)$ when G is one of the other groups.

- For an interval J, the integral operator $K_{J,G(N)}$ is defined by

$$(K_{J,G(N)}f)(x) = \int_J K_{G(N)}(x, y) f(y) \, dy$$

for functions f integrable on J, and similarly the operator $K_{J,G}$ is defined by

$$(K_{J,G}f)(x) = \int_J K_G(x, y) f(y) \, dy$$

These operators have eigenvalues denoted by $\lambda_{j,G(N)}(J)$ $(j = 1, 2, \ldots, N)$ and $\lambda_{j,G}(J)$, $(j = 1, 2, 3, \ldots)$ respectively.

- The Chebyshev polynomials are $T_n(x), U_n(x)$, and $V_n(x)$ where

$$T_n(\cos\theta) = \cos n\theta$$

$$U_n(\cos\theta) = \frac{\sin(n+1)\theta}{\sin\theta}.$$

$$V_n(\cos\theta) = U_{2n}\left(\cos\frac{\theta}{2}\right) = \frac{\sin(n+\frac{1}{2})\theta}{\sin\frac{1}{2}\theta}.$$

- We let $\mu_{G(N),j}(s)$ be the density function for the jth nearest neighbor spacing for eigenangles of $G(N)$ and $\mu_{G,j}(s)$ is the large–N–scaled–limit of this density function. Similarly, $\nu_{G(N),j}(s)$ is the density of the jth lowest eigenangle for $G(N)$ and $\nu_{G,j}(s)$ is its scaled limit.

- We let $E_{G(N)}(J,n)$ be the measure of the set of matrices $X \in G(N)$ which have precisely n eigenangles in the set J.

2   Introduction

In 1972 the fortuitous introduction of Montgomery and Dyson served also as an introduction of the worlds of analytic number theory and random matrix theory. The symbiosis between these two subjects developed slowly for the next 25 years with the principal developments being the numerical work of Odlyzko (see [O1] and [O2]) and the calculations of the third and higher correlations of the Riemann zeta-function (and other L-functions) by Hejhal [H], and Rudnick-Sarnak [RS].

Around 1998, there were two very important developments that have stimulated a great deal of subsequent work. One was the theory of symmetry types associated to families of L-functions by Katz and Sarnak [KaSa]. The other was the relationship between moments of characteristic polynomials and moments of the Riemann zeta-function and of families of L-functions found by Keating and Snaith [KS].

While we still do not understand why there is such a strong connnection between random matrix theory and families of L-functions, we do realize that random matrix theory provides models for a wide range of statistical behavior of these families. Consequently, we can now confidently predict the answer to any number of difficult questions about L-functions which 10 years ago seemed hopelessly impossible.

The purpose of these notes is to provide an introduction to the random matrix aspects of the book [KaSa] by Katz and Sarnak on symmetry types associated with families of L-functions. In particular, we will develop here some of the basic tools needed to understand the beginnings of computing statistics of eigenvalues of unitary, orthogonal, and symplectic groups of matrices. The four statistics we are interested in computing are n-correlation, n–level density, jth nearest neighbor, and jth lowest eigenvalue.

The main goals of these notes are (a) to show how to rewrite the basic Weyl integration formula for each of our groups $G(N)$ as a determinant of a "kernel" function $K_{G(N)}$ (derived in sections 2 – 5, equations (4.8), (4.16), (4.17), and (4.18)); (b) to use Gaudin's lemma to compute level densities and correlations (derived in sections 6 – 8, equations (7.3), (7.4), and (7.5)); (c) to use the combinatorial identity (9.3) to deduce the mth nearest neighbor statistic from the correlations (derived in section 9.1, equation (9.4)); and (d) to use Gram's identity to write the neighbor and lowest eigenvalue statistics in terms of derivatives of infinite products of eigenvalues of simple operators (derived in sections 9.2 – 9.5, equations (9.16) and (9.19)).

3 Definitions and Haar measures

3.1 Unitary

If X is an $N \times N$ matrix with complex entries $X = (x_{jk})$, we let X^* be its conjugate transpose, i.e. $X^* = (x_{jk}^*)$ where $x_{jk}^* = \overline{x_{kj}}$. X is said to be unitary if $XX^* = I$. We let $U(N)$ denote the group of all $N \times N$ unitary matrices. This is a compact Lie group and has a Haar measure which allows us to do analysis.

All of the eigenvalues of $X \in U(N)$ have absolute value 1; we write them as
$$e^{i\theta_1}, e^{i\theta_2}, \ldots, e^{i\theta_N}$$
with
$$0 \le \theta_j < 2\pi. \tag{3.1}$$
The eigenvalues of X^* are $e^{-i\theta_1}, \ldots, e^{-i\theta_N}$. Clearly, the determinant, $\det X = \prod_{n=1}^{N} e^{i\theta_n}$ of a unitary matrix is a complex number with absolute value equal to 1.

For any sequence of N points on the unit circle there are matrices in $U(N)$ with these points as eigenvalues. The collection of all matrices with the same set of eigenvalues constitutes a conjugacy class in $U(N)$. Thus, the set of conjugacy classes can be identified with the collection of sets of N points on the unit circle.

We are interested in computing various statistics about these eigenvalues. Consequently, we identify all matrices in $U(N)$ that have the same set of eigenvalues. Weyl's integration formula gives a simple way to perform averages over $U(N)$ for functions f that are constant on conjugacy classes. Such functions are called 'class functions'. Note that f being constant on conjugacy classes entails that $f(\theta_1,\ldots,\theta_N)$ is necessarily symmetric in its N variables. Weyl's formula asserts that for such an f,

$$\int_{U(N)} f(X)\, dX = \int_{[0,2\pi]^N} f(\theta_1,\ldots,\theta_N) \prod_{1\le j<k\le N} \left|e^{i\theta_k} - e^{i\theta_j}\right|^2 \frac{d\theta_1\ldots d\theta_N}{N!(2\pi)^N}.$$

Notice that we have used X to represent a variable element of $U(N)$ and dX to denote the Haar measure. If we want to emphasize the group $U(N)$ we will designate the Haar measure by $dX_{U(N)}$.

3.2   Orthogonal and Symplectic

A unitary matrix X is said to be *orthogonal* if $XX^t = I$, where X^t denotes the transpose of X. Orthogonality for a unitary matrix implies that $X^t = X^*$ or $\overline{X} = X$. In other words any real unitary matrix is orthogonal. We let $SO(N)$ denote the subgroup of $U(N)$ consisting of $N \times N$ orthogonal matrices with determinant 1.

We want to distinguish these two cases. Thus, we consider $SO(2N)$ (even orthogonal) and $SO(2N+1)$ (odd orthogonal).

For any complex eigenvalue of an orthogonal matrix, its complex conjugate is also an eigenvalue. The eigenvalues of $X \in SO(2N)$ can be written as

$$e^{\pm i\theta_1},\ldots,e^{\pm i\theta_N}$$

with

$$0 \le \theta_j \le \pi.$$

The Weyl integration formula for integrating a symmetric function $f(X) = f(\theta_1,\ldots,\theta_N)$ over $SO(2N)$ is

$$\int_{SO(2N)} f(X)\, dX = \frac{2^{(N-1)^2}}{\pi^N N!} \int_{[0,\pi]^N} f(\theta_1,\ldots,\theta_N)$$

$$\times \prod_{1\le j<k\le N} (\cos\theta_k - \cos\theta_j)^2 d\theta_1 \ldots d\theta_N.$$

The eigenvalues of $X \in SO(2N+1)$ can be written as

$$1, e^{\pm i\theta_1},\ldots,e^{\pm i\theta_N}$$

with

$$0 \le \theta_j \le \pi.$$

The Weyl integration formula for integrating a symmetric function $f(X) = f(\theta_1, \ldots, \theta_N)$ over the space $SO(2N+1)$ is

$$\int_{SO(2N+1)} f(X)\, dX = \frac{2^{N^2}}{\pi^N N!} \int_{[0,\pi]^N} f(\theta_1, \ldots, \theta_N) \prod_{1 \le j < k \le N} (\cos\theta_k - \cos\theta_j)^2$$

$$\times \prod_{h=1}^{N} \sin^2 \frac{\theta_h}{2}\, d\theta_1 \ldots d\theta_N.$$

A unitary matrix X is said to be *symplectic* if $XZX^t = Z$ where

$$Z = \begin{pmatrix} 0 & I_N \\ -I_N & 0 \end{pmatrix}.$$

A symplectic matrix necessarily has determinant equal to 1. The *symplectic group* $\mathrm{USp}(2N)$ is the subgroup of $2N \times 2N$ symplectic matrices. The eigenvalues of a symplectic matrix are

$$e^{\pm i\theta_1}, \ldots, e^{\pm i\theta_N}$$

with

$$0 \le \theta_j \le \pi.$$

The Weyl integration formula for integrating a symmetric function $f(X) = f(\theta_1, \ldots, \theta_N)$ over $USp(2N)$ is

$$\int_{USp(2N)} f(X)\, dX = \frac{2^{N^2}}{\pi^N N!} \int_{[0,\pi]^N} f(\theta_1, \ldots, \theta_N) \prod_{1 \le j < k \le N} (\cos\theta_k - \cos\theta_j)^2$$

$$\prod_{h=1}^{N} \sin^2 \theta_h\, d\theta_1 \ldots d\theta_N.$$

4 Vandermonde determinants and orthogonal polynomials

We use the notation $(f(j,k))_{1 \le j,k \le N}$ to denote the $N \times N$ matrix whose j, k entry is $f(j,k)$. If the notation is otherwise clear, we will often drop the subscript $1 \le j, k \le N$.

We recall the basic fact about Vandermonde determinants. For any N-tuple of complex numbers $(x_1, \ldots, x_N)$ let

$$\Delta(x_1, \ldots, x_N) = \det_{N \times N} \left(x_k^{j-1} \right)_{j,k}. \tag{4.1}$$

Then

$$\Delta(x_1, \ldots, x_N) = \prod_{1 \le j < k \le N} (x_k - x_j). \tag{4.2}$$

To prove this, one observes that both sides are homogeneous polynomials of total degree $N(N-1)/2$ which vanish whenever $x_j = x_k$. This fact identifies the two sides up to a constant factor. That the coefficient of $x_N^{N-1} x_{N-1}^{N-2} \ldots x_2$ is 1 in both expressions completes the proof.

Observe that

$$\prod_{1 \le j < k \le N} |e^{i\theta_k} - e^{i\theta_j}|^2 = |\Delta(e^{i\theta_1}, \ldots, e^{i\theta_N})|^2 \tag{4.3}$$

and

$$\prod_{1 \le j < k \le N} (\cos\theta_k - \cos\theta_j)^2 = \Delta(\cos\theta_1, \ldots, \cos\theta_N)^2. \tag{4.4}$$

Useful in our calculations will be the

Lemma 1 (Transposing Lemma). We have

$$\det_{N \times N} (\phi_{j-1}(x_k)) \det_{N \times N} (\psi_{j-1}(x_k)) = \det_{N \times N} \left(\sum_{n=1}^{N} \phi_{n-1}(x_j)\psi_{n-1}(x_k) \right). \tag{4.5}$$

This identity just follows by using the fact that the determinant of a matrix and its transpose are the same, and matrix multiplication. Specifically,

$$\det_{N \times N} (\phi_{j-1}(x_k)) \det_{N \times N} (\psi_{j-1}(x_k)) = \det_{N \times N} (\phi_{n-1}(x_j))_{j,n} \det_{N \times N} (\psi_{n-1}(x_k))_{n,k}$$

$$= \det_{N \times N} \left(\sum_{n=1}^{N} \phi_{n-1}(x_j)\psi_{n-1}(x_k) \right)_{j,k}.$$

4.1   An alternate formula for the Haar measure on $U(N)$

In order to compute the statistics we desire, we require an alternate formula for the Haar measure. The Transposing Lemma implies the identity

$$\prod_{1 \le j < k \le N} |e^{i\theta_k} - e^{i\theta_j}|^2 = \det_{N \times N} \left(S_N(\theta_k - \theta_j) \right) \tag{4.6}$$

where

$$S_N(\theta) = \frac{\sin\frac{N\theta}{2}}{\sin\frac{\theta}{2}}. \tag{4.7}$$

From this identity we have

$$dX_{U_N} = \frac{d\theta_1 \ldots d\theta_N}{N!} \det_{N \times N} \left(S_N^*(\theta_k - \theta_j) \right) \tag{4.8}$$

where $S_N^*(x) = \frac{1}{2\pi} S_N(x)$. To prove this we apply the Transposing Lemma with $\phi_j(x_k) = e^{ij\theta_k}$ and $\psi_j(x_k) = e^{-ij\theta_k}$ and use the fact that

$$\sum_{n=1}^{N} e^{i(n-1)\theta} = \frac{e^{iN\theta} - 1}{e^{i\theta} - 1} = \frac{e^{iN\theta/2}}{e^{i\theta/2}} \frac{e^{iN\theta/2} - e^{-iN\theta/2}}{e^{i\theta/2} - e^{-i\theta/2}} = e^{i(N-1)\theta/2} S_N(\theta)$$

from which

$$
\begin{aligned}
|\det(e^{i(j-1)\theta_k})|^2 &= \det\Big(\sum_{n=1}^{N} e^{i(n-1)(\theta_j-\theta_k)}\Big)_{j,k} \\
&= \det\big(e^{iN(\theta_j-\theta_k)/2} S_N(\theta_j-\theta_k)\big) \\
&= \det\big(S_N(\theta_j-\theta_k)\big);
\end{aligned}
$$

the last line holds by factoring out $e^{iN\theta_j/2}$ from the jth row and $e^{-iN\theta_k/2}$ from the kth column and observing that the product of all of these factors is 1.

For future reference we introduce the notation

$$
S(x) = \frac{\sin \pi x}{\pi x}. \tag{4.9}
$$

4.2 Alternate formulas for orthogonal and symplectic Haar measures

Now we give alternate formulas for our other measures. To accomplish this, it is helpful to first recall the basic properties of the Tchebychev polynomials. Let $T_n(x)$ be the (Chebyshev) polynomial of degree n for which

$$
T_n(\cos\theta) = \cos n\theta \tag{4.10}
$$

and $U_n(x)$ is the polynomial of degree n for which

$$
U_n(\cos\theta) = \frac{\sin(n+1)\theta}{\sin\theta}. \tag{4.11}
$$

Thus, $T_0(x) = 1, T_1(x) = x, T_2(x) = 2x^2 - 1, T_3(x) = 4x^3 - 3x$ and so on and $U_0(x) = 1, U_1(x) = 2x, U_2(x) = 4x^2 - 1, U_3(x) = 8x^3 - 4x$, and so on. From $\cos(n+1)\theta = \cos n\theta \cos\theta - \sin n\theta \sin\theta$ and $\sin(n+1)\theta = \sin n\theta \cos\theta + \cos n\theta \sin\theta$ it is easy to see that

$$
T_{n+1}(x) = xT_n(x) - (1-x^2)U_{n-1}(x)
$$

and

$$
U_n(x) = xU_{n-1}(x) + T_n(x).
$$

Thus,

$$
\begin{aligned}
T_{n+2}(x) &= xT_{n+1}(x) - (1-x^2)U_n(x) \\
&= xT_{n+1}(x) - (1-x^2)(xU_{n-1}(x) + T_n(x)) \\
&= xT_{n+1}(x) - (1-x^2)T_n(x) + x(T_{n+1}(x) - xT_n(x)) \\
&= 2xT_{n+1}(x) - T_n(x).
\end{aligned}
$$

Similarly, $U_{n+2}(x) = 2xU_{n+1}(x) - U_n(x)$. Notice that the leading coefficient in $T_n(x)$ is 2^{n-1} and in $U_n(x)$ it is 2^n.

Finally, we let $V_n(x)$ be the polynomial of degree n for which

$$V_n(\cos\theta) = U_{2n}\left(\cos\frac{\theta}{2}\right) = \frac{\sin(n+\frac{1}{2})\theta}{\sin\frac{1}{2}\theta}. \qquad (4.12)$$

It can be shown that $V_n(x) = 2^n x^n + \ldots$ has leading coefficient 2^n.

Now $\prod_{1 \le j < k \le N}(\cos\theta_k - \cos\theta_j) = \Delta(\cos\theta_1, \ldots, \cos\theta_N)$. Let $x_j = \cos\theta_j$ for convenience. Then, by elementary row operations, $\Delta(x_1, \ldots, x_N)$

$$= \det\begin{pmatrix} 1 & 1 & \cdots & 1 \\ x_1 & x_2 & \cdots & x_N \\ \vdots & \vdots & \vdots & \vdots \\ x_1^{N-2} & x_2^{N-2} & \cdots & x_N^{N-2} \\ x_1^{N-1} & x_2^{N-1} & \cdots & x_N^{N-1} \end{pmatrix}$$

$$= \frac{1}{2^{N-2}}\det\begin{pmatrix} 1 & 1 & \cdots & 1 \\ x_1 & x_2 & \cdots & x_N \\ \vdots & \vdots & \vdots & \vdots \\ x_1^{N-2} & x_2^{N-2} & \cdots & x_N^{N-2} \\ 2^{N-2}x_1^{N-1} & 2^{N-2}x_2^{N-1} & \cdots & 2^{N-2}x_N^{N-1} \end{pmatrix}$$

$$= \frac{1}{2^{N-2}}\det\begin{pmatrix} 1 & 1 & \cdots & 1 \\ x_1 & x_2 & \cdots & x_N \\ \vdots & \vdots & \vdots & \vdots \\ x_1^{N-2} & x_2^{N-2} & \cdots & x_N^{N-2} \\ T_{N-1}(x_1) & T_{N-1}(x_2) & \cdots & T_{N-1}(x_N) \end{pmatrix}$$

by adding appropriate multiples of the first $N-1$ rows to the last row. Now we do the same thing to all of the rows, except the first which we leave alone, working our way from the bottom to the top. In this way, we find that

$$\Delta(\cos\theta_1, \ldots, \cos\theta_N) = 2^{-(N-1)(N-2)/2} \det_{N \times N}(T_{j-1}(\cos\theta_k)). \qquad (4.13)$$

For the Haar measure on $SO(2N)$, we have

$$\begin{aligned} dX_{SO(2N)} &= \frac{2^{(N-1)^2}}{\pi^N N!}\Delta(\cos\theta_1, \ldots, \cos\theta_N)^2 d\theta_1 \ldots d\theta_N \\ &= \frac{2^{N-1}}{\pi^N N!}\det_{N \times N}(T_{j-1}(\cos\theta_k))^2 d\theta_1 \ldots d\theta_N. \end{aligned}$$

If we multiply each row except the first by $\sqrt{2}$ we find that

$$\Delta(\cos\theta_1, \ldots, \cos\theta_N) = 2^{-(N-1)^2/2} \det_{N \times N}(T^*_{j-1}(\cos\theta_k))$$

where we let $T_j^* = \sqrt{2}T_j$ for $j \geq 1$ and $T_0^* = T_0 = 1$. Then,

$$dX_{SO(2N)} = \frac{1}{\pi^N N!} \det_{N \times N} \left(T_{j-1}^*(\cos\theta_k) \right)^2 d\theta_1 \ldots d\theta_N. \qquad (4.14)$$

By the Transposing Lemma,

$$\Delta(\cos\theta_1,\ldots,\cos\theta_N)^2 = 2^{-(N-1)^2} \det_{N \times N} \left(1 + 2\sum_{n=1}^{N-1} \cos n\theta_j \cos n\theta_k \right) \qquad (4.15)$$

Recall that $S_N(\theta) = \dfrac{\sin\frac{N\theta}{2}}{\sin\frac{\theta}{2}}$ and $S(x) = \frac{\sin\pi x}{\pi x}$. Now

$$\sum_{n=-N}^{N} \cos nx = \Re \sum_{n=-N}^{N} e^{inx} = \Re \frac{e^{i(N+1)x} - e^{-iNx}}{e^{ix} - 1}$$

$$= \Re \frac{e^{i(N+1/2)x} - e^{-i(N+1/2)x}}{e^{ix/2} - e^{-ix/2}} = \frac{\sin(N+1/2)x}{\sin x/2} = S_{2N+1}(x).$$

Consequently,

$$\sum_{n=1}^{N} \cos nx = \frac{S_{2N+1}(x) - 1}{2}$$

so that

$$1 + 2\sum_{n=1}^{N-1} \cos nx \cos ny = 1 + \sum_{n=1}^{N-1} \left(\cos n(x-y) + \cos n(x+y) \right)$$

$$= \frac{S_{2N-1}(x-y) + S_{2N-1}(x+y)}{2}.$$

Consequently, a basic identity is

$$2^{(N-1)^2} \prod_{1 \leq j < k \leq N} (\cos\theta_k - \cos\theta_j)^2 = \det_{N \times N} \left(\frac{S_{2N-1}(\theta_k - \theta_j) + S_{2N-1}(\theta_k + \theta_j)}{2} \right)$$

from which we deduce by (4.14) that

$$dX_{SO(2N)} = \frac{1}{N!} \det_{N \times N} \left(S_{2N-1}^*(\theta_k - \theta_j) + S_{2N-1}^*(\theta_k + \theta_j) \right) d\theta_1 \ldots d\theta_N$$

$$= \frac{1}{N!} \det_{N \times N} \left(K_{SO(2N)}(\theta_j, \theta_k) \right) d\theta_1 \ldots d\theta_N, \qquad (4.16)$$

where we define

$$K_{SO(2N)}(x,y) = S_{2N-1}^*(y-x) + S_{2N-1}^*(y+x)$$

and, for use in a moment,

$$K_{USp(2N)}(x,y) = S_{2N+1}^*(y-x) - S_{2N+1}^*(y+x)$$

and

$$K_{SO(2N+1)}(x,y) = S_{2N}^*(y-x) - S_{2N}^*(y+x).$$

Now we do the same for the Haar measure of the symplectic group. Again, by elementary row operations on the determinant $\Delta(x_1,\ldots,x_N)$, we find that (recall that the leading term of the Chebyshev polynomial $U_N(x)$ is $(2x)^N$),

$$\Delta(x_1,\ldots,x_N) = 2^{-N(N-1)/2} \det_{N\times N}(U_{j-1}(x_k)).$$

Then $dX_{USp(2N)}$, the Haar measure on USp($2N$), satisfies

$$
\begin{aligned}
dX_{USp(2N)} &= \frac{2^{N^2}}{\pi^N N!} \Delta(\cos\theta_1,\ldots,\cos\theta_N)^2 \prod_{n=1}^{N} \sin^2\theta_n d\theta_1 \ldots d\theta_N \\
&= \frac{2^N}{\pi^N N!} \det_{N\times N}\left(U_{j-1}(\cos\theta_k)\right)^2 \prod_{n=1}^{N} \sin^2\theta_n d\theta_1 \ldots d\theta_N.
\end{aligned}
$$

Now, by the Transposing Lemma

$$
\begin{aligned}
\prod_{1\le j<k\le N} (\cos\theta_k - \cos\theta_j)^2 \prod_{n=1}^{N} \sin^2\theta_n &= 2^{-N(N-1)} \det_{N\times N}\left(\sum_{n=1}^{N} \sin n\theta_j \sin n\theta_k\right) \\
&= 2^{-N(N-1)} \det_{N\times N}\left(\frac{S_{2N+1}(\theta_k - \theta_j) - S_{2N+1}(\theta_k + \theta_j)}{2}\right)
\end{aligned}
$$

since

$$
\begin{aligned}
2\sum_{n=1}^{N} \sin nx \sin ny &= \sum_{n=1}^{N}\left(\cos n(x-y) - \cos n(x+y)\right) \\
&= \frac{S_{2N+1}(x-y) - S_{2N+1}(x+y)}{2}.
\end{aligned}
$$

Therefore,

$$
\begin{aligned}
dX_{USp(2N)} &= \frac{1}{\pi^N N!} \det_{N\times N}\left(\frac{S_{2N+1}(\theta_k - \theta_j) - S_{2N+1}(\theta_k + \theta_j)}{2}\right) d\theta_1 \ldots d\theta_N \\
&= \frac{1}{N!} \det_{N\times N}\left(K_{USp(2N)}(\theta_k,\theta_j)\right) d\theta_1 \ldots d\theta_N. \qquad (4.17)
\end{aligned}
$$

Finally, $dX_{SO(2N+1)}$, the Haar measure on $SO(2N+1)$, satisfies

$$
\begin{aligned}
dX_{SO(2N+1)} &= \frac{2^{N^2}}{\pi^N N!} \Delta(\cos\theta_1,\ldots,\cos\theta_N)^2 \prod_{n=1}^{N} \sin^2\frac{\theta_n}{2} d\theta_1 \ldots d\theta_N \\
&= \frac{2^N}{\pi^N N!} \det_{N\times N}\left(V_{j-1}(\cos\theta_k)\right)^2 \prod_{n=1}^{N} \sin^2\frac{\theta_n}{2} d\theta_1 \ldots d\theta_N.
\end{aligned}
$$

By the Transposing Lemma,

$$
\begin{aligned}
dX_{SO(2N+1)} &= \frac{2^N}{\pi^N N!} \det_{N \times N} \Big(\sum_{n=1}^{N} \sin(n - \tfrac{1}{2})\theta_j \sin(n - \tfrac{1}{2})\theta_k \Big) d\theta_1 \ldots d\theta_N \\
&= \frac{1}{\pi^N N!} \det_{N \times N} \Big(\frac{S_{2N}(\theta_k - \theta_j) - S_{2N}(\theta_k + \theta_j)}{2} \Big) d\theta_1 \ldots d\theta_N \\
&= \frac{1}{N!} \det_{N \times N} \big(S_{2N}^*(\theta_k - \theta_j) - S_{2N}^*(\theta_k + \theta_j) \big) d\theta_1 \ldots d\theta_N.
\end{aligned}
$$

Therefore,

$$
dX_{SO(2N+1)} = \frac{1}{N!} \det_{N \times N} \big(K_{SO(2N+1)}(\theta_k, \theta_j) \big) \, d\theta_1 \ldots d\theta_N. \tag{4.18}
$$

5   Andréief's identity

As a consistency check, we now deduce that our measures have total mass one. To do this we use a formula of Andréief:

Lemma 2 (Andréief's identity). For any interval J and integrable functions ϕ_j and ψ_j:

$$
\frac{1}{N!} \int_{J^N} \det_{N \times N}(\phi_j(\theta_k)) \det_{N \times N}(\psi_j(\theta_k)) \, d\theta_1 \ldots d\theta_N = \det_{N \times N} \Big(\int_J \phi_j(\theta)\psi_k(\theta) \, d\theta \Big). \tag{5.1}
$$

5.1   Proof of Andréief's identity

We use the definition of determinant for a matrix $X = (x_{jk})$:

$$
\det X = \sum_{\sigma \in \pi_N} \operatorname{sgn}(\sigma) \prod_{j=1}^{N} x_{j,\sigma j}
$$

where π_N denotes the collection of the $N!$ permutations of $[1, N] := \{1, 2, \ldots, N\}$, and where we have written σj in place of $\sigma(j)$ to make the notation easier.

Thus,

$$
\int_{J^N} \det_{N\times N} \left(\phi_j(\theta_k)\right) \det_{N\times N} \left(\psi_j(\theta_k)\right) d\theta_1 \ldots \, d\theta_N
$$

$$
= \int_{J^N} \sum_\sigma \operatorname{sgn}(\sigma) \prod_{j=1}^N \phi_j(\theta_{\sigma j}) \sum_\tau \operatorname{sgn}(\tau) \prod_{k=1}^N \psi_k(\theta_{\tau k}) \prod_{i=1}^N d\theta_i
$$

$$
\underset{\tau \to \sigma\tau}{=} \int_{J^N} \sum_{\sigma,\tau} \operatorname{sgn}(\tau) \prod_{j=1}^N \phi_j(\theta_{\sigma j}) \prod_{k=1}^N \psi_k(\theta_{\sigma\tau k}) \prod_{i=1}^N d\theta_i
$$

$$
\underset{k \to \tau^{-1}k}{=} \int_{J^N} \sum_{\sigma,\tau} \operatorname{sgn}(\tau) \prod_{j=1}^N \phi_j(\theta_{\sigma j}) \prod_{k=1}^N \psi_{\tau^{-1}k}(\theta_{\sigma k}) \prod_{i=1}^N d\theta_i
$$

$$
= \int_{J^N} \sum_{\sigma,\tau} \operatorname{sgn}(\tau) \prod_{j=1}^N \phi_j(\theta_{\sigma j}) \psi_{\tau^{-1}j}(\theta_{\sigma j}) \prod_{i=1}^N d\theta_i.
$$

Letting $\tau \to \tau^{-1}$, we see that the above is

$$
= \int_{J^N} \sum_{\sigma,\tau} \operatorname{sgn}(\tau) \prod_{j=1}^N \phi_j(\theta_{\sigma j}) \psi_{\tau j}(\theta_{\sigma j}) \prod_{i=1}^N d\theta_i
$$

$$
= \sum_{\sigma,\tau} \operatorname{sgn}(\tau) \prod_{j=1}^N \int_J \phi_j(\theta) \psi_{\tau j}(\theta) \, d\theta
$$

$$
= N! \sum_\tau \operatorname{sgn}(\tau) \prod_{j=1}^N \int_J \phi_j(\theta) \psi_{\tau j}(\theta) \, d\theta
$$

$$
= \det_{N\times N} \left(\int_J \phi_j(\theta) \psi_k(\theta) \, d\theta \right).
$$

Note that virtually the same proof leads to the slightly more general result

$$
\frac{1}{N!} \int_{J^N} \prod_{i=1}^N f(\theta_i) \det_{N\times N}(\phi_j(\theta_k)) \det_{N\times N}(\psi_j(\theta_k)) \, d\theta_1 \ldots d\theta_N
$$

$$
= \det_{N\times N} \left(\int_J f(\theta)\phi_j(\theta)\psi_k(\theta) \, d\theta \right).
$$

5.2   Verification that the Haar measures have total mass 1

Using

$$
\phi_j(\theta) = e^{i(j-1)\theta}
$$

we see that

$$
\begin{aligned}
\int_{[0,2\pi]^N} dX_{U(N)} &= \int_{[0,2\pi]^N} |\det_{N\times N}(e^{i(j-1)\theta_k})|^2 \frac{d\theta_1 \ldots d\theta_N}{N!(2\pi)^N} \\
&= \frac{1}{(2\pi)^N} \det_{N\times N} \Big(\int_0^{2\pi} e^{i(j-1)\theta} e^{-i(k-1)\theta} \, d\theta \Big) \\
&= \frac{1}{(2\pi)^N} \det_{N\times N} \big(2\pi I \big) = 1.
\end{aligned}
$$

This shows that the total mass of the Haar measure of $U(N)$ is 1.

We observe further that (4.14) and Andréief's identity together imply that

$$
\int_{[0,\pi]^N} dX_{SO(2N)} = \frac{2^{N-1}}{\pi^N} \det_{N\times N} \Big(\int_0^\pi T_{j-1}(\cos\theta) T_{k-1}(\cos\theta) \, d\theta \Big) = 1,
$$

since

$$
\begin{aligned}
\int_0^\pi T_{j-1}(\cos\theta) T_{k-1}(\cos\theta) \, d\theta &= \int_0^\pi \cos(j-1)\theta \cos(k-1)\theta \, d\theta \\
&= \frac{1}{2} \int_0^\pi \big(\cos(j+k-2)\theta + \cos(j-k)\theta \big) \, d\theta \\
&= \frac{\pi}{2} \delta_{j,k}(1 + \delta_{1,j})
\end{aligned}
$$

because the integral is 0 unless $j = k$ in which case it is π if $j > 1$ and 2π if $j = 1$. Also,

$$
\begin{aligned}
\int_{USp(2N)} dX &= \frac{2^N}{\pi^N} \det_{N\times N} \Big(\int_0^\pi \sin^2\theta \; U_{j-1}(\cos\theta) \; U_{k-1}(\cos\theta) \, d\theta \Big) \\
&= \frac{2^N}{\pi^N} \det_{N\times N} \Big(\int_0^\pi \sin j\theta \sin k\theta \, d\theta \Big) \\
&= \frac{2^N}{\pi^N} \det_{N\times N} \Big(\frac{1}{2} \int_0^\pi \big(\cos(j-k)\theta - \cos(j+k)\theta \big) \, d\theta \Big).
\end{aligned}
$$

Since the integral is π when $j = k$ and 0 otherwise, this confirms that the total measure of $USp(2N)$ is 1.

Similarly, we can calculate that the total mass of $SO(2N+1)$ is 1.

6   Gaudin's Lemma

The following Lemma is the key to begin computing the statistics of interest.[1]

---

[1] Editors' comment: This lemma is also applied in the lectures of Y.V. Fyodorov, page 31, Section 4.

Lemma 3 (Gaudin's Lemma). Suppose that we have a function f and a measurable set J such that

$$\int_J f(x,\theta)f(\theta,y)\,d\theta = Cf(x,y) \tag{6.1}$$

for all x and y where $C = C(J,f)$ is a constant. Suppose also that

$$\int_J f(x,x)\,dx = D, \tag{6.2}$$

where $D = D(J,f)$ is constant. Then

$$\int_J \det_{M \times M} \big(f(\theta_j,\theta_k)\big)\,d\theta_M = (D - (M-1)C)\det_{M-1}\big(f(\theta_j,\theta_k)\big). \tag{6.3}$$

This Lemma allows us to "integrate out" variables not under consideration when computing some statistic. We apply Gaudin's Lemma with $f(x,y) = S_N^*(x-y)$ and $J = [0,2\pi]$, so that $D = S_N^*(0) = N$. Reëxpressing S_N as a geometric series and integrating term-by-term, we find that

$$\int_0^{2\pi} S_N^*(\theta_j - \theta)S_N^*(\theta - \theta_k)\,d\theta = S_N^*(\theta_k - \theta_j),$$

so that $C = 1$. Thus, for example,

$$\int_{[0,2\pi]} \det_{N \times N}\big(S_N^*(\theta_k - \theta_j)\big)\,d\theta_N = \det_{(N-1)\times(N-1)}\big(S_N^*(\theta_k - \theta_j)\big).$$

Applying the Lemma repeatedly gives

$$\int_{[0,2\pi]^{N-n}} \det_{N \times N}\big(S_N^*(\theta_k - \theta_j)\big)\,d\theta_{n+1}\ldots\,d\theta_N$$
$$= (N-n)!\det_{n \times n}\big(S_N^*(\theta_k - \theta_j)\big).$$

In particular, by integrating out all but n variables, we have

$$\int_{U(N)} \sum_{\substack{J \subset \{1,\ldots,N\} \\ J = \{j_1,\ldots,j_n\}}} f(\theta_{j_1},\ldots,\theta_{j_n})\,dX_{U(N)}$$
$$= \frac{1}{n!}\int_{[0,2\pi]^n} f(\theta_1,\ldots,\theta_n)\det_{n \times n} S_N^*(\theta_k - \theta_j)\,d\theta_1 \ldots d\theta_n. \tag{6.4}$$

Proof of Gaudin's Lemma. Let π_M be the symmetric group on $\{1,\ldots,M\}$. Then,

$$\det_{M \times M}\big(f(\theta_j,\theta_k)\big) = \sum_{\sigma \in \pi_M} \text{sgn}(\sigma)\prod_{j=1}^{M} f(\theta_j,\theta_{\sigma j}).$$

If $\sigma M \neq M$, then

$$
\int_J \prod_{j=1}^{M} f(\theta_j, \theta_{\sigma j}) d\theta_M = \prod_{\substack{j=1 \\ \sigma j \neq M}}^{M-1} f(\theta_j, \theta_{\sigma j}) \int_J f(\theta_{\sigma^{-1}M}, \theta_M) f(\theta_M, \theta_{\sigma M}) \, d\theta_M
$$

$$
= C f(\theta_{\sigma^{-1}M}, \theta_{\sigma M}) \prod_{\substack{j=1 \\ \sigma j \neq M}}^{M-1} f(\theta_j, \theta_{\sigma j}). \tag{6.5}
$$

For a permutation $\sigma \in \pi_M$ with $\sigma M \neq M$ define a permutation $\sigma' \in \pi_{M-1}$ by

$$
\sigma' j = \begin{cases} \sigma j & \text{if } \sigma j \neq M \\ \sigma M & \text{if } \sigma j = M \end{cases}.
$$

Then (6.5) may be reëxpressed as

$$
\int_J \prod_{j=1}^{M} f(\theta_j, \theta_{\sigma j}) d\theta_M = C \prod_{j=1}^{M-1} f(\theta_j, \theta_{\sigma' j}).
$$

Clearly, each permutation σ' arises from $(M-1)$ different σ. Note also that $\operatorname{sgn}(\sigma') = -\operatorname{sgn}(\sigma)$. Thus, we have

$$
\int_J \sum_{\substack{\sigma \in \pi_M \\ \sigma M \neq M}} \operatorname{sgn}(\sigma) \prod_{j=1}^{M} f(\theta_j, \theta_{\sigma j}) d\theta_M = -(M-1)C \sum_{\sigma' \in \pi_{M-1}} \operatorname{sgn}(\sigma') \prod_{j=1}^{M-1} f(\theta_j, \theta_{\sigma' j})
$$

$$
= -(M-1)C \det_{M-1} \big(f(\theta_j, \theta_k) \big).
$$

Now consider the σ for which $\sigma M = M$; let σ' be defined by $\sigma' j = \sigma j$ for $j \leq M - 1$. Then, for these σ, we have

$$
\int_J \prod_{j=1}^{M} f(\theta_j, \theta_{\sigma j}) d\theta_M = \prod_{j=1}^{M-1} f(\theta_j, \theta_{\sigma j}) \int_J f(\theta_M, \theta_M) \, d\theta_M
$$

$$
= D \prod_{j=1}^{M-1} f(\theta_j, \theta_{\sigma' j}).
$$

These σ' have the same sign as the σ they came from. Therefore,

$$
\int_J \sum_{\substack{\sigma \in \pi_M \\ \sigma M = M}} \operatorname{sgn}(\sigma) \prod_{j=1}^{M} f(\theta_j, \theta_{\sigma j}) d\theta_M = D \sum_{\sigma' \in \pi_{M-1}} \operatorname{sgn}(\sigma') \prod_{j=1}^{M-1} f(\theta_j, \theta_{\sigma' j})
$$

$$
= D \det_{M-1} \big(f(\theta_k, \theta_j) \big).
$$

Combining the two cases we obtain the Lemma.

6.1   Calculation for orthogonal and symplectic cases

Recall that

$$\pi K_{SO(2N)}(x,y) = \frac{S_{2N-1}(x-y) + S_{2N-1}(x+y)}{2} = 1 + 2 \sum_{n=1}^{N-1} \cos nx \cos ny$$

and

$$\pi K_{USp(2N)}(x,y) = \frac{S_{2N+1}(x-y) - S_{2N+1}(x+y)}{2} = 2 \sum_{n=1}^{N} \sin nx \sin ny.$$

Then, Gaudin's Lemma for these groups is expressed as

$$\int_{[0,\pi]^{N-n}} \det_{N \times N} \left(K_{G(N)}(\theta_j, \theta_k) \right) d\theta_{n+1} \dots d\theta_N = (N-n)! \det_{n \times n} \left(K_{G(N)}(\theta_j, \theta_k) \right)$$

where $G(N)$ can stand for $USp(2N)$, $SO(2N)$, or $SO(2N+1)$. This allows us to "integrate out" variables in the orthogonal and symplectic settings.

Note, for future reference, that

$$\lim_{N \to \infty} \frac{1}{2N} S_{2N+1}(\pi x/N) = \lim_{N \to \infty} \frac{\sin \frac{(N+1/2)\pi x}{N}}{2N \sin \frac{\pi x}{2N}} = \frac{\sin \pi x}{\pi x} = S(x)$$

so that

$$K_G(x,y) := \lim_{N \to \infty} \frac{\pi K_{G(N)}(\pi x/N, \pi y/N)}{N} = S(y-x) \pm S(y+x).$$

To prove Gaudin's Lemma in this situation, it again suffices to prove the $n = N-1$ case, since the general case follows by a repeated application of this case:

$$\int_{[0,\pi]} \det_{N \times N} \left(K_{G(N)}(\theta_j, \theta_k) \right) d\theta_N = \det_{(N-1) \times (N-1)} \left(K_{G(N)}(\theta_j, \theta_k) \right).$$

The key formulae are

$$\int_0^\pi K_{G(N)}(\theta_j, \theta) K_{G(N)}(\theta, \theta_k) \, d\theta = K_{G(N)}(\theta_j, \theta_k)$$

and

$$\int_0^\pi K_{G(N)}(\theta, \theta) \, d\theta = N.$$

Knowing these, the rest of the proof is the same; so we now verify these formulae. We calculate

$$\int_0^\pi K_{USp(2N)}(x,\theta)K_{USp(2N)}(\theta,y)\,d\theta$$

$$= \frac{1}{\pi^2}\int_0^\pi \sum_{m=1}^N 2\sin mx \sin m\theta \sum_{n=1}^N 2\sin n\theta \sin ny\,d\theta$$

$$= \frac{4}{\pi^2}\sum_{m,n=1}^N \sin mx \sin ny \int_0^\pi \sin m\theta \sin n\theta\,d\theta$$

$$= \frac{2}{\pi^2}\sum_{m,n=1}^N \sin mx \sin ny \int_0^\pi (\cos(m-n)\theta - \cos(m+n)\theta)\,d\theta$$

$$= \frac{2}{\pi}\sum_{n=1}^N \sin nx \sin ny = K_{USp(2N)}(x,y).$$

Similarly,

$$\int_0^\pi K_{SO(2N)}(x,\theta)K_{SO(2N)}(\theta,y)\,d\theta$$

$$= \frac{1}{\pi^2}\int_0^\pi \left(1+2\sum_{m=1}^{N-1}\cos mx \cos m\theta\right)\left(1+2\sum_{n=1}^{N-1}\cos n\theta \cos ny\right)d\theta$$

$$= \frac{1}{\pi} + \frac{4}{\pi^2}\sum_{m,n=1}^{N-1}\cos mx \cos ny \int_0^\pi \cos m\theta \cos n\theta\,d\theta$$

$$= \frac{1}{\pi} + \frac{2}{\pi^2}\sum_{m,n=1}^{N-1}\cos mx \cos ny \int_0^\pi (\cos(m-n)\theta + \cos(m+n)\theta)\,d\theta$$

$$= \frac{1}{\pi} + \frac{2}{\pi}\sum_{n=1}^{N-1}\cos nx \cos ny = K_{SO(2N)}(x,y).$$

Similarly for $K_{SO(2N+1)}$.

7 n-level density

7.1 Unitary

We can use Gaudin's Lemma to compute an integral of the sort

$$\int_{U(N)}\sum_{j=1}^N f(\theta_j)dX,$$

or

$$\int_{U(N)} \sum_{\substack{1 \le j,k \le N \\ j \ne k}} f(\theta_j, \theta_k) dX,$$

or

$$\int_{U(N)} \sum_{\substack{1 \le j_1,\dots,j_n \le N \\ j_h \ne j_i}} f(\theta_{j_1}, \dots, \theta_{j_n}) dX.$$

These are precisely the definitions of the 1-, 2-, and n-level densities. By Gaudin's Lemma, these integrals are, respectively,

$$\frac{N}{2\pi} \int_0^{2\pi} f(\theta) \, d\theta,$$

$$\int_{[0,2\pi]^2} f(\theta_1, \theta_2) \det_{2\times 2} S_N^*(\theta_k - \theta_j) \, d\theta_1 \, d\theta_2,$$

and [2]

$$\int_{[0,2\pi]^n} f(\theta_1, \dots, \theta_n) \det_{n\times n} S_N^*(\theta_k - \theta_j) \, d\theta_1 \dots \, d\theta_n. \tag{7.1}$$

7.2   Normalized eigenangles and large N limits

For a matrix $X \in U(N)$ with eigenvalues

$$e^{i\theta_1}, \dots, e^{i\theta_N}$$

we let

$$\tilde{\theta}_j = \theta \frac{N}{2\pi} \tag{7.2}$$

be the normalized eigenangles. They satisfy

$$0 \le \tilde{\theta}_j < N.$$

If arranged in increasing order, the sequence of $\tilde{\theta}$ have mean spacing 1 and so give a way to compare statistics for different N. Thus, for the n-level density, we have (for a rapidly decaying smooth f)

$$\lim_{N \to \infty} \int_{U(N)} \sum_{\substack{1 \le j_1,\dots,j_n \le N \\ j_h \ne j_i}} f(\tilde{\theta}_{j_1}, \dots, \tilde{\theta}_{j_n}) dX$$

$$= \lim_{N \to \infty} \int_{[0,2\pi]^n} f(\tilde{\theta}_1, \dots, \tilde{\theta}_n) \det_{n\times n} S_N^*(\theta_k - \theta_j) \, d\theta_1 \dots \, d\theta_n$$

$$= \lim_{N \to \infty} \int_{[0,N]^n} f(x_1, \dots, x_n) \det_{n\times n} \frac{1}{N} S_N \Big(\frac{2\pi(x_k - x_j)}{N} \Big) \, dx_1 \dots \, dx_n$$

$$= \int_{\mathbf{R}_+^n} f(x_1, \dots, x_n) \det_{n\times n} S(x_k - x_j) \, dx_1 \dots dx_n. \tag{7.3}$$

---

[2]Editors' comment: Up to a constant factor $f(\theta_1, \dots, \theta_n)$ is being integrated against the quantity defined in the lectures of Y.V. Fyodorov, page 31, Section 3, as the n-point correlation function.

We say that $\det_{n \times n} S(x_k - x_j)$ is the n-level density function for U.

7.3 Orthogonal and symplectic

We can use Gaudin's Lemma to compute

$$\int_{SO(2N)} \sum_{\substack{1 \leq j_1, \dots, j_n \leq N \\ j_h \neq j_i}} f(\theta_{j_1}, \dots, \theta_{j_n}) dX$$

and

$$\int_{USp(2N)} \sum_{\substack{1 \leq j_1, \dots, j_n \leq N \\ j_h \neq j_i}} f(\theta_{j_1}, \dots, \theta_{j_n}) dX.$$

By Weyl's integration formula and Gaudin's Lemma, these integrals are

$$\int_{[0,\pi]^n} f(\theta_1, \dots, \theta_n) \det_{n \times n} \left(K_{SO(2N)}(\theta_k, \theta_j) \right) d\theta_1 \dots d\theta_n$$

and

$$\int_{[0,\pi]^n} f(\theta_1, \dots, \theta_n) \det_{n \times n} \left(K_{USp(2N)}(\theta_k, \theta_j) \right) d\theta_1 \dots d\theta_n$$

respectively.

For eigenangles of matrices in $SO(2N)$ and $USp(2N)$ we let

$$\tilde{\theta}_j = \frac{\theta_j N}{\pi}$$

be the normalized eigenangles. Thus, for the n-level density, we have (for a rapidly decaying smooth f)

$$\lim_{N \to \infty} \int_{SO(2N)} \sum_{\substack{1 \leq j_1, \dots, j_n \leq N \\ j_h \neq j_i}} f(\tilde{\theta}_{j_1}, \dots, \tilde{\theta}_{j_n}) dX$$

$$= \lim_{N \to \infty} \int_{[0,\pi]^n} f(\tilde{\theta}_1, \dots, \tilde{\theta}_n) \det_{n \times n} \left(K_{SO(2N)}(\theta_k, \theta_j) \right) d\theta_1 \dots d\theta_n$$

$$= \int_{\mathbf{R}_+^n} f(\theta_1, \dots, \theta_n) \det_{n \times n} \left(K_{SO}(\theta_k, \theta_j) \right) d\theta_1 \dots d\theta_n \qquad (7.4)$$

for the n-level density for SO even and

$$\lim_{N \to \infty} \int_{USp(2N)} \sum_{\substack{1 \leq j_1, \dots, j_n \leq N \\ j_h \neq j_i}} f(\tilde{\theta}_{j_1}, \dots, \tilde{\theta}_{j_n}) dX$$

$$= \lim_{N \to \infty} \int_{[0,\pi]^n} f(\tilde{\theta}_1, \dots, \tilde{\theta}_n) \det_{n \times n} \left(K_{USp(2N)}(\theta_k, \theta_j) \right) d\theta_1 \dots d\theta_n$$

$$= \int_{\mathbf{R}_+^n} f(\theta_1, \dots, \theta_n) \det_{n \times n} \left(K_{USp}(\theta_k, \theta_j) \right) d\theta_1 \dots d\theta_n \qquad (7.5)$$

for the n-level density for USp. The n-level density for $SO(2N+1)$ is slightly complicated by the fact that the matrices in this ensemble always have an eigenangle equal to 0. This fact leads to the presence of a δ-function in the formulation of the n-level density function.

8 Correlations

8.1 Pair correlation for $U(N)$

Let f be a suitable test function and consider

$$Q_N(f) := \int_{U(N)} \sum_{j \neq k} f(\tilde{\theta}_j - \tilde{\theta}_k) dX.$$

Applying Gaudin's Lemma we find that

$$Q_N(f) = \int_{[0,2\pi]^2} f(\tilde{\theta}_1 - \tilde{\theta}_2) \det \left(\begin{matrix} N & S_N(\theta_1 - \theta_2) \\ S_N(\theta_1 - \theta_2) & N \end{matrix} \right) \frac{d\theta_1 \, d\theta_2}{(2\pi)^2}.$$

After a change of variables, this is

$$= \int_{[0,N]^2} f(\theta_1 - \theta_2) \det_{2 \times 2} \frac{1}{N} S_N \left(\frac{2\pi(\theta_k - \theta_j)}{N} \right) d\theta_1 \, d\theta_2.$$

After expanding the determinant and performing another change of variables, we have

$$Q_N(f) = \int_{[-N,N]} f(v) \left(1 - \left(\frac{S_N(2\pi v/N)}{N} \right)^2 \right) (N - |v|) \, dv.$$

Now

$$\lim_{N \to \infty} \frac{1}{N} S_N \left(\frac{2\pi v}{N} \right) = \lim_{N \to \infty} \frac{1}{N} \frac{\sin \pi v}{\sin \frac{\pi v}{N}} = \frac{\sin \pi v}{\pi v} = S(v).$$

Now it follows, with a little bit of analysis, that

$$\lim_{N \to \infty} \frac{1}{N} Q_N(f) = \int_{-\infty}^{\infty} f(v) \left(1 - \left(\frac{\sin \pi v}{\pi v} \right)^2 \right) dv.$$

This is the same as the pair correlation for zeros of $\zeta(s)$ found by Montgomery[3]. This important fact was fortuitously discovered at tea at the Institute for Advanced Study one afternoon in 1971 when Chowla introduced Hugh Montgomery and Freeman Dyson to each other.

---

[3] Editors' comment: See lectures by D.A. Goldston, page 79, equation 6.7.

8.2 *n*-correlation for $U(N)$

Let $f(\theta_1, \ldots, \theta_n)$ be a test-function which is translation invariant i.e. $f(\theta_1 + t, \ldots, \theta_n + t) = f(\theta_1, \ldots, \theta_n)$. Let's suppose, for convenience, that $f(0, \theta_2, \ldots, \theta_n)$ is compactly supported, say on $[-A, A]$. We seek to evaluate

$$Q_N(f) = \int_{U(N)} \sum_{\substack{1 \leq j_1, j_2, \ldots, j_n \leq N \\ j_h \neq j_i}} f(\tilde{\theta}_{j_1}, \ldots, \tilde{\theta}_{j_n}) dX.$$

By Gaudin's Lemma and a change of variables, this is

$$= \int_{[0,N]^n} f(\theta_1, \ldots, \theta_n) \det_{n \times n} \frac{1}{N} S_N(2\pi(\theta_j - \theta_k)/N) d\theta_1 \ldots d\theta_n.$$

We make the change of variable $x_1 = \theta_1$, $x_2 = \theta_2 - \theta_1$, ... $x_n = \theta_n - \theta_1$ and the integral becomes

$$\int_0^N \int_{[-x_1, N-x_1]^{n-1}} g(x_1, x_2 + x_1, \ldots, x_n + x_1) dx_2 \ldots dx_n \, dx_1$$

where $g(x_1, \ldots, x_n) = f(x_1, \ldots, x_n) \det\left(\frac{S_N(2\pi(x_k - x_j)/N)}{N}\right)$. Since g is translation invariant and $g(0, x_2, \ldots, x_n)$ is compactly supported, for sufficiently large N this is

$$= \int_0^N \int_{[-x_1, N-x_1]^{n-1}} g(0, x_2, \ldots, x_n) dx_2 \ldots dx_n \, dx_1$$

$$= \int_{[-A,A]^{n-1}} g(0, x_2, \ldots, x_n) \left(\int_{\max\{-x_j\}}^{\min\{N-x_j\}} dx_1 \right) dx_2 \ldots dx_n.$$

Thus,

$$\lim_{N \to \infty} \frac{Q_N(f)}{N} = \tag{8.1}$$

$$\int_{\mathbf{R}^{n-1}} f(x_1, x_2, \ldots, x_n) \det_{n \times n} S(x_j - x_k)\big|_{x_1 = 0} dx_2 \ldots dx_n.$$

We note that if f is symmetric in all of its variables, then we also have

$$\lim_{N \to \infty} \frac{Q_N(f)}{N} = \tag{8.2}$$

$$n \int_{\mathbf{R}_+^{n-1}} f(x_1, x_2, \ldots, x_n) \det_{n \times n} S(x_j - x_k)\big|_{x_1 = 0} dx_2 \ldots dx_n.$$

8.3 n-correlation for orthogonal and symplectic

Let $f(\theta_1, \ldots, \theta_n)$ be a test-function as in the last section. We seek to evaluate

$$Q_N(f) = \int_{USp(N)} \sum_{\substack{1 \leq j_1, j_2, \ldots, j_n \leq N \\ j_h \neq j_i}} f(\tilde{\theta}_{j_1}, \ldots, \tilde{\theta}_{j_n}) dX.$$

By Gaudin's Lemma and a change of variables, this is

$$= \int_{[0,N]^n} f(\theta_1, \ldots, \theta_n) \det_{n \times n} \frac{1}{2N} \bigg(S_{2N-1}(\pi(\theta_k - \theta_j)/N)$$

$$+ S_{2N-1}(\pi(\theta_k + \theta_j)/N) \bigg) d\theta_1 \ldots d\theta_n.$$

We make the change of variable $x_1 = \theta_1$, $x_2 = \theta_2 - \theta_1$, ... $x_n = \theta_n - \theta_1$ and the integral becomes, for sufficiently large N,

$$\int_{[-A,A]^{n-1}} f(0, x_2, \ldots, x_n) \int_{\max\{-x_j\}}^{\min\{N-x_j\}} \det_{n \times n} \frac{1}{2N} \bigg(S_{2N-1}(\pi(x_k - x_j)/N)\big|_{x_1=0}$$

$$+ S_{2N-1}(\pi(x_k^* + x_j^* + 2x_1)/N) \bigg) dx_1 \, dx_2 \ldots dx_n$$

where $x_j^* = x_j$ if $j \neq 1$ whereas $x_1^* = 0$. Now we claim that

$$\lim_{N \to \infty} \frac{1}{N} \int_{\max\{-x_j\}}^{\min\{N-x_j\}} \det_{n \times n} \frac{1}{2N} \big(S_{2N-1}(\pi(x_k - x_j)/N)\big|_{x_1=0}$$

$$+ S_{2N-1}(\pi(x_k^* + x_j^* + 2x_1)/N)) dx_1$$

$$= \det_{n \times n} S(x_k - x_j)\big|_{x_1=0}.$$

To see this claim, note that in the expansion of the determinant there are $n!$ terms each of which is a product of n factors $\frac{1}{2N} S_{2N-1}(\pi(x_k - x_j))\big|_{x_1=0} + \frac{1}{2N} S_{2N-1}(\pi(x_k^* + x_j^* + 2x_1))$. If we multiply out each term, there are 2^n terms, all but one of which will contain at least one factor with $\frac{1}{2N} S_{2N-1}(\pi(x_k + x_j + 2x_1))$. Any of the terms with at least one factor like this will tend to 0 after integrating with respect to x_1 and dividing by N; for letting

$$c(a, b, N)(x) = \frac{\sin(ax + b)}{N \sin(ax/N + b/N)},$$

it is not difficult to see that $c(a, b, N)(x) \leq \frac{2}{\pi} \frac{\sin(ax+b)}{ax+b}$ provided that $ax + b < \frac{\pi N}{2}$, and $|c(a, b, N)(x)| \leq 1$ for all x and integer N. Therefore, using the fact that $\int_0^B \left(\frac{\sin x}{x}\right)^j dx$ is uniformly bounded in B for each fixed j, we see that

$$\frac{1}{N} \int_0^{N-B} \prod_{j=1}^J c(a_j, b_j, N)(x) \, dx \to 0$$

as $N \to \infty$ through integers. This leaves only the term with all $\frac{1}{2N}S_{2N-1}(\pi(x_k - x_j))$ factors which tend to $S(x_k - x_j)$ as $N \to \infty$.

Thus, just as in the case of $U(N)$, we find that

$$\lim_{N \to \infty} \frac{Q_N(f)}{N} =$$
$$\int_{\mathbf{R}^{n-1}} f(x_1, x_2, \ldots, x_n) \det_{n \times n} S(x_j - x_k)\big|_{x_1=0} dx_2 \ldots dx_n.$$

In particular, the scaled limit of the n-correlation functions are the same for all of unitary, orthogonal, and symplectic groups.

9 Neighbor spacings

9.1 Nearest neighbor for $U(N)$

We derive the combinatorial relation between nearest neighbor spacings and n-correlations (see [KS], Lemma 2.3.8). Let $Y = (\theta_1, \ldots, \theta_N)$ be an N-tuple with non-decreasing entries: $\theta_1 \leq \theta_2 \leq \cdots \leq \theta_N$. Let

$$\mathcal{S}_n(s, Y) := \#\{j : \theta_{j+n} - \theta_j \leq s\},$$

and let

$$\mathcal{C}_m(s, Y) := \#\{B \subset \{1, \ldots, N\} : |B| = m, \max_{j,k \in B} |\theta_j - \theta_k| \leq s\}.$$

These are related to the Sep and Clump functions used in Katz-Sarnak.

Lemma 4 (Combinatorial Lemma). For any Y,

$$\mathcal{C}_{m+2}(s, Y) = \sum_{n \geq m} \binom{n}{m} \mathcal{S}_{n+1}(s, Y). \tag{9.1}$$

Proof. Take an $m+2$-tuple of indices $i_0 < i_1 < \cdots < i_{m+1}$ whose endpoints satisfy $\theta_{i_{m+1}} - \theta_{i_0} \leq s$. Let $n = i_{m+1} - i_0$ so that the pair of endpoints is counted in $\mathcal{S}_n(s, Y)$. Then there are $\binom{n-1}{m}$ sets of points of size m between these endpoints, which, taken with the endpoints can be counted in $\mathcal{C}_{m+2}(s, Y)$. Therefore, $\mathcal{C}_{m+2} = \sum \binom{n-1}{m} \mathcal{S}_n$. Adjusting the index n by one gives the result.

In general, the relation $a_m = \sum_{n \geq m} \binom{n}{m} b_n$ can be inverted to give

$$b_m = \sum_{n \geq m} (-1)^{n-m} \binom{n}{m} a_n.$$

This follows from the identity for binomial coefficients

$$\sum_{\ell=m}^{n}(-1)^{\ell}\binom{\ell}{m}\binom{n}{\ell} = \left\{ \begin{array}{l} (-1)^m \text{ if } m = n \\ \quad 0 \text{ if } m \neq n \end{array} \right. .$$

Thus, with $a_m = \mathcal{C}_{m+2}$ and $b_n = \mathcal{S}_{n+1}$ we have

Corollary 1.

$$\mathcal{S}_{m+1}(s, Y) = \sum_{n \geq m}(-1)^{n-m}\binom{n}{m}\mathcal{C}_{n+2}(s, Y) \tag{9.2}$$

or, after adjusting the indices,

$$\mathcal{S}_m(s, Y) = \sum_{n \geq m+1}(-1)^{n-m-1}\binom{n-2}{m-1}\mathcal{C}_n(s, Y). \tag{9.3}$$

For a unitary matrix X let $\tilde{Y}_X$ be the N-tuple of normalized eigenangles of X arranged in increasing order. We want to compute

$$\int_0^s \mu_1(x)\,dx: \quad = \quad \text{Prob \{Neighboring eigenangles are } < s \text{ apart\}}$$

$$= \quad \lim_{N \to \infty} \frac{1}{N}\int_{U(N)} \mathcal{S}_1(s, \tilde{Y}_X)dX$$

and more generally

$$\int_0^s \mu_m(x)\,dx: \quad = \quad \text{Prob \{mth neighboring eigenangles are } < s \text{ apart\}}$$

$$= \quad \lim_{N \to \infty} \frac{1}{N}\int_{U(N)} \mathcal{S}_m(s, \tilde{Y}_X)dX.$$

We apply the combinatorial identity just derived and so are led to calculate the n-correlation function associated with the translation invariant function

$$f(\theta_1, \ldots, \theta_n) = \prod_{1 \leq j < k \leq n} \chi_{[0,s]}(|\theta_j - \theta_k|).$$

Note that f is symmetric in all of its variables. Using the calculations from the last chapter, we have

$$\lim_{N \to \infty} \frac{1}{N}\int_{U(N)} \mathcal{C}_n(s, \tilde{Y}_X)dX = \lim_{N \to \infty} \frac{1}{N}\int_{U(N)} \sum_{\substack{B \subset \{1,\ldots,N\} \\ |B|=n}} f(\theta_B)dX$$

$$= \frac{1}{(n-1)!}\int_{[0,s]^{n-1}} \det_{n \times n} S(x_j - x_k)\Big|_{x_1=0} dx_2 \ldots dx_n.$$

Thus, by (9.2)

$$\int_0^s \mu_m(x)\, dx = \sum_{n=m+1}^{\infty} \frac{(-1)^{n-m-1}}{(n-1)!} \binom{n-2}{m-1}$$
$$\times \int_{[0,s]^{n-1}} \det_{n\times n} S(x_j - x_k)\big|_{x_1=0} dx_2 \ldots dx_n.$$

In particular, for the nearest neighbor spacing, we have

$$\mu_1(s) = \frac{d}{ds} \sum_{n=2}^{\infty} \frac{(-1)^n}{(n-1)!} \int_{[0,s]^{n-1}} \det_{n\times n} S(x_j - x_k)\big|_{x_1=0} dx_2 \ldots dx_n.$$

Now, for any symmetric, even, translation invariant function g,

$$\frac{d}{ds} \int_{[0,s]^m} g(x_1, \ldots, x_m) dx_1 \ldots dx_m = m \int_{[0,s]^{m-1}} g(s, x_2, \ldots, x_m) dx_2 \ldots dx_m.$$

Therefore,

$$\begin{aligned}
\mu_1(s) &= \frac{d^2}{ds^2} \sum_{n=2}^{\infty} \frac{(-1)^n}{n!} \int_{[0,s]^n} \det_{n\times n} S(x_j - x_k)\, dx_1 dx_2 \ldots dx_n \\
&= \frac{d^2}{ds^2} \sum_{n=0}^{\infty} \frac{(-1)^n}{n!} \int_{[0,s]^n} \det_{n\times n} S(x_j - x_k)\, dx_1 dx_2 \ldots dx_n.
\end{aligned}$$

Also, temporarily letting

$$F(z) = \sum_{n=0}^{\infty} \frac{z^n}{n!} \int_{[0,s]^n} \det_{n\times n} S(x_j - x_k)\, dx_1 dx_2 \ldots dx_n,$$

we have

$$\begin{aligned}
\mu_m(s) &= \frac{d^2}{ds^2} \frac{d^{m-1}}{dz^{m-1}} \left(\frac{F(z) - 1 - z \int_0^s \det_{1\times 1} S\, d\theta}{z^2} \right)\Bigg|_{z=-1} \\
&= \frac{d^2}{ds^2} \frac{d^{m-1}}{dz^{m-1}} \frac{F(z)}{z^2}\bigg|_{z=-1}.
\end{aligned} \tag{9.4}$$

In the next few sections we will work toward relating the right side of this formula to another simple function.

9.2   Gram's identity

An identity of Gram is helpful for our further considerations.

Lemma 5 (Gram's Identity). For an interval J and integrable functions ϕ_j and ψ_j,

$$\det_{N\times N}\left(I + z\int_J \phi_j(x)\psi_k(x)\,dx\right) = \sum_{n=0}^{N}\frac{z^n}{n!}\int_{J^n}\left(\det_{n\times n}\sum_{h=1}^{N}\phi_h(x_j)\psi_h(x_k)\right)\,dx_1\ldots dx_n.$$

$$(9.5)$$

Proof. The left-hand-side of (9.5) is

$$\det_{N\times N}\left(I + z\int_J \phi_j(x)\psi_k(x)\,dx\right)$$

$$= \sum_{\sigma\in\pi_N}\operatorname{sgn}(\sigma)\prod_{j=1}^{N}\left(\delta_{j,\sigma j} + z\int_J \phi_j(x)\psi_{\sigma j}(x)\,dx\right)$$

$$= \sum_{\sigma\in\pi_N}\operatorname{sgn}(\sigma)\sum_{A\subset N}\prod_{j\notin A}\delta_{j,\sigma j}\prod_{j\in A} z\int_J \phi_j(x)\psi_{\sigma j}(x)\,dx$$

$$= \sum_{A\subset N}z^{|A|}\sum_{\sigma\in\pi_A}\operatorname{sgn}(\sigma)\prod_{j\in A}\int_J \phi_j(x)\psi_{\sigma j}(x)\,dx$$

$$= \sum_{n=0}^{N}z^n\sum_{\substack{A\subset N\\|A|=n}}\det_A\int_J \phi_j(x)\psi_k(x)\,dx$$

and the right-hand-side is

$$= \sum_{n=0}^{N}\frac{z^n}{n!}\int_{J^n}\sum_{\sigma\in\pi_n}\operatorname{sgn}(\sigma)\prod_{j=1}^{n}\sum_{h=1}^{N}\phi_h(x_j)\psi_h(x_{\sigma j})\,dx_1\ldots dx_n$$

$$= \sum_{n=0}^{N}\frac{z^n}{n!}\int_{J^n}\sum_{\sigma\in\pi_n}\operatorname{sgn}(\sigma)\sum_{\lambda:[1,n]\to[1,N]}\prod_{j=1}^{n}\phi_{\lambda j}(x_j)\psi_{\lambda j}(x_{\sigma j})\,dx_1\ldots dx_n$$

$$= \sum_{n=0}^{N}\frac{z^n}{n!}\sum_{\sigma\in\pi_n}\operatorname{sgn}(\sigma)\sum_{\lambda:[1,n]\to[1,N]}\prod_{j=1}^{n}\int_J \phi_{\lambda j}(x)\psi_{\lambda\sigma^{-1}j}(x)\,dx$$

$$= \sum_{n=0}^{N}\frac{z^n}{n!}\sum_{\lambda:[1,n]\to[1,N]}\sum_{\sigma\in\pi_n}\operatorname{sgn}(\sigma)\prod_{j=1}^{n}\int_J \phi_{\lambda j}(x)\psi_{\lambda\sigma^{-1}j}(x)\,dx$$

$$= \sum_{n=0}^{N}\frac{z^n}{n!}\sum_{\lambda:[1,n]\to[1,N]}\det\left(\int_J \phi_{\lambda j}(x)\psi_{\lambda k}(x)\,dx\right).$$

If λ is not one-to-one, then the inner determinant is 0. If λ is one-to-one, call the image A. Each such set A appears $n!$ times and we get the left-hand-side.

Remark. This proof is reminiscent of the proof that that the determinant of a product is the product of the determinants. Thus,

$$
\det_{N\times N}(AB) = \det(a_{jh})(b_{hk}) = \det\left(\sum_{h=1}^{N} a_{jh}b_{hk}\right)
$$

$$
= \sum_{\sigma\in\pi_N} \operatorname{sgn}(\sigma)\prod_{j=1}^{N}\sum_{h=1}^{N} a_{jh}b_{h,\sigma j}
$$

$$
= \sum_{\sigma\in\pi_N} \operatorname{sgn}(\sigma)\sum_{\lambda:[1,N]\to[1,N]}\prod_{j=1}^{N} a_{j,\lambda j}b_{\lambda j,\sigma j}
$$

where the sum over λ is over all of the N^N functions from $[1, N]$ to itself. Now, we claim that for each fixed λ the sum

$$
\sum_{\sigma\in\pi_N}\operatorname{sgn}(\sigma)\prod_{j=1}^{N} a_{j,\lambda j}b_{\lambda j,\sigma j}
$$

is 0 unless λ is actually a permutation. For suppose that $\lambda u = \lambda v$ for some $u \neq v \in [1, N]$. Then, for each σ let σ' be the permutation defined by $\sigma'j = \sigma j$ if j is not equal to u or v, whereas

$$
\sigma'j = \begin{cases} \sigma u \text{ if } j = v \\ \sigma v \text{ if } j = u \end{cases}.
$$

In this way π_N splits up into pairs (σ, σ') of permutations. Note that $\operatorname{sgn}(\sigma) = -\operatorname{sgn}(\sigma')$. Then

$$
\prod_{j=1}^{N} a_{j,\lambda j}b_{\lambda j,\sigma j} = a_{u,\lambda u}b_{\lambda u,\sigma u}a_{v,\lambda v}b_{\lambda v,\sigma v}\prod_{j\neq u,v} a_{j,\lambda j}b_{\lambda j,\sigma j}
$$

$$
= a_{u,\lambda u}b_{\lambda v,\sigma'v}a_{v,\lambda v}b_{\lambda u,\sigma'u}\prod_{j\neq u,v} a_{j,\lambda j}b_{\lambda j,\sigma'j}
$$

Thus, the contribution from σ cancels that from σ' in the case that $\lambda u = \lambda v$. Therefore,

$$
\det(AB) = \sum_{\sigma,\tau\in\pi_N}\operatorname{sgn}(\sigma)\prod_{j=1}^{N} a_{j,\tau j}b_{\tau j,\sigma j}
$$

$$
= \sum_{\sigma,\tau\in\pi_N}\operatorname{sgn}(\sigma)\prod_{j=1}^{N} a_{j,\tau j}\prod_{k=1}^{N} b_{\tau k,\sigma k}
$$

$$
\overset{=}{\scriptstyle\sigma\to\sigma\tau} \sum_{\sigma,\tau\in\pi_N}\operatorname{sgn}(\sigma\tau)\prod_{j=1}^{N} a_{j,\tau j}\prod_{k=1}^{N} b_{\tau k,\sigma\tau k}
$$

$$
\overset{=}{\scriptstyle k\to\tau^{-1}k} \sum_{\sigma,\tau\in\pi_N}\operatorname{sgn}(\sigma)\operatorname{sgn}(\tau)\prod_{j=1}^{N} a_{j,\tau j}\prod_{k=1}^{N} b_{k,\sigma k}
$$

$$
= \det A \det B.
$$

9.3 Intervals with precisely n eigenvalues

Let $E_{G(N)}(n, J)$ be the measure of the set of matrices $A \in G(N)$ which have precisely n eigenvalues in the interval J. Here $G(N)$ can be $U(N)$, $SO(2N)$, $SO(2N+1)$, or $USp(2N)$; we denote the Haar measure by dX. Then we have a series of identities related to $E_{G(N)}(n, J)$ which will provide a basis for obtaining tractable expressions for our functions μ_m, and later for ν_j (to be introduced in the near future). Let χ_J be the characteristic function of the interval J. First of all,

$$\sum_{n=0}^{N} (1+z)^n E_{G(N)}(n, J) = \int_{G(N)} \prod_{j=1}^{N} (1 + z\chi_J(\theta_j)) dX \qquad (9.6)$$

since for any $X \in G(N)$ which has precisely n eigenvalues in J, the integrand is $(1+z)^n$. Expanding out the product on the right side gives

$$\int_{G(N)} \prod_{j=1}^{N} (1 + z\chi_J(\theta_j)) dX = \sum_{n=0}^{N} z^n \binom{N}{n} \int_{G(N)} \prod_{j=1}^{n} \chi_J(\theta_j) dX. \quad (9.7)$$

Next by Gaudin's Lemma, (6.4)

$$\binom{N}{n} \int_{G(N)} \prod_{j=1}^{n} \chi_J(\theta_j) dX = \frac{1}{n!} \int_{J^n} \det_{n \times n} K_{G(N)}(\theta_j, \theta_k) \, d\theta_1 \ldots d\theta_n$$

where $K_{G(N)}(x, y)$ is the appropriate kernel for the group $G(N)$. Thus,

$$\sum_{n=0}^{N} z^n \binom{N}{n} \int_{G(N)} \prod_{j=1}^{n} \chi_J(\theta_j) dX = \sum_{n=0}^{N} \frac{z^n}{n!} \int_{J^n} \det_{n \times n} K_{G(N)}(\theta_j, \theta_k) \, d\theta_1 \ldots d\theta_n.$$
$$(9.8)$$

Now, for each $G(N)$ we can express

$$K_{G(N)}(x, y) = \sum_{h=1}^{N} \phi_{h,G}(x) \psi_{h,G}(y) \qquad (9.9)$$

for appropriate ϕ and ψ. Therefore, by Gram's identity

$$\sum_{n=0}^{N} \frac{z^n}{n!} \int_{J^n} \det_{n \times n} K_{G(N)}(\theta_j, \theta_k) \, d\theta_1 \ldots d\theta_n \qquad (9.10)$$

$$= \det_{N \times N} \left(I + z \int_J \phi_{j,G}(\theta) \psi_{k,G}(\theta) \, d\theta \right).$$

Let $M_{J,G(N)}$ denote the $N \times N$ matrix with entries

$$m_{j,k} = \int_J \phi_{j,G}(\theta) \psi_{k,G}(\theta) \, d\theta. \qquad (9.11)$$

Then

$$\det_{N \times N} \left(I + z \int_J \phi_{j,G}(\theta) \psi_{k,G}(\theta) \, d\theta \right) = \prod_{j=1}^{N} \left(1 + z\lambda_{j,G(N)}(J) \right) \tag{9.12}$$

where the $\lambda_{j,G(N)}(J)$ are the eigenvalues of $M_{J,G(N)}$.

We claim that if the kernel is symmetric (i.e. $K_{G(N)}(x,y) = K_{G(N)}(y,x)$), then the eigenvalues of $M_{J,G(N)}$ are also the eigenvalues of the integral operator $K_{J,G(N)}$ defined by

$$(K_{J,G(N)}f)(\theta) = \int_J K_{G(N)}(\theta,\mu)f(\mu) \, d\mu \tag{9.13}$$

acting on the N-dimensional space generated by $\{\psi_j(x) : 1 \le j \le N\}$.

Proof. Suppose that λ is an eigenvalue of $M_{J,G(N)}$ corresponding to an eigenvector $\vec{v} = (b_1, \dots, b_N)'$ where the prime indicates transpose. Then, for each j,

$$\lambda b_j = \sum_{k=1}^{N} m_{jk}b_k = \sum_{k=1}^{N} b_k \int_J \phi_j(\theta)\psi_k(\theta) \, d\theta.$$

Multiplying both sides by $\psi_j(\mu)$ and summing over j, we obtain

$$\lambda \sum_{j=1}^{N} b_j \psi_j(\mu) = \int_J \left(\sum_{j=1}^{N} \phi_j(\theta)\psi_j(\mu) \right) \left(\sum_{k=1}^{N} b_k\psi_k(\theta) \right) d\theta$$

$$= \int_J K_{G(N)}(\theta,\mu) \left(\sum_{k=0}^{N} b_k\psi_k(\theta) \right) d\theta$$

$$= \int_J K_{G(N)}(\mu,\theta) \left(\sum_{k=1}^{N} b_k\psi_k(\theta) \right) d\theta = K_{J,G(N)} \sum_{k=1}^{N} b_k\psi_k(\mu)$$

so that λ is an eigenvalue of $K_{J,G(N)}$ corresponding to the eigenfunction $f(\mu) = \sum_{k=1}^{N} b_k\psi_k(\mu)$.

Recapitulating, we have found that

$$\sum_{n=0}^{N} (1+z)^n E_{G(N)}(n,J) = \sum_{n=0}^{N} \frac{z^n}{n!} \int_{J^n} \det_{n \times n} K_{G(N)}(\theta_j,\theta_k) \, d\theta_1 \dots d\theta_n$$

$$= \prod_{j=1}^{N} \left(1 + z\lambda_{j,G(N)}(J) \right) \tag{9.14}$$

where the $\lambda_{j,G(N)}(J)$ are the eigenvalues of the integral operator $K_{J,G(N)}$ defined by

$$(K_{J,G(N)}f)(\theta) = \int_J K_{G(N)}(\theta,\mu)f(\mu) \, d\mu.$$

It can be shown that this equation scales appropriately for each G so that the large N limit can be taken. This results in (with an obvious notation E_G)

$$\sum_{n=0}^{\infty}(1+z)^n E_G(n,J) \;=\; \sum_{n=0}^{\infty}\frac{z^n}{n!}\int_{J^n}\det_{n\times n}K_G(\theta_j,\theta_k)\,d\theta_1\dots d\theta_n \quad (9.15)$$

$$= \prod_{j=1}^{\infty}\left(1+z\lambda_{j,G}(J)\right)$$

where the $\lambda_{j,G}(J)$ are the eigenvalues of the integral operator $K_{J,G}$ defined by

$$(K_{J,G}f)(\theta) \;=\; \int_J K_G(\theta,\mu)f(\mu)\,d\mu.$$

The function $F(z)$ of (9.4) is equal to each of the above with G=U. Thus, we find for $\mu_m(s)$ that

$$\mu_m(s)=\frac{d^2}{ds^2}\frac{d^{m-1}}{dz^{m-1}}\left(z^{-2}\prod_{j=1}^{\infty}(1+z\lambda_{j,U}([0,s]))\right)\Bigg|_{z=-1}. \quad (9.16)$$

9.4 jth lowest eigenvalue

Let

$$\nu_{G(N)}(j,s)$$

be the density function for the jth lowest eigenvalue so that

$$\text{meas}\{A\in G(N): \text{ the } j\text{th eigenvalue } \theta_j \text{ is smaller than } s\} \quad (9.17)$$

$$=\int_0^s \nu_{G(N)}(j,x)\,dx.$$

Then the set of $A\in G(N)$ with $\theta_j>s$ is the disjoint union over $n=0,1,\dots,j-1$ of the set of A with exactly n eigenangles in $[0,s]$. Thus,

$$\int_s^{\infty}\nu_{G(N)}(j,x)dx=\sum_{n=0}^{j-1}E_{G(N)}(n,[0,s]).$$

Therefore, by (9.14), we have

$$\nu_{G(N)}(j,s)=-\frac{d}{ds}\sum_{n=0}^{j-1}\frac{d^n}{dz^n}\prod_{n=1}^{N}(1+z\lambda_{G(N),n}([0,s]))\Big|_{z=-1}. \quad (9.18)$$

In the large N limit, this becomes

$$\nu_G(j,s)=-\frac{d}{ds}\sum_{n=0}^{j-1}\frac{d^n}{dz^n}\prod_{n=1}^{\infty}(1+z\lambda_{G,n}([0,s]))\Big|_{z=-1}. \quad (9.19)$$

For example,

$$\nu_G(1,s)=-\frac{d}{ds}\prod_{n=1}^{\infty}(1-\lambda_{G,n}([0,s])). \quad (9.20)$$

9.5 Relations between the eigenvalues

In this section, we develop a relationship between the eigenvalues λ_U and the eigenvalues λ_O and λ_S. (9.11). In the case that $J = [-s, s]$, note that if $\psi(\theta)$ is an eigenfunction of $M_{[-s,s],U(N)}$ with eigenvalue λ then $\psi(-\theta)$ is also an eigenfunction with eigenvalue λ, since

$$\lambda\psi(\theta) = \int_{-s}^{s} S_N(\theta - \mu)\psi(\mu)\, d\mu$$

implies that

$$
\begin{aligned}
\lambda\psi(-\theta) &= \int_{-s}^{s} S_N(-\theta - \mu)\psi(\mu)\, d\mu \\
&= \int_{-s}^{s} S_N(\theta + \mu)\psi(\mu)\, d\mu \\
&= \int_{-s}^{s} S_N(\theta - \mu)\psi(-\mu)\, d\mu.
\end{aligned}
$$

Therefore, if $\psi(\theta) + \psi(-\theta) \neq 0$, then it is also an eigenfunction with eigenvalue λ. A similar comment holds for $\psi(\theta) - \psi(-\theta)$. Consequently, each eigenfunction can be taken to be even or odd. The even eigenfunctions are also eigenfunctions of the integral equation with kernel

$$\frac{S_N(\mu - \theta) + S_N(\mu + \theta)}{2}$$

and the odd eigenfunctions are also eigenfunctions of the integral equation with kernel

$$\frac{S_N(\mu - \theta) - S_N(\mu + \theta)}{2}.$$

In general, if a matrix b is a "checkerboard" matrix, then the determinant of b factors. Specifically, if $b_{j,k} = 0$ whenever $i + j$ is odd, then

$$\det_{N \times N}(b_{j,k}) = \det_{[(N+1)/2]}(b_{2i-1,2j-1}) \det_{[N/2] \times [N/2]}(b_{2i,2j})$$

where $[x]$ is the greatest integer less than or equal to x.

We have such a factorization for $\det(I - M_{[-s,s],U(N)})$. Using the fact that

$$\sum_{h}(\delta_{jh} - \cos(j\theta)\cos(h\theta))(\delta_{hk} - \sin(h\theta)\sin(k\theta)) = \delta_{jk} - \cos(k - j)\theta$$

we deduce from (9.11) (see also Mehta (10.2.6)) that

$$\det(I - M_{[-s,s],U(N)}) = \det(I - M_{[-s,s],SO(2N)})\det(I - M_{[-s,s],USp(2N)}).$$

This gives a factorization

$$\prod_{n=1}^{2N}(1 - \lambda_{n,U(2N)}(s)) = \prod_{n=1}^{N}(1 - \lambda_{n,SO(2N)}(s))(1 - \lambda_{n,USp(2N)}(s)) \qquad (9.21)$$

into even and odd eigenvalues. In particular, in the limit we have

$$\prod_{n=1}^{\infty}(1 - \lambda_{n,U}(s)) = \prod_{n=1}^{\infty}(1 - \lambda_{n,Sp}(s))(1 - \lambda_{n,SO,even}(s)). \qquad (9.22)$$

Alternatively, we have

$$\prod_{n=1}^{\infty}(1 - \lambda_{n,Sp}(s)) = \prod_{n=1}^{\infty}(1 - \lambda_{2n,U}(s)) \qquad (9.23)$$

and

$$\prod_{n=1}^{\infty}(1 - \lambda_{n,SO,even}(s)) = \prod_{n=1}^{\infty}(1 - \lambda_{2n-1,U}(s)) \qquad (9.24)$$

provided that the $\lambda_{n,U}(s)$ are indexed so that an even index n corresponds to an even eigenfunction and an odd index n is for an odd eigenfunction of the integral operator (9.13) with kernel $K_U(x,y) = S(x-y)$. These formulae can be used to give expressions for $\nu_G(j,s)$ in terms of the eigenvalues $\lambda_{n,U}(s)$.

10   Conclusion

These notes have introduced four of the basic eigenvalue statistics for the groups of particular interest to number theorists and have shown the preliminary steps needed to make these statistics somewhat tractable. There are many directions to go after this basic introduction. Topics such as Painlevé equations, Toeplitz operators, the Szego limit theorems, and averages of characteristic polynomials are among the many developments that are essential for a more full introduction to random matrix theory. Many of these are covered elsewhere in this volume.

References

[C]   Conrey, J. Brian: L-functions and random matrices. In: Mathematics unlimited—2001 and beyond. Springer, Berlin Heidelberg New York (2001) (arXiv math.NT/0005300).

[Dei]   Deift, P. A.: Orthogonal polynomials and random matrices: a
             Riemann-Hilbert approach. Courant Lecture Notes in Mathemat-
             ics, 3. New York University, Courant Institute of Mathematical
             Sciences, New York. American Mathematical Society, Providence,
             RI (1999).

[H]   Hejhal, D.A.: On the triple correlation of zeros of the zeta function.
             Inter. Math. Res. Notices, 7 (1994) 293–302.

[ILS]   Iwaniec, Henryk ; Luo, Wenzhi ; Sarnak, Peter. Low lying zeros of
             families of *L*-functions. Inst. Hautes Études Sci. Publ. Math. No.
             91, (2000), 55–131 (2001).

[KaSa]   Katz, Nicholas M. and Sarnak, Peter: Random matrices, Frobe-
             nius eigenvalues, and monodromy. American Mathematical Soci-
             ety Colloquium Publications, 45. American Mathematical Society,
             Providence, RI (1999).

[K]   Keating, J. P. Random matrices and the Riemann zeta-function.
             Highlights of mathematical physics (London, 2000), 153–163,
             Amer. Math. Soc., Providence, RI, 2002.

[KS]   Keating, J. P.; Snaith, N. C. Random matrices and *L*-functions.
             Random matrix theory. J. Phys. A 36 (2003), no. 12, 2859–2881.

[Meh]   Mehta, Madan Lal: Random matrices. Second edition. Academic
             Press, Inc., Boston, MA (1991).

[M]   Montgomery, H. L. The pair correlation of zeros of the zeta func-
             tion. Analytic number theory (Proc. Sympos. Pure Math., Vol.
             XXIV, St. Louis Univ., St. Louis, Mo., 1972), pp. 181–193. Amer.
             Math. Soc., Providence, R.I., 1973.

[O1]   Odlyzko, A. M. On the distribution of spacings between zeros of
             the zeta function. Math. Comp. 48 (1987), no. 177, 273–308.

[O2]   Odlyzko, A. M. The 10^{22}-nd zero of the Riemann zeta function. Dy-
             namical, spectral, and arithmetic zeta functions (San Antonio, TX,
             1999), 139–144, Contemp. Math., 290, Amer. Math. Soc., Provi-
             dence, RI, 2001.

[RS]   Rudnick, Zeév ; Sarnak, Peter. Zeros of principal *L*-functions and
             random matrix theory. A celebration of John F. Nash, Jr. Duke
             Math. J. 81 (1996), no. 2, 269–322.

[TW]   Tracy, Craig A. and Widom, Harold. Introduction to Random Ma-
             trices. ITD 92/93–10. Springer Lecture Notes in Physics 424 (1993)
             103-130. arXiv hep-th/9210073.

[W]   Weyl, Hermann. The classical groups. Their invariants and representations. Princeton University Press, Princeton, Eighth Printing, 1973.

American Institute of Mathematics
360 Portage Ave.
Palo Alto CA 94306 USA

conrey@aimath.org

Compound Nucleus Resonances, Random Matrices, Quantum Chaos

Oriol Bohigas

> *It is important for him who wants to discover*
> *not to confine himself to one chapter of science,*
> *but to keep in touch with various others.*
> Jacques Hadamard ([Had45], quoted in [Dy72])

1 Introduction

The field 'Random Matrix Theories and Number Theory', the core of a Newton Institute Workshop, is a meeting area for some mathematicians and theoretical physicists. The purpose of this contribution, based on a talk given at the Institute and which was followed by another also of historical character by Michael Berry, is to give, particularly to the mathematicians, a flavour of the initial motivations which lead some physicists to introduce and study ensembles of random matrices. I was asked by the organizers to emphasize developments, to which I partly contributed, which lead to the merging of two at first sight seemingly disconnected fields, namely random matrix theories (RMT) and quantum chaos (QC), the study of quantum systems whose classical analogue is chaotic. Mention of directly related number theoretic problems will also be made, as well as, towards the end, of some more recent developments. To make this contribution more coherent, in section 5 I have touched aspects which were developed by Berry in his talk.

This general introductory overview is admittedly biased and incomplete, partly because it reflects my personal scientific itinerary, partly because it excludes important developments not directly related to spectral fluctuations of (quasi) bound systems, the bottom line adopted here. In this respect, the absence of the Novosibirsk-Milano-Atlanta activity on quantum suppression of chaos and localization deserves a special mention. Since the mid 70's Boris Chirikov, Joseph Ford, Giulio Casati and some of their associates played an essential role in the birth and development of what QC would become. Remarkably, this soviet-american-italian intense cooperation was started vigorously by these authors at a time when the cold war was not a nightmare of the past, but still a rather efficiently operating reality. Important questions of current interest concerning wave functions, scattering and more generally problems dealing with open systems, though closely related to the ones discussed here, will also be ignored.

There are books and monographs which cover the different aspects of the material touched here. Concerning RMT and their applications let me mention [Po65], which contains an introduction as well as reprints of the important papers in RMT prior to 1963, the classical book of Mehta [Me90], the forthcoming book [Fo], the monographs [BFFMP81] and [GMW98]. Concerning QC (as well as in most cases connections with RMT) one can consult the books [Oz88], [Gu90], [Ha91], [CC95], [St99], [Cv], as well as the Proceedings of the Les Houches Lectures [GVZ91]. A condensed matter physics perspective of several of the problems discussed here can be found in various contributions to [AMPZ95] and in [Ef97], [Ri00]. With few exceptions and because of the non-technical character of the present article, no effort has been made in quoting original material (nonetheless, references from paragraphs or figures literally reproduced are given). It can be found in the books and monographs referred to at the end. Occasionally, subjective statements are made and metaphoric language used, departing from the sober and auster prose in common use in academic journals.

Nobody better than Freeman Dyson can be chosen to exemplify the ideas and problems, the general landscape described here. This contribution is dedicated to him at least for three reasons: (i) he has played a prominent role in the setting and development of RMT with a series of classical and seminal papers [Po65]; (ii) he has provided inspiration, making connections and remarks which have substantially contributed to enlarge this general field (fortunately he has never completely recovered from the 'random matrix disease' he contracted in the early 60's); (iii) on a personal basis, some of his comments or remarks have been invaluable for me.

Dyson is sometimes qualified as a 'theoretical physicist', or as a 'mathematician', or as an 'applied mathematician'. In the physics community he is known to everyone, not because of RMT but because of Quantum Electrodynamics (QED). QED is the most precise physical theory ever invented by the human mind. Presently some of its predictions are discussed, in connection with experiment, at the level of 10^{-11} precision! And Dyson has been one of the four men (with Feynman, Schwinger and Tomonaga) who constructed QED in the late 40's and beginning of the 50's [Sc94].

Some number theorists know Dyson's name as well. Is he a physicist, a mathematician, is he both? In the foreword of [Hav03] he writes: *'I fell in love with mathematics at school and have been a professional mathematician ever since'.* However in [Dy96]:*'I am half mathematician and half physicist'.* And in [Dy72]: *'I happen to be a physicist who started life as a mathematician'.* In [Dy96] he gives a more detailed account:

'... Davenport was the first mathematician I met who had a group of young research students around him and regularly supplied them with problems... I told him my ideas about the Siegel conjecture and he encouraged me to continue my efforts. At that time I had tentatively decided to switch my activities from mathematics to physics... I thought that it would be more exciting to solve one of the basic mysteries of nature than to continue proving theorems that were of interest only to a small coterie of number theorists. But Davenport's friendliness tempted me to stay with mathematics. I decided to launch an all-out attack on the Siegel conjecture and to let the result determine my future. If I succeeded in proving it, I would be a mathematician. If I failed, I would be a physicist. After three months of intensive work, I admitted failure. I would after all be a physicist...

... It was easy for me to switch from mathematics to physics, because both number theory and physics are branches of applied mathematics. I define a pure mathematician to be somebody who creates mathematical ideas, and I define an applied mathematician to be somebody who uses existing mathematical ideas to solve problems. According to this definition, I was always an applied mathematician, whether I was solving problems in number theory or in physics... The main difference between number theory and physics is that in number theory the experimental data are more accurately known. In recent years the increasing use of computers has made number theory more than ever an experimental science...'

In Table 1, partly following Dyson's scheme, some of the landmarks of the present 'guided tour' extending till the mid 80's are given. The itinerary will not necessarily be straight but will comprise zigzags and chronological back and forths. It starts here, in Cambridge, at the Cavendish, with the discovery of the neutron by Chadwick, an event often considered as the beginning of nuclear physics. In Figure 1 is displayed the apparatus, of rustic appearence, which led him to this far reaching discovery.

	EXPERIMENT		THEORY		
Year	Physics	Mathematics	Physics	Mathematics	Year
1930					1930
	CHADWICK the neutron				
-			N. BOHR the compound nucleus		-
-	HAHN & STRASSMANN				-
1940	nuclear fission				1940
-					-
-	development of nuclear reactor technology				-
-					-
1950	TURING first use of a programmed electronic computer for Riemann's zeros	↓ V KAM theorem classical chaos	WIGNER random matrices (RMT)		1950
-					-
-					-
1960			development of RMT GAUDIN, MEHTA, DYSON		1960
-	RAINWATER systematic study of compound nucleus resonances with neutrons		GORKOV & ELIASHBERG disordered metallic grains		-
-		↓ V towards quantum chaos (QC)	GUTZWILLER periodic orbit theory (POT)		-
1970			PERCIVAL regular and irregular states	MONTGOMERY pair correlations of Riemann's zeros	1970
-					-
-			BERRY		-
1980	ODLYZKO Extended search of Riemann's zeros	BGS -random matrix conjecture merging of RMT and QC -universalities	EFETOV supersymmetric non-linear sigma-model understanding connection of RMT and QC with POT, universal versus system-specific properties	connections between QC and Riemann's zeros	1980
-					-
-					-
1990					1990

Table 1: Scheme of the 'guided tour' provided in this article.

Figure 1: Chadwick's neutron chamber (The Cavendish Laboratory, University of Cambridge).

The tour continues with the advent of the compound nucleus hypothesis by Niels Bohr [Bo36]. With charged particles a nuclear reaction is partly prevented by mutual electric repulsion. Such a mechanism does not operate for the neutron (a neutral particle) and in reactions with neutrons much larger probabilities (cross sections) are found. By studying neutron capture, Bohr was led to the hypothesis that a long lived compound system was formed, whose disintegration had no immediate connection with the first stage of the collision. Long lived means here 10^{-16}s, much longer than a typical nuclear time interval (10^{-21}s) derived from the known dimension of the nucleus. The compound nucleus resonances (long-lived quasi-bound states) can be seen, for instance, in the transmission measurements of neutrons through a nuclear target containing, say, A nucleons. Their energy appears as thin well located peaks (resonances) of the cross-section as a function of the neutron incident energy. The sequence of resonance peaks, which starts at the neutron threshold (typically 6-7 MeV above the ground state for a heavy nucleus corresponding to, say, the 10^6th level of the compound nucleus containing $(A+1)$ nucleons), depends on the target nucleus chosen. An example of such a cross section is displayed on Figure 2.

The next stop will be to signal the discovery of nuclear fission by Hahn and Strassmann. This occurred at the border of the period when the world, as well as the nucleus, would split, leading to some of the most ferocious events of contemporary history: among them, the manufacturing and dropping of atomic bombs. Both the discovery of fission and the perspective of atomic (nuclear) energy would have enormous consequences. D. Wilkinson has given

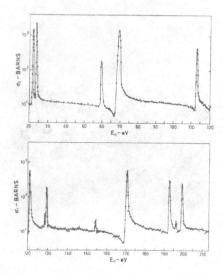

Figure 2: Total cross section for the reaction n + ^{232}Th as a function of the neutron energy (from the compilation 'Neutron Cross Sections', 1964)

succinctly an awe-striking view [Wil68]: *'Fission is a process of deadly fasci-nation: had nature chosen her constants just a little differently, we should have been deprived of its potential for social good and spared its power for social evil. Despite the former and despite the indeniable fact that the latter is responsible for nuclear and particle physics being decades in advance of what would other-wise have been their time, I know what my own choice of the constants would have been.'*

2   Random Matrix Theories (RMT): their origin and developments

After the end of the war, military as well as civilian purposes lead to an impressive effort in connection with nuclear fission. It was important, for instance for nuclear reactor purposes, to understand the properties of the compound nucleus resonances (see Figure 3). It is in this general context that RMT were introduced by Eugene Wigner. On Figures 4 and 5 are reproduced parts of Wigner's contributions to two conferences on neutron spectroscopy held in the mid 50's. On the former he gives a guess of the nearest neighbor spacing distribution $p(s)$,

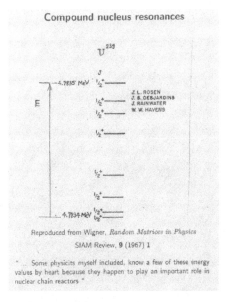

Figure 3: From Wigner's review in [Wi67].

$$p(s) = \frac{\pi}{2} s \exp\left(-\frac{\pi}{2} s^2\right),\qquad(2.1)$$

(what we call Wigner's surmise) which happens to be close to the asymptotic (large $N \times N$ matrix) result. He gives a (primitive) comparison with data. He writes (for what we call $\beta = 1$, real symmetric matrices, Gaussian Orthogonal Ensemble-GOE, Wishart distribution) the joint distribution of eigenvalues

$$P(E_1,\ldots,E_N) = C_{N\beta} \exp\left(-\sum_i E_i^2\right) \prod_{i<j} |E_i - E_j|^\beta \qquad(2.2)$$

($\beta = 2$ (4) for hermitean (quaternion real) matrices   - Gaussian Unitary (Symplectic) Ensemble; GUE (GSE)). In the latter, after (surprisingly to me) stating that the spacing of levels is not terribly important, he again refers to (2.1), and to his work with Von Neumann [VW29] on variation of eigenvalues and eigenfunctions in quantum mechanics under continuous changes of one or several parameters, as probably forgotten (which it is not, at least presently). He states that something like (2.1) is probably a universal function (independent on the details of the model) and finally he gives the eigenvalue distribution

Perhaps I am now too courageous when I try to guess the distribution of the distances between successive levels. I should re-emphasize that levels that have different J values are not connected at all with each other. They are entirely independent. So far experimental data are available only on even-even elements, and Dr. Hughes has projected a curve showing the probability of a spacing as a function of the spacing itself. The data with which I am familiar (Slide 7) come from Th^{232} and U^{238}. Many more data will become available as we shall learn more and more to distinguish levels, that is, to ascertain their J values.

Theoretically the situation is quite simple if one attacks it in a simple-minded fashion. The question is simply what are the distances of the characteristic values of a symmetric matrix with random coefficients? We know that the chance that two such energy levels coincide is infinitely unlikely. We can consider a two-dimensional matrix.

$$\begin{vmatrix} a_{11} & a_{12} \\ a_{21} & a_{22} \end{vmatrix}$$

in which case the distance between the two levels is $\sqrt{(a_{11} - a_{22})^2 + 4 a_{12}^2}$. This distance can be zero only if $a_{11} = a_{22}$ and $a_{12} = 0$. The difference between the two energy values is the distance of a point from the origin the two coordinates of which are $a_{11} - a_{22}$ and a_{12}. The probability that this distance be S is, for small values of S, always proportional to S itself, because the volume element of the plane in polar coordinates contains the radius as a factor.

Now Dr. Hughes, and several speakers before him, mentioned that one would expect an exponential distribution for the probability of a certain spacing as a function of the spacing. The reasoning which leads to this expectation is as follows. If we have a level at a certain point, then the probability of finding another level a distance S from it is independent of S and given by ρdS, where ρ is the mean level density. Therefore, the probability for finding the next adjacent level at distance S, in dS, is $e^{-\rho S} \rho dS$. I think that this is the basis for expecting an exponential distribution of the level spacing. However, the argument is erroneous, because the probability that there shall be a level at a distance S from a given level is not independent of this distance; for small values of S it is proportional to S. If this same law is assumed for large S also, the probability of finding the next level at a distance S becomes proportional to $S dS$. Hence this simplest assumption will give the probability $\frac{1}{2}\pi \rho^2 e^{-(1/4)\pi \rho^2 S^2} S dS$ for a spacing between S and $S + dS$.

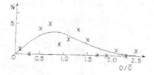

Slide 7. Distribution of Level Spacings.

A. M. LANE: I should like to ask if you think the distribution of the level spacing is a correlated distribution or a noncorrelated one. I think I am using the right term here. The correlation that I have in mind is the following: If you are given two levels, 1 and 2, is the spacing between 2 and 3 affected by the spacing between 1 and 2, or are these both chosen out of the set of samples corresponding to the distribution that you get at random? Are they both chosen at random, or is there a correlation between adjacent spacings?

E. P. WIGNER: There is one in the only case in which the mathematicians have calculated a formula. Let me give that formula. The mathematicians have considered the case of a real symmetric N-dimensional matrix in which every matrix element has a Gaussian distribution. Then the probability for the characteristic values $\lambda_1, \lambda_2, \dots, \lambda_N$ is:

$$P(\lambda_1, \lambda_2, \dots, \lambda_N) = e^{-(\lambda_1^2 + \lambda_2^2 + \dots + \lambda_N^2)} |(\lambda_2 - \lambda_1)(\lambda_3 - \lambda_1) \dots (\lambda_N - \lambda_{N-1})|$$

The product contains all $N(N - 1)/2$ differences between the λ. To calculate the probability of a spacing of two adjacent levels requires an integration over all the other levels, with the space between the two considered levels excluded from the region of integration, and as far as I know this has not been done. But it is pretty evident, I believe, that there are further correlations in this case at least, and I would think that is general.

E. GUTH: May I ask what the reference for this is?

E. P. WIGNER: This is called the Wishart distribution.

Figure 4: Part of Wigner's contribution to Gatlinburg Conference [Wi56].

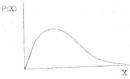

Figure 5: Part of Wigner's contribution to Columbia University Conference [Wi57].

corresponding to (2.1), namely what we call Wigner's semi-circle law:

$$\rho(E) = \begin{cases} \frac{1}{2\pi}(N - E^2)^{1/2} \text{ for } |E| < \sqrt{N} \\ \\ 0 \text{ otherwise.} \end{cases} \quad (2.3)$$

The basis of RMT can probably not be stated more succinctly and emphatically than in the quotation from Wigner

> *'...the Hamiltonian which governs the behavior of a complicated system is a random symmetric matrix with no particular properties except for its symmetric nature.'* [Wi61]

The three main ingredients specifying the Wigner-Dyson ensembles are: i) space-time symmetries, ii) isotropy in Hilbert space, iii) statistical independence of the matrix elements. i) and ii) are physical requirements, iii) is an (unessential) requirement of mathematical simplicity.

One major task that RMT theorists faced was to derive, from (2.2), the n-point correlation functions, spacing distribution etc. This has been the achievement of Gaudin, Mehta, Dyson and some others (see reprints in [Po65]) (we have heard about Gaudin's method several times in this workshop). Dyson intensively worked on it for more than one decade, starting from the early 60's. He gave important generalizations and interpretations. In particular, he introduced the circular ensembles (ensembles of unitary matrices asymptotically equivalent to the Gaussian ones), the Coulomb gas, the brownian motion ensembles. (In the physics literature the circular unitary ensemble, CUE, is the group $U(N)$ which appears in many of the number theoretical problems discussed in this workshop).

The beauty and depth of RMT were recognised and appreciated almost since their birth: *'... it is the one-dimensional theory par excellence! Not only does it have immediate usefulness and validity for real physical systems but, from the mathematical point of view, it has given rise to profound results and makes use of the deepest theorems of analysis. One might almost say that it is also a new branch of mathematics itself.'* [LM66]. Though a mature theory already in the mid 70's, RMT have known further and partly unexpected developments since. To discuss them goes beyond the aim of this contribution, but let us mention some of them. For systems depending on external parameters it has been shown that parametric correlations are universal. This extends significantly the universalities previously found and gives to the general physical picture a much broader view. From a more mathematical standpoint, Brézin and Zee, in the early 90's, found new results concerning universalities with respect to the confining potential, thereby starting new lines of research. In the early 80's, physicists from the Kyoto school, followed by Tracy, Widom, Mehta, Forrester and others, found new connections with Painlevé equations.

Two major developments have taken place more recently: the Riemann-Hilbert approach to random matrices [De98] and a general setting of measures of classical groups and spectral properties, with emphasis on applications to number theory [KS99]. Ref. [BI01] based on talks and lectures delivered at the MSRI Berkeley semester 'Random Matrix Models and Their Applications' in Spring 1999 includes contributions of physicists and mathematicians, with emphasis on recent developments.

As explained above, RMT were introduced in order to understand empirical facts. How far did the comparison with data go? Since the 60's a group lead by J. Rainwater including Havens and joined by Camarda and others, engaged, at the Nevis Laboratory of Columbia University, in a long term and systematic study of compound nucleus resonances by neutron time of flight detection. One main motivation strongly supported by Dyson was to test the RMT predictions. The task was difficult because what was ideally needed were long and pure sequences of consecutive resonances (no missing levels, no spurious levels) having the same exact quantum numbers. This patient and laborious enterprise lasted for many years. Among other things it provided for say 40 nucleides, the parameters of presumably pure and complete sequences of consecutive resonances each containing of the order of 100 levels (additional information coming from high resolution proton resonance data on medium nuclei were also obtained). These efforts did not momentarily change radically the rather disappointing conclusion reached some years before by Dyson and Mehta: *'We would be very happy if we could report that our theoretical model had been strikingly confirmed by the statistical analysis of neutron capture levels. We would be even happier if we could report that our theoretical model had been decisively contradicted... Unfortunately, our model is as yet neither proved nor disproved.'* [DM63]

Though RMT were born in a nuclear physics context, Gorkov and Eliashberg, in a seminal work well ahead of their time, applied them in describing the electromagnetic properties of weakly disordered small metallic grains. The dimension of the system is so small that the discrete character of the excitation spectrum, as well as its nature (presence or absence of level repulsion, for instance), must be taken into account. However it took more than ten years before this problem was fully understood theoretically, through Efetov's supersymmetric nonlinear sigma-model formulation of the problem [Ef97].

Let me now trace an itinerary, in several aspects remarkably parallel to the physics itinerary briefly depicted above, concerning this time some number theoretical problems related to the Riemann zeta function. I shall start with A. Turing's contribution (see Table 1). Turing was a distinguished British mathematician educated here in Cambridge, known worldwide for his concept developed in the 30's of universal machine (Turing machine). During World War II he greatly contributed in breaking the code used by the German Navy. As a consequence the North Atlantic traffic was safely controlled by the Allies during a long period of time.

Towards the end of his short, dense and exceptionally fruitful life [Ho00], Turing engaged himself in one of the first systematic efforts to devise and use software in an electronic computer for mathematical purposes, a highly unconventional activity at that time. It took place at the prototype Manchester computer (see Figure 6). In his last published mathematical paper 'Some calculations of the Riemann zeta-function', [Tu53] Turing writes: '... *The*

Figure 6: The prototype Manchester computer (Department of Computer Science, University of Manchester).

calculations were done in an optimistic hope that a zero would be found off the critical line, and the calculations were directed more towards finding such zeros than proving that none existed... If it had not been for the fact that the computer remained in serviceable condition for an unusually long period from 3 p.m. one afternoon to 8 a.m. the following morning it is probable that the calculation would never have been done at all... The general reliability of the machine was checked from time to time by repeating small sections... The interval $1414 < t < 1608$ was investigated and checked, but unfortunately at this point the machine broke down and no further work was done. Furthermore this interval was subsequently found to have been run with a wrong error value, and the most that can consequently be asserted with certainty is that the zeros lie on the critical line up to $t = 1540$, a negligible advance, Titchmarsh having investigated as far as 1468...' . A sort of uncertain heroic atmosphere emanates from this description. Despite the sad conclusion after several calamitous events, this work was one of the first announcing a new chapter in which experimental mathematics performed with computers would play an important role.

The comparison of Figures 1 and 6 provides a striking view of evolution in opposite directions. The ancestors of the modern gigantic particle physics accelerators and detectors needed for new discoveries on the constitution of matter (the CMS detector being presently assembled at CERN to operate at the Large Hadron Collider (LHC) in 2007 weighs more than 10.000 tonnes!) were relatively small primitive detector devices (Figure 1). In contrast the ancestors of the compact, ever more fast and powerful modern computers were the kind of comparatively dinosaur-looking apparatus reproduced in Figure 6.

The first contact between RMT and Riemann's zeta function has its origin in the well known encounter of Hugh Montgomery and Freeman Dyson. Montgomery has given a vivid description of the scene [Mo04]:

' *Because of possible applications to class numbers of imaginary quadratic fields, I was interested in showing that the Riemann zeta function has pairs of zeros that are close together. I worked on this in late 1971 and early 1972, in Cambridge, where I was a graduate student at the time. In March, 1972 I attended a conference on analytic number theory at Washington University in St. Louis, and spoke on my results. On my way back to Cambridge I made a stop in Princeton, in order to talk with Selberg and show him my results. I talked with Selberg in the morning, in his office. That afternoon, at tea, by chance I was talking with Chowla. S. Chowla was a number theorist at Penn State University, but he made frequent visits to the Institute for Advanced Study. Back in the 1930s, Chowla had worked with such people as Mordell, Davenport, Mahler, and Erdos. At that time I had known him personally for two or three years. Chowla noticed that Freeman Dyson was standing across the room. I knew Freeman by sight, as I had been a visitor at IAS the year before (i.e., the academic year 1970-1971). A number of times I had encountered Freeman, and we would smile and nod to each other. We had never spoken, and I doubt that he knew who I was, but I was familiar with his work in number theory dating from the 1940s, and I knew that he was an old friend of my supervisor Harold Davenport, who had died in June, 1969. On the day in question, Chowla noticed that Freeman was across the room, and suggested that he should introduce me to him. I demurred, as I saw no reason for bothering the great man. Chowla insisted, and I demurred again. This went back and forth for some time, until eventually Chowla won. We walked over to Freeman, and Chowla introduced me, explaining that I was a student of Davenport. Freeman was very nice and charming, and asked me what I was working on. I responded that it looks like the differences between zeros of the zeta function have a distribution with a density $1 - (\frac{\sin \pi r}{\pi r})^2$. Freeman immediately responded that this is the pair correlation density of random matrices. I don't remember much about the rest of the conversation, it didn't last too much longer.*

The next day, Selberg gave me a handwritten note from Freeman that read "Atle, tell Montgomery to look at p. 76 of Mehta's book on random matrix theory. Freeman.". This quote may not be 100% accurate, but it's pretty close...

In my talk in St. Louis, I did not use the term "pair correlation", because I did not know it at the time. When I prepared my paper for the conference proceedings [Mo73], I introduced terminology that I learned from Mehta's book. The mathematics didn't change, only the names and notation.

When I first found the results, I found them a little puzzling. I felt that there was a message in what I had found, but didn't know what that message was. This was resolved by my conversation with Dyson, because I learned then that it's simply a matter that the zeros look like eigenvalues.'

Montgomery's result is at the origin of much of the subsequent work concerning statistical properties of Riemann's zeros and RMT. Odlyzko, one of the main actors, puts it as follows (short answers to questions I asked him recently [Od04]): *' I had been interested in the RH ever since my college days. (My informal advisor, who supervised my senior thesis, was Tom Apostol, and my Ph. D. thesis supervisor was Harold Stark, both students of D.H. Lehmer, who had done some of the early computations of zeros, and both had lectured on the zeta function in the classes I took from them.) Furthermore, my Ph. D. thesis work on discriminants of number fields depended heavily on zeros of Dedekind zeta functions, in particular on their vertical distribution, so I had plenty of reasons to be interested in Montgomery's work.*

A major spur to my involvement in this project was the acquisition by Bell Labs of a Cray-1, which was the top supercomputer in the world at the time. As I recall, we were the first private company to get one. That was around 1979 or so.

Since the Cray-1 got its speed advantage largely from vectorization, which was a new technique at that time, it took a while for users to port their programs from other machines. Hence there was spare capacity, and so the Bell Labs Computer Center offered to give out 5-hour chunks of time for worthwhile scientific purposes. (There was internal accounting with charges for computer time, and this was for free, on a temporary basis.) I applied and got that 5 hours, and afterwards a few more. I asked for it to do the zeta computations, which I had been interested in doing ever since hearing Montgomery talk about his work at the Institute for Advance Study in Princeton a couple of years ago and verifying that the other people who had worked on zeros (Brent, te Riele, Schoenfeld) had not computed zeros, only verified they were on the critical line.'

Odlyzko has been intermittently digging deeper and deeper (climbing higher and higher on the critical line) in the Riemann ore ever since. This is because the asymptotic limit is often approached slowly when dealing with the zeta function. He has made calculations of various stretches of say 10^{10} consecutive zeros up to say the 10^{23}th zero (remember that stretches of consecutive compound nucleus resonances contain typically only of the order of hundred levels!). Once extracted, he has made the precious metal available to the users worldwide at his website [Od]. The work has been extended partly to other L-functions (see, for instance, Rubinstein's contribution to this workshop). And it has led, among other things, to what is referred to as the Montgomery-Odlyzko conjecture: Asymptotically Riemann's zeros behave locally as GUE (or CUE) eigenvalues.

3   My first two encounters with RMT: An (unanswered) question, an answer

I started my activity in theoretical physics working on the nuclear many-body problem, specifically in nuclear spectroscopy and nuclear structure problems. One of the more successful models in this field is the nuclear shell model. One chooses a finite Hilbert space generated by distributing say n valence nucleons (fermions) in a set of Ω single particle states, and one solves the Schrödinger equation in this subspace. The Hamiltonian is a (supposedly) known operator which describes the pairwise interaction among the nucleons. It is specified by the two-body matrix elements, namely the Hamiltonian matrix elements in the basis of the two-nucleon system (which has dimension $\binom{\Omega}{2}$). When dealing with more ($n > 2$) valence nucleons, the Hamiltonian matrix has a rapidly increasing dimensionality ($= \binom{\Omega}{n}$), and its matrix elements can be expressed as linear combinations of the two-body matrix elements. A shell-model calculation consists then in building and diagonalizing the Hamiltonian matrix. The model is considered successful when the low-lying part of the spectrum does reproduce with some accuracy the observed spectrum. Because of the exploding dimensionalities, this programme can be completed only for relatively light nuclei. Typical numbers corresponding to, say, silicium isotopes, are 10 valence nucleons, 50 two-body matrix elements, dimensionalities of the order of a few thousands, 'interesting' eigenvalues being at most the 100 smallest ones. One starts therefore from 50 numbers, one computes of the order of one million numbers (most of them zeros), and finally one pays attention to say 100 eigenvalues. This does not seem to be a very efficient procedure. Is it not possible to attack the problem differently? It is one of the important contributions of Bruce French to have addressed this type of question from a new and fresh point of view.

However, let me first tell something concerning French's scientific background. Like other promising theoreticians of his age, he started working in the not yet well formulated (see previous section) QED. Under the supervision of Weisskopf, he started a relativistic calculation of the at that time not yet measured hydrogen Lamb shift (the detailed description of the spectrum of the hydrogen atom, a bound proton and electron system, the simplest system one can imagine). Eventually, after complicated unconventional calculations, French produced a final result. Two other physicists, already famous at that time, Schwinger and Feynman, addressed the same question and produced another result, in some respects by different methods. Their result agreed among themselves and disagreed with the one obtained by French. It turned out at the end that the result obtained by French was the correct one (see [Sc94], for a detailed account).

After this brilliant start, French moved to problems of nuclear physics, were he made important contributions. Since the 60's, he was engaged in problems of developing and understanding shell model calculations. He put forward the idea of 'statistical spectroscopy', namely to examine shell model results not by observing individual levels but rather by studying them from a statistical point of view. It turned out that the spectral density of shell model calculations was showing a distribution close to a Gaussian (normal) distribution. This was in striking contrast with Wigner's semi-circle law obtained in RMT. On the other hand, the spectral fluctuations of shell-model spectra seemed to be consistent with RMT predictions. In order to try to identify this, and maybe other, qualitative similarities and differences, French and Wong, on the one side, and Flores and myself on the other, introduced independently and simultaneously in the early 70's what we called Two-Body Random Ensembles (TBRE): the two-body matrix elements were taken as independent random variables and from them the n-body matrix elements were computed. By construction, for $n = 2$, the ensemble coincided with the Wigner-Dyson ensemble and for $n > 2$ it differed. Indeed the Pauli principle as well as correlations among matrix elements, two important physical ingredients, were now explicitly taken into account. We performed numerical simulations and we reached the conclusion that the spectral density, with increasing n, tended to a Gaussian distribution and, tentatively and after some controversy and hesitations (see [Ga72]), that the spectral fluctuations were the same as for the GOE. This work had some impact in the random matrix community and Dyson wrote to Mehta [Dy71]: *'Dear Madan Lal... You have certainly thought about the recent work of Bohigas and Flores, French and Wong. This I find very good and interesting. But everything is numerical, nothing understood theoretically. So you have an obvious and important problem to work on – calculate rigorously the one and two-level distribution functions for the Bohigas-Flores ensemble in the limit of large numbers of levels. Au revoir – Yours. Freeman...'* As far as I know, Mehta never dealt with the problem. Mon and French, few years later, succeeded to a large extent in understanding analytically the one level distribution and the transition from semi-circle to Gaussian distribution. Despite several attempts, no further decisive progress has been made and the two and higher point correlations of the TBRE have resisted an analytical treatment. After more than 30 years, the problem is still waiting for a solution (see for instance [BFFMP81] or Benet and Weidenmüller in [FSV03]).

My second encounter with RMT came ten years after. We decided, with Rizwan Ul Haq and Akhilesh Pandey, to reconsider the analyses of the nuclear resonances. We introduced mainly two new ingredients: i) instead of making a comparison of individual nuclei, one by one, with theoretical predictions, we adopted the natural point of view that each resonance sequence of a given nucleus should be considered a realization and that the set of resonance sequences constitutes a sample of the ensemble which we called the Nuclear Data Ensemble (NDE); ii) we focused on partially new spectrally averaged measures. Both ingredients would greatly improve the statistical significance when comparing to theory. On Figure 7 are reproduced some of the results. In [HPB82] ('Fluc-

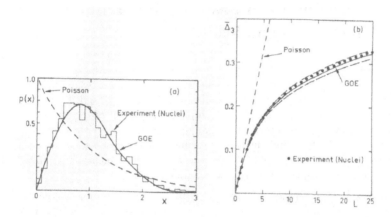

Figure 7: Left: nearest neighbour spacing histogram for the nuclear data ensemble (NDE) [BHP83]. Right: Dyson-Mehta spectral rigidity Δ_3 for NDE [HPB82]. GOE and Poisson predictions are plotted for the sake of comparison.

tuation Properties of Energy Levels: Do Theory and Experiment Agree?') we concluded: *'We have thus established an astonishingly good agreement between a parameter-free theory (GOE) and the data. We emphasize that, apart from rotation and time-reversal invariance resulting in the real symmetric nature of the matrices, the GOE takes no account of the specific properties of the nuclear Hamiltonian... The good agreement with experiment, coupled with the theoretical understanding which is slowly emerging, reinforces the belief that the GOE fluctuations are to be found in nature under very general conditions.'* Of course, I sent a preprint to Dyson. He replied [Dy81]: *'Thank you for your beautiful paper on the energy-level statistics... I am amazed by the exact agreement between theory and experiment. I think you succeeded, after everybody else had failed, for two reasons. One, you put all the data together. Two, you calculated the Δ_3 only for short stretches of levels instead of trying to cover* *long stretches. The second innovation was decisive in making the results in-sensitive to an occasional mistake in the data... your analysis is so perfect that I do not know how to improve it.'* Could we have expected a more rewarding answer? Not only we were receiving the 'nihil obstat' but a benediction from one of the highest authorities! I must say however that sometimes I felt as having behaved as a predator: after beautiful non-trivial developments, the theory was there, after painful long term work, the data were there; we had merely to assemble some pieces with due care in order to conclude and to take part in the final banquet.

4   Searching for a basis for RMT fluctuations: Chaotic Dynamics

Once the ability of RMT to predict fluctuation properties exhibited by the data was established, it still remained to understand in physical terms their origin, their domain of validity and their limitations. Wigner qualified a system to which RMT could be applied as a 'complicated system' (see quotation after equation (2.3)). Does this mean a many-particle system with many degrees of freedom, like the atomic nucleus, or could it also mean something else? It may be useful at this point to recall two major achievements, one related to classical mechanics, the other to quantum mechanics.

A new field of physics (and mathematics) has known impressive develop-ments since the 50's, namely the study of dynamical systems which, though deterministic, are unpredictable due to the presence of instabilities. Henri Poincaré may be considered as its founder. Here we will only refer to Hamil-tonian conservative systems. When parameters of the Hamiltonian are varied, the system may undergo very complicated changes, reflected in the geometrical properties of phase space. For instance, by changing a parameter, the system may undergo a transition from regular to chaotic motion, the regular situation being characterized by the presence of invariant tori in phase space, the fully chaotic by their absence. The effect of a small perturbation on an integrable system is the content of the Kolmogorov-Arnold-Moser (KAM) theorem, a landmark in the field (see Table 1). The problem of two bodies attracted by the gravitational force (the Kepler problem of the Sun-Earth system, the clas-sical analogue of the hydrogen atom mentioned above) is one of the simplest and more famous examples of regular systems. For the present discussion an important outcome in the study of dynamical systems must be mentioned: the motion may be very complicated, in fact fully chaotic, even if only few de-grees of freedom are present. Specifically, two degrees of freedom are sufficient for conservative Hamiltonian systems to exhibit hard chaos. On Figure 8 are displayed two examples, one of regular motion, the other of chaotic motion: a free particle inside a two-dimensional box (billiard) moves on straight line

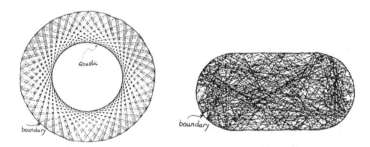

Figure 8: A typical trajectory in a billiard. Left: circular billiard (regular). Right: Bunimovich's stadium (chaotic).

segments and when hitting the boundary it bounces elastically undergoing a specular reflection. Generically, systems show mixed phase spaces: part of the phase space is occupied by invariant tori which are surrounded by chaotic seas. In summary, the study of conservative systems indicates a route which leads from integrable (regular) motion to chaotic (irregular, complicated) motion.

Let us now turn to quantum mechanics and to its relation with classical mechanics. A quantum description is necessary when dealing with processes involving atomic or smaller scales. On the other hand, macroscopic objects are well described by classical mechanics. Do quantum and classical descriptions conflict, does one contain the other? Bohr's correspondence principle addresses these questions. A. Pais, a particle theoretical physicist who, in his late years, has worked on the history of contemporary physics, writes [Pa91]: *'The correspondence principle is, I think, Bohr's greatest contribution to physics after his derivation of the Balmer formula (hydrogen atom). It is the first manifestation of what would remain the leading theme in his work: classical physics, though limited in scope, is indispensable for the understanding of quantum physics. In his own words: "Every description of natural processes must be based on ideas which have been introduced and defined by the classical theory."'* Sometimes the correspondence principle is stated as telling that in the classical limit, when interference effects can be neglected, the predictions of quantum and classical mechanics must agree. One further step is taken when asserting, as some physicists do, that anything classical mechanics can do, quantum mechanics can do better. Some cosmologists, Hawkings, for instance, don't even hesitate to consider the wave function of the universe. Dyson summarizes this wide span of views as follows [Dy04]: *'... The broad school says that quantum mechanics applies to all physical processes equally, while the strict school says that quantum mechanics covers only a small part of physics, namely the part dealing with events on a local or limited scale... The historic exponent of the strict view of quantum mechanics was Niels Bohr, who maintained that quantum mechanics can only describe processes occurring within a larger framework that must be defined classically.'*

Irrespective of different interpretations and philosophical preconceptions, the correspondence principle has been a powerful source of inspiration. Based on it and in the light of the developments in chaotic dynamics described above, Percival addressed questions concerning what we may call 'fingerprints' of chaos in quantum systems. For instance, in a short article suggestively entitled 'Regular and irregular spectra' [Pe73], he put forward, based on the correspondence principle and semiclassical reasoning, the notion that in the quantum discrete spectrum of a bound system to each level could be associated the character of being either regular or irregular, and he specified the properties characterizing regular and irregular states.

It was therefore natural to investigate the characteristics of the quantum spectra of systems known to be fully chaotic like appropriately chosen billiards (Sinai's, consisting in a square with a central circular disk, Bunimovich stadium, etc). One had to (numerically) compute the spectrum of the Laplacian on a flat region of the plane with the corresponding boundary, the wavefunction vanishing, say, on it. Some results of the first attempts in this direction are reproduced on Figure 9. Their authors concentrated on the nearest neighbour spacing distribution and, to a large extent, on its behaviour at the origin.

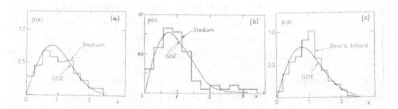

Figure 9: Nearest-neighbour spacing distribution of quantum billiards. (a) Bunimovich's stadium [MK79]; (b) same [CVG80]; (c) Sinai's billiard [Be81]. The GOE prediction has been superimposed for the sake of comparison.

Was it linear, as in RMT, or with a power different from one, as Zaslavski was suggesting? The numerical results, though highly suggestive, were not quite conclusive. Inspired by these works and equipped with some experience in RMT we decided, with M.-J. Giannoni and C. Schmit, to examine further the problem. We felt that only the tip of the iceberg was being considered, because RMT had much more in it than just, say, the linear level repulsion. What, for instance, about spacing correlations, spectral rigidity and other RMT predictions which we had previously considered when analyzing nuclear resonance data? Some of the results obtained, corresponding to Sinai's billiard (we also investigated Bunimovich's stadium) are reproduced in Figure 10. Though with present numerical capabilities and some developments in nu-

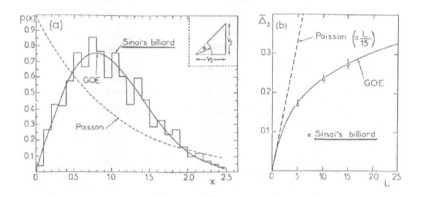

Figure 10: Spectral fluctuations of Sinai's billiard [BGS84a]. (a) nearest-neighbour spacing distribution; (b) Dyson-Mehta spectral rigidity. Poisson and GOE predictions for the sake of comparison.

merical methods they appear as primitive, we could make significant progress. In Ref [BGS84], entitled 'Characterization of Chaotic Quantum Spectra and Universality of Level Fluctuation Laws' we concluded: *'All fluctuation properties of Sinai's billiard investigated so far are fully consistent with GOE predictions... The present work should have further developments (for instance, when time-reversal invariance does not hold, the adequate model in RMT is the Gaussian unitary ensemble (GUE) and one should look for "simple" chaotic systems having GUE fluctuations). It is an attempt to put in close contact two areas – random matrix physics and the study of chaotic motion – that have remained disconnected so far. It indicates that the methods developed in RMT to study fluctuations provide the adequate tools to characterize chaotic spectra and that, conversely, the generality of GOE fluctuations is to be found in properties of chaotic systems. In summary, the question at issue is to prove or disprove the following conjecture: Spectra of time-reversal-invariant systems whose classical analogs are K systems show the same fluctuation properties as predicted by GOE (alternative stronger conjectures that cannot be excluded would apply to less chaotic systems, provided that they are ergodic). If the conjecture happens to be true, it will then have been established the universality of the laws of level fluctuations in quantal spectra already found in nuclei and to a lesser extent in atoms. Then, they should also be found in other quantal systems, such as molecules, hadrons, etc.'* This conjecture, or elaborate versions of it, is often referred to in the literature as the random-matrix or BGS conjecture (see Table 1).

Once again, I sent a preprint to Dyson, and once again he sent back an encouraging reply [Dy83a]: *'This is a beautiful piece of work. It is extraordinary that such a simple model shows the GOE behavior so perfectly. I agree completely with your conclusions. I would say the result is not quite surprising but certainly unexpected... I once suggested to a student at Haverford that he build a microwave cavity of irregular shape and observe the resonances to see whether they follow the GOE distribution. So far as I know, the experiment was never done... I always thought the cavity would have to be a complicated shape with many angles. I did not imagine that something as simple as the Sinai region would work...'*. It should be realized that there is a complete equivalence between the stationary Schrödinger equation for a particle in a 2-dimensional box (spectrum of the Laplacian) and the eigenmodes of an appropriately shaped microwave cavity (also equivalent to the Helmholtz equation governing the membrane vibrations of a drum). Some years later, in the early 90's, Stöckmann at Marburg, Sridhar at Boston and Richter at Darmstadt started experimental programmes with microwave cavities on these lines (see for instance, their contributions in [BA01], or [St99]) and many results were analysed on the light of RMT. This type of studies have been later extended to investigate acoustic waves in metallic blocks, quartz crystals, etc. Though the elastic wave equation is no longer equivalent to the Schrödinger equation but more complicated, still many properties seem to be well described by RMT (see, for instance, contribution of Ellegaard in [BA01] or [St99]). I find it remarkable that Porter clearly forsaw these possibilities almost thirty years in advance [Po65]: *'That matrix ensembles will most likely be relevant to the fields of acoustics and elasticity is rather evident. Some work in room acoustics has already pointed out in this direction... there is clearly further room in the context of this volume for at least some additional careful thought with respect to the utility of these ideas in other problems in which the system being considered is complex enough that many normal modes are present.'*

5   Towards a dynamical theory of quantum spectral fluctuations

To understand the progress made since the 70's in the description of quantum spectral fluctuations, it may be convenient to recall some basic facts related to the development of quantum mechanics since its early days. Before its modern formulation in the mid 20's with the revolutionary works of Schrödinger, Heisenberg and some others, there existed the 'old quantum mechanics'. One of the major achievements of the incipient quantum theory was the successful description of the spectrum of the hydrogen atom, the proton (nucleus)-electron system referred to above (Balmer formula, for instance). This was accomplished by applying a far from systematic quantization scheme, the Bohr-Sommerfeld quantization, later improved. It relied heavily on properties of the trajectories of the classical system. After this remarkable success, the next step was obvious, namely to generalize the treatment to the helium atom (the nucleus plus two electrons). But this lead to disasters. In [Pa91] this period is described as 'spring of hope, winter of despair'. Max Born, in 1925, would write: *'The systematic application of the principles of the quantum theory... gives results in agreement with experiment only on those cases were the motion of a single electron is considered; it fails even in the treatment of the motion of two electrons in the helium atom'*. Soon after, the quantum revolution took place with Schrödinger's wave equation and Heisenberg's matrix mechanics and ever since most of the approaches to determine the quantum spectrum of a bound system, for instance, take as starting point the time independent Schrödinger equation. The task consists then in devising analytic and/or numerical approximations to solve it, the interest in the 'old quantum mechanics' being left to the inquiry of science historians.

However, the understanding of the origin of the failure of the old quantum theory is at the origin of some of the most interesting developments related to our subject. The problems encountered in the quantization à la Bohr-Sommerfeld of systems with several degrees of freedom were clearly identified by Einstein. In a paper published in 1917 and practically unnoticed during forty years [Ei17], he realized that the application of the known quantization scheme was restricted to systems whose phase space exhibited only, using modern terminology, invariant tori (regular systems). Most of the systems, namely with mixed or fully chaotic phase space, were therefore excluded. Among them the famous three-body problem, and a particular case of it, the Sun-Earth-Moon problem, investigated intensively since Newton, passing through Poincaré, in which chaos is already lurking. The helium atom, a quantum three-body problem (nucleus and two electrons), can therefore not be treated in the frame of torus quantization, thereby explaining the failure referred to above of the old quantum mechanics.

Starting from the late 60's, due to a large extent to the solitary effort of Martin Gutzwiller, this situation changed dramatically (see Table 1). His method can be derived by starting from Feynman's path integral (exact) formulation of quantum mechanics and by using the stationary phase approximation [Gu90]. Gutzwiller succeeded in devising an approximate and systematic quantization scheme which applies to chaotic systems and which, like for regular systems, requires only the knowledge of the properties of the periodic orbits of the corresponding classical system. This approach is often referred to as the periodic orbit theory (POT). Work on closely related lines was performed independently and almost simultaneously by Balian and Bloch [BDVM75]. The physical motivations and emphasis were however different: Gutzwiller was mainly interested in the quantization of chaotic systems, Balian and Bloch on shell effects, namely strong departures from uniformity of the quantum spectrum, bunching of levels, effects which are mainly connected to regular motion.

For further discussion, it is convenient to recall that the quantum level density ρ can be separated in a smooth part $\bar{\rho}$ and a fluctuating, or more appropriately denoted oscillating part ρ_{osc}

$$\rho(E) = \bar{\rho}(E) + \rho_{osc}(E) \tag{5.1}$$

For a billiard, for instance, $\bar{\rho}$ is determined by global geometric properties (surface, perimeter,...). The oscillating part ρ_{osc} is given by

$$\rho_{osc}(E) = 2 \sum_p \sum_{r=1}^{\infty} A_{p,r}(E) \cos\left[r S_p(E)/\hbar + \mu_{p,r} \right] \tag{5.2}$$

The summation is over primitive periodic orbits, labelled by p, r are the repetitions, $\hbar$ is Planck's constant, S_p is the action of the corresponding periodic orbit and $\mu_{p,r}$ its Maslov index related to the number of conjugate points. The amplitudes $A_{p,r}$ have different properties depending on the system being considered, regular or chaotic. Here and in what follows (crucial) questions of convergence will be ignored.

For regular systems the number of periodic orbits increases with energy as a power law and periodic orbits appear in continuous families. This has far reaching physical consequences. Shell effects in nuclei responsible for the existence of the so called magic numbers (nuclei having particular numbers of protons or neutrons have exceptional properties), the dynamics of nuclear fission, for which shell effects are essential, the cosmic relative abundances of elements, the possible existence of super-heavy elements, the existence of shells (and supershells) in metallic clusters, are examples of phenomena to which the POT is applicable. Often it gives, in its almost simplest formulation, a clear physical insight and a semi-quantitative description [BB97] (for an excellent introduction to shell effects in nuclei from a modern perspective, in which chaotic features are also discussed, see [Le04]). In the 70's, Strutinsky and collaborators applied the ideas of POT thoroughly to several nuclear physics phenomena, from a qualitative as well as a quantitative point of view. Berry, who was working on semiclassical mechanics, realized since the beginning the relevance of Gutzwiller's and Balian-Bloch's works [Be97]. Based on them he studied, among many other things (see above and below), the regime of regular dynamics, with Tabor. Concerning spectral fluctuations of regular systems they demonstrated that, on a local scale, they behave as those of an uncorrelated (Poissonian) sequence, in strong contrast with RMT predictions.

The structure of the periodic orbit sum in (5.2) is very different for fully chaotic systems. In particular the periodic orbits don't appear in families but are isolated, their number proliferates with energy exponentially and not as a power law. Gutzwiller showed that the amplitude in (5.2) is given by

$$A_{p,r}(E) = \frac{\tau_p}{2\pi\hbar|\det(M_p^r - I)|^{\frac{1}{2}}};$$ (5.3)

in (5.3) τ_p is the period of the orbit and M_p the monodromy matrix, which contains information about its instability. One possible point of view is that the ultimate goal of POT is to determine individual highly excited states of a chaotic system. This goal has not been fully attained so far. On the other hand it may have limited practical interest because, unless one invents new methods, it is probably more difficult to determine the properties of an enormous number of periodic orbits than to solve the Schrödinger equation itself. A different point of view may be adopted, namely that POT provides a solid theoretical ground on which physical insight can be gained; however less 'microscopic' questions should be asked. It is with this strategy that Berry, who had a deep understanding of POT , addressed the question of spectral fluctuations in chaotic systems.

Before I proceed further let me make a short step back. In the previous section I reproduced part of a very supportive letter Dyson sent concerning our work on spectral universalities. A couple of months later I received another one, this time with warnings and remarks [Dy83b]: '...*After I had written to you in August [Dy83a] I began to read the paper of M.V. Berry [Be81] and realized that things are not so simple as you say in your preprint... This oscillation (in the counting function $N(E)$) comes from the effect in the quantum regime of the classical closed orbits. It implies that the long-range level-correlations should not agree precisely with the random-matrix model... perhaps you never had L large enough to show the Berry oscillations... It should still be true that* (the logarithmic increase of $\Delta_3(L)$) *holds for each fixed L as n* (level number) *goes to infinity* (semiclassical limit). *Can you prove it?*' In fact Dyson was anticipating basic properties of the theory that Berry was working out [Be85].

Berry started from the Gutzwiller periodic orbit sum which, to be more specific and for later comparison, takes for billiards the following form

$$\rho_{osc}(k) = \frac{1}{2\pi} \sum_p \sum_{r=1}^{\infty} \frac{l_p}{\sinh\frac{r\alpha_p}{2}} \cos\left[r(kl_p - \frac{\mu_p\pi}{2})\right];$$ (5.4)

for convenience, the wave number $k(= E^{1/2})$ instead of the energy E is used. l_p is the length of the primitive periodic orbit and α_p its instability exponent. To study the two-point correlation function, the same object encountered when referring to Montgomery's work, one has to work out the behaviour of the average of the product $\rho(E) \cdot \rho(E + \epsilon)$ or its Fourier transform with respect to ϵ, called the spectral form factor $K(\tau)$. It contains a double sum over periodic orbits p and p'. Schematically, one can summarize Berry's work as follows. Use the diagonal approximation (only terms $p = p'$) and the (classical)

Hannay-Ozorio sum-rule, which is based on the fact that in chaotic systems the phase-space distribution of the very long orbits is uniform on the energy surface, irrespective of the system considered. In this way one can not only recover in the semiclassical limit some of the basic results of RMT (linear behaviour of $K(\tau)$ at the origin, for instance), but also obtain departures from it, in particular due to the presence of short periodic orbits. The universality of RMT spectral fluctuations reflect the universality properties of long periodic orbits. However, short periodic orbits, which differ from system to system, give rise to long range oscillations in the fluctuations and have characteristic effects in several of the statistical measures considered previously, number variance or spectral rigidity, for instance. These effects are not contained in RMT, a one-parameter theory, parameter which fixes the scale (mean level spacing). For instance, by extending the range of L in Figure 10b, departures from GOE behaviour are predicted and found. And Dyson was right in his warnings and remarks.

In searching for a chaotic system for which all spectral correlations could be derived, free motion on constant negative curvature surfaces comes naturally to mind. One should restrict the motion to a compact surface and consider the spectrum of the Laplace-Beltrami operator. It is known that in this case the (approximate) Gutzwiller formula becomes the exact Selberg trace formula (same as equation (5.4) except that now $E = k^2 + 1/4$ and the sum p is over conjugacy classes of the primitive elements p). Numerical computations performed by Charles Schmit, for triangles with angles π/n, n integer, gave unexpected results. Sometimes the spectral fluctuations were found to be close to GOE predictions, as expected, sometimes far from them and close to Poissonian fluctuations. This 'anomaly' was understood when it was realized that, by accident, some of the triangles showing the anomaly were members of a set of 85 triangles possessing a special property, an arithmetic symmetry (arithmetic groups), known from mathematicians. Exploiting this arithmetic symmetry, the 'anomalous' spectral fluctuations could be derived analytically [BGGS97]. These studies have been continued and extended by Steiner and collaborators in Ulm.

For non-arithmetic polygons the question remains open. Sarnak gave rather recently a list of thirteen conjectures, chosen according to criteria of 'elegance, simplicity, concreteness and the problem be a central one' [Sa99]. He included, now phrased in elaborate mathematical language, *Conjecture XIII [BGS84b]: Fix $r \geq 2$ and let $R_r(X)$ denote the space of all Riemannian metrics on X in the C^r topology. For the generic (in the sense of measure as defined in [HSY92] for example) $g \in R_r(X)$ of negative curvature, the local spacing distributions between the eigenvalues of Δ_g in the large λ limit, follow the laws of the eigenvalue spacings of random matrices from the Gaussian Orthogonal Ensemble (see [Me90] for a description of the latter).*

Much effort has been devoted, since the seminal work of Berry, to go beyond the diagonal approximation in the computation of the form factor $K(\tau)$ using POT . The work of Sieber and Richter (see their contribution in [BA01]) and considerable extensions of it [MHBHA04] represent recent major advances.

Though not discussed here, let us mention the field theoretic supersymmetric non-linear σ-model derived in the context of studies of disordered conductors [Ef97]. Among many other things, it provides in several respects an alternative method to POT . In particular, it was shown by Efetov that the two-point correlation function of a disordered metal particle coincides with the corresponding function in RMT, thereby suggesting strong links between problems of disorder and problems of QC (Andreev, Agam, Simons, Altshuler). In fact, progress in one approach (field theoretic) has often stimulated progress in the other (POT) and vice-versa. Of particular interest has been the study of parametric correlation functions, for instance correlations at two energies corresponding to two different values of the parameter appearing in the Hamiltonian (see, for instance, Smolyarenko and Simons, in [FSV03]).

Before we leave quantum physics, let us revisit the hydrogen and helium atoms in the light of the previous discussion. The spectrum of hydrogen represented the success of torus quantization. However, when a magnetic field is applied, the classical mechanics changes (diamagnetic Kepler problem): there are tori which are destroyed and by increasing the magnetic field they eventually disappear, giving rise to a fully chaotic motion. Correspondingly, in the quantum regime, we can observe the transition from Poissonian to RMT spectral fluctuations. This physical system, studied theoretically (Wintgen, Delande,...) and experimentally (Rydberg atoms) provides the best example where many aspects of POT , RMT and their relationship have been thoroughly tested. Concerning the spectrum of helium, semiclassical quantization has taken its revenge. The failure of Pauli and Heisenberg reflected to some extent their lack of understanding of some subtleties of classical mechanics (role of conjugate points along classical trajectories) but mainly their complete ignorance of the importance of periodic orbits. Finally, Wintgen and collaborators, in 1991, successfully dealt with the spectrum of the helium atom in the frame of POT and semiclassical quantization.

Let us now go back to Riemann's zeta function in the light of RMT and POT , a subject which has known developments in contact with physicists, notoriously with Berry. Berry tells that he learned from me and Gutzwiller about the possible relevance to QC of Riemann's zeta [Be97]. I don't remember precisely in which circumstances. What I neatly remember is that, when I was working on compound nucleus resonances and billiards, I did some computations of cumulants of the number statistics for zeros on the critical line. The third and fourth cumulant tended rapidly to zero, with some small oscillations, as for eigenvalues of GUE matrices and consistent with the known GUE behaviour of the spacing distribution of Riemann's zeros. However the variance (second cumulant) increased logarithmically like in RMT only over a small range of L, and then showed a slower increase. There was a message in that discrepancy which I missed, discrepancy that I attributed to some unknown obvious reason (!). At that time, I was unprepared to realize that deviations from RMT and from asymptotic universalities were interesting.

Berry's approach to these problems, not surprisingly, was POT-based. Indeed, if one considers the density ρ of the imaginary part of the non-trivial zeros ($\zeta_n = 1/2 + iE_n$), one has

$$\rho(E) = \frac{1}{2\pi} \log \frac{E}{2\pi} + O(E^{-2}) - \frac{1}{\pi} \sum_p \sum_{r=1}^{\infty} \frac{\log p}{p^{r/2}} \cos(Er \log p), \qquad (5.5)$$

in complete analogy with (5.1) and (5.2). The 'quantum' spectrum (the sequence of E_n's) is related to the 'classical' dynamics (prime numbers). Periodic orbits are labelled by primes, their periods given by $\log p$, etc. (for a review, see [BK99]). Notice however that (5.2), (5.3), (5.4) are approximate, whereas (5.5) is exact. Some of the methods already used in QC and POT were further developed since the mid 80's by Berry, later joined by Keating, and a wealth of new and unexpected results concerning the statistical properties of Riemann's zeros followed.

Let me finally mention a beautiful illustration of the classical-quantum duality translated now in Riemann's zeta territory. In order to have RMT spectral fluctuations, subtle correlations among classical actions of periodic orbits must be present (see (5.2)). For the Riemann case, extending some of Montgomery's results, Bogomolny and Keating have shown that if the Hardy-Littlewood conjecture about prime correlations holds ('classical action correlations'), then the Montgomery-Odlyzko conjecture follows ('quantum spectral correlations').

6　Epilogue

At the end of the journey, after this long guided tour, let us take an image of some of the main figures encountered: Bohr, Riemann and his zeta function, Wigner, Poincaré, Dyson.

On Figure 11 is reproduced the illustration given by Bohr of the idea of the compound nucleus, at the origin of Wigner's RMT. Remarkably, Bohr used a billiard system that later played some role in studying chaotic systems.

Riemann's zeta function, with its zeros and its associated Riemann Hypothesis (RH), is behind many of the contributions at the workshop. Recall that there exists the Hilbert-Polya conjecture, which predicts that the RH is true because zeros of the zeta function correspond to eigenvalues of a hermitean operator. Polya, asked by Odlyzko almost 70 years later, made the following recollection [Pol82]: *'I spent two years in Göttingen ending around the begin of 1914. I tried to learn analytic number theory from Landau. He*

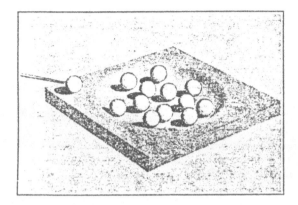

Figure 11: Illustration of the compound nucleus idea in a neutron-nucleus collision. The constituent nucleons are viewed as billiard balls and the nuclear binding as a shallow basin (after N. Bohr [Bo36]).

asked me one day: "You know some physics. Do you know a physical reason that the Riemann hypothesis should be true?" This would be the case, I answered, if the nontrivial zeros of the Xi-function were so connected with the physical problem that the Riemann hypothesis would be equivalent to the fact that all the eigenvalues of the physical problem are real. I never published this remark, but somehow it became known and it is still remembered.' But there is no indication of what operator would be involved.

The developments, concerning relations of zeta and RMT as well as the dynamical interpretations via POT , reinforce the indications on which many researchers base their belief that the Hilbert-Polya conjecture is the most promising approach to prove RH. The results by Keating and Snaith [KS00] deserve a special mention because they show that, in some cases, RMT is able to generate new and (presumably) exact mathematical results concerning the zeta and other L-functions. The precision attained by POT-based approaches to describe, beyond leading asymptotics, the pair correlation function $R_2(\epsilon)$ and the spacing distribution $p(s)$, two objects encountered many times, is illustrated on Figure 12. (observe the scale on the lower part of the figure). Let us also recall that Riemann's zeta is but one of the L-functions, that they are not connected by a continuous parameter, and that zeros of different L-functions are statistically independent, like eigenvalues of a quantum system possessing discrete symmetries [BL94]. This suggests that what may be the object to be found is an operator possessing discrete symmetries corresponding to different L-functions.

Oriol Bohigas

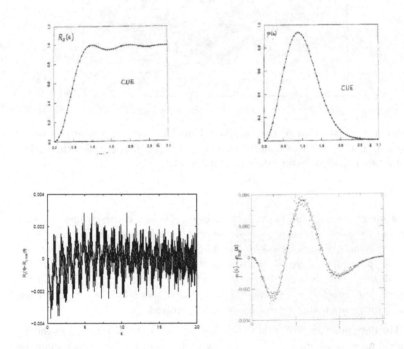

Figure 12: Statistical behaviour of Riemann's zeros. Data from [Od], around the 10^{23}th zero. Upper part: left, two-point correlation function $R_2(\epsilon)$; right, spacing distribution $p(s)$; continuous curves, asymptotic CUE distributions. Lower part: difference between Riemann's zeros values and asymptotic CUE distributions; continuous curves, theory [Bog03], [BBLM04].

One of my purposes in writing this contribution has been to show that, often, the paths followed by scientific ideas are interwoven and unpredictable, like trajectories of unstable orbits. I will consider my purpose attained if some mathematicians share the idea that what they are doing now takes partly its source in what some physicists and engineers were doing when studying the compound nucleus and designing nuclear reactors.

I could conclude with Wigner's statement *'The unreasonable effectiveness of mathematics in the natural sciences'* [Wi60] or its reverse 'The unreasonable effectiveness of the natural sciences in mathematics'. Or with Poincaré: *'La physique ne nous donne pas seulement l'occasion de résoudre des problèmes..., elle nous fait pressentir la solution'*. But I will conclude with Dyson, to whom this contribution is dedicated, with his characteristic style, always witty and sometimes controversial [Sc94]: *'Elegance comes first; if the problem that is solved is physically interesting or important, that's a bonus'.*

7   Acknowledgments:

Part of the work reported here is the outcome of collaborations with, in chronological order, Jorge Flores, Rizwan Ul Haq and Akhilesh Pandey, Marie-Joya Giannoni and Charles Schmit, and Eugene Bogomolny, Patricio Leboeuf and Alejandro Monastra. I thank all of them. In preparing the manuscript I have benefitted from exchanges with E. Bogomolny, P. Leboeuf, H. Montgomery, A. Odlyzko, I.E. Smolyarenko and B.D. Simons. I am most grateful to F. Mezzadri and N. Snaith, for encouragement, assistance and patience.

References

[AMPZ95]   Akkermans E., Montambaux G., Pichard J.-L., Zinn-Justin J., eds., *Mesoscopic quantum physics*, Les Houches Session LXI, 1994, Elsevier, 1995.

[Be81]   Berry M.V., 'Quantizing a Classically Ergodic System: Sinai's Billiard and the KKR Method ', Ann. Phys. (N.Y.) **131** (1981) 163-216.

[Be85]   Berry M.V., 'Semiclassical theory of spectral rigidity', Proc. R. Soc. Lond. **A400** (1985) 229-251

[Be97]   Berry M., 'Gutzwiller 70th birthday', www.phy.bris.ac.uk/ research/theory/Berry/the-papers/speeches.pdf

[Bo36]   Bohr N., 'Neutron Capture and Nuclear Constitution', Nature **137** (1936) 344-348; 351.

[Bog03]   Bogomolny E., 'Quantum and Arithmetical Chaos', lectures at workshop 'Frontiers in Number Theory, Physics and Geometry', Les Houches 2003, nlin.CD/031206

[Br92]   Britton J.L., editor, *Pure Mathematics. Collected Works of A.M. Turing*, North-Holland 1992.

[BA01]   Berggren K.-F., Aberg S., *Quantum Chaos Y2K*, Proceedings of Nobel Symposium **116**, Bäckaskog Castle, 2000, Physica Scripta **T90** (2001) 1-282.

[BB97]   Brack M., Bhaduri R.K., *Semiclassical Physics*, Addison-Wesley, 1997.

[BBLM04]   Bogomolny E., Bohigas O., Leboeuf P., Monastra A.G., 'On the spacing distribution of the Riemann zeros: corrections to the asymptotic result', J. Phys. A **39** (34) (2006) 10743–54.

[BDVM75]   Balian R., De Dominicis C., Gillet V., Messiah A., editors, *Claude Bloch. Scientific Works*, Vol. 2, North Holland, 1975.

[BFFMPW81] Brody T.A., Flores J., French J.B., Mello P.A., Pandey A. , Wong.S.S.M., 'Random-matrix physics: spectrum and strength fluctuations', Rev. Mod. Phys. **53** (1981) 385-479.

[BGGS97]   Bogomolny E.B., Georgot B., Giannoni M.-J., Schmit C., 'Arithmetic Chaos', Phys. Rep. **291** (1997) 219-326.

[BGS84a]   Bohigas O., Giannoni M.-J., Schmit C., 'Characterization of Chaotic Quantum Spectra and Universality of Level Fluctuation Laws', Phys. Rev. Lett. **52** (1984) 1-4.

[BGS84b]   Bohigas O., Giannoni M.-J., Schmit C., 'Spectral properties of the Laplacian and random matrix theories', J. Physique Lett. **45** (1984) L1015-L1022.

[BHP83]   Bohigas O., Haq R.U., Pandey A., 'Fluctuation properties of nuclear energy levels and widths: comparison of theory with experiment ', in *Nuclear Data for Science and Technology*, K.H. Böckhoff (ed.), Reidel, Dordrecht 1983, pp. 809-813.

[BI01]   Bleher P.M., Its A.R., editors, *Random Matrix Models and Their Applications*, MSRI Publications, Cambridge University Press, 2001.

[BK99]   Berry M.V., Keating J.P., 'The Riemann Zeros and Eigenvalue Asymptotics', SIAM Rev. **41** (1999) 236-266.

[BL94]   Bogomolny E., Leboeuf P., 'Statistical properties of the zeros of zetafunctions-beyond the Riemann case', Nonlinearity **7** (1994) 1155-1167.

[Cv]   Cvitanovic P. et al., *Classical and Quantum Chaos*, www.nbi.dk/ChaosBook/, to be published.

[CC95]   Casati G., Chirikov B., *Quantum Chaos*, Cambridge University Press, 1995.

[CVG80]   Casati G., Valz-Gris F., Guarneri I., 'On the Connection between Quantization of Nonintegrable Systems and Statistical Theory of Spectra', Lett. Nuov. Cim. **28** (1980) 279-282.

[De98]   Deift P., *Orthogonal Polynomials and Random Matrices: A Riemann-Hilbert Approach* , AMS, Courant Institute, 1998.

[Dy71]   Dyson F., letter to M.L. Mehta, 10/05/71.

[Dy72]   Dyson F., 'Missed oportunities', Bull. Am. Math. Soc. **78** (1972) 635-652., reproduced in [Dy96] .

[Dy81]   Dyson F., letter to the author, 16/12/81.

[Dy83a]   Dyson F., letter to the author, 24/08/83.

[Dy83b]   Dyson F., letter to the author, 25/10/83.

[Dy96]   Dyson F., *Selected Papers of Freeman Dyson with Commentary*, American Mathematical Society, Providence, 1996.

[Dy04]   Dyson F.J., 'Thought-experiments in honor of John Archibald Wheeler' in *Science and Ultimate Reality. Quantum Theory, Cosmology, and Complexity*, J.D. Barrow, P.C.W. Davies, C.L. Harper, Jr., editors, Cambridge University Press, 2004.

[DM63]   Dyson F.J., Mehta M.L., 'Statistical Theory of Energy Levels of Complex Systems. IV', J. Math. Phys. **4** (1963) 701-712.

[Ef97]   Efetov K.B., *Supersymmetry in Disorder and Chaos*, Cambridge University Press, Cambridge 1997.

[Ei17]   Einstein A., 'Zum Quantensatz von Sommerfeld und Epstein', Verh. Deutsch. Phys. Ges. **19** (1917) 82, english translation 'On the Quantum Theorem of Sommerfeld and Epstein' in *The Collected Papers of Albert Einstein*, Vol.**6** *The Berlin Years: Writings, 1914-1917*, Princeton University Press, 1997.

[Fo]   Forrester   P.,   *Log-gases   and   Random
              Matrices*, www.ms.unimelb.edu.au/~matpjf/matpjf.html, to
              be published.

[FSV03]   Forrester P. J., Snaith N.C., Verbaarschot J.J.M., editors,
              Special issue: Random Matrix Theory, J. Phys. A **36** (12)
              (2003) 2859-3646.

[Ga72]   Garg J.B., editor, Proceedings of the International Conference
              Statistical Properties of Nuclei, Albany, N.Y., 1971, Plenum
              Press, 1972.

[Gu90]   Gutzwiller M.C., *Chaos in Classical and Quantum Mechanics*,
              Springer-Verlag, 1990.

[GMW98]   Guhr T., Müller-Groeling A., Weidenmüller H.A., 'Random
              Matrix Theories in Quantum Physics: Common Concepts'.,
              Phys. Rep. **209** (1998) 189-425.

[GVZ91]   Giannoni M.-J., Voros A., and Zinn-Justin J., eds, *Chaos and
              Quantum Physics*, Les Houches, Session LII, 1989, Elsevier,
              1991.

[Had45]   Hadamard.J., *The psychology of invention in the mathe-
              matical field*, Princeton University Press 1945, revised edi-
              tion, Dover Publications, 1954; *Essai sur la psychologie de
              l'invention dans le domaine mathématique* , first French edi-
              tion, revised and extended, Albert Blanchard, Paris, 1959.

[Ha91]   Haake F., *Quantum Signatures of Chaos*, Springer Verlag,
              1991.

[Hav03]   Havil J., *Gamma. Exploring Euler's Constant*, Princeton Uni-
              versity Press, 2003.

[Ho00]   Hodges A., *Alan Turing: The Enigma*, Simon and Schuster,
              New York 1983; second edition Walker & Co., New York 2000.

[HPB82]   Haq R.U., Pandey A., Bohigas O., 'Fluctuation Properties of
              Energy Levels: Do Theory and Experiment Agree?', Phys.
              Rev. Lett. **48** (1982) 1086-1089.

[HSY92]   Hunt B., Sauer, Yorke J., 'Prevalence: A translational in-
              variant "almost everywhere" on infinite dimensional spaces',
              BAMS, vol. 27, **2** (1992) 217-238.

[KS99]   Katz N., Sarnak.P., *Random Matrices, Frobenius Eigenvalues,
              and Monodromy*, AMS, Providence, 1999.

[KS00]   Keating J.P., Snaith N.C., 'Random Matrix Theory and $\zeta(1/2 + it)$', Comm. Math. Phys. **214** (2000) 57-89.

[Le04]   Leboeuf P., 'Regularity and chaos in the nuclear masses', Lectures delivered at VIII Hispalensis International Summer School, Sevilla 2003, nucl-th/0406064, to appear in Lecture Notes in Physics, J.M. Arias, M. Lozano, eds Springer.

[LM66]   Lieb E.H. and Mattis D.C., *Mathematical Physics in One Dimension*, Academic Press, 1966.

[Me90]   Mehta M.L., *Random Matrices and the Statistical Theory of Energy Levels*, Academic Press 1967, second revised and enlarged edition, 1990, third enlarged edition, 2004.

[Mo73]   Montgomery H., 'The pair correlation of zeros of the zeta function', Washington University, St. Louis, Symposium on Analytic Number Theory (H.G. Diamond, ed.), Proc. Sympos. Pure Math. **24** (1973) 181-193, Amer. Math. Soc.

[Mo04]   Montgomery H., communication to the author, 04/08/04.

[MHBHA04]   Müller S., Heusler S., Braun P., Haake F., Altland A., 'Semiclassical Foundation of Universality in Quantum Chaos', Phys. Rev. Lett. **93** (2004) 014103-1-4.

[MK79]   McDonald S.W., Kaufman A.N., 'Spectrum and Eigenfunctions for a Hamiltonian with Stochastic Trajectories', Phys. Rev. Lett. **42** (1979) 1189-1191.

[Od]   Odlyzko A.M., www.dtc.umn.edu/~odlyzko/

[Od87]   Odlyzko A.M., 'On the distribution of spacings between zeros of the zeta function', Math. Comp. **48** (1987) 273-308.

[Od89]   Odlyzko A.M., 'The 10^{20}-th zero of the Riemann zeta function and 70 millions of its neighbors', ATT Bell Laboratories preprint, 1989.

[Od01]   Odlyzko A.M., 'The 10^{22}-nd zero of the Riemann zeta function', in *Dynamical, Spectral and Arithmetic Zeta Functions*, M. van Frankenhuysen and M.L. Lapidus (eds.), Amer. Math. Soc., Contemporary Math. Series, no **290** (2001) 139-144.

[Od04]   Odlyzko A.M., communication to the author, 24/07/04.

[Oz88]   Ozorio de Almeida A., *Hamiltonian Systems: Chaos and Quantization*, Cambridge University Press, 1988.

[Pa91]   Pais A., *Niels Bohr's Times: In physics, Philosophy and Polity*, Clarendon Press, Oxford, 1991.

[Pe73]   Percival I.C., 'Regular and irregular spectra', J. Phys. **B6** (1973) L229-L232.

[Po65]   Porter C.E., *Statistical Theories of Spectra: Fluctuations*, Academic Press 1965.

[Pol82]   Polya G., letter to A. Odlyzko, 03/01/82, in [Od].

[Ri00]   Richter K., *Semiclassical Theory of Mesoscopic Quantum Systems*, Springer, 2000.

[Sa99]   Sarnak P., in V. Arnold, M. Atiyah, P. Lax, B. Mazur, editors, *Mathematics: Frontiers and Perspectives*, Am. Math. Soc., 1999.

[Sc94]   Schweber.S., *QED and the men who made it: Dyson, Feynman, Schwinger, and Tomonaga*, Princeton University Press, 1994.

[St99]   Stöckmann H.-J., *Quantum Chaos, An Introduction*, Cambridge University Press, 1999.

[Tu53]   Turing A.M., 'Some Calculations of the Riemann Zeta-Function', Proc. London Math. Soc. **3** (1953) 99-117, reproduced in [Br92], with a comment by R. Heath-Brown.

[VW29]   Von Neumann J. and Wigner E., Physikalische Zeitschrift **30** (1929) 467-470, english translation 'On the Behavior of Eigenvalues in Adiabatic Processes'.

[Wi56]   Wigner E.P., 'Results and Theory of Resonance Absorption' in 'Conference on Neutron Physics by Time of Flight', Gatlinburg, 1-2 Nov. 1956, reproduced in [Po65].

[Wi57]   Wigner E.P., 'Interpretation of low energy neutron spectroscopy' in 'International Conference on Neutrons Interactions with the Nucleus', Columbia University, 9-13 Sept. 1957, reproduced in [Po65].

[Wi60]   Wigner E.P., 'The Unreasonable Effectiveness of Mathematics in the Natural Sciences', Comm. Pure Appl. Math. **13** (1960), reprinted in *Symmetries and Reflections*, see [Wi61]

[Wi61]   Wigner E.P., 'The Probability of the Existence of a Self-Reproducing Unit' in *The Logic of Personal Knowledge: Essays in Honor of Michael Polanyi*, London, Routledge and

Kegan Paul Ltd, 1961, reproduced in E.P. Wigner: *Symmetries and Reflections*, Indiana University Press, 1961, reedited by OxBow Press 1979.

[Wi67]   Wigner E.P., 'Random Matrices in Physics ', SIAM Review **9** (1967) 1-23.

[Wil68]   Wilkinson D.H., 'Shape Isomers', Comm. Nucl. Part. Phys. **2** (1968) 146-150.

Laboratoire Physique Théorique et Modèles Statistiques (LPTMS)
UMR Université Paris-Sud Associée au CNRS,
91405 Orsay Cedex,
France

Basic analytic number theory

David W Farmer *

Abstract

We give an informal introduction to the most basic techniques used
to evaluate moments on the critical line of the Riemann zeta-function
and to find asymptotics for sums of arithmetic functions.

1 Introduction

The simplest way to compute a moment of the zeta-function is to approximate
the zeta-function by Dirichlet polynomials and then compute the moment of
the polynomials. In this paper we describe the most rudimentary techniques in
this area. Along the way we discuss the basic methods for finding asymptotics
of sums of arithmetic functions, and we also compute the arithmetic factor
in the standard conjectures for moments of the zeta-function. The methods
described here are completely standard: our intention is to give a brief introduction to those who are new to the subject. The standard reference is
Titchmarsh [T] and we cite the specific sections where one can look for more
details.

---

*Work supported by the American Institute of Mathematics and by the Focused Research
Group grant (0244660) from the NSF.

We assume a knowledge of the calculus of one complex variable. Readers unfamiliar with the big-O, little-o, and $\ll$ notation (and physicists, who may use the $\ll$ notation differently), should consult the Appendix.

2   The "first moment"

The simplest approximation to the zeta-function[1] is

$$\zeta(s) = \sum_{1 \le n \le T} \frac{1}{n^s} - \frac{T^{1-s}}{1-s} + O(T^{-\sigma}), \qquad (2.1)$$

where $s = \sigma + it$ and the equation is valid for $|t| \le T$. See [T], Section 4.11. Specializing to $s = \frac{1}{2} + it$, we have

$$\zeta(\tfrac{1}{2} + it) = \sum_{1 \le n \le T} \frac{1}{n^{\frac{1}{2}+it}} + O\left(\frac{T^{\frac{1}{2}}}{1+|t|}\right). \qquad (2.2)$$

Now we can find the average value of the zeta-function on the critical line. The justification of the steps follows the calculation.

$$\begin{aligned}
\int_0^T \zeta(\tfrac{1}{2}+it)\,dt &= \sum_{1 \le n \le T} \frac{1}{\sqrt{n}} \int_0^T n^{-it}\,dt + O\left(T^{\frac{1}{2}} \int_0^T \frac{1}{1+|t|}\,dt\right) \\
&= T + \sum_{2 \le n \le T} \frac{1}{\sqrt{n}} \frac{1-n^{-iT}}{\log n} + O\left(T^{\frac{1}{2}} \log T\right) \\
&= T + O\left(T^{\frac{1}{2}} \log T\right). \qquad (2.3)
\end{aligned}$$

In the first step we switched the sum and the integral, which is justified because both are finite. The sum on the second line was estimated by

$$\sum_{2 \le n \le T} \frac{1}{\sqrt{n}} \frac{1-n^{-iT}}{\log n} \ll \sum_{2 \le n \le T} \frac{1}{\sqrt{n}\log n} \ll \int_2^T \frac{1}{\sqrt{x}\log x}dx \ll \frac{T^{\frac{1}{2}}}{\log T}, \qquad (2.4)$$

which is smaller than the other error term.

Thus, the zeta-function on the critical line is 1 on average. This is not a particularly useful piece of information, for it is the magnitude of the zeta-function which is of interest. So we consider the moments of $|\zeta(\tfrac{1}{2}+it)|$.

---

[1] Editors' comment: The Riemann zeta function is defined in the lectures by D.R. Heath-Brown starting on page 1, Section 3, equation (3.1).

3 The 2nd moment

The simple fact that $|z|^2 = z\overline{z}$ means that $|\zeta(\frac{1}{2}+it)|^K$ is much more amenable to methods of complex analysis when $K = 2k$ is an even integer. The easiest case is the 2nd moment, which we compute in this section.

For later use it will be helpful to first consider the 2nd moment of a general Dirichlet polynomial. Suppose

$$P(s) = \sum_{1 \le n \le N} \frac{a_n}{n^s}. \tag{3.1}$$

We have

$$
\begin{aligned}
\int_0^T |P(it)|^2 \, dt &= \int_0^T \left| \sum_{1 \le n \le N} \frac{a_n}{n^{it}} \right|^2 dt \\
&= \int_0^T \sum_n \frac{a_n}{n^{it}} \sum_m \frac{\overline{a_m}}{m^{-it}} \, dt \\
&= \sum_{n,m} a_n \overline{a_m} \int_0^T \left(\frac{m}{n} \right)^{it} dt \\
&= T \sum_n |a_n|^2 + \sum_{n \ne m} \frac{a_n \overline{a_m} \left(\left(\frac{m}{n} \right)^{iT} - 1 \right)}{\log(m/n)} \\
&= T\,\mathcal{M}(N) + \mathcal{E}(N,T), \tag{3.2}
\end{aligned}
$$

say. We think of $T\,\mathcal{M}(N)$ as the main term and $\mathcal{E}(N,T)$ as an error term, so we want to understand when $\mathcal{E}(N,T)$ will be smaller in magnitude than $T\,\mathcal{M}(N)$.

Setting $m = n + h$ we can rewrite the error term as

$$\mathcal{E}(N,T) = \sum_n \sum_{h \ne 0} \frac{a_n \overline{a_{n+h}} \left(\left(1 + \frac{h}{n}\right)^{iT} - 1 \right)}{\log(1 + h/n)}. \tag{3.3}$$

Now consider only the $h = 1$ term from the above sum:

$$\sum_{1 \le n \le N} a_n \overline{a_{n+1}} \cdot n \cdot (\text{something bounded}). \tag{3.4}$$

Without any information on a_n, the above sum, which is just one part of $\mathcal{E}(N,T)$, could be about the same size as $N\,\mathcal{M}(N)$. Thus, in general one should only expect $\mathcal{E}(N,T)$ to be smaller than $T\,\mathcal{M}(N)$ if $N < T$. And if $N > T$ then one may need some detailed information about the coefficients a_n in order to extract something meaningful from $\mathcal{E}(N,T)$. Goldston and Gonek [GG] have given a clear discussion of these issues.

One can carry through the above calculation to obtain a useful general mean value theorem for Dirichlet polynomials. See Titchmarsh [T], Section 7.20. However, a stronger result is provided by the mean value theorem of Montgomery and Vaughan [MV]:

$$\int_0^T \left| \sum_{1 \le n \le N} \frac{a_n}{n^{it}} \right|^2 dt = \sum_{1 \le n \le N} |a_n|^2 \left(T + O(n) \right). \tag{3.5}$$

This result is best possible for general sequences a_n. Note that if $N = o(T)$ then the error term is smaller than the main term.

To use the mean value theorem to compute the second moment of the zeta-function, first write (2.2) as $\zeta = S + E$. That is,

$$S = \sum_{1 \le n \le T} \frac{1}{n^{\frac{1}{2}+it}} \qquad \text{and} \qquad E = O\left(\frac{T^{\frac{1}{2}}}{1+|t|} \right). \tag{3.6}$$

Using

$$|\zeta|^2 = |S+E|^2 = (S+E)(\overline{S}+\overline{E}) = |S|^2 + 2\mathbb{R}e S\overline{E} + |E|^2, \tag{3.7}$$

we have

$$\int_0^T \left| \zeta(\tfrac{1}{2}+it) \right|^2 dt = \int_0^T |S|^2\, dt + 2\mathbb{R}e \int_0^T S\overline{E}\, dt + \int_0^T |E|^2\, dt. \tag{3.8}$$

We want to evaluate the $|S|^2$ integral as our main term and estimate the $|E|^2$ integral as our error term, but what to do about the cross term? By the Cauchy-Schwartz inequality,

$$2\mathbb{R}e \int_0^T S\overline{E}\, dt \ll \int_0^T |S||E|\, dt \ll \left(\int_0^T |S|^2 dt \right)^{\frac{1}{2}} \left(\int_0^T |E|^2 dt \right)^{\frac{1}{2}}. \tag{3.9}$$

If $\int_0^T |E|^2 dt$ is smaller than $\int_0^T |S|^2 dt$, then so is the right side of the inequality above. We have the general principle that if the "main error term" is smaller than the main term, then so are the cross terms. It remains only to evaluate $\int_0^T |S|^2 dt$ and estimate $\int_0^T |E|^2 dt$.

For the main term use (3.5) with $a_n = n^{-\frac{1}{2}}$:

$$
\begin{aligned}
\int_0^T |S|^2 dt &= \sum_{1 \le n \le T} \frac{1}{n}\left(T + O(n) \right) \\
&= T \left(\log T + O(1) \right) + \sum_{1 \le n \le T} O(1) \\
&= T \log T + O(T).
\end{aligned}
\tag{3.10}
$$

For the error term we have

$$\int_0^T |E|^2 dt \ll T \int_0^T \frac{1}{(1+|t|)^2}\, dt \ll T, \tag{3.11}$$

which is smaller than the main term. We have just proven the mean value result

$$\int_0^T |\zeta(\tfrac{1}{2}+it)|^2 dt = T \log T + O(T \log^{\frac{1}{2}} T). \tag{3.12}$$

Note that we do not obtain an error term of $O(T)$ in (3.12) by these methods. However, a much better error term can be obtained. The first step toward this is given in the next section.

Exercise. Approximate the sum by an integral to show that if $A > 1$ then

$$\sum_{n>T} \frac{1}{n^A} = \frac{T^{1-A}}{A-1} + O(T^{-A}). \tag{3.13}$$

Conclude that if $A > 1$ then

$$\sum_{n \leq T} \frac{1}{n^A} = \zeta(A) - \frac{T^{1-A}}{A-1} + O(T^{-A}). \tag{3.14}$$

Exercise. Show that if $\frac{1}{2} < \sigma < 1$ then

$$\int_0^T |\zeta(\sigma+it)|^2 dt = T\left(\zeta(2\sigma) - \frac{T^{1-2\sigma}}{2\sigma-1}\right) + O\left(\frac{T^{\frac{3}{2}-\sigma}}{1-\sigma} \log^{\frac{1}{2}} T\right), \tag{3.15}$$

where the implied constant in the big-O term is independent of σ.

Since the error term in (3.15) is uniform in sigma, we can let $\sigma \to \frac{1}{2}^+$ to recover (3.12). This makes use of the fact that $\zeta(s)$ has a simple pole with residue 1 at $s = 1$.

Note that if $\sigma > \frac{1}{2}$ is independent of T then the right side of (3.15) is of size $\approx T$. On the other hand, the second moment on the $\frac{1}{2}$-line is of size $T \log T$. Thus there is an abrupt change in the behavior of the zeta-function when one moves onto the critical line. Equation (3.15) illustrates that the transition occurs on the scale of $1/\log T$ from the $\frac{1}{2}$-line.

4 Better 2nd moment

The methods of the previous section are not sufficient to evaluate the main term of the 2nd moment with an error term $O(T^A)$ for $A < 1$, nor are those methods sufficient to evaluate the 4th moment of the zeta-function. Evaluating the 4th moment by squaring (2.2) gives a Dirichlet polynomial of length T^2, which cannot be handled by the Montgomery-Vaughan mean value theorem. So one needs either a shorter approximation to $\zeta(s)$, or an approximation to $\zeta^2(s)$ of length $\leq T$, or a way to handle longer polynomials. In preparation for the 4th moment, we first evaluate the 2nd moment with a better error term.

The "approximate functional equation" of Hardy and Littlewood expresses the ζ-function as a sum of *two* short Dirichlet polynomials:

$$\zeta(s) = \sum_{1 \leq n \leq x} \frac{1}{n^s} + \chi(s) \sum_{1 \leq n \leq y} \frac{1}{n^{1-s}} + O\left(x^{-\sigma} + |t|^{\frac{1}{2}-\sigma}y^{\sigma-1}\right), \qquad (4.1)$$

where $xy = t/2\pi$ and $\chi(s)$ is the usual factor in the functional equation[2] $\zeta(s) = \chi(s)\zeta(1-s)$. The name "approximate functional equation" comes from the fact that the right side looks like $\zeta(s)$ when x is large and like $\chi(s)\zeta(1-s)$ when y is large. On the $\frac{1}{2}$-line we have

$$\zeta(s) = \sum_{1 \leq n \leq N} \frac{1}{n^{\frac{1}{2}+it}} + \chi(s) \sum_{1 \leq n \leq N} \frac{1}{n^{\frac{1}{2}-it}} + O\left(N^{-\frac{1}{2}}\right), \qquad (4.2)$$

where we set $x = y = N = \sqrt{t/2\pi}$.

Hardy and Littlewood used the approximate functional equation to evaluate the second moment of the zeta-function with a better error term. We will not carry out the calculation, but just give the flavor. See Titchmarsh [T] Section 7.4 for details. Writing (4.2) as $\zeta = S + \chi(s)\overline{S} + E$, we have

$$\int_0^T |\zeta(\tfrac{1}{2}+it)|^2 dt = 2 \int_0^T |S|^2 dt + 2\mathbb{R}e \int_0^T \chi(\tfrac{1}{2}-it)S^2 dt + \text{error term.} \quad (4.3)$$

In that calculation we used the fact that $|\chi(\tfrac{1}{2}+it)| = 1$. Below we will use the more precise information

$$\chi(s) = \left(\frac{t}{2\pi}\right)^{\frac{1}{2}-s} e^{it+\pi i/4}\left(1 + O(t^{-1})\right). \qquad (4.4)$$

The main difficulty is evaluating the $\chi(\tfrac{1}{2}-it)S^2$ term, which equals

$$\sum_{n,m} \frac{1}{\sqrt{nm}} \int_0^T \chi(\tfrac{1}{2}-it)(nm)^{-it} dt. \qquad (4.5)$$

By (4.4), up to a negligible error that integral is of the form $\int e^{if(t)} dt$. If $f(t)$ has a stationary point in the range of integration then we can extract a main term, otherwise it will become an error term. In particular, that integral can be handled by the method of stationary phase. See Titchmarsh [T] Section 7.4

---

[2]Editors' comment: For more on the functional equation see the contribution of D.R. Heath-Brown, page 1, Theorem 8.

for details. Our point here is that by virtue of (4.4), integrals involving $\chi(\frac{1}{2}+it)$ and Dirichlet polynomials can be handled. The result that can be obtained by the above argument is

$$\int_0^T |\zeta(\tfrac{1}{2}+it)|^2 dt = T\log T + (2\gamma-1)T + O(T^{\frac{1}{2}+\varepsilon}). \tag{4.6}$$

The error term can be improved by more sophisticated methods.

There are many applications of mean values of the zeta-function multiplied by a Dirichlet polynomial. Suppose

$$M(s) = \sum_{1\le n<T^\theta} \frac{b_n}{n^s} \tag{4.7}$$

with $b_n \ll n^\varepsilon$. By the approximate functional equation, $\zeta(s)M(s)$ can be approximated by Dirichlet polynomials of length $T^{\frac{1}{2}+\theta}$. So by the above methods, if $\theta < \frac{1}{2}$ then one should be able to find an asymptotic formula for $\int_0^T |\zeta M(\frac{1}{2}+it)|^2 dt$. Evaluating such an integral, with $\theta = \frac{1}{2}-\varepsilon$, was key to Levinson's proof that more than one-third of the zeros of the zeta-function are on the critical line. Conrey made use of very deep and technical results to evaluate such an integral with $\theta = \frac{4}{7}-\varepsilon$, leading to the result that more than two-fifths of the zeros are on the critical line.

5   The 4th moment

To evaluate the 4th moment of the zeta-function by the methods described above, one requires an approximation to $\zeta(s)^2$ of length less than T. This is provided by the following approximate functional equation:

$$\zeta(s)^2 = \sum_{1\le n\le x} \frac{d(n)}{n^s} + \chi(s)^2 \sum_{1\le n\le y} \frac{d(n)}{n^{1-s}} + O(x^{\frac{1}{2}-\sigma}\log t), \tag{5.1}$$

where $xy = (t/2\pi)^2$ and $d(n)$ is the number of divisors of n. More generally one should expect an approximate functional equation of the form

$$\zeta(s)^k = \sum_{1\le n\le x} \frac{d_k(n)}{n^s} + \chi(s)^k \sum_{1\le n\le y} \frac{d_k(n)}{n^{1-s}} + \text{ error term} \tag{5.2}$$

where $xy \approx t^k$ and $d_k(n)$ is the k-fold divisor function

$$d_k(n) = \sum_{n_1\cdots n_k=n} 1, \tag{5.3}$$

which has generating function

$$\zeta(s)^k = \sum_{n=1}^{\infty} \frac{d_k(n)}{n^s}, \qquad \sigma > 1. \tag{5.4}$$

Plugging (5.1) into the Montgomery-Vaughan mean value theorem leads to

$$\int_0^T |\zeta(\tfrac{1}{2}+it)|^4 dt = \frac{1}{2\pi^2} T \log^4 T + \text{ error term.} \tag{5.5}$$

If you actually do the calculation, you will find that in order to determine the main term you need to evaluate sums like

$$\sum_{1\le n\le X} \frac{d(n)^2}{n}. \tag{5.6}$$

There is a standard technique for finding the leading-order asymptotics of such sums, which is given in the next section.

5.1 Comments on approximate functional equations

The error term in (5.1) is rather large and leads to an error term of size $O(T\log^2 T)$ in the 4th moment $\int_0^T |\zeta(\tfrac{1}{2}+it)|^4 dt$. The large size of the error term is due to the fact that our sums have a sharp cut-off. The error term can be much reduced by having a smooth weight in the sums. That is,

$$\zeta(s)^k = \sum_n \frac{d_k(n)}{n^s}\varphi(n,t) + \chi(s)^k \sum_n \frac{d_k(n)}{n^{1-s}}\varphi^*(n,t) + \text{ small error term,} \tag{5.7}$$

where φ and φ^* are particular functions that are approximately 1 for $n < t^{\frac{k}{2}}$ and decay for $n > t^{\frac{k}{2}}$. As our previous discussion should suggest, it is the length of the sums, and not the size of the error term, which provides the true difficulty when $k > 2$.

6 Perron's formula

Here is the problem: you have an arithmetical function a_n and you want to find the asymptotics of

$$S(X) = \sum_{1\le n\le X} a_n. \tag{6.1}$$

This problem can often be solved by the most basic methods of analytic number theory.

First note that for integers $N \ge 1$,

$$\frac{1}{2\pi i}\int_{1-iY}^{1+iY} A^s \frac{ds}{s^N} = \begin{cases} \frac{\log^{N-1} A}{(N-1)!} + \text{ error term} & \text{if } A > 1 \\ 0 + \text{ error term} & \text{if } A < 1 \end{cases} \tag{6.2}$$

To see this, consider the integral

$$\frac{1}{2\pi i}\int_{\mathcal{C}_1} A^s\, \frac{ds}{s^N} \tag{6.3}$$

where the integration is over the closed rectangular path connecting the points

$$\mathcal{C}_1 = \begin{cases} [1-iY, 1+iY, -B+iY, -B-iY], & \text{if } A > 1 \\ [1-iY, 1+iY, B+iY, B-iY], & \text{if } A < 1, \end{cases} \tag{6.4}$$

where B is a large positive number. In both cases the main term comes from the residue of the pole at 0, which is or is not inside the path of integration.

Exercise. Bound the error term in (6.2) by estimating the integral along the three segments of $\mathcal{C}_1$ other than $[1-iY, 1+iY]$. You should find that if $N \geq 2$, then you can let $Y \to \infty$ and the error vanishes. See Section 3.12 of [T] if you aren't sure how to begin.

To evaluate (6.1), let

$$F(s) = \sum_{n=1}^{\infty} \frac{a_n}{n^s}, \tag{6.5}$$

and suppose that the sum converges absolutely for $\sigma > \sigma_0$. Using (6.2) and supposing $\sigma > \sigma_0$, we have

$$
\begin{aligned}
\frac{1}{2\pi i}\int_{\sigma-iY}^{\sigma+iY} F(s)X^s\, \frac{ds}{s} &= \sum_{n=1}^{\infty} a_n \frac{1}{2\pi i}\int_{\sigma-iY}^{\sigma+iY} \left(\frac{X}{n}\right)^s \frac{ds}{s} \\
&= \sum_{n=1}^{\infty} a_n \begin{cases} 1 + \text{ error term} & \text{if } X > n \\ 0 + \text{ error term} & \text{if } X < n \end{cases} \\
&= S(X) + \text{ error term}. \tag{6.6}
\end{aligned}
$$

This is known as Perron's formula. If we can learn enough about the function $F(s)$ so that the integral in (6.6) can be evaluated in another way, then we will have a formula for $S(X)$.

Suppose (as is frequently the case) that $F(s)$ has a pole at σ_0 and no other poles in the half-plane $\sigma > \sigma_1$ for some $\sigma_1 < \sigma_0$. Then consider

$$\frac{1}{2\pi i}\int_{\mathcal{C}_2(\varepsilon)} F(s)X^s\, \frac{ds}{s}, \tag{6.7}$$

where for $\varepsilon > 0$ the integration is over the rectangular path with vertices

$$\mathcal{C}_2(\varepsilon) = [\sigma - iY, \sigma + iY, \sigma_1 + \varepsilon + iY, \sigma_1 + \varepsilon - iY]. \tag{6.8}$$

We can evaluate (6.7) by finding the residue at the pole $s = \sigma_0$ and at $s = 0$ (if 0 is inside the path of integration. And in the same way as you estimated the error term in (6.2), we find that (6.7) equals the integral in Perron's formula

194 David W. Farmer

plus an error term. The final step of bounding the integral on the 3 other segments requires the additional ingredient of a bound for $F(\sigma+it)$ as $t \to \infty$, uniformly for $\sigma > \sigma_1$.

For example, at the end of the previous section we wanted to evaluate the sum of $d(n)^2/n$.

Exercise. Check that

$$\sum_n \frac{d(n)^2}{n^s} = \frac{\zeta^4(s)}{\zeta(2s)}. \tag{6.9}$$

Hint: both sides have an Euler product. The factors on the right can be found from the Euler product for the zeta-function. Those on the left require summing $\sum_{j=0}^{\infty} d(p^j)^2 p^{-js}$.

Thus, we apply Perron's formula (6.6) with

$$F(s) = \frac{\zeta^4(s+1)}{\zeta(2s+2)}. \tag{6.10}$$

To determine the analytic properties of $F(s)$, use the fact that $\zeta(s)$ is entire except for a simple pole at $s = 1$, where we have the Laurent expansion

$$\zeta(s) = \frac{1}{s-1} + \gamma + \cdots . \tag{6.11}$$

Also $\zeta(s)$ has no zeros in $\sigma > 1$, and no zeros in $\sigma > \frac{1}{2}$ assuming the Riemann Hypothesis. In these calculations one frequently needs that $\zeta(2) = \pi^2/6$ and

$$\zeta(s) = -\frac{1}{2} - \frac{1}{2}\log(2\pi)s + \cdots \tag{6.12}$$

for s near 0.

To estimate the error terms, one can use the convexity estimate

$$\zeta(\sigma+it) \ll \begin{cases} 1 & \sigma > 1 \\ |t|^{\frac{1}{2}-\frac{1}{2}\sigma} & 0 < \sigma < 1 \\ |t|^{\frac{1}{2}-\sigma} & \sigma < 0, \end{cases} \tag{6.13}$$

along with $\zeta(s) \gg 1$ for $\sigma > 1$. Also, assuming RH we have $t^{-\varepsilon} \ll \zeta(\sigma+it) \ll t^{\varepsilon}$ for $\sigma > \frac{1}{2}$. All of these estimates are for fixed σ as $t \to \infty$.

Assembling the pieces we find

$$\sum_{1 \le n \le X} \frac{d(n)^2}{n} \sim \frac{\log^4 X}{4\pi^2}. \tag{6.14}$$

Exercise. Argue that (6.14) is of the form $XP_4(\log X) + O(X^B)$ where P_4 is a polynomial of degree 4 and $B < 1$. Find the next-to-leading coefficient of P_4, and estimate B both with and without assuming the Riemann Hypothesis.

Exercise. Deduce the following asymptotics:

$$\sum_{1 \le n \le X} d_k(n) \ \sim \ \frac{1}{k!} X \log^k X$$

$$\sum_{1 \le n \le X} \varphi(n) \ \sim \ \frac{3}{\pi^2} X^2, \tag{6.15}$$

where $\varphi(n)$ is the Euler totient function.[3] In addition to the generating function (5.4), you should use (and prove):

$$\sum_n \frac{\varphi(n)}{n^s} = \frac{\zeta(s-1)}{\zeta(s)}. \tag{6.16}$$

Exercise. Find the next-to-leading order terms in the previous exercise. Also determine the shape of the main terms and estimate the size of the error terms, both with and without the Riemann Hypothesis.

Note that there are interesting and important cases where the above analysis is inadequate. For example, in the proof of the prime number theorem $a_n = \Lambda(n)$, the von Mangoldt function.[4] Then $F(s)$ has a pole at $s = 1$ as well as poles at the zeros of the ζ-function and one must use a more complicated path of integration as well as nontrivial estimates for $\zeta(s)$ in the critical strip.

7 The conjecture for moments of the zeta-function

Much recent work on the relationship between L-functions and Random Matrix Theory was motivated by the problem of finding conjectures for the $2k$th moment of the Riemann zeta-function on the critical line. Conrey and Ghosh [CG] formulated it as follows: for each integer $k \ge 0$ there exists an integer g_k such that

$$\int_0^T |\zeta(\tfrac{1}{2} + it)|^{2k} dt \sim g_k \frac{a_k}{k^2!} T \log^{k^2} T, \tag{7.1}$$

where

$$a_k = \prod_p \left(1 - \frac{1}{p}\right)^{k^2} \sum_{m=0}^{\infty} \binom{k+m-1}{m}^2 p^{-m}. \tag{7.2}$$

In this conjecture the only missing ingredient is the *integer* g_k. Keating and Snaith [KS] computed the moments of characteristic polynomials of unitary

---

[3]Editors' comment: $\varphi(n)$ is defined in the contribution by D.R. Heath-Brown, page 1 after Theorem 20.

[4]Editors' comment: The von Mangoldt function is defined in the lectures by D.R. Heath-Brown, page 1, Corollary 1.

matrices and used the result to conjecture

$$g_k = k^2! \prod_{j=0}^{k-1} \frac{j!}{(k+j)!}. \tag{7.3}$$

It is not trivial to show that this g_k is actually an integer [CF].

Our last topic in this paper is to show how the factor a_k arises naturally. From the approximate functional equation (5.2) it is reasonable to consider

$$\int_0^T \left| \sum_{n<T} \frac{d_k(n)}{n^{\frac{1}{2}+it}} \right|^2 dt. \tag{7.4}$$

Note that the sum has length T. This is good because we can use the mean value theorem. But it is bad because the polynomial is not long enough to fully approximate $\zeta(\frac{1}{2}+it)^k$. We cannot expect this mean value to equal the $2k$th moment of the zeta-function, but how far will it be off? It would be nice if it were off by some simple factor, so one possible interpretation of g_k is "the number of length T polynomials needed to capture the $2k$th moment of the ζ-function." We do not claim that this was the original reasoning of Conrey and Ghosh.

By the Montgomery-Vaughan mean value theorem the above integral has main term

$$\sum_{n<T} \frac{d_k(n)^2}{n}. \tag{7.5}$$

By Perron's formula, to evaluate this we need to find the leading pole of

$$F(s) = \sum_{n<T} \frac{d_k(n)^2}{n^s}. \tag{7.6}$$

If $k > 2$ then $F(s)$ is not a simple expression involving known functions, but fortunately we do not require complete information about $F(s)$.

First note that

$$\begin{aligned} \zeta(s) &= \prod_p \left(1 + \frac{1}{p^s} + \cdots\right) \\ &= \frac{1}{s-1} + \cdots, \end{aligned} \tag{7.7}$$

so

$$\begin{aligned} \zeta(s)^N &= \prod_p \left(1 + \frac{N}{p^s} + \cdots\right) \\ &= \frac{1}{(s-1)^N} + \cdots. \end{aligned} \tag{7.8}$$

Thus, if the coefficients of p^{-js} in an Euler product are integers that only depend on j, then the coefficient of p^{-s} tells you the order of the pole at $s = 1$. Since

$$
\begin{aligned}
d_k(p) &= \sum_{n_1 \cdots n_k = p} 1 \\
&= k,
\end{aligned}
\tag{7.9}
$$

we see that $F(s)$ has a pole of order k^2 at $s = 1$, that is, $F(s) = a_k(s-1)^{-k^2} + \cdots$, where we will show that a_k is as given above. To see that $F(s)$ has no other poles in $\sigma > \frac{1}{2}$, note that

$$
\zeta^{-k^2}(s)F(s) = \prod_p \left(1 + \frac{\beta_2}{p^{2s}} + \frac{\beta_3}{p^{3s}} + \cdots \right),
\tag{7.10}
$$

where the β_j are certain integers that do not grow too fast. In particular, the above Euler product converges absolutely for $\sigma > \frac{1}{2}$ so it represents a regular function that is bounded in $\sigma > \frac{1}{2} + \varepsilon$. We have all of the pieces to apply the methods of the previous section, giving

$$
\int_0^T \left| \sum_{n < T} \frac{d_k(n)}{n^{\frac{1}{2}+it}} \right|^2 dt \sim \frac{a_k}{k^2!} T \log^{k^2} T,
\tag{7.11}
$$

where

$$
\begin{aligned}
a_k &= \lim_{s \to 1}(s-1)^{k^2} F(s) \\
&= \lim_{s \to 1} \zeta(s)^{-k^2} F(s) \\
&= \prod_p \left(1 - \frac{1}{p} \right)^{k^2} \sum_{m=0}^{\infty} d_k(p^m)^2 p^{-m}.
\end{aligned}
\tag{7.12}
$$

Note that the product converges because $d_k(p) = k$ and $d_k(n) \ll n^\varepsilon$.

Finally, $d_k(p^m) = \binom{k+m-1}{m}$, as can be seen by the following argument. Since $d_k(p^m)$ is the number of ways of writing $e_1 + \cdots + e_k = m$, we can select the e_j by writing down $m + k - 1$ circles $\circ$ and filling in $k - 1$ of them to make a dot $\bullet$. Then the e_j are the number of circles between the dots, including the circles before the first and after the last dot. For example, here is one configuration that arises from $d_5(p^3)$:

$$
\begin{array}{ccccccc}
\bullet & \circ & \circ & \bullet & \circ & \bullet & \bullet \\
\end{array}
$$
$$
p^3 = \quad 1 \quad p^2 \quad p \quad 1 \quad 1
\tag{7.13}
$$

8 The Estermann phenomenon

The idea behind formula (7.10) can be generalized to show that if $c(n)$ is a multiplicative function such that $c(n) \ll n^\varepsilon$ and $c(p^j)$ is an integer that is independent of the prime p, then

$$F(s) = \sum_{n=1}^{\infty} \frac{c(n)}{n^s}$$
$$= Z_J(s) \prod_{j<J} \zeta(js)^{C(j)}, \tag{8.1}$$

where the $C(j)$ are integers and $Z(s)$ is regular and bounded in $\sigma > 1/J$. Thus $F(s)$, which is originally defined for $\sigma > 1$, has a meromorphic continuation to $\sigma > 0$.

Note that $F(s)$ cannot be continued past $\sigma = 0$ unless $C(j) = 0$ for almost all j. This is because the zeros of the zeta-function lead to zeros or poles of $F(s)$ that accumulate along the $\sigma = 0$ line, giving a natural boundary. This is known as "the Estermann phenomenon".

9 Appendix

9.1 Big-O and $\ll$ notation

The statement
$$f(x) = O(g(x)) \qquad \text{as} \quad x \to \infty \tag{9.1}$$
is pronounced "$f(x)$ is big oh of $g(x)$." It is equivalent to

$$f(x) \ll g(x) \qquad \text{as} \quad x \to \infty, \tag{9.2}$$

which is pronounced "$f(x)$ is less than less than $g(x)$." The symbol $\ll$ is typed as \ll in TeX. Both of the above statements mean the following: there exists a constant $C > 0$ such that if x is sufficiently large then $|f(x)| \le C\, g(x)$. The number C is called "the implied constant."

Note:

- $g(x)$ must be positive if x is sufficiently large.

- $f(x) \ll g(x)$ does *not* mean that $f(x)$ is much smaller than $g(x)$. It is more accurate to say that $f(x)$ does not grow faster than $g(x)$.

- the above statements have the condition "as $x \to \infty$". It is also common to use the big-O and $\ll$ notation to describe the behavior of a function as $x \to 0$. Then the definition is modified to "if x is sufficiently small". Usually context makes it clear which behavior is being considered.

- Both notations are useful: the $\ll$ does not require parentheses, and the big-O can be used as one term in a formula.

Here are some examples. Below, A and ε are arbitrary fixed positive numbers.

Examples assuming $x \to \infty$:

$$
\begin{aligned}
x^3 &\ll x^4 \\
\log(x) &= O(x) \\
\log(x) &\ll x^\varepsilon \\
x^A &= O(e^x) \\
\sin(x) &\ll 1 \\
(x+2)^{10} &\ll x^{10} \\
(x+2)^{10} &= x^{10} + O(x^9).
\end{aligned}
$$

Examples assuming $x \to 0$:

$$
\begin{aligned}
x^4 &\ll x^3 \\
\log(1+x) &= O(x) \\
\log(1+x) &= x + O(x^2) \\
\sin(x) &\ll x \\
(x+2)^{10} &= 1024 + O(x).
\end{aligned}
$$

9.2 Little-o notation

The statement

$$
f(x) = o(g(x)) \qquad \text{as} \quad x \to \infty \tag{9.3}
$$

is pronounced "$f(x)$ is little oh of $g(x)$." It means

$$
\lim_{x \to \infty} \frac{f(x)}{g(x)} = 0. \tag{9.4}
$$

Equivalently, for all $C > 0$, if x is sufficiently large then $|f(x)| \le Cg(x)$. It is like big-O where the implied constant can be made arbitrarily small.

Note that $f(x) \sim g(x)$, "$f(x)$ is asymptotic to $g(x)$" is equivalent to $f(x) = (1 + o(1))g(x)$.

References

[CF] J. B. Conrey and D. W. Farmer, *Mean values of L-functions and symmetry*, Internat. Math. Res. Notices (2000) **17** pp. 883–908.

[CG] J. B. Conrey and A. Ghosh, *Mean values of the Riemann zeta-function*, Mathematika 31 (1984) 159–161

[GG] D.A. Goldston and S. M. Gonek, *Mean value theorems for long Dirichlet polynomials and tails of Dirichlet series*, Acta Arith. 84 (1998), no. 2, 155–192.

[KS] J. P. Keating and N. C. Snaith, *Random matrix theory and L-functions at* $s = \frac{1}{2}$, Comm. Math. Phys. **214** (2000) pp. 91–110.

[MV] H. L. Montgomery and R. C. Vaughan *The large sieve*, Mathematika, 1973 **20**, 119–134

[T] E. C. Titchmarsh, The theory of the Riemann zeta-function. Second edition. Edited and with a preface by D. R. Heath-Brown, The Clarendon Press, Oxford University Press New York (1986).

American Institute of Mathematics
360 Portage Ave.
Palo Alto, CA 94306

farmer@aimath.org

Applications of Mean Value Theorems to the Theory of the Riemann Zeta Function

S.M. Gonek *†

Abstract

In the first half of these lectures we discuss mean value theorems for functions representable by Dirichlet series and sketch several applications to the distribution of zeros of the Riemann zeta function. These include the clustering of zeros about the critical line, Levinson's result that a third of the zeros are on the critical line, and a conditional result on the number of simple zeros. The second half focuses on mean values of Dirichlet polynomials, particularly "long" ones. We then show how these can be used to investigate the pair correlation of the zeros of the zeta function and to conjecture the sixth and eighth power moments of the zeta function on the critical line.

*This work was supported by NSF grant DMS 0201457 and by an NSF Focused Research Group grant. The paper was written during the author's visit to the Isaac Newton Institute.

†The first six sections of this paper closely follow the author's lectures at a conference at Chonbuk National University in South Korea in 2002. The author is grateful to the organizers of that conference for their kind permission to use that material here.

1   What is a Mean Value Theorem?

By a mean value theorem we mean an estimate for the average of a function. When $F(s)$ has a convergent Dirichlet series expansion in some half–plane $\operatorname{Re} s > \sigma_0$ of the complex plane, we typically take the average over a vertical segment:

$$\int_0^T |F(\sigma + it)|^2 \, dt \qquad \text{or} \qquad \int_0^T F(\sigma + it) \, dt \, .$$

The path of integration here need *not* lie in this half–plane. For example, we would like to know the size of the integrals

$$I_k(\sigma, T) = \int_0^T |\zeta(\sigma + it)|^{2k} \, dt \, ,$$

for $\sigma \geq 1/2$ and k a positive integer. Here $F(s) = \zeta(s)^k$ and its Dirichlet series converges only for $\sigma > 1$.

There are many variations. For example, one can consider a discrete mean value

$$\sum_{r=1}^R |F(\sigma_r + it_r)|^2 \, ,$$

where the points $\sigma_r + it_r$ lie in $\mathbb{C}$. Or, one can estimate the mean value of a Dirichlet polynomial

$$F(s) = F_N(s) = \sum_{n=1}^N a_n n^{-s}$$

of "length" N.

2   Mean Values and Zeros

Mean value estimates are very useful for studying the zeros of the zeta function; this is one of the reasons so much effort has been expended on them. One link between means and zeros can be seen in Jensen's Formula from classical function theory.

Theorem 2.1. (**Jensen's Formula**) Let $f(z)$ be analytic for $|z| \leq R$ and suppose that $f(0) \neq 0$. If $r_1, r_2, \ldots, r_n$ are the moduli of all the zeros of $f(z)$ inside $|z| \leq R$, then

$$\log\left(\frac{|f(0)| R^n}{r_1 r_2 \cdots r_n}\right) = \frac{1}{2\pi} \int_0^{2\pi} \log |f(Re^{i\theta})| \, d\theta \, .$$

Here we see that the mean value of $\log |f(z)|$ around a circle is related to the distribution of the zeros of $f(z)$ inside that circle. There is an analogous result for rectangles, which is often more useful when working with Dirichlet series, namely,

Theorem 2.2. (**Littlewood's Lemma**) Let $f(s)$ be analytic and nonzero on the rectangle $\mathcal{C}$ with vertices σ_0, σ_1, $\sigma_1 + iT$, and $\sigma_0 + iT$, where $\sigma_0 < \sigma_1$. Then

$$2\pi \sum_{\rho \in \mathcal{C}} \mathrm{Dist}(\rho) = \int_0^T \log|f(\sigma_0 + it)|\, dt - \int_0^T \log|f(\sigma_1 + it)|\, dt$$

$$+ \int_{\sigma_0}^{\sigma_1} \arg f(\sigma + iT)\, d\sigma - \int_{\sigma_0}^{\sigma_1} \arg f(\sigma)\, d\sigma,$$

where the sum runs over the zeros ρ of $f(s)$ in $\mathcal{C}$ and "Dist(ρ)" is the distance from ρ to the left edge of the rectangle.

When we use Littlewood's Lemma below, only the first term on the right–hand side will be significant. In order not to be too technical, we will always use the result in the form

$$2\pi \sum_{\rho \in \mathcal{C}} \mathrm{Dist}(\rho) = \int_0^T \log|f(\sigma_0 + it)|\, dt + \mathcal{E},$$

where $\mathcal{E}$ is an error term that can be ignored and might be different on different occassions. The integral of the logarithm usually cannot be dealt with directly, so we often use the following trick:

$$\frac{1}{T}\int_0^T \log|f(\sigma + it)|\, dt = \frac{1}{2T}\int_0^T \log(|f(\sigma + it)|^2)\, dt$$

$$\leq \frac{1}{2}\log\left(\frac{1}{T}\int_0^T |f(\sigma + it)|^2\, dt\right),$$

where the inequality follows from the arithmetic–geometric mean inequality. In this way we see a direct connection between the location of the zeros within a rectangle and the type of mean values we have been considering.

3   A Sample of Important Estimates

Let

$$I_k(\sigma, T) = \int_0^T |\zeta(\sigma + it)|^{2k}\, dt.$$

When $k = 1$ we know that for each fixed $\sigma > 1/2$

$$I_1(\sigma, T) \sim c(\sigma)\, T,$$

as $T \to \infty$, where $c(\sigma)$ is a know function of σ. In 1918 Hardy and Littlewood [HL] proved that when $\sigma = 1/2$,

$$I_1(1/2, T) \sim T \log T.$$

What can such estimates tell us about the zeta function? Comparing the result for σ greater than $1/2$ with that for $\sigma = 1/2$, we see that the zeta function tends to assume, on average, much larger values on the critical line than to the right of it. Since it also has many zeros on the critical line, we should expect the zeta function to behave rather erratically there.

The next higher moment was determined in 1926 by Ingham [I], who proved that

$$I_2(1/2, T) \sim \frac{T}{2\pi^2} \log^4 T.$$

Unfortunately, no asymptotic estimate has been proved for any k greater than 2. It is known that for positive rational k[1],

$$I_k(1/2, T) \gg T \log^{k^2} T$$

(see Ramachandra [R] and Heath–Brown [H-B]). This is also known to hold for all positive k if the Riemann Hypothesis is true (see Ramachandra [R]). We expect that

$$I_k(1/2, T) \sim c_k T \log^{k^2} T,$$

but a proof seems a long way off. J. B. Conrey and A. Ghosh (unpublished) suggested that

$$c_k = \frac{a_k g_k}{\Gamma(k^2 + 1)},$$

where

$$a_k = \prod_p \left(\left(1 - \frac{1}{p}\right)^{k^2} \sum_{r=0}^{\infty} \frac{d_k^2(p^r)}{p^r} \right)$$

and g_k is an integer. Only recently has anyone put forth a plausible value for g_k. J. B. Conrey and A. Ghosh [CG] conjectured that $g_3 = 42$, and J. B. Conrey and the author [CGO] conjecured that $g_4 = 24024$. Then J. Keating and N. Snaith [KS], using random matrix theory, conjectured a value for g_k for every complex number k with $\operatorname{Re} k > -1/2$. For integer values of k, their conjecture takes the form $g_k = (k^2!) \prod_{j=0}^{k-1} j!/(j+k)!$.

Another type of mean value important for applications is

$$\int_0^T |\zeta^{(j)}(\sigma + it) M_N(\sigma + it)|^2 \, dt, \tag{3.1}$$

where[2]

$$M_N(s) = \sum_{1 \le n \le N} \frac{\mu(n)}{n^s} P\left(\frac{\log n}{\log N}\right)$$

---

[1] Editors' comment: See the Appendix of the lecture by D.W. Farmer, page 185, for a discussion of the $\gg$ notation.

[2] Editors' comment: The Möbius function, $\mu(n)$, is defined in the lectures of D.A. Goldston, page 79, equation 2.10.

and $P(x)$ is a polynomial. Since

$$\frac{1}{\zeta(s)} = \sum_{n=1}^{\infty} \frac{\mu(n)}{n^s} \qquad (\mathrm{Re}\, s > 1),$$

we can view $M_N(s)$ as an approximation to the reciprocal of $\zeta(s)$ in $\mathrm{Re}\, s > 1$. We might then expect the approximation to hold (in some sense) inside the critical strip as well. If that is the case, multiplying the zeta function by $M_N(s)$ should dampen (or mollify) the large values of zeta. Below we will see two applications of this idea. The most general estimates known for such integrals are due to Conrey, Ghosh, and the author [CGG2], who obtained asymptotic estimates for them when the length of the Dirichlet polynomial $M_N(s)$ is $N = T^\theta$ with $\theta < 1/2$. Later, Conrey [C] used Kloosterman sum techniques to show that these formulas also hold when $\theta < 4/7$.

We conclude this section by mentioning a few discrete mean value results. The author [G] proved asymptotic estimates for the sums

$$\sum_{0 \leq \gamma \leq T} |\zeta^{(j)}(\rho)|^2,$$

assuming the Riemann Hypothesis is true. Here γ runs over the ordinates of the zeros $\rho = 1/2 + i\gamma$ of $\zeta(s)$. Conrey, Ghosh, and Gonek [CGG2] proved discrete versions of the mollified mean values (3.1), namely

$$\sum_{0 < \gamma < T} |\zeta'(\rho) M_N(\rho)|^2,$$

under the assumption of the Riemann Hypothesis and the Generalized Lindelöf Hypothesis.

4   Application: A Simple Zero–Density Estimate

We want to show that there are relatively few zeros of the zeta function in the right half of the critical strip. Let $1/2 < \sigma_0 < 1$ be a fixed real number and let $\mathcal{C}$ be the rectangle in the complex plane with vertices at 2, $2 + iT$, $\sigma_0 + iT$, σ_0. Applying our (simplified) version of Littlewood's Lemma, we see that

$$\sum_{\rho \in \mathcal{C}} \mathrm{Dist}(\rho) = \frac{1}{2\pi} \int_0^T \log(|\zeta(\sigma_0 + it)|)\, dt + \mathcal{E},$$

Where $\mathrm{Dist}(\rho)$ is the distance of the zero $\rho = \beta + i\gamma$ of the zeta function from the line $\mathrm{Re}\, s = \sigma_0$. Now let σ be a fixed real number with $\sigma_0 < \sigma < 1$ and write $N(\sigma, T)$ for the number of zeros with $\sigma < \beta \le 2$ and $0 < \gamma < T$. On the one hand, we have

$$\sum_{\rho \in \mathcal{C}} \text{Dist}(\rho) \ge \sum_{\substack{\rho \in \mathcal{C} \\ \sigma \le \beta}} \text{Dist}(\rho) \ge (\sigma - \sigma_0) N(\sigma, T).$$

On the other hand,

$$\frac{1}{2\pi} \int_0^T \log(|\zeta(\sigma_0 + it)|)\, dt = \frac{1}{4\pi} \int_0^T \log(|\zeta(\sigma_0 + it)|^2)\, dt$$

$$\le \frac{T}{4\pi} \log\left(\frac{1}{T} \int_0^T |\zeta(\sigma_0 + it)|^2\right) dt$$

by the arithmetic–geometric mean inequality, as before. The integral on the last line is $I_k(\sigma_0, T)$, which we have seen is $\sim c(\sigma_0)T$, where $c(\sigma_0)$ is positive and independent of T. Thus, the last expression is $O(T)$. It follows that

$$N(\sigma, T) \ll T.$$

Since $N(T) \sim \frac{T}{2\pi} \log T$, we may interpret this as saying that the proportion of zeros to the right of any line $\text{Re}\, s = \sigma > 1/2$ is infinitesimal.

This, the first zero–density estimate, was proved by H. Bohr and E. Landau [BL] in 1914. Since then there have been much stronger results, typically of the form

$$N(\sigma, T) \ll T^{\lambda(\sigma)},$$

where $\lambda(\sigma) < 1$ for $\sigma > 1/2$. Nevertheless, the underlying idea in the proof of many of these results already appears here.

5   Application: Levinson's Method

Zero–density theorems tell us there are (relatively) few zeros to the right of the critical line. Our goal here is to sketch the method of Levinson [L], which shows that there are many zeros *on* it.

Recall that[3]

$$N(T) = \#\{\rho = \beta + i\gamma \mid \zeta(\rho) = 0, \quad 0 < \gamma < T\} \sim \frac{T}{2\pi} \log T$$

and let

$$N_0(T) = \# \{\rho = 1/2 + i\gamma \mid \zeta(\rho) = 0, \quad 0 < \gamma < T\}$$

denote the number of zeros on the critical line up to height T. The important estimations of $N_0(T)$ were:

---

[3]Editors' comment: See Section 7 of the lectures by D.R. Heath-Brown starting on page 1.

$N_0(T) \to \infty$ G. H. Hardy (1914)
$N_0(T) > cT$ G. H. Hardy-J. E. Littlewood (1921)
$N_0(T) > c'N(T)$ A. Selberg (1942)
$N_0(T) > \frac{1}{3}N(T)$ N. Levinson (1974)
$N_0(T) > \frac{2}{5}N(T)$ J. B. Conrey (1989)

Levinson's method begins with the following fact first proved by Speiser [Sp].

Theorem 5.1. (**Speiser**) The Riemann Hypothesis is equivalent to the assertion that $\zeta'(s)$ does not vanish in the left half of the critical strip.

In the early seventies, N. Levinson and H. L. Montgomery [LM] proved a quantitative version of this. Let

$$N'_-(T) = \#\{\rho' = \beta' + i\gamma' \mid \zeta'(\rho') = 0,\ -1 < \beta' < 1/2,\quad 0 < \gamma' < T\}$$

and

$$N_-(T) = \#\{\rho = \beta + i\gamma \mid \zeta(\rho) = 0,\ -1 < \beta < 1/2,\quad 0 < \gamma < T\}\ .$$

Theorem 5.2. (**Levinson-Montgomery**) We have $N_-(T) = N'_-(T) + O(\log T)$

.

The idea behind the proof is as follows. Let $0 < a < 1/2$ and let $\mathcal{C}$ denote the positively oriented rectangle with vertices $a + iT/2$, $a + iT$, $-1 + iT$, and $-1 + iT/2$. It is not difficult to show that

$$\Delta \arg \frac{\zeta'}{\zeta}(s)\bigg|_{\mathcal{C}} = O(\log T),$$

independently of a. Given this, we see that

$$2\pi(\#\text{ zeros of } \zeta'(s) \text{ in } \mathcal{C} - \#\text{ zeros of } \zeta(s) \text{ in } \mathcal{C}) = O(\log T).$$

The theorem now follows on observing that a was arbitrary, and by "adding" rectangles with top and bottom edges, respectively, at T and $T/2$, $T/2$ and $T/4, \ldots$.

We now sketch Levinson's method. We have just seen that $N_-(T) = N'_-(T) + O(\log T)$. Now, the nontrivial zeros of $\zeta(s)$ are symmetric about the critical line. Hence, the number of them lying to the right of the critical line up to height T is also $N_-(T)$. Therefore

$$N(T) = N_0(T) + 2N_-(T)$$
$$= N_0(T) + 2N'_-(T) + O(\log T),$$

or

$$N_0(T) = N(T) - 2N'_-(T) + O(\log T).$$

The size of the first term on the right–hand side of the last line is known, namely, $(1 + o(1))\frac{T}{2\pi}\log T$. Hence, if we can determine a sufficiently small upper bound for $N'_-(T)$, we can deduce a lower bound for $N_0(T)$.

To find such an upper bound it is convenient to first note that the zeros of $\zeta'(s)$ in the region $-1 < \sigma < 1/2$, $0 < t < T$, are identical to the zeros of $\zeta'(1-s)$ in the reflected region $1/2 < \sigma < 2$, $0 < t < T$. One can also show, by the functional equation for the zeta function, that $\zeta'(1-s)$ and $G(s) = \zeta(s) + \zeta'(s)/L(s)$ have the same zeros in $1/2 < \sigma < 2$, $0 < t < T$, where $L(s)$ is essentially $\frac{1}{2\pi}\log T$. It turns out to be technically advantageous to count the zeros of $G(s)$ rather than those of $\zeta'(1-s)$.

To bound the number of zeros of $G(s)$ in this region, we apply Littlewood's Lemma. Let $a = 1/2 - \delta/\log T$, with δ a small positive number, and let $\mathcal{R}_a$ denote the rectangle whose vertices are at $a, 2, 2+iT$, and $a+iT$. It would be natural to apply our abbreviated form of the lemma to obtain

$$\sum_{\rho^* \in \mathcal{R}_a} \mathrm{Dist}(\rho^*) = \frac{1}{2\pi}\int_0^T \log|G(a+it)|dt + \mathcal{E},$$

where ρ^* denotes a zero of $G(s)$ and $\mathrm{Dist}(\rho^*)$ is its distance to the left edge of $\mathcal{R}_a$. However, in the next step, when we apply the arithmetic–geometric mean inequality to the integral, we would lose too much. To avoid this loss, we first mollify $G(s)$ and then apply Littlewood's Lemma in the form

$$\sum_{\substack{\rho^{**} \in \mathcal{R}_a \\ GM(\rho^{**})=0}} \mathrm{Dist}(\rho^{**}) = \frac{1}{2\pi}\int_0^T \log|G(a+it)M(a+it)|dt + \mathcal{E}.$$

Here $M(s) = \sum_{n \leq T^\theta} a_n/n^s$, with $a_n = \mu(n)n^{a-1/2}\left(1 - \frac{\log n}{\log T^\theta}\right)$ and $\theta > 0$, approximates $1/\zeta(s)$. Note that among the zeros of $G(s)M(s)$ in $\mathcal{R}_a$ are all the zeros of $G(s)$ in $\mathcal{R}_a$. Therefore we have

$$\sum_{\substack{\rho^{**} \in \mathcal{R}_a \\ GM(\rho^{**})=0}} \mathrm{Dist}(\rho^{**}) \geq \sum_{\substack{\rho^* \in \mathcal{R}_a \\ G(\rho^*)=0}} \mathrm{Dist}(\rho^*)$$

$$\geq \sum_{\substack{\rho^* \in \mathcal{R}_a, \mathrm{Re}\rho^* > 1/2 \\ G(\rho^*)=0}} \mathrm{Dist}(\rho^*)$$

$$\geq (1/2 - a)N'_-(T).$$

We now see that

$$(1/2 - a)N'(T) \leq \frac{1}{2\pi} \int_0^T \log|GM(a+it)|dt + \mathcal{E}$$

$$= \frac{1}{4\pi} \int_0^T \log|GM(a+it)|^2 dt + \mathcal{E}$$

$$\leq \frac{T}{4\pi} \log\left(\frac{1}{T} \int_0^T |GM(a+it)|^2 dt\right) + \mathcal{E}.$$

Thus, we require an estimate for

$$\int_0^T |GM(a+it)|^2 dt.$$

This is similar to a mean value we saw in Section 3. Levinson was able prove an asymptotic estimate for this integral when $\theta = 1/2 - \epsilon$ with ϵ arbitrarily small. The resulting upper bound for $N'_-(T)$ then led to the lower bound

$$N_0(T) > (1/3 + o(1)) N(T).$$

Conrey later proved an asymptotic estimate for the integral when $\theta = 4/7 - \epsilon$. This led to

$$N_0(T) > (2/5 + o(1)) N(T).$$

The form of the asymptotic estimate in both cases is the same as a function of θ, and D. Farmer [F] has given heuristic arguments to suggest that this remains the case even when θ is arbitrarily large. From Farmer's conjecture it follows that

$$N_0(T) \sim N(T).$$

Before concluding this section, we remark that had we introduced a mollifier into our proof of the Bohr–Landau result in the previous section, we would have obtained a much stronger zero–density estimate of the form we alluded to previously, namely $N(\sigma, T) \ll T^{\lambda(\sigma)}$, with $\lambda(\sigma) < 1$.

6   Application: The Number of Simple Zeros

Our third application demonstrates the use of discrete mean value theorems.

Let

$$N_s(T) = \#\{\rho = \beta + i\gamma \mid \zeta(\rho) = 0, \zeta'(\rho) \neq 0, \quad 0 < \gamma < T\}$$

denote the number of simple zeros of the zeta function in the critical strip with ordinates between 0 and T. It is believed that all the nontrivial zeros are on the critical line and simple, in other words, that $N(T) = N_0(T) = N_s(T)$ for every $T > 0$. Unconditionally, it is known that at least 2/5 of the zeros are

simple (see Conrey [C]). In 1973, H. Montgomery [M], used his pair correlation
method to show that if the Riemann Hypothesis is true, then more than 2/3
of the zeros are simple. In other words,

$$N_s(T) > 2/3N(T)$$

provided that T is sufficiently large. We will outline his argument in section 8.
Here we briefly describe a different method of Conrey, Ghosh, and Gonek
[CGG1], which shows that on the stronger hypotheses of RH and the Gener-
alized Lindelöf Hypothesis, one can replace the 2/3 above by $19/27 = .703\ldots$.

By the Cauchy–Schwarz inequality, we have

$$\left| \sum_{0<\gamma<T} \zeta'(1/2+i\gamma)M_N(1/2+i\gamma) \right|^2 \le \left(\sum_{\substack{0<\gamma\le T \\ 1/2+i\gamma \text{ is simple}}} 1 \right)\left(\sum_{0<\gamma<T} |\zeta'(\rho)M_N(\rho)|^2 \right),$$

where $M_N(s)$ is a Dirichlet polynomial of length N with coefficients similar, but
not identical, to those of $M(s)$ in the last section. Its purpose is also similar: to
mollify $\zeta'(1/2+i\gamma)$ so as to minimize the loss in applyng the Cauchy–Schwarz
inequality. If one assumes RH, the sum on the left–hand side is easy to compute
and turns out to be $\sim 19/24N(T)\log T$. The sum on the right–hand side is
much more difficult to treat, but one can show that if RH and GLH are true,
then it is $\sim 57/64N(T)\log^2 T$. Inserting these estimates into the inequality
above and solving for $N_s(T)$, we obtain the stated result. An elaboration of
the method leads to the conclusion that, on the same hypotheses, over 95.5%
of the zeros of $\zeta(s)$ are either simple or double.

7   Mean Values of Dirichlet Polynomials

From now on we will focus on mean values of Dirichlet polynomials. Let

$$A(s) = A_N(s) = \sum_{n=1}^{N} a_n n^{-s}$$

be a Dirichlet polynomial of length N and let $s = \sigma + it$. The Classical Mean
Value Therorem for Dirichlet polynomials is

Theorem 7.1. (**Classical Mean Value Theorem**)

$$\int_0^T |\sum_{n=1}^{N} a_n n^{-s}|^2 dt = \left(T + O(N\log N) \right)\sum_{n=1}^{N} |a_n|^2 n^{-2\sigma}.$$

A more precise version due to H. L. Montgomery and R. C. Vaughan is

Theorem 7.2. (**Montgomery–Vaughan**)

$$\int_0^T |\sum_{n=1}^N a_n n^{-s}|^2 dt = \sum_{n=1}^N |a_n|^2 n^{-2\sigma}\left(T + O(n)\right).$$

From this we see that if $N = o(T)$, then

$$\int_0^T |\sum_{n=1}^N a_n n^{-\sigma-it}|^2 dt \sim T \sum_{n=1}^N |a_n|^2 n^{-2\sigma}.$$

On the other hand, if $N \gg T$ the O–term dominates and we have only

$$\int_0^T |\sum_{n=1}^N a_n n^{-\sigma-it}|^2 dt \ll N \sum_{n=1}^N |a_n|^2 n^{-2\sigma}.$$

It is natural to ask whether this is the actual size of the mean when N is larger than T. The following example answers this question.

Example. Let each $a_n = 1$ and take $\sigma = 1/2$. Montgomery and Vaughan's mean value formula gives

$$\int_0^T |\sum_{n=1}^N n^{-\frac{1}{2}-it}|^2 dt = \sum_{n=1}^N \frac{1}{n}\left(T + O(n)\right)$$

$$= T\left(\log N + O(1)\right) + O(N)$$

$$= \begin{cases} (1+o(1))T \log N & \text{if } N = O(T), \\ O(N) & \text{if } N > T^\alpha \ (\alpha > 1). \end{cases}$$

We can also evaluate this using a crude form of the approximate functional equation for the zeta function (see Titchmarsh [T], p.77), namely

$$\zeta(s) = \sum_{1 \le n \le N} n^{-s} + \frac{N^{1-s}}{s-1} + O(N^{-\sigma}).$$

Taking $\sigma = 1/2$, we obtain

$$\int_0^T \left|\sum_{n=1}^N n^{-1/2-it}\right|^2 dt = \int_0^T \left|\zeta(1/2+it) + \frac{N^{1/2-it}}{1/2-it} + O(N^{-1/2})\right|^2 dt.$$

Now, we know (see Titchmarsh [T]) that

$$\int_0^T |\zeta(1/2+it)|^2 dt \sim T \log T.$$

Furthermore, it is easy to see that

$$\int_0^T \left| \frac{N^{1/2-it}}{1/2 - it} \right|^2 dt = N \int_0^T \frac{1}{1/4 + t^2} dt \sim \pi N \qquad \text{and} \qquad \int_0^T N^{-1} dt = T/N \, .$$

Hence, we find that

$$\int_0^T \left| \sum_{n=1}^N n^{-1/2-it} \right|^2 dt \sim \pi N$$

if $N \gg T^\alpha$ and $\alpha > 1$. Therefore, the $O-$ term cannot be reduced in this case and, in fact, we can extract a new main term. So it makes sense to ask whether there is a useful general asymptotic formula for

$$\int_0^T \left| \sum_{n=1}^N a_n n^{-s} \right|^2 dt$$

when $N = T^\alpha$, $\alpha > 1$. In order to answer this, let us consider the proof of the Classical Mean Value Theorem. Squaring out and integrating, we obtain

$$\int_0^T \left| \sum_{n=1}^N a_n n^{-\sigma - it} \right|^2 dt = \sum_{n=1}^N \sum_{m=1}^N \frac{a_n \bar{a}_m}{(nm)^\sigma} \int_0^T (m/n)^{it} \, dt$$

$$= T \sum_{n=1}^N \frac{|a_n|^2}{n^{2\sigma}} + \sum_{\substack{1 \le m,n \le N \\ m \ne n}} \frac{a_n \bar{a}_m}{(nm)^\sigma} \left(\frac{e^{iT \log(m/n)} - 1}{i \log(m/n)} \right) .$$

The second sum consists of "off–diagonal" terms and is

$$\ll \sum_{\substack{1 \le m,n \le N \\ m \ne n}} \frac{|a_n a_m|}{(nm)^\sigma |\log(m/n)|} \le \frac{1}{2} \sum_{\substack{1 \le m,n \le N \\ m \ne n}} \left(\frac{|a_n|^2}{n^{2\sigma}} + \frac{|a_m|^2}{m^{2\sigma}} \right) \frac{1}{|\log(m/n)|}$$

$$= \sum_{\substack{1 \le m,n \le N \\ m \ne n}} \frac{|a_n|^2}{n^{2\sigma} |\log(m/n)|} = \sum_{1 \le n \le N} \frac{|a_n|^2}{n^{2\sigma}} \left(\sum_{\substack{1 \le m \le N \\ m \ne n}} \frac{1}{|\log(m/n)|} \right) .$$

The inner sum is

$$\ll \left(\sum_{\substack{m \le N \\ |m-n| \le n/2}} + \sum_{\substack{m \le N \\ n/2 < |m-n|}} \right) \frac{1}{|\log(m/n)|}$$

$$\ll \sum_{1 \le h \le n/2} \frac{1}{|\log((n \pm h)/n)|} + \sum_{\substack{m < n/2 \, \text{or} \\ 3n/2 < m \le N}} \frac{1}{|\log(m/n)|}$$

$$\ll \sum_{1 \le h \le N} \frac{n}{h} + \sum_{1 \le m \le N} 1 \ll N \log N + N \ll N \log N \, .$$

Hence, the off–diagonal terms are

$$\ll N \log N \sum_{1 \leq n \leq N} \frac{|a_n|^2}{n^{2\sigma}}.$$

We therefore find that

$$\int_0^T \Big| \sum_{n=1}^N a_n n^{-\sigma - it} \Big|^2 dt = (T + O(N \log N)) \sum_{n=1}^N \frac{|a_n|^2}{n^{2\sigma}}.$$

From this it is clear that if we want a precise formula when N is much larger than T, we need to estimate the off–diagonal terms more carefully.

Returning to our initial expression for these terms, we see that

$$\sum_{\substack{1 \leq m,n \leq N \\ m \neq n}} \frac{a_n \bar{a}_m}{(nm)^\sigma} \left(\frac{e^{iT \log(m/n)} - 1}{i \log(m/n)} \right)$$

$$= 2\mathrm{Re} \sum_{1 \leq n < m \leq N} \frac{a_n \bar{a}_m}{(nm)^\sigma} \left(\frac{e^{iT \log(m/n)} - 1}{i \log(m/n)} \right)$$

$$= 2\mathrm{Re} \sum_{1 \leq n < N} \sum_{1 \leq h \leq N-n} \frac{a_n \bar{a}_{n+h}}{(n(n+h))^\sigma} \left(\frac{e^{iT \log((n+h)/n)} - 1}{i \log((n+h)/n)} \right)$$

$$= 2\mathrm{Re} \sum_{1 \leq h < N} \sum_{1 \leq n \leq N-h} \frac{a_n \bar{a}_{n+h}}{n^{2\sigma}} (1 + h/n)^{-\sigma} \left(\frac{e^{iT \log(1+h/n)} - 1}{i \log(1+h/n)} \right).$$

For the sake of simplicity, consider only the terms with $h/n < 1/2$. In these $\log(1 + h/n)$ is approximately h/n and $(1 + h/n)^\sigma$ is approximately 1. These terms therefore contribute about

$$2\mathrm{Re} \sum_{1 \leq h < N} \sum_{2h < n \leq N-h} \frac{a_n \bar{a}_{n+h}}{n^{2\sigma-1}} \left(\frac{e^{iTh/n} - 1}{ih} \right).$$

To simplify further, we restrict our attention to the terms with $Th < n/2$. In these $(e^{iTh/n} - 1)/ih$ is approximately T/n, so their contribution is about

$$2T \, Re \sum_{h \neq 0} \sum_n a_n \bar{a}_{n+h} n^{-2\sigma}.$$

We can clearly estimate this if we have good estimates for the sums

$$\sum_{n=1}^N a_n \bar{a}_{n+h}.$$

In fact, this would be sufficient to estimate the terms we ignored as well. To state the final result obtained, we assume the a_n satisfy the following conditions (see Goldston and Gonek [GG] for the details):

1. (Normalization)

$$a_n \ll n^\epsilon.$$

2. There is a function $M(x)$ and a real number θ with $0 < \theta < 1$ such that

$$\sum_{n \le x} a_n = M(x) + O(x^\theta),$$

$M'(x) \ll x^\epsilon$, and $M''(x) \ll x^{-1+\epsilon}$.

3. There is a function $M(x, h)$, real numbers ϕ and η with $0 < \phi, \eta < 1$, such that

$$\sum_{n \le x} a_n \bar{a}_{n+h} = M(x, h) + O(x^\phi)$$

uniformly for $h \le x^\eta$, and $M'(x, h) \ll (hx)^\epsilon$.

In applications it is often more convenient to estimate

$$\int_0^T |\sum_{n=1}^N a_n n^{-s} - \int_1^N M'(x) x^{-s} dx|^2 dt$$

rather than

$$\int_0^T |\sum_{n=1}^N a_n n^{-s}|^2 dt .$$

Here the integral involving $M'(x)$ may be thought of as an expected value. Also, it is much easier to work with a weighted mean

$$\int_{-\infty}^\infty \Psi_U(\frac{t}{T}) |\sum_{n=1}^N a_n n^{-s} - \int_1^N M'(x) x^{-s} dx|^2 dt ,$$

where $\Psi_U(x)$ is nonnegative, has support in $[1 - U^{-1}, 1 + U^{-1}]$ with $U = \log^A T$, and satisfies $\Psi_U^{(j)}(x) \ll \log^j T$ and $\Psi_U(x) = 1$ in $[1 + U^{-1}, 1 - U^{-1}]$. It follows that the Fourier transform $\widehat{\Psi}_U(v)$ is approximately 1 for $|v| \le 1$ and drops off rapidly as $|v|$ increases. Thus $\Psi_U(t/T)$ is a smooth approximation to the characteristic function of the interval $[0, T]$. Our result is

Theorem 7.3. (**Goldston–Gonek**) Let $\epsilon > 0$ be arbitraily small, $\sigma < 1$, and

θ, ϕ, η as above. Then for $T \ll N \ll T^{(1-\epsilon)/(1-\eta)}$ we have

$$\int_{-\infty}^{\infty} \Psi(\frac{t}{T}) \left| \sum_{n=1}^{N} a_n n^{-s} - \int_1^N M'(x) x^{-s} dx \right|^2 dt$$

$$= \widehat{\Psi}(0) T \sum_{n \leq N} |a_n|^2 n^{-2\sigma}$$

$$+ 4\pi (\frac{T}{2\pi})^{2-2\sigma} \text{Re} \int_{T/2\pi N}^{\infty} \left(\sum_{1 \leq h \leq 2\pi Nv/T} M'(\frac{hT}{2\pi v}, h) h^{1-2\sigma} \right) \frac{\widehat{\Psi}(v)}{v^{2-2\sigma}} dv$$

$$- 4\pi (\frac{T}{2\pi})^{2-2\sigma} \text{Re} \int_{T^\epsilon/2\pi N}^{\infty} \left(\int_0^{2\pi Nv/T} |M'(\frac{uT}{2\pi v})|^2 u^{1-2\sigma} du \right) \frac{\widehat{\Psi}(v)}{v^{2-2\sigma}} dv$$

$$+ O(N^{1-2\sigma+max(\theta,\phi)+5\epsilon}) + O(N^{2-2\sigma+5\epsilon}T^{-1}) + O(N^{2\epsilon}).$$

A similar formula can be proved for the tails of Dirichlet series minus their expected value, that is, for $\sum_{n>N} a_n n^{-s} - \int_N^{\infty} M'(x) x^{-s} dx$. One can also estimate the "mixed" means

$$\int_{-\infty}^{\infty} \Psi(\frac{t}{T}) \left(\sum_{n=1}^{N} a_n n^{-s} - \int_1^N M'_a(x) x^{-s} dx \right) \overline{\left(\sum_{n=1}^{N} b_n n^{-s} - \int_1^N M'_b(x) x^{-s} dx \right)} dt,$$

where $M'_a(x)$ and $M'_b(x)$ have an obvious meaning. Finally, one can show that the integrals of "crossed" expressions consisting of a Dirichlet polynomial times the complex conjugate of the tail of a Dirichlet series (minus their expected values in both cases) are generally of smaller order than means involving a polynomial times a polynomial or a tail times a tail.

We now turn to applications of long mean value theorems.

8 Application: A Lower Bound for $F(\alpha)$

H. L. Montgomery [M] studied the function[4]

$$F(\alpha) = (\frac{T}{2\pi} \log T)^{-1} \sum_{0 < \gamma, \gamma' \leq T} T^{i\alpha(\gamma-\gamma')} \frac{4}{4 + (\gamma - \gamma')^2}.$$

It is known that $F(\alpha)$ is even and nonnegative, and Montgomery showed that if the Riemann Hypothesis is true, then

$$F(\alpha) = (1 + o(1)) T^{-2\alpha} \log T + \alpha + o(1) \qquad (8.1)$$

for $|\alpha| \leq 1$. He also conjectured that

$$F(\alpha) = (1 + o(1)) \qquad (8.2)$$

---

[4]Editors' comment: The form factor, $F(\alpha)$, is also discussed in Sections 4 and 6 of the lectures of D.A. Goldston, page 79.

when $1 \leq |\alpha| \leq A$ with A arbitrarily large. The only known nontrivial lower bound for $F(\alpha)$ when $|\alpha| \geq 1$ is given by

Theorem 8.1. **(Goldston–Gonek–Ozluk–Snyder)** Assume the Generalized Riemann Hypothesis. Then

$$F(\alpha) \geq 3/2 - |\alpha| - \epsilon$$

uniformly for $1 \leq |\alpha| \leq 3/2 - 2\epsilon$ and $T \geq T_0(\epsilon)$.

See [GGOS].

Sketch of the proof. First we sketch the derivation of Montgomery's results (8.1) and (8.2).

We begin with the explicit formula

$$-2 \sum_{0 < \gamma \leq T} x^{i(\gamma - t)} \frac{1}{1 + (t - \gamma)^2} = x^{-1} \left(\sum_{n \leq x} \Lambda(n) n^{1/2 - it} - \int_1^x u^{1/2 - it} du \right)$$
$$+ x \left(\sum_{n > x} \Lambda(n) n^{-3/2 - it} - \int_x^\infty u^{-3/2 - it} du \right) + \mathcal{E},$$

where $\mathcal{E}$, as usual, denotes an ignorable error term. Integrating the modulus squared of both sides (see Montgomery [M] for details), we see that the left–hand side is

$$\int_0^T \left| 2 \sum_{0 < \gamma \leq T} x^{i(\gamma - t)} \frac{1}{1 + (t - \gamma)^2} \right|^2 dt = 2\pi \sum_{0 < \gamma, \gamma' \leq T} x^{i(\gamma - \gamma')} \frac{4}{4 + (\gamma - \gamma')^2} + \mathcal{E}$$
$$= 2\pi F(x, T) + \mathcal{E},$$

where we write

$$F(x, T) = \sum_{0 < \gamma, \gamma' \leq T} x^{i(\gamma - \gamma')} \frac{4}{4 + (\gamma - \gamma')^2}.$$

Note that

$$F(\alpha) = (\frac{T}{2\pi} \log T)^{-1} F(T^\alpha, T).$$

Equating this with the mean squared modulus of the right–hand side, we find that

$$2\pi F(x, T) = \int_0^T \left| x^{-1} \left(\sum_{n \leq x} \Lambda(n) n^{\frac{1}{2} - it} - \int_1^x u^{\frac{1}{2} - it} du \right) \right.$$
$$\left. + x \left(\sum_{n > x} \Lambda(n) n^{-\frac{3}{2} - it} - \int_x^\infty u^{-\frac{3}{2} - it} du \right) \right|^2 dt + \mathcal{E}.$$
$$(8.3)$$

Case 1. $x = T^\alpha$, $\alpha < 1$. Applying the Montgomery–Vaughan mean value theorem in a straightforward way, we obtain (8.1).

Case 2. $x = T^\alpha$, $1 \leq \alpha < A$. Applying Theorem 7.3 and a strong form of the Twin Prime Conjecture, we obtain the conjecture 8.2. More precisely, if we assume that

$$\sum_{n \leq y} \Lambda(n)\Lambda(n+h)) = c(h)x + O(x^{1/2+\epsilon}),$$

where $c(h)$ is defined in Theorem 8.2 below, we obtain 8.2 with $A = 2$. If we also assume there is significant cancellation among the O–terms when averaged over h, we obtain (8.2) with A arbitrarily large.

Theorem 8.1 is proved as follows. We have no proof of the Twin Prime Conjecture, but we have its analogue for the functions

$$\lambda_Q(n) = \sum_{q \leq Q} \frac{\mu^2(q)}{\phi(q)} \sum_{\substack{d|q \\ d|n}} d\mu(d),$$

which approximate the $\Lambda(n)$'s. Let us rewrite (8.3) as

$$2\pi \, F(x,T) = \int_0^T \left|\mathbb{A}(x,t) + \mathbb{A}^*(x,t)\right|^2 dt + \mathcal{E}.$$

Also, let $\mathbb{A}_Q(x,t)$ and $\mathbb{A}_Q^*(x,t)$ be the same as $\mathbb{A}(x,t)$ and $\mathbb{A}^*(x,t)$, respectively, but with the $\Lambda(n)$'s replaced by $\lambda_Q(n)$'s. Clearly we have

$$0 \leq \int_{-\infty}^{\infty} \Psi_U(\tfrac{t}{T}) \left| \left(\mathbb{A}(x,t) + \mathbb{A}^*(x,t)\right) - \left(\mathbb{A}_Q(x,t) + \mathbb{A}_Q^*(x,t)\right) \right|^2 dt.$$

It follows that

$$2\mathrm{Re} \int_{-\infty}^{\infty} \Psi_U(\tfrac{t}{T}) \left(\mathbb{A}\overline{\mathbb{A}_Q} + \mathbb{A}^*\overline{\mathbb{A}_Q^*} + \mathbb{A}\overline{\mathbb{A}_Q^*} + \mathbb{A}^*\overline{\mathbb{A}_Q} - \mathbb{A}_Q\overline{\mathbb{A}_Q^*} \right) dt$$

$$- \int_{-\infty}^{\infty} \Psi_U(\tfrac{t}{T}) \left(\mathbb{A}_Q\overline{\mathbb{A}_Q} + \mathbb{A}_Q^*\overline{\mathbb{A}_Q^*} \right) dt$$

$$\leq \int_{-\infty}^{\infty} \Psi_U(\tfrac{t}{T}) \left| \left(\mathbb{A}(x,t) + \mathbb{A}^*(x,t)\right) \right|^2 dt$$

$$= 2\pi \, F(x,T).$$

The coefficient correlation sums needed to estimate the long mean values here are

$$\sum_{n \leq y} \Lambda(n)\lambda_Q(n+h) \qquad \text{for} \qquad \mathbb{A}\overline{\mathbb{A}_Q} \quad \text{and} \quad \mathbb{A}^*\overline{\mathbb{A}_Q^*}$$

and

$$\sum_{n \leq y} \lambda_Q(n)\lambda_Q(n+h) \qquad \text{for} \qquad \mathbb{A}_Q\overline{\mathbb{A}_Q} \quad \text{and} \quad \mathbb{A}_Q^*\overline{\mathbb{A}_Q^*}.$$

These are available from

Theorem 8.2. (**J. Friedlander–D. Goldston**) Assume the Generalized Riemann Hypothesis. Let $Q = y^\delta$ with $1/4 \le \delta \le 1/2$. Then

$$\sum_{n \le y} \Lambda(n)\lambda_Q(n+h) \qquad \text{and} \qquad \sum_{n \le y} \lambda_Q(n)\lambda_Q(n+h)$$

are both $= c(h)y + O(y^{\frac{1}{2}+\delta+\epsilon})$ uniformly for $1 \le h \le y^{1-\epsilon}$, where

$$c(h) = \begin{cases} 2\prod_{p>2}\left(1 - \frac{1}{(p-1)^2}\right)\prod_{\substack{p>2 \\ p|h}}\left(\frac{p-1}{p-2}\right) & \text{if h is even}, \\ 0 & \text{if h is odd}. \end{cases}$$

Applying this to our terms, taking $X = T^\alpha$, and choosing δ optimally as a function of α leads to

$$F(\alpha, T) \ge \frac{3}{2} - |\alpha| - \epsilon$$

for $1 \le \alpha \le 3/2 - 2\epsilon$.

9 Application: The 6th and 8th Power Moments of the Zeta Function

In Section 3 we defined

$$I_k(1/2, T) = \int_0^T |\zeta(1/2 + it)|^{2k}\, dt$$

for positive values of k. Recall that Hardy and Littlewood showed that

$$\int_0^T |\zeta(1/2 + it)|^2\, dt \sim T \log T,$$

Ingham showed that

$$\int_0^T |\zeta(1/2 + it)|^4\, dt \sim \frac{1}{2\pi^2} T \log^4 T,$$

and no other asymptotic formula has ever been proved. In the mid 1990's J. B. Conrey and A. Ghosh [CG] made the following

Conjecture 1. (**Conrey–Ghosh**) As $T \to \infty$,

$$\int_0^T |\zeta(1/2 + it)|^6\, dt \sim \frac{42}{9!} \prod_p \left(\sum_{r=0}^\infty \frac{d_3(p^r)^2}{p^r}\right) T \log^9 T,$$

where $d_3(n)$ denotes the number of ways to write n as a product of three positive integers.

J. B. Conrey and I [CGO] followed this a few years later with

Conjecture 2. (**Conrey–Gonek**) As $T \to \infty$,

$$\int_0^T |\zeta(1/2 + it)|^8 \, dt \sim \frac{24024}{16!} \prod_p \left(\sum_{r=0}^{\infty} \frac{d_4(p^r)^2}{p^r} \right) T \log^{16} T \,,$$

where $d_4(n)$ is the four–fold divisor function.

All these results and conjectures relied on estimating mean values of Dirichlet polynomial approximations to powers of the zeta function. It should be mentioned that the Keating–Snaith [KS] conjecture previously refered to used an entirely different method, namely, they modeled the zeta function by characteristic polynomials of random unitary matrices.

Here we sketch our method for the 6th and 8th moment conjectures. It gives the 2nd and 4th moment asymptotics as well. We begin with a discussion of the approximate functional equation.

For $s = \sigma + it$ and $\sigma > 1$, $\zeta^k(s)$ has the Dirichlet series expansion

$$\zeta^k(s) = \prod_p \left(1 - p^{-s}\right)^{-k} = \prod_p \left(1 + \frac{d_k(p)}{p^s} + \frac{d_k(p^2)}{p^{2s}} + \cdots \right) = \sum_{n=1}^{\infty} \frac{d_k(n)}{n^s} \,,$$

where $d_k(p^j) = (-1)^j \binom{-k}{j}$ is the kth divisor function. The series does not converge when $\sigma \le 1$, but we can approximate $\zeta^k(s)$ in this region by an expression of the form

$$\zeta(s)^k = \sum_{n=1}^{N} \frac{d_k(n)}{n^s} + \chi(s)^k \sum_{n=1}^{M} \frac{d_k(n)}{n^{1-s}} + \mathcal{E}_k(s) \,,$$

where $\mathcal{E}_k(s)$ denotes an error term. This is an approximate functional equation. We write it as

$$\zeta(s)^k = \mathcal{D}_{k,N}(s) + \chi(s)^k \mathcal{D}_{k,M}(1 - s) + \mathcal{E}_k(s) \,,$$

where

$$\mathcal{D}_{k,N}(s) = \sum_{n=1}^{N} \frac{d_k(n)}{n^s} \,,$$

$MN = \left(\frac{t}{2\pi}\right)^k$, and $\chi(s) = \pi^{s-1/2} \Gamma(\frac{1-s}{2})/\Gamma(\frac{s}{2})$ is the factor from the functional equation for the zeta function, $\zeta(s) = \chi(s)\zeta(1 - s)$. Taking $s = 1/2 + it$, we find that

$$\zeta(1/2 + it)^k = \mathcal{D}_{k,N}(1/2 + it) + \chi(1/2 + it)^k \mathcal{D}_{k,M}(1/2 - it) + \mathcal{E}_k(1/2 + it) \,.$$

Assuming the error term is negligible, we obtain

$$\int_T^{2T} |\zeta(1/2+it)|^{2k}\, dt \sim \int_T^{2T} |\mathcal{D}_{k,N}(1/2+it)|^2\, dt + \int_T^{2T} |\mathcal{D}_{k,M}(1/2+it)|^2\, dt$$

$$+ 2\mathrm{Re} \int_T^{2T} \chi(1/2-it)^k \mathcal{D}_{k,N}(1/2+it)\mathcal{D}_{k,M}(1/2+it)\, dt\, .$$

There is reason to believe that the cross term is smaller than the main term and that $MN = (t/2\pi)^k$ may be replaced by $MN = (T/2\pi)^k$. Thus, we expect that

$$\int_T^{2T} |\zeta(1/2+it)|^{2k}\, dt \sim \int_T^{2T} |\mathcal{D}_{k,N}(1/2+it)|^2\, dt + \int_T^{2T} |\mathcal{D}_{k,M}(1/2+it)|^2\, dt.$$

$$(9.1)$$

where $MN = (T/2\pi)^k$ and $M, N \geq 1$. We can prove this when $k = 1$ or $k = 2$, provided that M and N are both $\ll T$. When $k \geq 3$, the known bounds for $\mathcal{E}_k(s)$ are too large and it is difficult to show that the cross term really is small. (However, it might be possible to overcome these problems by appealing to a more complicated form of the approximate functional equation developed by A. Good [GD].) Our problem now is to determine an asymptotic estimate for the mean square of the Dirichlet polynomials $\mathcal{D}_{k,N}(1/2+it)$ and $\mathcal{D}_{k,M}(1/2+it)$.

Montgomery and Vaughan's mean value theorem, Theorem 7.2, gives

$$\int_T^{2T} \left|\mathcal{D}_{k,N}(1/2+it)\right|^2\, dt = \sum_{n \leq N} \frac{d_k(n)^2}{n}(T + O(n))\, .$$

By standard techniques one can show that

$$\sum_{n \leq N} d_k(n)^2 \sim \frac{a_k}{\Gamma(k^2)} N \log^{k^2-1} N$$

and that

$$\sum_{n \leq N} \frac{d_k(n)^2}{n} \sim \frac{a_k}{\Gamma(k^2+1)} \log^{k^2} N\, ,$$

where

$$a_k = \prod_p \left(\left(1 - \frac{1}{p}\right)^{k^2} \sum_{r=0}^{\infty} \frac{d_k(p^r)^2}{p^r} \right)\, .$$

Thus, for $N \ll T$, we deduce that

$$\int_T^{2T} \left|\mathcal{D}_{k,N}(1/2+it)\right|^2\, dt \sim \frac{a_k}{\Gamma(k^2+1)} T \log^{k^2} N\, .$$

Using this with $M, N \ll T$ and $MN = (T/2\pi)^k$, we obtain the classical estimates for $I_1(T)$ and $I_2(T)$.

If $k \geq 3$, the condition $MN = (T/2\pi)^k$ forces at least one of M or N to be $\gg T$, so we need Theorem 7.3, the mean value for long Dirichlet polynomials. This requires good uniform estimates for the additive divisor sums

$$D_k(x, h) = \sum_{n \leq x} d_k(n) d_k(n+h) \, .$$

No such formula has been proved when $k > 2$, but a precise formula for the main term of $D_k(x, h)$ can be conjectured by a heuristic application of the circle method. This leads us to guess that

$$D_k(x, h) = m_k(x, h) + O(x^{1/2+\epsilon})$$

uniformly for $1 \leq h \leq x^{1/2}$, where $m_k(x, h)$ is a certain smooth function of x. Using this in Theorem 7.3, we obtain the

Conjecture 1. Let $N = (T/2\pi)^{1+\eta}$ with $0 \leq \eta \leq 1$. Then

$$\int_T^{2T} |\mathcal{D}_{k,N}(1/2 + it)|^2 \; dt \sim w_k(\eta) \frac{a_k}{\Gamma(k^2+1)} TL^{k^2} \, ,$$

where a_k is the product over primes defined previously and

$$w_k(\eta) = (1+\eta)^{k^2} \left(1 - \sum_{n=0}^{k^2-1} \binom{k^2}{n+1} \gamma_k(n) \left(1 - (1+\eta)^{-(n+1)} \right) \right) \, ,$$

with

$$\gamma_k(n) = (-1)^n \sum_{i=0}^{k} \sum_{j=0}^{k} \binom{k}{i} \binom{k}{j} \binom{n-1}{i-1, j-1, n-i-j+1}$$

when $n \geq 1$ and $\gamma_k(0) = k$.

The conjecture restricts us to $N \ll T^2$. Thus, M and N in (9.1) must satisfy

$$M \ll T^2, \quad N \ll T^2, \quad \text{and} \quad MN = (T/2\pi)^k \, .$$

Writing $N = (T/2\pi)^{1+\eta}$ and $M = (T/2\pi)^{k-1-\eta}$, we find that

$$\int_T^{2T} |\zeta(1/2 + it)|^{2k} \; dt \sim \int_T^{2T} |\mathcal{D}_{k,(T/2\pi)^{1+\eta}}(1/2+it)|^2 \; dt$$

$$+ \int_T^{2T} |\mathcal{D}_{k,(T/2\pi)^{k-1-\eta}}(1/2+it)|^2 \; dt \, ,$$

with $0 \leq \eta \leq 1$. Hence,

$$\int_T^{2T} |\zeta(1/2 + it)|^{2k} \; dt \sim \left(w_k(\eta) + w_k(k-2-\eta) \right) \frac{a_k}{\Gamma(k^2+1)} TL^{k^2}$$

Example:The 6th moment.   Take $k = 3$. Then

$$\int_T^{2T} |\zeta(1/2 + it)|^6 \, dt \sim \left(w_3(\eta) + w_3(1 - \eta)\right)\frac{a_3}{\Gamma(10)}TL^9$$

for $0 \leq \eta \leq 1$. We find from the conjecture that

$$w_3(\eta) = 1 + 9\eta + 36\eta^2 + 84\eta^3 + 126\eta^4 - 630\eta^5 + 588\eta^6 + 180\eta^7 - 9\eta^8 + 2\eta^9 \,,$$

and one can verify that

$$w_3(\eta) + w_3(1 - \eta) = 42$$

for $0 \leq \eta \leq 1$. Therefore

$$\int_T^{2T} |\zeta(1/2 + it)|^6 \, dt \sim 42\frac{a_3}{9!}TL^9 \,.$$

Example:The 8th moment.   Take $k = 4$. Then

$$\int_T^{2T} |\zeta(1/2 + it)|^8 \, dt \sim \left(w_4(\eta) + w_4(2 - \eta)\right)\frac{a_4}{\Gamma(17)}TL^{16} \,,$$

where η and $2 - \eta$ must be in $[0, 1]$. This forces $\eta = 1$. Now

$$w_4(1) = 12012 \,.$$

Hence

$$\int_T^{2T} |\zeta(1/2 + it)|^8 \, dt \sim 24024\frac{a_4}{16!}TL^{16} \,.$$

Originally we thought we would be able to take $N > T^2$ in our formulas. In other words, we expected the error terms in

$$D_k(x, h) = m_k(x, h) + O_h(x^{1/2+\epsilon}) \,,$$

when used in conjunction with the long mean value theorem and averaged over h up to $x^{1-\epsilon}$, would cancel. We were surprised to find, however, that they accumulate once h exceeds $x^{1/2-\epsilon}$ and contribute to the main term. It would be very interesting to understand this behavior better.

References

[BL]   H. Bohr and E. Landau, *Ein Satz über Dirichletsche Reihen mit Anwendung auf die ζ–Funktion und die L–Funktionen*, Rend. di Palermo **37** (1914), 269–272.

[C]   J. B. Conrey, *More than two-fifths of the zeros of the Riemann zeta-function are on the critical line,* J. Reine Angew. Math. **399** (1989), 1–26.

[CG]   J. B. Conrey and A. Ghosh, *A conjecture for the sixth power moment of the Riemann zeta-function,* Int. Math. Res. Not., **15** (1998), 775–780.

[CGG1] J. B. Conrey, A. Ghosh, and S. M. Gonek, *Simple zeros of the Riemann zeta-function,* Proc. London Math. Soc. (3) **76** (1998), 497–522.

[CGG2] J. B. Conrey, A. Ghosh, and S. M. Gonek, *Mean values of the Riemann zeta-function with Application to the distribution of zeros,* in: Number Theory, Trace Formulas and Discrete Groups (Oslo, 1987), Academic Press, Boston (1989), 185–199.

[CGO]   J. B. Conrey and S. M. Gonek, *High moments of the Riemann zeta function,* Duke Math. J. **107** (2001), 577–604.

[F]   D. W. Farmer, *Long mollifiers of the Riemann zeta-function,* Mathematika, **40** (1993), 71–87.

[G]   S. M. Gonek, *Mean values of the Riemann zeta-function and its derivatives,* Invent. Math., **75** (1984), 123–141.

[GD]   A. Good, *Approximative Approximative Funktionalgleichungen und Mittelwertsätze für Dirichletreihen, die Spitzenformen assoziiert sind,* Comment. Math. Helv., **50** (1975), 327–361.

[GG]   D. A. Goldston and S. M. Gonek, *Mean value theorems for long Dirichlet polynomials and tails of Dirichlet series,* Acta Arithmetica, **LXXXIV** 2 (1998), 155- 192.

[GGOS] D. A. Goldston, S. M. Gonek, A. Ozluk, and C. Snyder, *On the pair correlation of zeros of the Riemann zeta-function,* Proc. London Math. Soc. (3) **80** (2000), 31-49.

[H]   G. H. Hardy, *Sur les zéros de la fonction $\zeta(s)$ de Riemann,* C. R. **158** (1914), 1012–1014.

[H-B]   D. R. Heath–Brown, *Fractional moments of the Riemann zeta-function,* J. London Math. Soc. (2) **24** (1981), 65–78.

[HL]   G. H. Hardy and J. E. Littlewood, *Contributions to the theory of the Riemann zeta-function and the theory of the distribution of primes,* A. M. **41** (1918), 119–196.

[I]   A. E. Ingham, *Mean-value theorems in the theory of the Riemann zeta-function,* Proc. London Math. Soc. (92) **27** (1926), 273–300.

[KS]   J.P. Keating and N.C. Snaith, *Random matrix theory and* $\zeta(1/2 + \imath t)$, Commun. Math. Phys. **214** (2000), 57–89.

[L]   N. Levinson, *More than one third of the zeros of Riemann's zeta-function are on* $\sigma = \frac{1}{2}$, Adv. Math. **13** (1974), 383–436.

[LM]   N. Levinson and H. L. Montgomery, *Zeros of the derivative of the Riemann zeta-function*, Acta Math. **133** (1974), 49–65.

[M]   H. L. Montgomery, *The pair correlation of zeros of the zeta function*, in: Proc. Sympos. Pure Math. **24**, Amer. Math. Soc., Providence, R. I. (1973), 181–193.

[R]   K. Ramachandra, *Some remarks on the mean value of the Riemann zeta-function and other Dirichlet series–III*, Ann. Acad. Sci. Fenn. Ser. AI Math. **5** (1980), 145–158.

[Sp]   Speiser, *Geometrisches zur Riemannschen Zetafunktion*, M.A. **110** (1934), 514–521.

[T]   E. C. Titchmarsh, *The Theory of the Riemann Zeta-Function,* 2nd ed., revised by D. R. Heath-Brown, Clarendon (Oxford), (1986).

Families of L-functions and 1-level densities

Brian Conrey *

Abstract

In these notes we will describe some of the families of L-functions with (conjectured) symmetry types unitary, orthogonal, and symplectic. Then we will indicate a general method to compute the 1-level density functions and compare the results for our families with the 1-level density functions that can be computed for the scaled limits of $U(N)$, $USp(2N)$, $SO(2N)$, and $SO(2N+1)$.

1   The Selberg Class of L-functions

All of the L-functions in the families we will discuss belong to the Selberg class (at least conjecturally; see [S], [CG], or [KP] for more details about this section) whose definition we now give. For detailed information about the L-functions in these families see [IK].

---

*This paper was written at the Isaac Newton Institute and was partially supported by an FRG grant from the NSF

Let $s = \sigma + it$ with σ and t real. An *L-function* is a Dirichlet series

$$L(s) = \sum_{n=1}^{\infty} \frac{\lambda_n}{n^s},$$

satisfying the Ramanujan bound[1] $\lambda_n \ll_\epsilon n^\epsilon$ for every $\epsilon > 0$, which has three additional properties.

- Analytic continuation: $L(s)$ continues to a meromorphic function of finite order, with at most finitely many poles, and all poles are located on the $\sigma = 1$ line.

- Functional equation: There is a number ε with $|\varepsilon| = 1$, and a function $\gamma_L(s)$ of the form

$$\gamma_L(s) = P(s)Q^s \prod_{j=1}^{k} \Gamma(w_j s + \mu_j)$$

where P is a polynomial whose only zeros in $\sigma > 0$ are at the poles of $L(s)$, $Q > 0$, $w_j > 0$, and $\mathrm{Re}\,\mu_j \geq 0$, such that

$$\xi_L(s) := \gamma_L(s)L(s)$$

is entire, and

$$\xi_L(s) = \varepsilon \overline{\xi_L}(1 - s),$$

where $\overline{\xi_L}(s) = \overline{\xi_L(\bar{s})}$.

ϵ is often called the "sign" of the functional equation (especially when it is ± 1) or the root number. It is sometimes convenient to write the functional equation in asymmetric form:

$$L(s) = \varepsilon X_L(s)\overline{L}(1 - s),$$

where $X_L(s) = \dfrac{\overline{\gamma_L}(1 - s)}{\gamma_L(s)}$.

- Euler product: For $\sigma > 1$ we have

$$L(s) = \prod_p L_p(1/p^s),$$

where the product is over the primes p, and

$$L_p(1/p^s) = \sum_{k=0}^{\infty} \frac{\lambda_{p^k}}{p^{ks}} = \exp\left(\sum_{k=1}^{\infty} \frac{b_{p^k}}{p^{ks}}\right),$$

where $b_n \ll n^\theta$ with $\theta < \frac{1}{2}$.

---

[1]See the Appendix of the lectures of D.W. Farmer, page 185, for a discussion of the $\ll$ notation.

Note that $L(s) \equiv 1$ is the only constant L-function, the set of L-functions is closed under products, and if $L(s)$ is an L-function then so is $L(s + iy)$ for any real y. An L-function is called *primitive* if it cannot be written as a nontrivial product of L-functions, and it can be shown, assuming Selberg's orthonormality conjectures, that any L-function has a unique representation as a product of primitive L-functions.

The *degree* of the L-function is the sum of the w_j. In all known examples, one may take $w_j = 1/2$ for all j; this may require using the duplication formula for the Gamma-function. This is a useful thing to do, because then $\gamma_L(s)$ is uniquely determined. Also, in all known examples the Euler product is expressible as the reciprocal of a polynomial whose degree is equal to d for all primes that do not divide the level q which (in the formulation with all $w_j = 1/2$) is given by $q = \pi^d Q^2$; for the primes dividing q it is the case in all known examples that the Euler factor is the reciprocal of a polynomial of degree at most d.

For each L-function we define its *log-conductor* $c(L, t)$ at the point $1/2 + it$ by

$$c(L, t) = \frac{X'_L}{X_L}(1/2 + it). \tag{1.1}$$

The density of zeros near the point $1/2 + it$ is given by

$$\frac{2\pi}{c(L, t)}.$$

2 Degree 1 L-functions

2.1 The Riemann zeta-function

The Riemann zeta-function is given by

$$\zeta(s) := 1 + \frac{1}{2^s} + \frac{1}{3^s} + \cdots = \sum_{n=1}^{\infty} \frac{1}{n^s}.$$

The series converges in the half-plane where the real part of s is larger than 1. Riemann proved that $\zeta(s)$ has an analytic continuation to the whole plane apart from a simple pole at $s = 1$. Moreover, he proved that $\zeta(s)$ satisfies a *functional equation* which in its symmetric form is given by

$$\xi(s) := \tfrac{1}{2}s(s - 1)\pi^{-\frac{s}{2}}\Gamma\left(\frac{s}{2}\right)\zeta(s) = \xi(1 - s)$$

where $\Gamma(s)$ is the usual Gamma-function. Euler had earlier proved that

$$\zeta(s) = \left(1 + \frac{1}{2^s} + \frac{1}{4^s} + \frac{1}{8^s} + \cdots\right)\left(1 + \frac{1}{3^s} + \frac{1}{9^s} + \cdots\right)\left(1 + \frac{1}{5^s} + \cdots\right)\cdots$$

$$= \prod_p \left(1 - \frac{1}{p^s}\right)^{-1}$$

where the infinite product (called the *Euler product*) is over all the prime numbers. The product converges when the real part of s is greater than 1. The Euler product implies that there are no zeros of $\zeta(s)$ with real part greater than 1; the functional equation implies that there are no zeros with real parts less than 0, apart from the *trivial zeros* at $s = -2, -4, -6, \ldots$. Thus, all of the complex zeros are in the *critical strip* $0 \leq \mathrm{Re}\, s \leq 1$. The functional equation shows that the complex zeros are symmetric with respect to the line $\mathrm{Re}\, s = \frac{1}{2}$. Riemann calculated the first few complex zeros $\frac{1}{2} + i14.134\ldots, \frac{1}{2} + i21.022\ldots$ and proved that the number $N(T)$ of zeros with imaginary parts between 0 and T is

$$N(T) = \frac{T}{2\pi}\log\frac{T}{2\pi e} + \frac{7}{8} + S(T) + O(1/T)$$

where $S(T) = \frac{1}{\pi}\arg\zeta(1/2 + iT)$ is computed by continuous variation starting from $\arg\zeta(2) = 0$ and proceeding along straight lines, first up to $2 + iT$ and then to $1/2 + iT$. Riemann also proved that $S(T) = O(\log T)$. Note for future reference that at a height T the average gap between zero heights is $\sim 2\pi/\log T$. Riemann suggested that the number $N_0(T)$ of zeros of $\zeta(1/2 + it)$ with $0 < t \leq T$ seemed to be about

$$\frac{T}{2\pi}\log T;$$

and then made his conjecture that all of the zeros of $\zeta(s)$ in fact lie on the 1/2-line; this is the Riemann Hypothesis.

The family $\{\zeta(1/2 + it) : t \in \mathcal{R}\}$ parametrized by t is a unitary family.

2.2   Dirichlet L-functions

The simplest L-function after the ζ-function is the Dirichlet L-function for the non-trivial character of conductor 3:

$$L(s, \chi_{-3}) = 1 - \frac{1}{2^s} + \frac{1}{4^s} - \frac{1}{5^s} + \frac{1}{7^s} - \frac{1}{8^s} + - \ldots.$$

This can be written as an Euler product

$$L(s, \chi_{-3}) = \prod_{p \equiv 1 \bmod 3} (1 - p^{-s})^{-1} \prod_{p \equiv 2 \bmod 3} (1 + p^{-s})^{-1},$$

satisfies the functional equation

$$\xi(s,\chi_{-3}) := \left(\tfrac{\pi}{3}\right)^{-\frac{s}{2}} \Gamma(\tfrac{s+1}{2}) L(s,\chi_3) = \xi(1-s,\chi_{-3}),$$

and is expected to have all of its non-trivial zeros on the $1/2$-line. In general, a Dirichlet character is a completely multiplicative periodic function $\chi : \mathbb{N} \to \mathbb{C}$; i.e. $\chi(mn) = \chi(m)\chi(n)$ for all m,n and $\chi(m+q) = \chi(m)$ for some integer q. It is the *primitive* characters which lead to the L-functions in the Selberg class . For each $q \geq 1$ there are precisely

$$\psi(q) = \sum_{d|q} \mu(d)\varphi(q/d) \tag{2.1}$$

primitive characters to the modulus q.[2] If q has the factorization $q = p_1^{e_1} \dots p_r^{e_r}$, then any primitive character χ mod q has a unique representation as a product $\chi = \chi_1 \dots \chi_r$ where χ_j is a primitive character modulo $p_j^{e_j}$. We now describe how to construct the primitive characters modulo p^e. If p is odd, then the number of integers less than or equal to p^e and relatively prime to p^e is given by $\varphi(p^e) = p^e - p^{e-1}$. These reduced residues modulo p^e form a multiplicative group which is cyclic; let g be a generator of this group (i.e. a *primitive root of* p^e.) We can specify any character χ modulo p^e by saying what the value of $\chi(g)$ is (clearly this value must be a $\varphi(p^e)$ root of unity). The primitive characters are those for which $\chi(g) = \exp(2\pi i a/\varphi(p^e))$ where $(a, \varphi(p^e)) = 1$. For $p = 2$, the reduced residues modulo 2^e do not form a cyclic group unless $e = 1$ or 2. If $e \geq 3$ then the reduced residues are given by $\pm 5^j$ with $j = 0, 1, \dots 2^{e-2}$. The primitive characters χ modulo 2^e are determined by the value of $\chi(5) = \exp(2\pi i a/2^{e-2})$ with $1 \leq a \leq 2^{e-2}$ odd and by the value of $\chi(-1) = \pm 1$. This describes all primitive characters.

For each primitive character χ mod q the *Gauss sum* is given by

$$\tau(\chi) = \sum_{n=1}^{q} \chi(n)e(n/q). \tag{2.2}$$

where $e(x) = \exp(2\pi i x)$. It satisfies $|\tau(\chi)| = \sqrt{q}$; we write $\tau(\chi) = \varepsilon_\chi \sqrt{q}$. The Dirichlet L-function is given by

$$L(s,\chi) = \sum_{n=1}^{\infty} \frac{\chi(n)}{n^s} = \prod_p \left(1 - \frac{\chi(p)}{p^s}\right)^{-1}$$

for $\sigma > 1$. Odd characters are those for which $\chi(-1) = -1$; even characters have $\chi(-1) = 1$. The functional equation for an even character is

$$\xi(s,\chi) := (\pi/\sqrt{q})^{-s/2}\Gamma(s/2)L(s,\chi) = \varepsilon_\chi \xi(1-s,\overline{\chi}). \tag{2.3}$$

---

[2]Editors' comment: Here φ is Euler's phi-function, defined in the lectures by D.R. Heath-Brown (page 1) after Theorem 20. μ is defined by D.A. Goldston (page 79) in equation 2.10 and is called the Möbius function.

For an odd character, the functional equation is

$$\xi(s,\chi) := (\pi/\sqrt{q})^{-s/2}\Gamma((s+1)/2)L(s,\chi) = \varepsilon_\chi\xi(1-s,\overline{\chi}). \qquad (2.4)$$

We have described the primitive characters above. Imprimitive characters arise in two ways. First, the principal character χ_0 modulo p defined by $\chi_0(n)$ $= 0$ if $p \mid n$ and $= +1$ if $p \nmid n$ is an imprimitive character. Second, a primitive character modulo p^e regarded as a character modulo p^f where $f > e$ is an imprimitive character. Finally, the product of a primitive character with an imprimitive character is an imprimitive character. Any character χ (primitive or imprimitive) which satisfies $\chi(m+q) = \chi(m)$ is called a character modulo q. There are $\varphi(q)$ characters modulo q.

2.2.1   Orthogonality relations

The basic orthogonality relation is expressed by: if $(mn,q) = 1$, then

$$\sum_{\chi \bmod q} \chi(m)\overline{\chi(n)} = \begin{cases} \varphi(q) \text{ if } m = n \bmod q \\ 0 \text{ if } m \neq n \bmod q \end{cases} \qquad (2.5)$$

For primitive characters, this takes the shape: if $(mn,q) = 1$, then

$$\sum_{\chi \bmod q}^{*} \chi(m)\overline{\chi(n)} = \sum_{d\mid(q,m-n)} \varphi(d)\mu(q/d). \qquad (2.6)$$

The Polya-Vinogradov inequality asserts that

$$|\sum_{n=1}^{N}\chi(n)| \ll q^{1/2}\log q \qquad (2.7)$$

for any non-principal character $\chi \bmod q$.

The family $\{L(1/2, \chi_q) : q \text{ is primitive}\}$ is also a unitary family.

2.3   Real primitive characters

A special role is played by the real or quadratic Dirichlet characters. These we denote by χ_d where d is a fundamental discriminant: d can be positive or negative, is either odd, squarefree, and congruent to 1 modulo 4, or is 4 times a squarefree integer congruent to 2 or 3 modulo 4. Thus, the sequence of positive fundamental discriminants begins $d = 1, 5, 8, 12, 13, 17, 21, 24, 28, \ldots$ and the sequence of negative fundamental discriminants begins $d=-3, -4, -7,$ $-8, -11, -15, -19, -20, -23, -24, \ldots$. The character χ_d only takes on the values $+1, 0, -1$; it is primitive with the modulus $|d|$. If $d > 0$, then χ_d is an even character and if $d < 0$ it is an odd character. The character χ_d is the character associated with the quadratic field $Q(\sqrt{d})$. In particular, the prime p *splits* or factors in this field if $\chi_d(p) = +1$; it remains prime if $\chi_d(p) = -1$; and it *ramifies* (has a square factor) if $\chi_d(p) = 0$. The real characters χ_d can be decomposed into a product of characters $\chi_{-4}, \chi_8, \chi_{-8}, \chi_p$ ($p \equiv 1 \bmod 4$), and χ_{-p}, ($p \equiv 3 \bmod 4$) for odd primes p where $\chi_{\pm p}(n) = \left(\frac{n}{p}\right)$ is the Legendre symbol ($= 1$ if n is a non-zero square modulo p, and $= -1$ if n is a non-zero non-square modulo p, $= 0$ if $p \mid n$). The character $\chi_{-4}(n)$ is 0 for even n, is $+1$ for n congruent to 1 modulo 4, and is -1 for n congruent to 3 modulo 4. The character $\chi_8(n)$ is 0 for even n, is $+1$ for n congruent to ± 1 modulo 8, and is -1 for n congruent to ± 3 modulo 8. Finally, $\chi_{-8}(n)$ is 0 for even n, is $+1$ for n congruent to 1 or 3 modulo 8, and is -1 for n congruent to 5 or 7 modulo 8.

2.3.1 Orthogonality

First of all, the number $N_q^+(x)$ of fundamental discriminants d with $0 < d \leq x$ and $(d, q) = 1$ satisfies $N_q^+(x) \sim \frac{3}{\pi^2} \frac{\varphi(q)}{q} x$ and similarly the number $N_q^-(x)$ of negative fundamental discriminants d with $0 < -d < x$ and $(d, q) = 1$ satisfies $N_q^-(x) \sim \frac{3}{\pi^2} \frac{\varphi(q)}{q} x$.

By the Generalized Riemann Hypothesis,

$$\sum_{0 < d \leq x} \chi_d(n) = \begin{cases} N_n^+(x) & \text{if } n \text{ is a square} \\ O(x^{1/2}(nx)^\epsilon) & \text{if } n \text{ is not a square.} \end{cases} \tag{2.8}$$

Elliott, Jutila and Heath-Brown have proven unconditional estimates for averages of real characters. In particular, Heath-Brown [H-B] has shown that

$$\sum_m \left| \sum_n a_n \left(\frac{n}{m}\right) \right|^2 \ll (MN)^\varepsilon (M + N) \sum_n |a_n|^2,$$

where n and m are restricted to odd square-free numbers in the intervals $[1, M]$ and $[1, N]$, respectively, and the a_n are any complex numbers.

The family $\{L(1/2, \chi_d) : d \text{ is a fundamental discriminant }\}$ is a symplectic family.

3   Degree two L-functions

3.1   Modular L-functions

A first example of a degree 2 L-function arises from Ramanujan's tau-function, defined implicitly by

$$x \prod_{n=1}^{\infty} (1 - x^n)^{24} = \sum_{n=1}^{\infty} \tau(n)x^n.$$

The Fourier series

$$\Delta(z) := \sum_{n=1}^{\infty} \tau(n)e(nz),$$

where $e(z) = \exp(2\pi i z)$, satisfies

$$\Delta\left(\frac{az + b}{cz + d}\right) = (cz + d)^{12}\Delta(z)$$

for all integers a, b, c, d with $ad - bc = 1$. A function satisfying these equations is called a *modular form* of weight 12. The associated L-function is

$$L_\Delta(s) := \sum_{n=1}^{\infty} \frac{\tau(n)/n^{11/2}}{n^s} = \prod_p \left(1 - \frac{\tau(p)/p^{11/2}}{p^s} + \frac{1}{p^{2s}}\right)^{-1};$$

it satisfies the functional equation

$$\xi_\Delta(s) := (2\pi)^{-s}\Gamma(s + 11/2)L_\Delta(s) = \xi_\Delta(1 - s).$$

Note that, by the duplication formula, this can be written in the form

$$\xi_\Delta(s) = \pi^{-s}\Gamma\left(\frac{s}{2} + \frac{11}{4}\right)\Gamma\left(\frac{s}{2} + \frac{13}{4}\right)L_\Delta(s).$$

It is expected that all of the complex zeros of $L_\Delta(s)$ are on the $1/2$-line. In general a cusp form of weight k for the full modular group is a holomorphic function f on the upper half-plane which satisfies

$$f\left(\frac{az + b}{cz + d}\right) = (cz + d)^k f(z)$$

for all integers a, b, c, d with $ad - bc = 1$ and also has the property that $\lim_{y\to\infty} f(iy) = 0$. Cusp forms for the whole modular group exist only for even integers $k = 12$ and $k \geq 16$. The cusp forms of a given weight k of this form make a complex vector space S_k of dimension $[k/12]$ if $k \neq 2 \bmod 12$ and of dimension $[k/12] - 1$ if $k = 2 \bmod 12$. Each such vector space has a special basis H_k of Hecke eigenforms which consist of functions $f(z) = \sum_{n=1}^{\infty} a_f(n)e(nz)$ for which

$$a_f(m)a_f(n) = \sum_{d|(m,n)} d^{k-1}a_f(mn/d^2). \tag{3.1}$$

The Fourier coefficients $a_f(n)$ are real algebraic integers of degree at most the dimension $\#H_k$ of the vector space. Thus, when $k = 12, 16, 18, 20, 22, 26$ the spaces are one dimensional and the coefficients are ordinary integers. We can express these explicitly in terms of the Eisenstein series

$$E_4(z) = 1 + 240 \sum_{n=1}^{\infty} \sigma_3(n)e(nz)$$

and

$$E_6(z) = 1 - 504 \sum_{n=1}^{\infty} \sigma_5(n)e(nz)$$

where $\sigma_r(n)$ is the sum of the rth powers of the positive divisors of n:

$$\sigma_r(n) = \sum_{d|n} d^r.$$

Then, $\Delta(z)E_4(z)$ gives the unique Hecke form of weight 16; $\Delta(z)E_6(z)$ gives the unique Hecke form of weight 18; $\Delta(z)E_4(z)^2$ is the Hecke form of weight 20; $\Delta(z)E_4(z)E_6(z)$ is the Hecke form of weight 22; and $\Delta(z)E_4(z)^2E_6(z)$ is the Hecke form of weight 26. The two Hecke forms of weight 24 are given by

$$\Delta(z)E_4(z)^3 + x\Delta(z)^2$$

where $x = -156 \pm 12\sqrt{144169}$. To define the L-function, we scale the coefficients and write

$$\lambda_f(n) = \frac{a_f(n)}{n^{(k-1)/2}}.$$

Then the L-function associated with a Hecke form f of weight k is given by

$$L_f(s) = \sum_{n=1}^{\infty} \lambda_f(n)n^s = \prod_p \left(1 - \frac{\lambda_f(p)}{p^s} + \frac{1}{p^{2s}}\right)^{-1}. \tag{3.2}$$

By Deligne's theorem $\lambda_f(p) = 2\cos\theta_f(p)$ for a real $\theta_f(p)$. It is conjectured (Sato-Tate) that for each f the $\{\theta_f(p) : p \text{ prime}\}$ is uniformly distributed on $[0, \pi)$ with respect to the measure $\frac{2}{\pi}\sin^2\theta\,d\theta$. We write $2\cos\theta_f(p) = \alpha_f(p) + \overline{\alpha_f(p)}$ where $\alpha_f(p) = e^{i\theta_f(p)}$; then

$$L_f(s) = \prod_p \left(1 - \frac{\alpha_f(p)}{p^s}\right)^{-1}\left(1 - \frac{\overline{\alpha_f(p)}}{p^s}\right)^{-1}. \tag{3.3}$$

The functional equation satisfied by $L_f(s)$ is

$$\xi_f(s) = (2\pi)^{-s}\Gamma(s + (k-1)/2)L_f(s) = (-1)^{k/2}\xi_f(1-s). \tag{3.4}$$

Note that the *sign* $(-1)^{k/2}$ of the functional equation is $+1$ when $k \equiv 0 \bmod 4$ and is -1 when $k \equiv 2 \bmod 4$.

3.1.1   Orthogonality relations

The Petersson inner product on the space S_k is defined by

$$\langle f,g \rangle = \iint_D f(z)\overline{g(z)} y^k \frac{dxdy}{y^2}. \tag{3.5}$$

Here the integration is over the *fundamental domain*

$$\mathcal{D} := \{(x,y) : -1/2 \le x \le 1/2, y \ge \sqrt{1-x^2}\}.$$

Let $\mathcal{F}$ be an orthogonal basis of S_k with respect to this inner product. The Petersson formula tells us that

$$\frac{\Gamma(k-1)}{(4\pi\sqrt{mn})^{k-1}} \sum_{f \in \mathcal{F}} \frac{a_f(m)\overline{a_f(n)}}{\langle f,f \rangle} = \delta_{m,n} + 2\pi i^{-k} \sum_{c=1}^{\infty} \frac{S(m,n,c)}{c} J_{k-1}\left(\frac{4\pi\sqrt{mn}}{c}\right) \tag{3.6}$$

where J_{k-1} is the Bessel function of index $k-1$ and $S(m,n,c)$ is the Kloosterman sum

$$S(m,n,c) = \sum_{(x,c)=1} e((mx+n\overline{x})/c) \tag{3.7}$$

where the sum is over a set of reduced residue classes modulo c and where $\overline{x}$ satisfies $x\overline{x} \equiv 1 \bmod c$. By a theorem of Weil, $|S(m,n,c)| \le (m,n,c)^{1/2} d(c)\sqrt{c}$ where $d(c)$ is the number of positive divisors of c.

The family $\{L_f(1/2) : f$ is a primitive form of weight $k\}$ is an orthogonal family. If we restrict to $k \equiv 0 \bmod 4$ then it is an even orthogonal family and if we restrict to $k \equiv 2 \bmod 4$ then it is an odd orthogonal family.

3.2   Higher level modular forms

An example of a higher level modular form is the modular form $\sum_{n=1}^{\infty} a(n)e(nz)$ associated to an elliptic curve $E : y^2 = x^3 + Ax + B$ where A, B are integers. The associated L-function, called the Hasse-Weil L-function, is

$$L_E(s) = \sum_{n=1}^{\infty} \frac{a(n)/n^{1/2}}{n^s} = \prod_{p\nmid q}\left(1 - \frac{a(p)/p^{1/2}}{p^s} + \frac{1}{p^{2s}}\right)^{-1} \prod_{p|q}\left(1 - \frac{a(p)/p^{1/2}}{p^s}\right)^{-1}$$

where q is the conductor of the curve. The coefficients $a(n)$ are constructed easily from $a(p)$ for prime p; in turn the $a(p)$ are given by $a(p) = p - N_p$ where N_p is the number of solutions of E when considered modulo p. The work of Wiles and others proved that these L-functions are associated to modular forms of weight 2. This modularity implies the functional equation

$$\xi_E(s) := (2\pi/\sqrt{q})^{-s}\Gamma(s+1/2)L_E(s) = w_E\xi_E(1-s) \tag{3.8}$$

where $w_E = \pm 1$ is the sign of the functional equation. It is believed that all of the complex zeros of $L_E(s)$ are on the 1/2-line.

3.2.1 Level q cusp forms

We let $\Gamma_0(q)$ denote the group of matrices $\begin{pmatrix} a & b \\ c & d \end{pmatrix}$ with integers a, b, c, d satisfying $ad - bc = 1$ and $q \mid c$. This group is called the *Hecke congruence group of level q*. A function f holomorphic on the upper half plane satisfying

$$f\left(\frac{az+b}{cz+d}\right) = (cz+d)^k f(z) \tag{3.9}$$

for all matrices in $\Gamma_0(q)$ and $\lim_{y\to 0} f(\frac{a}{q} + iy) = 0$ for all rational numbers a/q is called a cusp form for $\Gamma_0(q)$; the space of these is a finite dimensional vector space $S_k(q)$. The space S_k above is the same as $S_k(1)$. Again, these spaces are empty unless k is an even integer. If k is an even integer, then

$$\dim S_k(q) = \frac{(k-1)}{12}\nu(q) + \left(\left[\frac{k}{4}\right] - \frac{k-1}{4}\right)\nu_2(q)$$
$$+ \left(\left[\frac{k}{3}\right] - \frac{k-1}{3}\right)\nu_3(q) - \frac{\nu_\infty(q)}{2}$$

where $\nu(q)$ is the index of the subgroup $\Gamma_0(q)$ in the full modular group $\Gamma_0(1)$:

$$\nu(q) = q\prod_{p\mid q}\left(1+\frac{1}{p}\right);$$

$\nu_\infty(q)$ is the number of *cusps* of $\Gamma_0(q)$:

$$\nu_\infty(q) = \sum_{d\mid q}\varphi((d, q/d));$$

$\nu_2(q)$ is the number of inequivalent *elliptic points* of order 2:

$$\nu_2(q) = \begin{cases} 0 \text{ if } 4\mid q \\ \prod_{p\mid q}(1+\chi_{-4}(p)) \text{ otherwise} \end{cases}$$

and $\nu_3(q)$ is the number of inequivalent *elliptic points* of order 3:

$$\nu_3(q) = \begin{cases} 0 \text{ if } 9\mid q \\ \prod_{p\mid q}(1+\chi_{-3}(p)) \text{ otherwise} \end{cases}.$$

It is clear from this formula that the dimension of $S_k(q)$ grows approximately linearly with q and k. For the spaces $S_k(q)$ the issue of primitive forms and imprimitive forms arise, much as the situation with characters. In fact, one should think of the Fourier coefficients of cusp forms as being a generalization of characters. They are not periodic, but they act as harmonic detectors, much as characters do, through their orthogonality relations (below). Imprimitive cusp forms arise in two ways. Firstly , if $f(z) \in S_k(q)$, then $f(z) \in S_k(dq)$ for any integer $d > 1$. Secondly, if $f(z) \in S_k(q)$, then $f(dz) \in S_k(\Gamma_0(dq))$ for any $d > 1$. The dimension of the space $S_k^{\mathrm{new}}(q)$ generated by primitive forms is given by

$$\dim S_k^{\mathrm{new}}(q) = \sum_{d|q} \mu_2(d) \dim S_k(q/d)$$

where $\mu_2(n)$ is the multiplicative function defined for prime powers by $\mu_2(p^e) = -2$ if $e = 1$, $= 1$ if $e = 2$, and $= 0$ if $e > 2$. The set of primitive forms (or Hecke forms) which generate this space is denoted $H_k(q)$. The elements f of this set have a Fourier series

$$f(z) = \sum_{n=1}^{\infty} a_f(n)e(nz)$$

where the $a_f(n) = \lambda_f(n)n^{(k-1)/2}$ have the property that the associated L-function has an Euler product

$$\begin{aligned} L_f(s) &= \sum_{n=1}^{\infty} \frac{\lambda_f(n)}{n^s} \\ &= \prod_{p \nmid q} \left(1 - \frac{\lambda_f(p)}{p^s} + \frac{1}{p^{2s}}\right)^{-1} \prod_{p|q} \left(1 - \frac{\lambda_f(p)}{p^s}\right)^{-1}. \end{aligned}$$

We can express this as

$$L_f(s) = \prod_p \left(1 - \frac{\alpha_f(p)}{p^s}\right)^{-1} \left(1 - \frac{\alpha'_f(p)}{p^s}\right)^{-1}$$

where if $p \nmid q$ then $\alpha'_f(p) = \overline{\alpha_f(p)}$ whereas if $p \mid q$ then $\alpha'_f(p) = 0$. The Hecke relations, equivalent to the Euler product, are given by

$$\lambda_f(m)\lambda_f(n) = \sum_{\substack{d|(m,n) \\ (d,q)=1}} \lambda_f(mn/d^2).$$

The functional equation of the L-function is

$$\xi_f(s) := (2\pi/\sqrt{q})^{-s}\Gamma(s + (k-1)/2)L_f(s) = \pm\xi_f(1-s).$$

Now the $\pm$ depends on more than the weight k.

The family $\{L_f(1/2) : f$ is a primitive form of weight 2 and level $q\}$ is an orthogonal family. If we restrict to those f with a plus in the functional equation, then it is an even orthogonal family and if we restrict to those f with a minus in the functional equation then it is an odd orthogonal family.

3.3   Twists of modular L-functions by real characters

If $L_f(s) = \sum \lambda_f(n)n^{-s}$ is an L-function associated with a primitive form f, then we can form the twisted L-function

$$L_f(s, \chi) = \sum_{n=1}^{\infty} \frac{\lambda_f(n)\chi(n)}{n^s}$$

where $\chi(n)$ is a primitive character with modulus q. In general, if L_f has level N and $(N, q) = 1$, then $L_f(s, \chi)$ will have level Nq^2. If the functional equation of L_f is

$$\xi_f(s) = \left(\frac{\sqrt{N}}{2\pi}\right)^s \Gamma(s + (k-1)/2)L_f(s) = w_f\xi_f(1-s)$$

then the functional equation of the twist by a real quadratic character $L_f(s, \chi_d)$ is

$$\xi_f(s, \chi_d) = \left(\frac{\sqrt{N}|d|}{2\pi}\right)^s \Gamma(s + (k-1)/2)L_f(s, \chi_d) = w_f\chi_d(-N)\xi_f(1-s, \chi_d).$$

$$(3.10)$$

3.4   Maass forms

There is another kind of cusp form associated with the group $\Gamma_0(q)$. This is a function $f(z)$ which is real analytic on the upper half-plane. It transforms like a weight 0 cusp form and is an eigenfunction of the Laplace operator:

$$\Delta := y^2 \left(\frac{\partial^2}{\partial x^2} + \frac{\partial^2}{\partial y^2}\right).$$

It has a Fourier expansion as a linear combination of terms $e(nx)$ in which the dependence on y is expressed through K-Bessel functions. The prototype for these is given by the Eisenstein series (for the full modular group)

$$E(z, s) = \sum_{\gamma \in \Gamma_\infty \backslash \Gamma_0(1)} y(\gamma z)^s = \sum_{(c,d)=1} \frac{y^s}{|cz + d|^{2s}}$$

where $y(z)$ denotes the imaginary part of z and where Γ_∞ is the group which fixes ∞, i.e. the group of matrices $\begin{pmatrix} 1 & b \\ 0 & 1 \end{pmatrix}$ for integer b. This is not a cusp form (because it doesn't vanish at iy as $y \to \infty$.) However, its Fourier expansion is similar to that of the Maass cusp forms for which no explicit construction is known (apart from some forms with eigenvalue $1/4$). Let

$$\theta(s) := \pi^{-s}\Gamma(s)\zeta(2s) = \theta(1-s).$$

Then $\theta(s)E(z,s) =$

$$\theta(s)y^s + \theta(1-s)y^{1-s} + 4y^{1/2}\sum_{n=1}^{\infty}\sum_{ab=n}(a/b)^{s-1/2}K_{s-1/2}(2\pi ny)\cos(2\pi nx).$$

Since $\theta(s)$, $\sum_{ab=n}(a/b)^{s-1/2}$ and $K_{s-1/2}(2\pi ny)$ are all invariant under $s \to 1-s$, we see that $\theta(s)E(z,s) = \theta(1-s)E(z,1-s)$.

A Maass form f with eigenvalue $\lambda = 1/4 + \kappa^2$ satisfies $(\Delta + \lambda)f = 0$ and has Fourier expansion

$$f(z) = y^{1/2}\sum_{n=1}^{\infty}\lambda_f(n)K_{i\kappa}(2\pi ny)\cos(2\pi nx)$$

for an even Maass form and

$$f(z) = y^{1/2}\sum_{n=1}^{\infty}\lambda_f(n)K_{i\kappa}(2\pi ny)\sin(2\pi nx)$$

for an odd Maass form.

The L-function $L_f(s) = \sum_{n=1}^{\infty}\lambda_f(n)N^{-s}$ associated with a Maass form is entire, has an Euler product, and satisfies the functional equation

$$\xi_f(s) := \pi^{-s}\Gamma((s+i\kappa)/2)\Gamma((s-i\kappa)/2)L_f(s) = \xi_f(1-s)$$

for even Maass forms and

$$\xi_f(s) := \pi^{-s}\Gamma((s+1+i\kappa)/2)\Gamma((s+1-i\kappa)/2)L_f(s) = \xi_f(1-s)$$

for odd Maass forms.

Selberg's trace formula provides us with a kind of Weyl law for the number of Maass forms with eigenvalue less than a given quantity.

Ramanujan's conjecture for Maass forms is that $|\lambda_f(p)| \le 2$. However, this has not yet been proven. The best result is $\lambda_f(p) \ll p^{1/9}$. Thus, we don't know for sure that the Maass-form L-functions are in the Selberg class.

4   Higher degree L-functions

4.1   Symmetric square L-functions

Recall that the Euler product for a level q modular form has the shape

$$L_f(s) = \prod_p\left(1 - \frac{\alpha_f(p)}{p^s}\right)^{-1}\left(1 - \frac{\alpha'_f(p)}{p^s}\right)^{-1}.$$

We can form the symmetric square L-function associated to f as

$$L_f(\text{sym}^2, s) = \prod_p \left(1 - \frac{\alpha_f^2(p)}{p^s}\right)^{-1} \left(1 - \frac{\alpha_f(p)\alpha_f'(p)}{p^s}\right)^{-1} \left(1 - \frac{\alpha_f'(p)^2}{p^s}\right)^{-1}.$$

Note that this L-function has a degree three Euler product associated with it. Shimura proved that this is an entire function which satisfies the functional equation

$$\begin{aligned} \xi_f(\text{sym}^2, s) : &= \pi^{-3s/2} q^s \Gamma((s+1)/2) \Gamma((s+k-1)/2) \Gamma((s+k)/2) L_f(\text{sym}^2, s) \\ &= \xi_f(\text{sym}^2, 1-s). \end{aligned} \qquad (4.1)$$

The family $\{L_f(\text{sym}^2, 1/2) : f$ is a primitive form of weight $k\}$ is a symplectic family. Similarly for $\{L_f(\text{sym}^2, 1/2) : f$ is a primitive form of weight 2 and level $q\}$.

More generally, for an L-function with Euler product

$$L(s) = \prod_p \prod_{j=1}^{d} \left(1 - \frac{\alpha_j(p)}{p^s}\right)^{-1}$$

one can form its symmetric square L-function

$$L(\text{sym}^2, s) = \prod_p \prod_{1 \le i \le j \le d} \left(1 - \frac{\alpha_i(p)\alpha_j(p)}{p^s}\right)^{-1};$$

this is expected to be a degree $k(k+1)/2$ L-function in the Selberg class which may or may not be primitive. One can also form the exterior square L-function

$$L(\text{ext}^2, s) = \prod_p \prod_{1 \le i < j \le d} \left(1 - \frac{\alpha_i(p)\alpha_j(p)}{p^s}\right)^{-1};$$

this is expected to be an L-function in the Selberg class of degree $k(k-1)/2$, which is not necessarily primitive. See the nice survey of Bump [B] for more information about these L-functions.

4.2   Convolution L-functions

Given two L-functions

$$L_f(s) = \prod_p \left(1 - \frac{\alpha_f(p)}{p^s}\right)^{-1} \left(1 - \frac{\alpha_f'(p)}{p^s}\right)^{-1}$$

where $f \in H_k(q_1)$ and

$$L_g(s) = \prod_p \left(1 - \frac{\beta_g(p)}{p^s}\right)^{-1} \left(1 - \frac{\beta_g'(p)}{p^s}\right)^{-1}$$

where $g \in H_\ell(q_2)$ with $(q_1, q_2) = 1$ we form the convolution L-function

$$L_{f \times g}(s) = \prod_p \left(1 - \frac{\alpha_f(p)\beta_g(p)}{p^s}\right)^{-1} \left(1 - \frac{\alpha_f(p)\beta'_g(p)}{p^s}\right)^{-1} \times$$
$$\left(1 - \frac{\alpha'_f(p)\beta_g(p)}{p^s}\right)^{-1} \left(1 - \frac{\alpha'_f(p)\beta'_g(p)}{p^s}\right)^{-1}.$$

If $f \neq g$, then this L-function is entire – an Euler product of degree 4 – and satisfies the functional equation

$$\xi_{f \times g}(s) : = (2\pi)^{-2s}(q_1 q_2)^s \Gamma(s + (|k - \ell|)/2)\Gamma(s - 1 + (k + \ell - 1)/2)L_{f \times g}(s)$$
$$= \pm \xi_{f \times g}(1 - s).$$

One can form a convolution between any two L-functions

$$L_1(s) = \prod_p \prod_{i=1}^{d_1} \left(1 - \frac{\alpha_i(p)}{p^s}\right)^{-1}$$

and

$$L_2(s) = \prod_p \prod_{j=1}^{d_2} \left(1 - \frac{\beta_j(p)}{p^s}\right)^{-1};$$

it is given by

$$L(s) = \prod_p \prod_{i=1}^{d_1} \prod_{j=1}^{d_2} \left(1 - \frac{\alpha_i(p)\beta_j(p)}{p^s}\right)^{-1}.$$

In general, this convolution L-function is expected to be an L-function in the Selberg class of degree $d_1 d_2$.

5   One-level densities

Now we indicate how to compute the one level density functions for these families. Since one-level densities have to do with low-lying zeros, we will focus attention on zeros near the point $1/2$. Suppose that we have a family $\mathcal{F}$ ordered by log-conductor $c(L) = c(L, 0)$. We assume the Riemann Hypothesis for L for convenience, and let γ_L denote a generic ordinate of a zero of L. We want to consider

$$D(f, \mathcal{F}, Y) := \sum_{\substack{L \in \mathcal{F} \\ c(L) \leq \log Y}} \sum_{\gamma_L} f(c(L)\gamma_L). \qquad (5.1)$$

The density conjecture is that

$$\lim_{Y \to \infty} \frac{D(f, \mathcal{F}, Y)}{\sum_{\substack{L \in \mathcal{F} \\ c(L) \leq \log Y}} 1} = \int_{-\infty}^{\infty} f(x) W_{\mathcal{F}}(x) \, dx$$

where $W_{\mathcal{F}}(x) = W_G(x)$ is the one-level density function for the (scaled) limit of $U(N), O(N), USp(2N), SO(2N)$, or $SO(2N+1)$.

Recall that

$$
\begin{aligned}
W_U(x) &= 1, \\
W_{SO^+}(x) &= 1 + \frac{\sin 2\pi x}{2\pi x}, \\
W_{SO^-}(x) &= \delta_0(x) + 1 - \frac{\sin 2\pi x}{2\pi x}, \\
W_O(x) &= 1 + \frac{1}{2}\delta_0(x), \\
W_{USp}(x) &= 1 - \frac{\sin 2\pi x}{2\pi x}.
\end{aligned}
$$

By Plancherel's formula (and because f is even),

$$
\int_{-\infty}^{\infty} f(x)W_G(x)\,dx = \int_{-\infty}^{\infty} \hat{f}(x)\hat{W}_G(x)\,dx.
$$

So, it is useful to record that

$$
\begin{aligned}
\hat{W}_U(x) &= \delta_0(x), \\
\hat{W}_{SO^+}(x) &= \delta_0(x) + \frac{1}{2}\chi_{[-1,1]}(x), \\
\hat{W}_{SO^-}(x) &= \delta_0(x) - \frac{1}{2}\chi_{[-1,1]}(x) + 1, \\
\hat{W}_O(x) &= \delta_0(x) + \frac{1}{2}, \\
\hat{W}_{USp}(x) &= \delta_0(x) - \frac{1}{2}\chi_{[-1,1]}(x).
\end{aligned}
$$

The fundamental tool for beginning any calculations is the explicit formula.

5.1 Explicit formulae

We describe an explicit formula of the type initially found by Riemann, and later studied especially by by Guinand and Weil. We suppose that L is entire and in the Selberg class. Let $\Lambda_L(n)$ be defined as the Dirichlet series coefficients of $-L'/L$:

$$
\frac{L'}{L}(s) = -\sum_{n=1}^{\infty} \frac{\Lambda_L(n)}{n^s}. \tag{5.2}
$$

We further assume that the function $\phi(t)$ is even and decays rapidly and that the Fourier transform

$$
\hat{\phi}(x) = \int_{-\infty}^{\infty} \phi(t)e^{-2\pi i x t}\,dt
$$

has compact support. Then

$$\sum_{\gamma_L} \phi(\gamma_L) = -\frac{1}{2\pi} \int_{-\infty}^{\infty} \frac{X_L'}{X_L}(1/2 + it)\phi(t)\, dt - \sum_{n=1}^{\infty} \frac{\Lambda_L(n)}{\sqrt{n}} \hat{\phi}\left(\frac{\log n}{2\pi}\right) \quad (5.3)$$

Idea of proof. If we pretend that $F(s)$ is a holomorphic function with $F(1/2 + it) = \phi(t)$ then we can easily see the terms of the explicit formula emerging. We consider the integral

$$\frac{1}{2\pi i} \int_{2-i\infty}^{2+i\infty} \frac{L'}{L}(s)F(s)\, ds.$$

Expanding L'/L into its Dirichlet series and integrating term-by term we find one-half of the sum over n on the right side of the explicit formula. Moving the path of integration to $\mathrm{Re}\,s = 1/4$, we obtain the sum over the γ_L on the left-side from the residues of the poles of L'/L at its zeros. Then, change variables $s \to 1 - s$ and use the functional equation $L'/L(1 - s) = X'/X(s) - L'/L(s)$ and consider these terms separately. The integral-term in the explicit formula follows by moving the path of integration in the X'/X term to $\mathrm{Re}\,s = 1/2$. For the other term, we move the path back into a region where the Dirichlet series converges absolutely and then integrate term-by-term. Because of the evenness of $\hat{\phi}$ we obtain the other half of the sum over n on the right side of the explicit formula.

5.2   The X'/X term

We now want to substitute the explicit formula (5.3) into the D formula (5.1). To scale things we let

$$\phi(t) = f\left(\frac{t \log Y}{2\pi}\right);$$

then

$$\hat{\phi}(t) = \frac{2\pi}{\log Y} \hat{f}\left(\frac{2\pi t}{\log Y}\right).$$

The explicit formula then becomes

$$\sum_{\gamma_L} f\left(\frac{\gamma \log Y}{2\pi}\right) = -\frac{1}{2\pi} \int_{-\infty}^{\infty} \frac{X_L'}{X_L}(1/2 + it)f\left(\frac{t \log Y}{2\pi}\right) dt$$

$$- \frac{4\pi}{\log Y} \sum_{n=1}^{\infty} \frac{\Lambda_L(n)}{\sqrt{n}} \hat{f}\left(\frac{\log n}{\log Y}\right)$$

If we assume that all of the $w_j = 1/2$, then we have

$$X_L(s) = \frac{Q^{1-s} \prod_{j=1}^{d} \Gamma\left(\frac{1-s}{2} + \overline{\mu_j}\right)}{Q^s \prod_{j=1}^{d} \Gamma\left(\frac{s}{2} + \mu_j\right)}$$

and

$$-\frac{X_L'}{X_L}(1/2 + it) = 2 \log Q + \operatorname{Re} \sum_{j=1}^{d} \frac{\Gamma'}{\Gamma} \left(\frac{1}{4} + \frac{it}{2} + \mu_j \right)$$

Therefore, in the situation that $|\mu_j| \ll 1$, so that Q is the main parameter,

$$
\begin{aligned}
-\frac{1}{2\pi} \int_{-\infty}^{\infty} \frac{X_L'}{X_L}(1/2 + it) f\left(\frac{t \log Y}{2\pi} \right) dt &= \frac{2 \log Q}{\log Y} \hat{f}(0) + O\left(\frac{\log \log Y}{\log Y} \right) \\
&= \frac{\log q}{\log Y} \hat{f}(0) + O\left(\frac{\log \log Y}{\log Y} \right).
\end{aligned}
$$

In the family of L-functions associated with primitive forms of level 1 and large weight k, the parameter k appears in the shifts μ_j in the gamma factors of the functional equation. In this case

$$-\frac{1}{2\pi} \int_{-\infty}^{\infty} \frac{X_L'}{X_L}(1/2 + it) f\left(\frac{t \log Y}{2\pi} \right) dt = \frac{\log k}{\log Y} \hat{f}(0) + O\left(\frac{\log \log Y}{\log Y} \right).$$

A similar phenomenon happens with Maass forms in which case the μ_j are complex numbers with large imaginary part. Also, if we wanted to consider the one level density of zeros of $\zeta(s + iT)$ with a large T then we would have a $\log T$ term in place of $2 \log Q$.

To handle the sum with the $\Lambda_L(n)$ we assume that the support of $\hat{f}$ is contained in the interval $[-a, a]$. Then our sum over n is truncated at $n = Y^a$. The idea is to have a as large as possible.

The $\Lambda_L(n)$ are 0 unless n is a power of a prime. The terms with $n = p^k$ for some $k \geq 3$ clearly converge. Thus,

$$
\begin{aligned}
\sum_{n=1}^{\infty} \frac{\Lambda_L(n)}{\sqrt{n}} \hat{f}\left(\frac{\log n}{\log Y} \right) &= \sum_{p \leq Y^a} \frac{\Lambda_L(p)}{\sqrt{p}} \hat{f}\left(\frac{\log p}{\log Y} \right) \\
&+ \sum_{p \leq Y^{a/2}} \frac{\Lambda_L(p^2)}{p} \hat{f}\left(\frac{2 \log p}{\log Y} \right) + O(1).
\end{aligned}
$$

Let's obtain a formula for $\Lambda_L(p^k)$. Suppose that

$$L(s) = \prod_p L_p(1/p^s) = \prod_p \prod_{j=1}^{d} \left(1 - \frac{\alpha_j(p)}{p^s} \right)^{-1}.$$

and suppose also (for convenience and as is generally believed) that $|\alpha_j(p)| = 1$ or 0. Then a brief calculation using the power series expansion for $\log(1 - x)$ yields the formula

$$\Lambda_L(p^k) = \log p \sum_{j=1}^{k} \alpha_{j,L}(p)^k.$$

On the other hand, we have

$$L_p(x) = 1 + \lambda(p)x + \lambda(p^2)x^2 + \cdots = \prod_{j=1}^{d}(1 + \alpha_j(p)x + \alpha_j(p)^2x^2 + \dots)$$

so that

$$\sum_{j=1}^{d} \alpha_j(p) = \lambda(p)$$

and

$$\sum_{j=1}^{d} \alpha_j(p)^2 = \lambda(p^2) - \sum_{1 \leq i < j \leq d} \alpha_i(p)\alpha_j(p)$$

In a typical situation there will be a number δ such that

$$\sum_{j=1}^{d} \alpha_j(p)^2 - \delta$$

is small on average; the value of δ depends on the behavior of

$$\prod_p (1 + \frac{1}{p^s} \sum_{j=1}^{d} \alpha_j(p)^2)$$

near $s = 1$; if it is analytic, then $\delta = 0$; if it has a simple pole, then $\delta = 1$, and if it has a simple zero, then $\delta = -1$. It should be the case that δ is constant throughout $L \in \mathcal{F}$. Then it is not hard to show that

$$\sum_{p \leq Y^{a/2}} \frac{\Lambda_L(p^2)}{p} \hat{f}\left(\frac{2\log p}{\log Y}\right) \sim \delta \sum_{p \leq Y^{a/2}} \frac{\log p}{p} \hat{f}\left(\frac{2\log p}{\log Y}\right)$$

$$\sim \delta \int_1^{Y^{a/2}} \hat{f}(2(\log u)/\log Y) \frac{du}{u}$$

$$= \frac{\delta}{2} \int_{-a}^{a} \hat{f}(v) \, dv.$$

This should be valid for any bounded a. Thus, we arrive at

$$\sum_{\gamma_L} f\left(\frac{\gamma \log Y}{2\pi}\right) = \frac{\log q}{\log Y} \hat{f}(0) - \frac{\delta}{2} \int_{-a}^{a} \hat{f}(v) \, dv$$

$$- \frac{4\pi}{\log Y} \sum_{p \leq Y^a} \frac{a_{p,L} \log p}{\sqrt{p}} \hat{f}\left(\frac{\log p}{\log Y}\right) + o(1)$$

This leads to

$$
\frac{D(f, \mathcal{F}, Y)}{\sum_{\substack{L \in \mathcal{F} \\ c(L) \leq \log Y}} 1} = \hat{f}(0) - \frac{\delta}{2} \int_{-a}^{a} \hat{f}(v) \, dv
$$

$$
- \frac{4\pi}{\log Y} \frac{\sum_{\substack{L \in \mathcal{F} \\ c(L) \leq \log Y}} \sum_{p \leq Y^{a/2}} \frac{a_L(p) \log p}{\sqrt{p}} \hat{f}\left(\frac{\log p}{\log Y}\right)}{\sum_{\substack{L \in \mathcal{F} \\ c(L) \leq \log Y}} 1} + o(1)
$$

$$
= \int_{-a}^{a} \hat{f}(x)\left(-\frac{\delta}{2} + \delta_0(x)\right) \, dx
$$

$$
- \frac{4\pi}{\log Y} \frac{\sum_{\substack{L \in \mathcal{F} \\ c(L) \leq \log Y}} \sum_{p \leq Y^{a/2}} \frac{a_L(p) \log p}{\sqrt{p}} \hat{f}\left(\frac{\log p}{\log Y}\right)}{\sum_{\substack{L \in \mathcal{F} \\ c(L) \leq \log Y}} 1} + o(1).
$$

What remains is to calculate the contribution from the primes. This is the subtle part. From comparing the formula above with what is predicted we see that in the unitary case, we expect $\delta = 0$ and the sum over primes should never contribute to the main term for any a. In the symplectic case, we expect $\delta = 1$ and that the sum over primes is small when $a < 1$ but for $a \geq 1$ the sum over primes should contribute

$$
\int_{1}^{a} \hat{f}(x) \, dx.
$$

For the case of symmetry type SO^+, we expect $\delta = -1$ and that the sum over primes contributes $-\int_{1}^{a} \hat{f}(x) \, dx$ when $a > 1$. For the case of SO^- we expect $\delta = -1$ and that the sum over primes should give $\int_{1}^{a} \hat{f}(x) \, dx$ for $a > 1$. Finally, in the case of symmetry type O, we expect $\delta = -1$ and that the primes never contribute to the main term.

5.3 Sample calculations

5.3.1 Zeta.

The simplest example is one level density of zeros of $\{\zeta(1/2+it) : T \leq t \leq 2T\}$. (Our explicit formula assumed that our L-function was entire, and so needs to be modified for $\zeta(s)$; the following is intended to capture the spirit of the calculation.) In this case, we have that $\Lambda_L(p) = \log p \, p^{-it}$ and the "sum" over L is an integral with respect to t over $[T, 2T]$. Since $\int_T^{2T} p^{-it} \, dt \ll 1/\log p$, we have

$$
\frac{1}{T} \int_T^{2T} \sum_{p \leq T^a} \frac{\log p}{p^{1/2+it}} \ll \frac{1}{T} \sum_{p \leq T^a} p^{-1/2} \ll T^{a/2-1}/\log T = o(1)
$$

as long as $a \leq 2$. Moreover, $\Lambda_L(p^2) = \log p \, p^{-2it}$ so

$$\frac{1}{T} \int_T^{2T} \sum_{p \leq T^{a/2}} \frac{\log p}{p^{1+2it}} \ll a(\log T)/T = o(1)$$

for any fixed a. Thus, we obtain the one-level density for $\zeta(s + it)$ for support of $\hat{f}$ in $[-2, 2]$, and we see that it agrees with the predicted one-level density for a unitary symmetry type.

Being more careful in this analysis, we note that the Riemann Hypothesis implies that

$$\sum_{p \leq T^a} \frac{p^{-2iT}}{p^{1/2}} \ll_\epsilon T^{a\epsilon}$$

so that in fact we obtain the one level density for this family for any finite range $[-a, a]$.

5.3.2   All Dirichlet L-functions $L(s, \chi)$.

A similar argument can be carried out for the unitary family of all Dirichlet characters modulo q; Hughes and Rudnick [HR] have obtained one-level density for this family for any test function whose Fourier transform has support in $[-a, a] = [-2, 2]$.

To extend this range, it seems that one would need a result of the sort

$$\sum_{\substack{n \equiv 1 \bmod q \\ n \leq X}} \frac{\Lambda(n)}{\sqrt{n}} = \frac{X}{\varphi(q)} + O_\epsilon \left(\frac{X^{1/2+\epsilon}}{q^\theta} \right)$$

for some positive θ. Such error terms for primes in arithmetic progressions (with θ as large as $1/2$) have been conjectured by various people, including Montgomery.

5.3.3   Dirichlet L-functions with real character $L(s, \chi_d)$.

For the family of real quadratic characters we see, by the Polya-Vinogradov inequality, that

$$\frac{1}{X^*} \sum_{d \leq X} \sum_{p \leq X^a} \frac{\log p \, \chi_d(p)}{\sqrt{p}} \ll X^{-1} \sum_{p \leq X^a} \log^2 p \ll X^{a-1} \log X = o(1)$$

provided that $a < 1$; here X^* denotes $\sum_{d \leq X} 1$. The term involving squares of primes is

$$\frac{1}{X^*} \sum_{d \leq X} \sum_{p \leq X^{a/2}} \frac{\log p \; \chi_d(p^2)}{p} \hat{f}\left(\frac{2 \log p}{\log d}\right) \sim \sum_{p \leq X^{a/2}} \frac{\log p}{p} \hat{f}\left(\frac{2 \log p}{\log X}\right)$$

$$\sim \int_1^{X^{a/2}} \hat{f}(2(\log u)/\log X) \frac{du}{u}$$

$$= \frac{1}{2} \int_{-a}^{a} \hat{f}(v) \; dv.$$

This is in agreement with the prediction for one-level density for a symplectic family.

Ozluk and Snyder [OS], assuming GRH, have proven one-level density for the family of real Dirichlet L-functions $L(s, \chi_d)$ for $\hat{f}$ supported in $[-2, 2]$. To detect the contributions from the prime sum when $a > 1$, they use a transformation property of sums

$$\sum_d \chi_d(p) g(d)$$

for smooth g, where the point is that $\chi_d(p) = \left(\frac{d}{p}\right)$ (the Legendre symbol for p) is periodic in the d variable with period p. This transformation is just an application of the Poisson summation formula ($\sum_d g(d) = \sum_d \hat{g}(d)$) to each residue class modulo p separately; if one thinks of the above sum as being over all d (and not just fundamental discriminants d), then then one has

$$\sum_d \left(\frac{d}{p}\right) g(d) = \frac{1}{\sqrt{p}} \sum_d \left(\frac{d}{p}\right) \hat{g}(d/p).$$

Now we bring in the sum over p and get a contribution from the d which are squares (for which $\left(\frac{d}{p}\right) = 1$). This techniques allows them to take $a = 2$.

6 Statements of further results

Iwaniec, Luo, and Sarnak [ILS] have obtained one-level density theorems for the orthogonal families of L-functions of newforms of level 1 and large weight k, when $\hat{f}$ has support in $[-2, 2]$; similarly for L-functions of newforms of weight 2 and large (prime) level q and also for the symmetric squares of these families, all for $\hat{f}$ supported in $[-2, 2]$. The first examples can be separated into even orthogonal and odd orthogonal or combined to have orthogonal. The symmetric square examples have a symplectic symmetry type. With an additional hypothesis about the behavior of a certain exponential sum over primes, they can obtain the larger support $[-7/3, 7/3]$. The extra main terms from their prime sums arise from extracting main terms out of the Kloosterman sums that arise in the Petersson formula (3.6).

Matthew Young [Y] has looked at families of elliptic curve L-functions. Let

$$\mathcal{F}_1 = \{L_E(s) \text{ where } E : y^2 = x^3 + ax + b^2\}$$

be a family of rank (at least) one elliptic curves. Young can do one-level density for this family for $\hat{f}$ supported in $[-7/9, 7/9]$. Let

$$\mathcal{F}_2 = \{L_E(s) \text{ where } E : y^2 = x^3 + ax + b\}$$

be the family of all elliptic curves. Young can do one-level density for $\hat{f}$ supported in $[-23/48, 23/48]$. Note, that in this case, it is expected that, $\hat{W}_{\mathcal{F}_2}(x) = 1 + \frac{3}{2}\delta_0(x)$ (see [Sn]).

Royer[Ro] has proven one-level density results for families of L-functions associated with (fixed) weight k and (large) level N newforms where the space is restricted to N with a fixed number ℓ of prime divisors and such that the newforms have prescribed behavior under the *Atkin -Lehner* operators. These results are for test functions with support of $\hat{f}$ in $[-2, 2]$.

Finally, we mention the interesting work of Fouvry and Iwaniec [FI] on low lying zeros of dihedral L-functions. These are L-functions of characters of the class group of $Q(\sqrt{d})$ for a fundamental discriminant d. The L-functions are associated with modular forms of weight 1 but instead of having an orthogonal symmetry type, it is expected that they have a symplectic symmetry type, because the symmetric square L-functions have a pole at $s = 1$. The authors obtain the one-level density for $\hat{f}$ supported in $[-1, 1]$.

References

[B]   Bump, Daniel. The Rankin-Selberg method: a survey. Number theory, trace formulas and discrete groups (Oslo, 1987), 49–109, Academic Press, Boston, MA, 1989.

[CG]   Conrey, J. B.; Ghosh, A. On the Selberg class of Dirichlet series: small degrees. Duke Math. J. 72 (1993), no. 3, 673–693.

[FI]   Fouvry, E.; Iwaniec, H. Low-lying zeros of dihedral L-functions. Duke Math. J. 116 (2003), no. 2, 189–217.

[H-B]   Heath-Brown, D. R. A mean value estimate for real character sums. Acta Arith. 72 (1995), no. 3, 235–275.

[HR]   Hughes, C. P.; Rudnick, Z. Linear statistics of low-lying zeros of L-functions. Q. J. Math. 54 (2003), no. 3, 309–333.

[IK]   Iwaniec, Henryk; Kowalski, Emmanuel. Analytic Number Theory. American Mathematical Society, 2004.

[ILS]   Iwaniec, Henryk ; Luo, Wenzhi; Sarnak, Peter. Low lying zeros of families of L-functions. Inst. Hautes Études Sci. Publ. Math. No. 91, (2000), 55–131 (2001).

[KP]   Kaczorowski, J.; Perelli, A. The Selberg class: a survey. Number theory in progress, Vol. 2 (Zakopane-Kościelisko, 1997), 953–992, de Gruyter, Berlin, 1999.

[KS]   Katz, Nicholas M.; Sarnak, Peter: Random matrices, Frobenius eigenvalues, and monodromy. American Mathematical Society Colloquium Publications, 45. American Mathematical Society, Providence, RI (1999).

[OS]   Özlük, A. E.; Snyder, C. On the distribution of the nontrivial zeros of quadratic L-functions close to the real axis. Acta Arith. 91 (1999), no. 3, 209–228.

[Ro]   Royer, Emmanuel. Petits zéros de fonctions L de formes modulaires. Acta Arith. 99 (2001), no. 2, 147–172.

[R]   Rubinstein, Michael. Evidence for the spectral interpretation of the zeros of L-functions. PhD thesis, Princeton, 1998.

[S]   Selberg, Atle. Old and new conjectures and results about a class of Dirichlet series. Proceedings of the Amalfi Conference on Analytic Number Theory (Maiori, 1989), 367–385, Univ. Salerno, Salerno, 1992. Also in Collected papers. Vol. II. With a foreword by K. Chandrasekharan. Springer-Verlag, Berlin, 1991.

[Sn]   Snaith, Nina. Derivatives of random matrix characteristic polynomials with applications to elliptic curves. J. Phys. A. 38(48) (2005), 10345–60.

[Y]   Young, Matthew. Random Matrix Theory and Families of Elliptic Curves. PhD thesis, Rutgers University, 2004.

American Institute of Mathematics
360 Portage Ave.
Palo Alto CA 94306 USA

conrey@aimath.org

L-functions and the Characteristic Polynomials of Random Matrices

J. P. Keating

1 Introduction

My purpose in these lecture notes is to review and explain the basic ideas underlying the connection between random matrix theory and the moments of L-functions. Both of these subjects are introduced separately, and at length, in other lectures. I will focus on their intersection – specifically on the way in which random matrix theory can be used to predict values of the moments and on some applications of the resulting conjectures to other important problems in number theory.

The ideas I shall be reviewing were introduced in [17], [18] and [3], and the applications in [8] and [9]. More recent results and developments, such as those related to calculating complete asymptotic expansions of the moments [4, 5], are described in other lecture notes.

2   The Circular Unitary Ensemble (CUE) of Random Matrices

According to the Katz-Sarnak philosophy [14, 15], the connections between random matrix theory and the statistical properties of L-functions within families take their simplest and most general form when they are expressed in terms of random matrices drawn from the classical compact groups. The simplest example is the unitary group $U(N)$ when the probability measure is taken to be Haar measure. In the random matrix literature, this is called the *circular unitary ensemble* (CUE) of random matrices. In this section, as an introductory example, I will describe the calculation of the two-point correlation function of the eigenvalues of matrices in the CUE, originally due to Dyson [10].

Let A be an $N \times N$ unitary matrix, so that $A(A^T)^* = AA^\dagger = I$. The eigenvalues of A lie on the unit circle; that is, they may be expressed in the form $e^{i\theta_n}$, $\theta_n \in \mathbb{R}$.

Definition 2.1. $f(A) = f(\theta_1, \theta_2, \ldots, \theta_N)$ is called a *class function* if f is symmetric in all of its variables.

Weyl [27] gave an explicit formula for averaging class functions over the CUE:

$$\int_{U(N)} f(A) d\mu_{Haar}(A) = \frac{1}{(2\pi)^N N!} \int_0^{2\pi} \cdots \int_0^{2\pi} f(\theta_1, \ldots, \theta_N)$$
$$\times \prod_{1 \leq j < k \leq N} |e^{i\theta_j} - e^{i\theta_k}|^2 d\theta_1 \cdots d\theta_N. \quad (2.1)$$

One way to understand this formula is to note that, by definition, $d\mu_{Haar}(A)$ is invariant under $A \to \tilde{U} A \tilde{U}^\dagger$ where $\tilde{U}$ is any $N \times N$ unitary matrix, and that A can always be diagonalized by a unitary transformation; that is, it can be written as

$$A = U \begin{pmatrix} e^{i\theta_1} & \cdots & 0 \\ \vdots & \ddots & \vdots \\ 0 & \cdots & e^{i\theta_N} \end{pmatrix} U^\dagger, \quad (2.2)$$

where U is an $N \times N$ unitary matrix. Therefore the integral over A can be written as an integral over the matrix elements of U and the eigenphases θ_n. Because the measure is invariant under unitary transformations, the integral over the matrix elements of U can be evaluated straightforwardly, leaving the integral over the eigenphases (2.1).

Henceforth, to simplify the notation, I shall drop the subscript on the measure $d\mu(A)$ – in all integrals over compact groups the measure may be taken to be the Haar measure on the group.

3 Two-Point Correlations

As an example of a calculation involving an average over the CUE I will now describe Dyson's evaluation of the two-point correlation function of the eigenphases θ_n.

The first step is to scale the eigenphases so that they have unit mean spacing:

$$\phi_n = \theta_n \frac{N}{2\pi}. \tag{3.1}$$

The two-point correlation function for a given matrix A is then defined as

$$R_2(A; x) = \frac{1}{N} \sum_{n=1}^{N} \sum_{m=1}^{N} \sum_{k=-\infty}^{\infty} \delta(x + kN - \phi_n + \phi_m), \tag{3.2}$$

so that

$$\frac{1}{N} \sum_{n,m} f(\phi_n - \phi_m) = \int_0^N R_2(A; x) f(x) dx. \tag{3.3}$$

$R_2(A; x)$ is clearly periodic in x, so it can be expressed as a Fourier series:

$$R_2(A; x) = \frac{1}{N^2} \sum_{k=-\infty}^{\infty} |\mathrm{Tr} A^k|^2 e^{2\pi i kx/N}. \tag{3.4}$$

The goal is to calculate the average of $R_2(A; x)$ with respect to the CUE of $N \times N$ unitary matrices. It follows from (3.4) that this can be achieved by computing the corresponding average of the Fourier coefficients $|\mathrm{Tr} A^k|^2$. This was done by Dyson:

Theorem 3.1. (Dyson 1963 [10])

$$\int_{U(N)} |\mathrm{Tr} A^k|^2 d\mu(A) = \begin{cases} N^2 & k = 0 \\ |k| & |k| \leq N \\ N & |k| > N \end{cases} \tag{3.5}$$

To prove this we use the following lemma, due originally to Heine.

Lemma 3.2. For f a class function

$$\int_{U(N)} f(\theta_1, \ldots, \theta_N) d\mu(A) \tag{3.6}$$

$$= \frac{1}{(2\pi)^N} \int_0^{2\pi} \cdots \int_0^{2\pi} f(\theta_1, \ldots, \theta_N) \det(e^{i\theta_n (n-m)}) d\theta_1 \cdots d\theta_N.$$

This lemma may be proved by using the Weyl integration formula (2.1) to write

$$\int_{U(N)} f(\theta_1,\ldots,\theta_N)d\mu(A) = \frac{1}{(2\pi)^N N!}\int_0^{2\pi}\cdots\int_0^{2\pi} f(\theta_1,\ldots,\theta_N)$$
$$\times \prod_{1\leq j<k\leq N}|e^{i\theta_j}-e^{i\theta_k}|^2 d\theta_1\cdots d\theta_N,$$

and then by noting that

$$\prod_{1\leq j<k\leq N}|e^{i\theta_j}-e^{i\theta_k}|^2$$

$$= \det\left[\begin{pmatrix} 1 & 1 & \cdots \\ e^{i\theta_1} & e^{i\theta_2} & \cdots \\ \vdots & \ddots & \cdots \\ e^{i(N-1)\theta_1} & e^{i(N-1)\theta_2} & \cdots \end{pmatrix}\begin{pmatrix} 1 & e^{-i\theta_1} & e^{-2i\theta_1} & \cdots \\ 1 & e^{-i\theta_2} & e^{-2i\theta_2} & \cdots \\ \vdots & \vdots & \ddots & \cdots \\ 1 & e^{-i\theta_N} & e^{-2i\theta_N} & \ddots \end{pmatrix}\right]$$

$$= \det[\sum_{\ell=1}^N e^{i\theta_\ell(n-m)}]. \tag{3.7}$$

Hence

$$\int_{U(N)} f(\theta_1,\ldots,\theta_N)d\mu(A) = \frac{1}{(2\pi)^N N!}\int_0^{2\pi}\cdots\int_0^{2\pi} f(\theta_1,\ldots,\theta_N)$$

$$\times \begin{vmatrix} \sum_{\ell=1}^N 1 & \sum_{\ell=1}^N e^{-i\theta_\ell} & \cdots & \sum_{\ell=1}^N e^{-(N-1)i\theta_\ell} \\ \sum_{\ell=1}^N e^{i\theta_\ell} & \sum_{\ell=1}^N 1 & \cdots & \sum_{\ell=1}^N e^{-i(N-2)\theta_\ell} \\ \vdots & \vdots & \ddots & \cdots \\ \sum_{\ell=1}^N e^{i(N-1)\theta_\ell} & \sum_{\ell=1}^N e^{i(N-2)\theta_\ell} & \cdots & \sum_{\ell=1}^N 1 \end{vmatrix} d\theta_1\cdots d\theta_N \tag{3.8}$$

and so, using the fact that f is assumed to be symmetric in its arguments,

$$\int_{U(N)} f(\theta_1,\ldots,\theta_N)d\mu(A) = \frac{N}{(2\pi)^N N!}\int_0^{2\pi}\cdots\int_0^{2\pi} f(\theta_1,\ldots,\theta_N)$$

$$\times \begin{vmatrix} 1 & e^{-i\theta_1} & \cdots & e^{-(N-1)i\theta_1} \\ \sum_{\ell=1}^N e^{i\theta_\ell} & \sum_{\ell=1}^N 1 & \cdots & \sum_{\ell=1}^N e^{-i(N-2)\theta_\ell} \\ \vdots & \vdots & \ddots & \cdots \\ \sum_{\ell=1}^N e^{i(N-1)\theta_\ell} & \sum_{\ell=1}^N e^{i(N-2)\theta_\ell} & \cdots & \sum_{\ell=1}^N 1 \end{vmatrix} d\theta_1\cdots d\theta_N. \tag{3.9}$$

Subtracting $e^{i\theta_1}$ times the first row from the second row then gives

$$\int_{U(N)} f(\theta_1,\ldots,\theta_N)d\mu(A) = \frac{N}{(2\pi)^N N!}\int_0^{2\pi}\cdots\int_0^{2\pi} f(\theta_1,\ldots,\theta_N)$$

$$\times \begin{vmatrix} 1 & e^{-i\theta_1} & \cdots & e^{-(N-1)i\theta_1} \\ \sum_{\ell=2}^N e^{i\theta_\ell} & \sum_{\ell=2}^N 1 & \cdots & \sum_{\ell=2}^N e^{-i(N-2)\theta_\ell} \\ \vdots & \vdots & \ddots & \cdots \\ \sum_{\ell=2}^N e^{i(N-1)\theta_\ell} & \sum_{\ell=2}^N e^{i(N-2)\theta_\ell} & \cdots & \sum_{\ell=2}^N 1 \end{vmatrix} d\theta_1\cdots d\theta_N. \tag{3.10}$$

This process may be continued, reducing the second row to $e^{i\theta_2}, 1, e^{-i\theta_2}, \ldots,$ $e^{-(N-2)i\theta_2}$ and so pulling out a factor of $N-1$, then doing the same to the third row and so on. The factor of $N!$ resulting from these row manipulations cancels the $N!$ in the normalization constant of Weyl's formula. This proves the lemma.

We now return to

$$\int_{U(N)} |\mathrm{Tr} A^k|^2 d\mu(A) = \frac{1}{(2\pi)^N} \int_0^{2\pi} \cdots \int_0^{2\pi} \sum_m \sum_n e^{ik(\theta_m - \theta_n)} \qquad (3.11)$$

$$\times \begin{vmatrix} 1 & e^{-i\theta_1} & e^{-2i\theta_1} & \cdots & e^{-i(N-1)\theta_1} \\ e^{i\theta_2} & 1 & e^{-i\theta_2} & \cdots & e^{-i(N-2)\theta_2} \\ \vdots & \vdots & \vdots & \ddots & \cdots \\ e^{i(N-1)\theta_N} & e^{i(N-2)\theta_N} & e^{i(N-3)\theta_N} & \cdots & 1 \end{vmatrix} d\theta_1 \cdots d\theta_N.$$

The net contribution from the diagonal $(m = n)$ terms is N, because the measure is normalized and there are N diagonal terms. Using the fact that

$$\frac{1}{2\pi} \int_0^{2\pi} e^{in\theta} d\theta = \begin{cases} 1 & n = 0 \\ 0 & n \neq 0 \end{cases}, \qquad (3.12)$$

if $k \geq N$ then the integral of the off-diagonal terms is zero, because, for example, when the determinant is expanded out and multiplied by the prefactor there is no possibility of θ_1 cancelling in the exponent. If $k = N - j$, $j = 1, \ldots, N-1$, then the off-diagonal terms contribute $-j$; for example, when $j = 1$ only one non-zero term survives when the determinant is expanded out, multiplied by the prefactor, and integrated term-by-term – this is the term coming from multiplying the bottom-left entry by the top-right entry and all of the diagonal entries on the other rows. Thus the combined diagonal and off-diagonal terms add up to give the expression in Theorem 3.1, bearing in mind that when $k = 0$ the total is just N^2, the number of terms in the sum over m and n.

We thus have that

$$\int_{U(N)} R_2(A; x) d\mu(A) = \frac{1}{N^2} \sum_{k=-\infty}^{\infty} e^{2\pi i k x/N} \begin{cases} N^2 & k = 0 \\ |k| & |k| < N \\ N & |k| \geq N \end{cases} \qquad (3.13)$$

$$= \sum_{j=-\infty}^{\infty} \delta(x - jN) + 1 - \frac{\sin^2(\pi x)}{N^2 \sin^2\left(\frac{\pi x}{N}\right)}. \qquad (3.14)$$

Hence, for test functions f such that $f(x) \to 0$ as $|x| \to \infty$,

$$\lim_{N \to \infty} \int_{U(N)} \int_{-\infty}^{\infty} f(x) R_2(A; x) dx d\mu(A)$$

$$= \int_{-\infty}^{\infty} f(x) \left(\delta(x) + 1 - \frac{\sin^2(\pi x)}{\pi^2 x^2}\right) dx. \qquad (3.15)$$

For example,

$$\lim_{N\to\infty} \int_{U(N)} \frac{1}{N} \#\{\phi_n, \phi_m : \alpha \le \phi_n - \phi_m \le \beta\} d\mu(A)$$

$$= \int_\alpha^\beta (\delta(x) + 1 - \frac{\sin^2(\pi x)}{\pi^2 x^2}) dx. \tag{3.16}$$

The key point for us is that this is exactly the two-point correlation function conjectured by Montgomery [21] for the complex zeros of the Riemann zeta function $\zeta(s)$[1]. Let us denote the nth nontrivial zero of the Riemann zeta function by $1/2 + it_n$. For convenience, we shall assume the truth of the Riemann Hypothesis, that $t_n \in \mathbb{R} \ \forall n$. Then defining

$$w_n = t_n \frac{1}{2\pi} \log \frac{t_n}{2\pi} \tag{3.17}$$

(so that the w_ns have unit mean density) Montgomery's conjecture asserts that

$$\lim_{W\to\infty} \frac{1}{W} \#\{w_n, w_m \in [0, W] : \alpha \le w_n - w_m < \beta\} = \int_\alpha^\beta (\delta(x) + 1 - \frac{\sin^2(\pi x)}{\pi^2 x^2}) dx, \tag{3.18}$$

or, more generally, that

$$\lim_{N\to\infty} \frac{1}{N} \sum_{n,m\le N} f(w_n - w_m) = \int_{-\infty}^\infty f(x)(\delta(x) + 1 - \frac{\sin^2(\pi x)}{\pi^2 x^2}) dx. \tag{3.19}$$

There is considerable computational and theoretical support for this conjecture and for its generalizations to correlations between n-tuples of zeros [22, 24, 1, 2].

4   Value distribution of $\zeta(\frac{1}{2} + it)$ and $\log \zeta(\frac{1}{2} + it)$

Rather than pursuing further the question of how the nontrivial zeros of the zeta function are distributed, I turn now to the distributional properties of $\zeta(1/2 + it)$ and $\log \zeta(1/2 + it)$.

$\log \zeta(1/2 + it)$ satisfies the following central limit theorem.

Theorem 4.1. (Selberg) For any rectangle B in the complex plane,

$$\lim_{T\to\infty} \frac{1}{T} \text{meas.}\{T \le t \le 2T : \frac{\log \zeta(\frac{1}{2} + it)}{\sqrt{\frac{1}{2} \log \log \frac{t}{2\pi}}} \in B\}$$

$$= \int\int_B e^{-\frac{1}{2}(x^2 + y^2)} dx dy. \tag{4.1}$$

---

[1]Editors' comment: The result in equation (3.16) is also derived, by a different method, in the lectures of J.B. Conrey, page 111, Section 8.1. See also equation (6.7) in the lectures of D.A. Goldston, page 79, for the pair correlation of the Riemann zeros

Odlyzko has investigated the value distribution of $\log \zeta(1/2 + it)$ numerically for values of t around the height of the 10^{20}th zero of the zeta function. Surprisingly, he found a distribution that differs markedly from the limiting Gaussian. His data are plotted in Figures 1 and 2. The CUE curve will be discussed later.

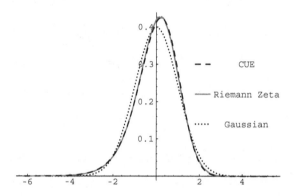

Figure 1: Odlyzko's data for the value distribution of $\operatorname{Re} \log \zeta(1/2+it)$ near the 10^{20}th zero (taken from [22]), the value distribution of $\operatorname{Re} \log Z$ with respect to matrices taken from $U(42)$, and the standard Gaussian, all scaled to have unit variance. (Taken from [17].)

In order to quantify the discrepancy illustrated in Figures 1 and 2, I list in Table 1 the moments of $\operatorname{Re} \log \zeta(1/2 + it)$, normalized so that the second moment is equal to one, calculated numerically by Odlyzko in [22]. The data in the second and third columns relate to two different ranges near the height of the 10^{20}th zero. The difference between them is therefore a measure of the fluctuations associated with computing over a finite range near this height. The data labelled $U(42)$ will be explained later.

Next let us turn to the value distribution of $\zeta(1/2 + it)$ itself. Its moments satisfy the following long-standing and important conjecture.

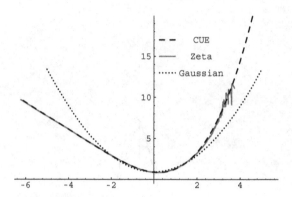

Figure 2: The logarithm of the inverse of the value distribution plotted in Figure 1. (Taken from [17].)

Moment	ζ a)	ζ b)	$U(42)$	Normal
1	0.0	0.0	0.0	0
2	1.0	1.0	1.0	1
3	-0.53625	-0.55069	-0.56544	0
4	3.9233	3.9647	3.89354	3
5	-7.6238	-7.8839	-7.76965	0
6	38.434	39.393	38.0233	15
7	-144.78	-148.77	-145.043	0
8	758.57	765.54	758.036	105
9	-4002.5	-3934.7	-4086.92	0
10	24060.5	22722.9	25347.77	945

Table 1: Moments of $\mathrm{Re}\log\zeta(1/2+it)$, calculated by Odlyzko over two ranges (labelled a and b) near the 10^{20}th zero ($t \simeq 1.520 \times 10^{19}$) (taken from [22]), compared with the moments of $\mathrm{Re}\log Z$ for $U(42)$ and the Gaussian (normal) moments, all scaled to have unit variance.

Conjecture 4.2.

$$\lim_{T \to \infty} \frac{1}{(\log \frac{T}{2\pi})^{\lambda^2}} \frac{1}{T} \int_0^T |\zeta(\tfrac{1}{2} + it)|^{2\lambda} dt$$

$$= f_\zeta(\lambda) \prod_p \left[(1 - \tfrac{1}{p})^{\lambda^2} \sum_{m=0}^{\infty} \left(\frac{\Gamma(\lambda + m)}{m! \Gamma(\lambda)} \right)^2 p^{-m} \right] \quad (4.2)$$

This can be viewed in the following way. It asserts that the moments grow like $(\log \frac{T}{2\pi})^{\lambda^2}$ as $T \to \infty$. Treating the primes as being statistically independent of each other gives the right hand side with $f_\zeta(\lambda) = 1$. $f_\zeta(\lambda)$ thus quantifies deviations from this simple-minded ansatz.

The conjecture is known to be correct in only two non-trivial cases, when $\lambda = 1$ and $\lambda = 2$. It was shown by Hardy and Littlewood in 1918 that $f_\zeta(1) = 1$ [11] and by Ingham in 1926 that $f_\zeta(2) = \frac{1}{12}$ [12]. On number-theoretical grounds, Conrey and Ghosh have conjectured that $f_\zeta(3) = \frac{42}{9!}$ [6] and Conrey and Gonek that $f_\zeta(4) = \frac{24024}{16!}$ [7].

We shall now look to random matrix theory to see what light, if any, it can shed on these issues.

5 Characteristic polynomials of random unitary matrices

Our goal is to understand the value distribution of $\zeta(1/2 + it)$. Recalling that the zeros of this function are believed to be correlated like the eigenvalues of random unitary matrices, we take as our model the functions whose zeros are these eigenvalues, namely the characteristic polynomials of the matrices in question.

Let us define the characteristic polynomial of a matrix A by

$$\begin{aligned} Z(A, \theta) &= \det(I - Ae^{-i\theta}) \\ &= \prod_n (1 - e^{i(\theta_n - \theta)}). \end{aligned} \quad (5.1)$$

Consider first the function

$$P_N(s, t) = \int_{U(N)} |Z(A, \theta)|^t e^{is \text{Im} \log Z(A, \theta)} d\mu(A). \quad (5.2)$$

This is the moment generating function of $\log Z$: the joint moments of $\text{Re} \log Z$

and Im log Z are obtained from derivatives of P at $s = 0$ and $t = 0$ and

$$\int_{U(N)} \delta(x - \operatorname{Re}\log Z)\delta(y - \operatorname{Im}\log Z)d\mu(A) \tag{5.3}$$

$$= \frac{1}{4\pi^2} \int_{-\infty}^{\infty}\int_{-\infty}^{\infty} e^{-itx-isy}P(s,it)dsdt. \tag{5.4}$$

Written in terms of the eigenvalues,

$$P_N(s,t) = \int_{U(N)} \prod_{n=1}^{N}|1 - e^{i(\theta_n-\theta)}|^t e^{-is\sum_{m=1}^{\infty}\frac{\sin[(\theta_n-\theta)m]}{m}}d\mu(A). \tag{5.5}$$

Since the integrand is a class function, we can use Weyl's integration formula (2.1) to write

$$P_N(s,t) = \frac{1}{(2\pi)^N N!}\int_0^{2\pi}\cdots\int_0^{2\pi}\prod_{n=1}^{N}|1 - e^{i(\theta_n-\theta)}|^t e^{-is\sum_{m=1}^{\infty}\frac{\sin[(\theta_n-\theta)m]}{m}}$$

$$\times \prod_{1\le j<k\le N}|e^{i\theta_j} - e^{i\theta_k}|^2 d\theta_1\cdots d\theta_N. \tag{5.6}$$

This integral can then be evaluated using a form of Selberg's integral described in [20], giving

Theorem 5.1. Keating & Snaith [17]

$$P_N(s,t) = \prod_{j=1}^{N}\frac{\Gamma(j)\Gamma(t+j)}{\Gamma(j+\frac{t}{2}+\frac{s}{2})\Gamma(j+\frac{t}{2}-\frac{s}{2})} \tag{5.7}$$

Note that the result is independent of θ. This is because the average over $U(N)$ includes rotations of the spectrum and is itself therefore rotationally invariant.

5.1   Value distribution of $\log Z$

Consider first the Taylor expansion

$$P_N(s,t) = e^{\alpha_{00}+\alpha_{10}t+\alpha_{01}s+\alpha_{20}t^2/2+\alpha_{11}ts+\alpha_{02}s^2/2+\cdots}. \tag{5.8}$$

The α_{m0} are the cumulants of $\operatorname{Re}\log Z$ and the α_{0n} are i^n times the cumulants of $\operatorname{Im}\log Z$. Expanding (5.7) gives:

$$\alpha_{10} = \alpha_{01} = \alpha_{11} = 0; \tag{5.9}$$
$$\alpha_{20} = -\alpha_{02} = \tfrac{1}{2}\log N + \tfrac{1}{2}(\gamma+1) + O(\tfrac{1}{N^2}); \tag{5.10}$$
$$\alpha_{mn} = O(1) \text{ for } m+n \ge 3; \tag{5.11}$$

and more specifically,

$$\alpha_{m0} = (-1)^m(1 - \tfrac{1}{2^{m-1}})\Gamma(m)\zeta(m-1) + O(\tfrac{1}{N^{m-2}}), \text{ for } m \ge 3. \tag{5.12}$$

This leads to

Theorem 5.2. (Keating & Snaith [17]) For any rectangle B in the complex plane

$$\lim_{N \to \infty} \text{meas.} \left\{ A \in U(N) : \frac{\log Z(A, \theta)}{\sqrt{\frac{1}{2} \log N}} \in B \right\}$$
$$= \frac{1}{2\pi} \int \int_B e^{-\frac{1}{2}(x^2 + y^2)} dx dy. \tag{5.13}$$

We see that $\log \zeta(1/2 + it)$ and $\log Z$ both satisfy a central limit theorem when, respectively, $t \to \infty$ and $N \to \infty$. Note that the scalings in theorems 4.1 and 5.2, corresponding to the asymptotic variances, are the same if we make the identification

$$N = \log \tfrac{t}{2\pi}. \tag{5.14}$$

This is the same as identifying the mean eigenvalue density with the mean zero density; c.f. the unfolding factors in (3.1) and (3.17).

The identification (5.14) provides a connection between matrix sizes and heights up the critical line. The central limit theorems imply that when both of these quantities tend to infinity $\log \zeta(1/2 + it)$ and $\log Z$ have the same limit distribution. This supports the choice of Z as a model for the value distribution of $\zeta(1/2 + it)$ when $t \to \infty$. It is natural then to ask if it also constitutes a useful model when t is large but finite; that is, whether it can explain the deviations from the limiting Gaussian seen in Odlyzko's data.

The value of t corresponding to the height of the 10^{20}th zero should be associated, via (5.14), to a matrix size of about $N = 42$. The moments and value distribution of $\log Z$ for any size of matrix can be obtained directly from the formula for the moment generating function (5.7). The value distribution when $N = 42$ is the CUE curve plotted in Figures 1 and 2. Values of the moments are listed in Table 1. The obvious agreement between the results for random 42×42 unitary matrices and Odlyzko's data provides significant further support for the model. It suggests that random matrix theory models not just the limit distribution of $\log \zeta(1/2 + it)$, but the rate of approach to the limit as $t \to \infty$.

5.2   Moments of $|Z|$

We now turn to the more important problem of the moments of $|\zeta(1/2 + it)|$.

It is natural to expect these moments to be related to those of the modulus of the characteristic polynomial Z, which are defined as

$$\int_{U(N)} |Z(A,\theta)|^{2\lambda} d\mu(A) = P(0, 2\lambda)$$

$$= \prod_{j=1}^{N} \frac{\Gamma(j)\Gamma(j+2\lambda)}{(\Gamma(j+\lambda))^2} \tag{5.15}$$

$$= e^{\sum_{m=0}^{\infty} \alpha_{m0}(2\lambda)^n/n!}. \tag{5.16}$$

Therefore

$$\lim_{N\to\infty} \frac{1}{N^{\lambda^2}} \int_{U(N)} |Z(A,\theta)|^{2\lambda} d\mu(A) = e^{\lambda^2(\gamma+1)+\sum_{m=3}^{\infty}(-2\lambda)^m \frac{2^{m-1}-1}{2^{m-1}} \frac{\zeta(m-1)}{m}}, \tag{5.17}$$

for $|\lambda| < \frac{1}{2}$. Note that since we are identifying Z with $\zeta(1/2 + it)$ and N with $\log\frac{t}{2\pi}$, the expression on the left-hand side of (5.17) corresponds precisely to that in Conjecture 4.2.

We now recall some properties of the Barnes' G-function. This is an entire function of order 2 defined by

$$G(1+z) = (2\pi)^{z/2} e^{-[(1+\gamma)z^2+z]/2} \prod_{n=1}^{\infty} \left[(1+z/n)^n e^{-z+z^2/(2n)}\right]. \tag{5.18}$$

It satisfies

$$G(1) = 1, \tag{5.19}$$

$$G(z+1) = \Gamma(z)G(z) \tag{5.20}$$

and

$$\log G(1+z) = (\log 2\pi - 1)\frac{z}{2} - (1+\gamma)\frac{z^2}{2} + \sum_{n=3}^{\infty}(-1)^{n-1}\zeta(n-1)\frac{z^n}{n}. \tag{5.21}$$

Thus we have that

$$\lim_{N\to\infty} \frac{1}{N^{\lambda^2}} \int_{U(N)} |Z(A,\theta)|^{2\lambda} d\mu(A) = f_U(\lambda), \tag{5.22}$$

with

$$f_U(\lambda) = \frac{G^2(1+\lambda)}{G(1+2\lambda)}. \tag{5.23}$$

Using (5.20) we further have that for positive integers k

$$f_U(k) = \prod_{j=0}^{k-1} \frac{j!}{(j+k)!}. \tag{5.24}$$

In particular, $f_U(1) = 1$, $f_U(2) = \frac{1}{12}$, $f_U(3) = \frac{42}{9!}$ and $f_U(4) = \frac{24024}{16!}$, which match the values of f_ζ listed after Conjecture 4.2. This then motivates the following conjecture.

Conjecture 5.3. (Keating & Snaith [17]) For $\text{Re}\lambda > -\frac{1}{2}$

$$f_\zeta(\lambda) = f_U(\lambda) \tag{5.25}$$

5.3   Value distribution of $|Z|$

Let us now define the value distribution of $|Z(A,0)|$ by

$$\int_{U(N)} \delta(|Z(A,0)| - w)d\mu(A) = \rho_U(w, N). \qquad (5.26)$$

Obviously,

$$\int_{U(N)} |Z|^t d\mu(A) = \int_0^\infty \rho_U(w, N)w^t dw; \qquad (5.27)$$

that is, the moments of $|Z|$ are given by the Mellin transform of the value distribution we seek to evaluate. Therefore, using (5.15),

$$\rho_U(w, N) = \frac{1}{2\pi i} \int_{c-i\infty}^{c+i\infty} \prod_{j=1}^{N} \frac{\Gamma(j)\Gamma(j+t)}{(\Gamma(j+\frac{t}{2}))^2} \frac{1}{w^{t+1}} dt, \qquad (5.28)$$

where $c > 0$. As $N \to \infty$ this can be approximated using the method of stationary phase. In the limit as $w \to 0$ an expansion in increasing powers of w can be formed by considering the residues from the poles at the negative integers. The rightmost pole is at $t = -1$, and so $\rho_U(w, N) \to$ constant as $w \to 0$.

Values of $\rho_U(w, N)$ when $N = 12$ are plotted and compared with the value distribution of $|\zeta(\frac{1}{2} + it)|$ when $t = 10^6$ in Figure 3.

6   Other Compact Groups

We have seen so far that the characteristic polynomials of random unitary matrices may be used to model the moments and value distribution of the Riemann zeta function on its critical line. Katz and Sarnak [14, 15] have shown that the distribution of the zeros within families of L-functions is related to averages over the various classical compact groups, the particular group in question being determined by symmetries of the family. This suggests that the moments and value distribution of L-functions within a family may be understood by extending the calculations for the unitary group described above to the other classical compact groups.

Consider a matrix $A \in USp(2N)$ or $A \in O(N)$. In both cases there is a symmetry in the spectrum not present for general unitary matrices: the complex eigenvalues come in complex conjugate pairs, $e^{\pm i\theta_n}$. For example, in the case of $USp(2N)$ and $O(2N)$ the characteristic polynomial is

$$Z(A, \theta) = \prod_{n=1}^{N} (1 - e^{i(\theta_n - \theta)})(1 - e^{i(-\theta_n - \theta)}). \qquad (6.1)$$

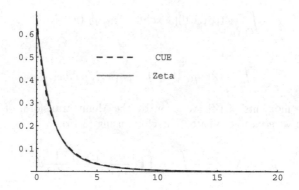

Figure 3: The CUE value distribution of $|Z|$ corresponding to $N = 12$ (that is, $U(12)$) (dashed), with numerical data for the value distribution of $|\zeta(1/2+it)|$ (solid) near $t = 10^6$.

Our goal now is to determine the moments and value distribution of the characteristic polynomials with respect to averages over these groups. Like for $U(N)$, averages are understood to be computed with respect to the relevant Haar measure. Unlike for $U(N)$, in both cases the symmetry in the spectrum means that the results depend on θ. We will focus on the symmetry point $\theta = 0$, as this is where the differences are greatest.

6.1   Moments

To calculate the moments of the characteristic polynomials with respect to averages over $O(N)$ or $USp(2N)$ we need the two key ingredients used in the calculation for $U(N)$: the Weyl integration formula for these groups [27] and appropriate forms of the Selberg integral (c.f. [20], chapter 17). Following the steps detailed above we then find for the symplectic group that

$$
\begin{aligned}
\int_{USp(2N)} Z(A,0)^s d\mu(A) &= 2^{2Ns} \prod_{j=1}^{N} \frac{\Gamma(1+N+j)\Gamma(\frac{1}{2}+s+j)}{\Gamma(\frac{1}{2}+j)\Gamma(1+s+N+j)} \\
&\equiv M_{Sp}(s;N).
\end{aligned}
\tag{6.2}
$$

It follows that $\log Z$ again satisfies a central limit theorem and that

$$\lim_{N\to\infty} \frac{1}{N^{s(s+1)/2}} \int_{USp(2N)} Z(A,0)^s d\mu(A) \tag{6.3}$$

$$= 2^{s^2/2} \frac{G(1+s)\sqrt{\Gamma(1+s)}}{\sqrt{G(1+2s)\Gamma(1+2s)}} \equiv f_{Sp}(s). \tag{6.4}$$

For positive integers n

$$f_{Sp}(n) = \frac{1}{\prod_{j=1}^{n}(2j-1)!!} = \frac{1}{(2n-1)(2n-3)^2(2n-5)^3\cdots}. \tag{6.5}$$

The distribution of values of $Z(A,0)$ for $A \in USp(2N)$ is given by

$$\int_{USp(2N)} \delta(Z(A,0)-w)d\mu(A) = \frac{1}{2\pi i} \int_{c-i\infty}^{c+i\infty} M_{Sp}(s) \frac{ds}{w^{s+1}}. \tag{6.6}$$

By way of illustration, this distribution is plotted in Figure 4 when $N = 6$ and $N = 42$. As $w \to 0$ it vanishes like $w^{1/2}$, because, from (6.2), the rightmost pole of $M_{Sp}(s; N)$ is at $s = -3/2$.

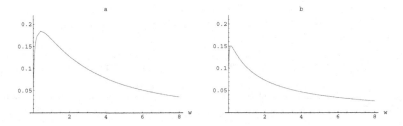

Figure 4: Value distribution of $Z(U,0)$ for matrices in $USp(2N)$ (6.6), when a)$N = 6$, b) $N = 42$.

As a second example we take the orthogonal group $SO(2N)$. In this case

$$\int_{SO(2N)} Z(A,0)^s d\mu(A) = 2^{2Ns} \prod_{j=1}^{N} \frac{\Gamma(N+j-1)\Gamma(s+j-1/2)}{\Gamma(j-1/2)\Gamma(s+j+N-1)}$$

$$\equiv M_O(s;N), \tag{6.7}$$

$\log Z$ again satisfies a central limit theorem, and

$$\lim_{N\to\infty} \frac{1}{N^{s(s-1)/2}} \int_{SO(2N)} Z(A,0)^s d\mu(A) \tag{6.8}$$

$$= 2^{s^2/2} \frac{G(1+s)\sqrt{\Gamma(1+2s)}}{\sqrt{G(1+2s)\Gamma(1+s)}} \equiv f_O(s). \tag{6.9}$$

For positive integers n we have

$$f_O(n) = 2^n f_{Sp}(n-1). \tag{6.10}$$

(I note in passing the following relationship between the leading order moment coefficients for the three compact groups discussed:

$$f_O(s) f_{Sp}(s) = 2^{s^2} f_U(s).) \tag{6.11}$$

The value distribution of the characteristic polynomials is again given by

$$\int_{SO(2N)} \delta(Z(A,0) - w) d\mu(A) = \frac{1}{2\pi i} \int_{c-i\infty}^{c+i\infty} M_O(s;N) \frac{1}{w^{s+1}} ds. \tag{6.12}$$

In this case it diverges like $w^{-1/2}$ as $w \to 0$. The distribution is illustrated in Figure 5.

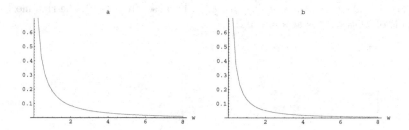

Figure 5: Value distribution of $Z(U,0)$ for matrices in $SO(2N)$ (6.12), when a)$N = 6$, b) $N = 42$.

7 Families of L-functions and Symmetry

I now describe how the results listed above may be applied to L-functions within families. The main ideas will be illustrated by focusing on two representative examples.

7.1 Example 1: Dirichlet L-functions

Let

$$\chi_d(p) = \left(\frac{d}{p}\right) = \begin{cases} +1 & \text{if } p \nmid d \text{ and } x^2 \equiv d \pmod{p} \text{ solvable} \\ 0 & \text{if } p|d \\ -1 & \text{if } p \nmid d \text{ and } x^2 \equiv d \pmod{p} \text{ not solvable;} \end{cases} \tag{7.1}$$

denote the Legendre symbol. That is, $\chi_d(p)$ is a real quadratic Dirichlet character. Then define

$$L_D(s, \chi_d) = \prod_p (1 - \frac{\chi_d(p)}{p^s})^{-1}$$

$$= \sum_{n=1}^{\infty} \frac{\chi_d(n)}{n^s}, \tag{7.2}$$

where the product is over the prime numbers. These functions form a family of L-functions parameterized by the integer index d.

7.2   Example 2: *L*-functions associated with elliptic curves

Consider the function

$$f(z) = e^{2\pi i z} \prod_{n=1}^{\infty} (1 - e^{2\pi i n z})^2 (1 - e^{22\pi i n z})^2$$

$$= \sum_{n=1}^{\infty} a_n e^{2\pi i n z}, \tag{7.3}$$

where the integers a_n are the Fourier coefficients of f. This function may be shown to satisfy

$$f(\frac{az + b}{cz + d}) = (cz + d)^2 f(z) \tag{7.4}$$

for every $\begin{pmatrix} a & b \\ c & d \end{pmatrix} \in SL_2(\mathbb{Z})$ with $11|c$. That is, $f(z)$ is a cusp form of weight 2 for $\Gamma_0(11)$. It is important to note that the weight is an integer.

Now consider the elliptic curve

$$E_{11} : \quad y^2 = 4x^3 - 4x^2 - 40x - 79. \tag{7.5}$$

Let

$$N_p = \#\{(x, y) \in \mathbb{F}_p^2 : y^2 = 4x^3 - 4x^2 - 40x - 79\}. \tag{7.6}$$

Then

$$a_p = p - N_p; \tag{7.7}$$

that is, the Fourier coefficients of f determine the number of solutions of E_{11}. One can construct a zeta function

$$\zeta_{E_{11}}(s) = \sum_{n=1}^{\infty} \frac{a_n}{n^s} \tag{7.8}$$

and then a family of L-functions by twisting with the Dirichlet characters defined in (7.1):

$$L_{E_{11}, d}(s) = \sum_{n=1}^{\infty} \frac{a_n \chi_d(n)}{n^s}. \tag{7.9}$$

This family is again parameterized by the integer index d. The L-functions satisfy the following functional equation:

$$
\begin{aligned}
\Phi_{E_{11},d}(s) &\equiv \left(\frac{2\pi}{\sqrt{11|d|}}\right)^{-s} \Gamma(s) L_{E_{11},d}(s) \\
&= \chi_d(-11)\Phi_{E_{11},d}(2-s).
\end{aligned}
\tag{7.10}
$$

We will here focus on those L-functions associated with characters satisfying

$$
\chi_d(-11) = +1,
\tag{7.11}
$$

i.e. those that do not vanish trivially at $s = 1$.

In both of the above examples the L-functions satisfy a Riemann Hypothesis. In the first example, this places their complex zeros on the (critical) line $\mathrm{Re}\,s = 1/2$; in the second, it places them on the line $\mathrm{Re}\,s = 1$ (this is merely a matter of conventional normalization rather than a significant difference). In each case the zeros high up on the critical line are believed to be distributed like the eigenvalues of random unitary matrices [24], and so the results obtained for the Riemann zeta function extend, conjecturally, to every individual (principal) L-function.

Rather than fixing the L-function and averaging along the critical line, we can instead fix a height on the critical line and average through the family; that is, average with respect to d. In this way one can therefore examine the distribution of the zeros nearest the critical point, $s = 1/2$ or $s = 1$, within these families.

It was conjectured by Katz and Sarnak [14, 15] that the zero statistics around the critical point are related to the eigenvalue statistics of one of the compact groups described above near to a spectral symmetry point (if one exists). The particular group in question is determined by symmetries of the family. There is now extensive numerical and theoretical evidence in support of this [23].

The first example of a family of L-functions given above (the Dirichlet L-functions) is conjectured to have symplectic symmetry and so the zeros behave like the eigenvalues of matrices from $USp(2N)$. The family of elliptic curve L-functions in the second example is conjectured to have orthogonal symmetry. Their zeros behave like the eigenvalues of $SO(2N)$ matrices.

Following the Katz-Sarnak philosophy, it is natural to believe that random matrix theory can predict the moments of L-functions in families like those described here.

Conjecture 7.1. (Conrey & Farmer [3], Keating & Snaith [18])

The moments

$$
\frac{1}{X^*} \sum_{0<d<X}^{*} \left(L_D(\tfrac{1}{2}, \chi_d)\right)^s
$$

(where the sum is over fundamental discriminants d, and X^* is the number of terms in the sum) are modelled by

$$\int_{USp(2N)} (Z(A,0))^s dA,$$

whereas the moments

$$\frac{1}{X^*} \sum_{\substack{0 < d < X \\ \chi_{-d}(-11)=+1}}^{*} (L_{E_{11},d}(1))^s$$

are modelled by

$$\int_{SO(2N)} (Z(A,0))^s dA.$$

For example, the factors corresponding to f_ζ in the moments of the L-functions are conjectured to be given by $f_{Sp}(s)$ and $f_O(s)$. This agrees with all previous results and conjectures for the integer moments (see, for example, [3, 18]). (These factors must be multiplied by arithmetical contributions to give the moments.) Furthermore, the value distributions of the L-functions with respect to varying d are expected to be related to the value distributions of the associated characteristic polynomials. Numerical evidence in support of this is illustrated in Figure 6 for the two families we are here focusing on.

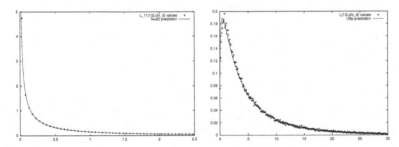

Figure 6: The first picture depicts the value distribution of $L_{E_{11},d}(1)$, for prime $|d|$, $-788299808 < d < 0$, even functional equation, compared to equation (6.12), with $N = 20$. In the second picture we depict the value distribution of $L_D(1/2, \chi_d)$ (Dirichlet L-functions) for all fundamental $800000 < |d| < 1000000$. Here, the Katz-Sarnak philosophy predicts a Unitary Symplectic family, and so we compare the data with equation (6.6), $N = 5$. In these pictures the L-function values have been normalized so that they have the same means as the random matrix value distributions. (From [8].)

The key question is obviously: what use can be made of the random matrix model for the value distribution of L-functions? I will now outline some applications that are currently being explored for the L-functions associated with elliptic curves. These exploit certain explicit formulae for the values at the central point $s = 1$. The approach is general, but for simplicity I shall describe it in the specific context of the family defined in example 2.

The formula for $L_{E_{11},d}(1)$ that we shall exploit is an example of a general class of formulae developed by Shimura [25], Waldspurger [26] and Kohnen-Zagier [19]. For $d < 0$ and $\chi_d(-11) = +1$ it asserts that

$$L_{E_{11},d}(1) = \frac{\kappa c_{|d|}^2}{\sqrt{|d|}}, \tag{7.12}$$

where

$$\kappa = \text{constant} \ (= 2.91763\ldots)$$

and

$$g(z) = \sum_{n=1}^{\infty} c_n e^{2\pi i n z} \tag{7.13}$$

satisfies

$$g\left(\frac{az+b}{cz+d}\right) = \epsilon(a,b,c,d)(cz+d)^{3/2} g(z) \tag{7.14}$$

for every $\begin{pmatrix} a & b \\ c & d \end{pmatrix} \in SL_2(\mathbb{Z})$ such that $44|c$; that is the numbers c_n are the Fourier coefficients of a three-halves-weight form for $\Gamma_0(44)$. Note that the L-functions were originally defined using the Fourier coefficients a_n of an integer-weight form (weight-two in our example), but that at the central point their values are related to the Fourier coefficient of a half-integer weight form. One important point to notice is that

$$c_n \in \mathbb{Z}.$$

I will now describe two conjectural implications that follow from combining this formula with the random-matrix model.

7.3 Generalization of the Sato-Tate law to half-integer weight modular forms

The Sato-Tate law describes the value distribution of the Fourier coefficients a_p defined in (7.3). According to the theorems of Hasse and Deligne these satisfy $|a_p| \leq 2\sqrt{p}$ and so may be written

$$\frac{a_p}{\sqrt{p}} = 2\cos\theta_p, \ \ 0 \leq \theta \leq \pi.$$

The question then is: how are the angles θ_p distributed as the prime p varies? This is the subject of the Sato-Tate law.

Conjecture 7.2. Sato-Tate

$$\lim_{x\to\infty} \frac{1}{\pi(x)} \#\{p < x : \alpha < \theta_p \leq \beta\} = \frac{2}{\pi} \int_{\alpha}^{\beta} \sin^2\theta d\theta, \qquad (7.15)$$

where $\pi(x) = \#\{p < x\}$.

Given that the Fourier coefficients of integer-weight forms satisfy a simple limit distribution law, it is natural to ask whether the Fourier coefficients of half-integer-weight forms do as well.

Combining (7.12) with the random matrix model for the value distribution of $L_{E_{11},d}(1)$ provides a conjectural answer to this question [9]. For example, it follows from the central limit theorem for the logarithm of the characteristic polynomial that

Conjecture 7.3. (Conrey, Keating, Rubinstein & Snaith [9])

$$\lim_{D\to\infty} \frac{1}{D^*} \#\{2 < d \leq D : \chi_{-d}(-11) = 1,$$

$$\frac{2\log|c_{|d|}| - \frac{1}{2}\log d + \frac{1}{2}\log\log d}{\sqrt{\log\log d}} \in (\alpha, \beta)\}$$

$$= \frac{1}{\sqrt{2\pi}} \int_{\alpha}^{\beta} e^{-x^2/2} dx, \qquad (7.16)$$

where

$$D^* = \#\{2 < d \leq D : \chi_{-d}(-11) = 1\}$$

Furthermore, the value distribution of $c_{|d|}$ should be related to that illustrated in Figure 5.

7.4   Frequency of vanishing of L-functions

I now turn to the question of the frequency of vanishing of L-functions at the central point. In the light of the Birch & Swinnerton-Dyer conjecture, this is an issue of considerable importance.

The formula (7.12) for $L_{E_{11},d}(1)$ implies a discretization (or quantization) of its values. So if $L_{E_{11},d}(1) < \frac{\kappa}{\sqrt{|d|}}$ then in fact $L_{E_{11},d}(1) = 0$. Pushing the random matrix model to the very limits of the range where it can be justified (and hopefully not beyond), the probability that $L_{E_{11},d}(1) < \frac{\kappa}{\sqrt{|d|}}$ may be estimated by integrating the probability density (6.12) from 0 to $\frac{\kappa}{\sqrt{|d|}}$. Using the fact that the probability density has a square-root singularity at the origin then motivates the following two conjectures.

Conjecture 7.4. (Conrey, Farmer, Keating, Rubinstein & Snaith [8])

$$\#\{p \leq D : \chi_{-p}(-11) = 1, L_{E_{11},-p}(1) = 0\} \asymp \frac{D^{3/4}}{(\log D)^{5/8}}. \qquad (7.17)$$

Conjecture 7.5. (Conrey, Farmer, Keating, Rubinstein & Snaith [8]) Let

$$R_p(D) = \frac{\#\{d < D : \chi_{-d}(-11) = 1, \chi_{-d}(p) = 1, L_{E_{11},d}(1) = 0\}}{\#\{d < D : \chi_{-d}(-11) = 1, \chi_{-d}(p) = -1, L_{E_{11},d}(1) = 0\}}. \quad (7.18)$$

Then

$$R_p = \lim_{D \to \infty} R_p(D) = \sqrt{\frac{p + 1 - a_p}{p + 1 + a_p}}. \quad (7.19)$$

Data relating to the first conjecture are plotted in Figure 7. These would appear to support the dependence on $D^{3/4}$, but do not cover a large enough range to determine the power of $\log D$.

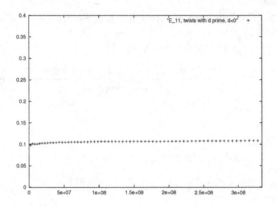

Figure 7: Figure in support of Conjecture 7.4. This depicts the l.h.s. of (7.17) divided by $D^{3/4}(\log D)^{-5/8}$. The calculations include only twists with $d < 0$, d prime, and cases with even functional equation. While the picture is reasonably flat, $\log(D)$ is almost constant for most of the interval in question. The flatness observed therefore reflects the main dependence on $D^{3/4}$. (From [8].)

Data in support of the second conjecture are listed in Table 2 and are plotted in Figure 8. In this case the agreement with the conjecture is striking.

7.5   Extension to other compact Lie groups

It is interesting that the ideas reviewed above concerning connections between the value distribution of L-functions and averages over the classical compact groups extend to other Lie groups, such as the exceptional Lie groups [16]. For example, consider G_2. This is a 14-dimensional group of rank 2 (it is the automorphism group of the octonions), with an embedding into $SO(7)$. In the 7-dimensional representation, the characteristic polynomial associated

p	conjectured R_p for E_{11}	data for E_{11}	conjectured R_p for E_{19}	data for E_{19}	conjectured R_p for E_{32}	data for E_{32}
3	1.2909944	1.2774873	1.7320508	1.7018241	1	0.99925886
5	0.84515425	0.84938811	0.57735027	0.57825622	1.4142136	1.4113424
7	1.2909944	1.288618	1.1338934	1.134852	1	1.0003445
11		0	0.77459667	0.76491219	1	1.0001457
13	0.74535599	0.73266305	1.3416408	1.3632977	0.63245553	0.61626177
17	1.118034	1.1282072	1.183216	1.196637	0.89442719	0.88962298
19	1	1.000864		0	1	1.0006726
23	1.0425721	1.0470095	1	0.99857962	1	1.0000812
29	1	0.99769402	0.81649658	0.80174375	1.4142136	1.4615854
31	0.80064077	0.78332934	1.1338934	1.143379	1	1.0008405
37	0.92393644	0.91867671	0.9486833	0.94311279	1.0540926	1.0603105
41	1.2126781	1.2400086	1.1547005	1.1683113	0.78446454	0.76494748
43	1.1470787	1.1642671	1.0229915	1.0229106	1	1.0006774
47	0.84515425	0.82819492	1.0645813	1.0708874	1	0.99951502
53	1.118034	1.1332312	0.79772404	0.77715638	0.76696499	0.74137107
59	0.91986621	0.91329134	1.1055416	1.1196252	1	0.99969828
61	0.82199494	0.79865031	1.0162612	1.0199932	1.1766968	1.1996892
67	1.1088319	1.1216776	1.0606602	1.0705574	1	1.0002831
71	1.0425721	1.0497774	0.91986621	0.90939741	1	0.99992715
73	0.94733093	0.94345043	1.099525	1.1110782	1.0846523	1.0950853
79	1.1338934	1.1562237	0.90453403	0.8922209	1	0.99882039
83	1.0741723	1.0854551	0.8660254	0.84732408	1	0.99979996
89	0.84515425	0.82410673	0.87447463	0.85750248	0.89442719	0.88154899
97	1.0741723	1.0877289	0.92144268	0.90867892	0.8304548	0.80811684
101	0.98058068	0.97846254	0.94280904	0.93032086	1.0198039	1.0229108
103	1.1677484	1.1976448	0.87333376	0.855721	1	1.0004009
107	0.84515425	0.82186438	1.183216	1.2153554	1	1.0009282
109	0.91287093	0.89933354	1.1577675	1.1844329	0.94686415	0.94015124
113	0.92393644	0.9146531	0.9486833	0.93966595	1.1313708	1.1534106
127	0.93933644	0.93052596	0.98449518	0.98005032	1	0.99904006
131	1.1470787	1.171545	1.1208971	1.1413931	1	0.99916309
137	1.052079	1.0603352	1.0219806	1.0285831	1.1744404	1.2066518
139	0.93094934	0.91532106	1.0975994	1.1176423	1	1.0000469
149	1.069045	1.0833831	0.86855395	0.84844439	0.91064169	0.89706709

Table 2: A table in support of Conjecture 7.5, comparing R_p v.s. $R_p(D)$ for three elliptic curves E_{11}, E_{19}, E_{32} (D equal to 333605031, 263273979, 930584451 respectively). More of this data, for $p < 2000$, is depicted in the Figure 8. The 0 entries for $p = 11$ and $p = 19$ are explained by the fact that we are restricting ourselves to twists with even functional equation. Hence for E_{11} and E_{19}, we are only looking at twists with $\chi_d(11) = \chi_d(19) = -1$. (From [8].)

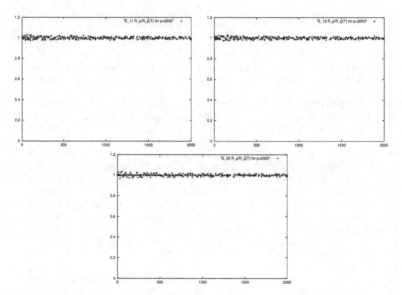

Figure 8: Pictures depicting $R_p/R_p(D)$, for $p < 2000$, D as in Table 2. (From [8].)

with the corresponding unitary matrix U factorizes as

$$Z(U,\theta) = \det(I - Ue^{-i\theta}) = (1 - e^{-i\theta})\tilde{Z}(U,\theta). \qquad (7.20)$$

The moments of $\tilde{Z}(U,\theta)$ with respect to an average over the group can be calculated as for the classical compact groups using the corresponding Weyl integration formula and one of MacDonald's constant term identities (which plays the role of the Selberg integral). The result is that [16]

$$\int_{G_2} |\tilde{Z}(U,0)|^s d\mu(U) = \frac{\Gamma(3s+7)\Gamma(2s+3)}{\Gamma(2s+6)\Gamma(s+4)\Gamma(s+3)\Gamma(s+2)}. \qquad (7.21)$$

We note in this context that Katz [13] has found a one-parameter family of L-functions over a finite field whose value distribution in the limit as the size of the field grows is related to G_2. Thus the random matrix moments (7.21) determine the value distribution of these L-functions.

The random matrix calculations extend straightforwardly to all of the exceptional Lie groups. It would be very interesting to know whether the others also describe families of L-functions over finite fields.

Acknowledgements

I am very grateful to the Isaac Newton Institute for hospitality and support during the programme on Random Matrix Approaches in Number Theory. I am also very grateful to Nina Snaith for her help in preparing these notes.

References

[1] E.B. Bogomolny and J.P. Keating Random matrix theory and the Riemann zeros I: three- and four-point correlations, *Nonlinearity*, **8**: 1115–1131, 1995.

[2] E.B. Bogomolny and J.P. Keating Random matrix theory and the Riemann zeros II: *n*-point correlations, *Nonlinearity*, **9**: 911–935, 1996.

[3] J.B. Conrey and D.W. Farmer, Mean values of *L*-functions and symmetry, *Int. Math. Res. Notices*, **17**: 883–908, 2000, arXiv:math.nt/9912107.

[4] J.B. Conrey, D.W. Farmer, J.P. Keating, M.O. Rubinstein, and N.C. Snaith, Autocorrelation of random matrix polynomials, *Commun. Math. Phys.*, **237**: 365–395, 2003.

[5] J.B. Conrey, D.W. Farmer, J.P. Keating, M.O. Rubinstein, and N.C. Snaith, Integral moments of *L*-functions, *Proc. Lond. Math. Soc*, **91**(1):33–104, 2005.

[6] J.B. Conrey and A. Ghosh, On mean values of the zeta-function, iii, *Proceedings of the Amalfi Conference on Analytic Number Theory, Università di Salerno*, 1992.

[7] J.B. Conrey and S.M. Gonek, High moments of the Riemann zeta-function, *Duke Math. J.*, **107**: 577–604, 2001.

[8] J.B. Conrey, J.P. Keating, M.O. Rubinstein, and N.C. Snaith, On the frequency of vanishing of quadratic twists of modular *L*-functions, In *Number Theory for the Millennium I: Proceedings of the Millennial Conference on Number Theory*, editor, M.A. Bennett et al., pages 301–315. A K Peters, Ltd, Natick, 2002, arXiv:math.nt/0012043.

[9] J.B. Conrey, J.P. Keating, M.O. Rubinstein, and N.C. Snaith, Random matrix theory and the Fourier coefficients of half-integral weight forms, *Experimental Mathematics*, **15**(1):67–82, 2006.

[10] F.J. Dyson, Statistical theory of the energy levels of complex systems, i, ii and iii, *J. Math. Phys.*, **3**: 140–175, 1962.

[11] G.H. Hardy and J.E. Littlewood, Contributions to the theory of the Riemann zeta-function and the theory of the distribution of primes, *Acta Mathematica*, **41**: 119–196, 1918.

[12] A.E. Ingham, Mean-value theorems in the theory of the Riemann zeta-function, *Proceedings of the London Mathematical Society (2)*, **27**: 273–300, 1926.

[13] N. Katz, *Exponential sums and differential equations (Annals of Mathematics Studies 124)*, Princeton, NJ: Princeton University Press, 1990.

[14] N.M. Katz and P. Sarnak, *Random Matrices, Frobenius Eigenvalues and Monodromy*, American Mathematical Society Colloquium Publications, 45. American Mathematical Society, Providence, Rhode Island, 1999.

[15] N.M. Katz and P. Sarnak, Zeros of zeta functions and symmetry, *Bull. Amer. Math. Soc.*, **36**: 1–26, 1999.

[16] J.P. Keating, N. Linden, and Z. Rudnick, Random matrix theory, the exceptional Lie groups, and L-functions, *J. Phys. A-Math. Gen.*, **36**(12): 2933–44, 2003.

[17] J.P. Keating and N.C. Snaith, Random matrix theory and $\zeta(1/2 + it)$, *Commun. Math. Phys.*, **214**: 57–89, 2000.

[18] J.P. Keating and N.C. Snaith, Random matrix theory and L-functions at $s = 1/2$, *Commun. Math. Phys*, **214**: 91–110, 2000.

[19] W. Kohnen and D. Zagier, Values of L-series of modular forms at the center of the critical strip, *Invent. Math.*, **64**: 175–198, 1981.

[20] M.L. Mehta, *Random Matrices*, Academic Press, London, second edition, 1991.

[21] H.L Montgomery, The pair correlation of zeros of the zeta function, *Proc. Symp. Pure Math.*, **24**: 181–193, 1973.

[22] A.M. Odlyzko, The 10^{20}th zero of the Riemann zeta function and 70 million of its neighbors, *Preprint*, 1989.

[23] M.O. Rubinstein, *Evidence for a spectral interpretation of the zeros of L-functions*, Ph.D. thesis, Princeton University, 1998.

[24] Z. Rudnick and P. Sarnak Zeros of principal L-functions and random matrix theory, *Duke Math. J.*, **81**: 269–322, 1996.

[25] G. Shimura, On modular forms of half integral weight, *Ann. Math.*, **97**(2): 440–481, 1973.

[26] J.-L. Waldspurger, Sur les coefficients de Fourier des formes modulaires de poids demi-entier, *J. Math. Pures Appl.*, **60**(9): 375–484, 1981.

[27] H. Weyl, *Classical Groups*, Princeton University Press, 1946.

School of Mathematics,
University of Bristol,
Bristol BS8 1TW, UK

Spacing Distributions in Random Matrix Ensembles

Peter J. Forrester

1 Introduction

1.1 Motivation and definitions

The topic of spacing distributions in random matrix ensembles is almost as old as the introduction of random matrix theory into nuclear physics. Both events can be traced back to Wigner in the mid 1950's [37, 38]. Thus Wigner introduced the model of a large real symmetric random matrix, in which the upper triangular elements are independently distributed with zero mean and constant variance, for purposes of reproducing the statistical properties of the highly excited energy levels of heavy nuclei. This was motivated by the gathering of experimental data on the spectrum of isotopes such as ^{238}U at energy levels beyond neutron threshold. Wigner hypothesized that the statistical properties of the highly excited states of complex nuclei would be the same as those of the eigenvalues of large random real symmetric matrices. For the random matrix model to be of use at a quantitative level, it was necessary to deduce analytic forms of statistics of the eigenvalues which could be compared against statistics determined from experimental data.

What are natural statistics for a sequence of energy levels, and can these statistics be computed for the random matrix model? Regarding the first question, let us think of the sequence as a point process on the line, and suppose for simplicity that the density of points is uniform and has been normalized to unity. For any point process in one dimension a fundamental quantity is the probability density function for the event that given there is a point at the origin, there is a point in the interval $[s, s + ds]$, and further there are n points somewhere in between these points and thus in the interval $(0, s)$. Let us denote the probability density function by $p(n; s)$. In the language of energy levels, this is the spacing distribution between levels n apart.

Another fundamental statistical quantity is the k-point distribution function $\rho_{(k)}(x_1, \ldots, x_k)$. This can be defined recursively, starting with $\rho_{(1)}(x)$, by the requirement that

$$\rho_{(k)}(x_1, \ldots, x_k)/\rho_{(k-1)}(x_1, \ldots, x_{k-1}) \tag{1.1}$$

is equal to the density of points at x_k, given there are points at $x_1, \ldots, x_{k-1}$. One sees from the definitions that

$$\frac{\rho_{(2)}(0, s)}{\rho_{(1)}(0)} = \sum_{n=0}^{\infty} p(n; s). \tag{1.2}$$

From empirical data of a long energy level sequence, the quantity $p(n; s)$ for small values of n at least is readily estimated (the statistical uncertainty gets worse as n increases). Use of (1.2) then allows for an estimation of $\rho_{(2)}(0; s)$.

We thus seek the theoretical determination of $p(n; s)$ for matrix ensembles.

1.2 Spacing between primes

Before taking up the problem of determining $p(n; s)$ for matrix ensembles, which is the theme of these lectures, let us digress a little and follow the line of introduction to spacing distributions given by Porter in the review he wrote as part of the book [31], which collected together the major papers written in the field up to 1965. Porter's introduction is particularly relevant to the theme of the present school because it uses the prime numbers as an example of a deterministic sequence which, like energy levels of heavy nuclei, nevertheless exhibit pronounced stochastic features.

It turns out the spacing distributions between primes relate to perhaps the simplest example of a point process. This is when the probability that there is a point in the interval $[s, s + ds]$ is equal to ds, independent of the location of the other points. This generates the so called Poisson process with unit density, or in the language of statistical mechanics, a perfect gas. By definition of the process the ratio (1.1) is unity for all k and thus

$$\rho_{(k)}(x_1, \ldots, x_k) = 1. \tag{1.3}$$

To compute $p(n; s)$, we think of the Poisson process as the $N \to \infty$ limit of a process in which each unit interval on the line is broken up into N equal sub-intervals, with the probability of there being a particle in any one of the subintervals equal to $1/N$. Thus

$$p(s; n) = \lim_{N \to \infty} (1 - \frac{1}{N})^{sN-n} N^{-n} \binom{sN}{n} = \frac{s^n}{n!} e^{-s}. \tag{1.4}$$

In the first equality of (1.4), the first factor is the probability that $sN - n$ subintervals do not contain a particle, the second factor is the probability that n subintervals do contain a particle, while the final factor is the number of ways of choosing n occupied sites amongst sN sites in total. The probability density in the final equality of (1.4) is the Poisson distribution. Substituting (1.4) in (1.2) gives $\rho_{(2)}(0, x) = 1$, as required by (1.3).

The distribution (1.4) ties in with prime numbers through Cramér's model.[1] In this approximation, statistically the primes are regarded as forming a Poisson process on the positive integer lattice. The probability of occupation of the Nth site is taken to equal $1/\log N$, so as to be consistent with the prime number theorem. Cramér's model predicts that as an approximation

$$p^{(N)}(n; s) = \frac{s^n}{n!} e^{-s}, \qquad s = t/\log N \tag{1.5}$$

where $p^{(N)}(n; s)$ refers to the probability that for primes p in the neighbourhood of a prime N, there is a prime at $p + t$, and furthermore there are exactly n primes between p and $p + t$.

To compare the prediction (1.5) against empirical data, we choose a value of N, say 10^9, and for the subsequent M primes (say $M = 2,000$) record the

---

[1] Editor's comment: See the lectures by D.R. Heath-Brown, page 1, after Conjecture 2.

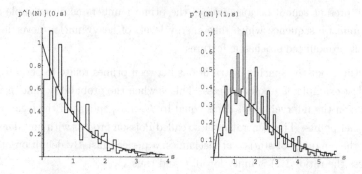

Figure 1: Distribution of the spacing t between primes (leftmost graph) and the spacing t between every second prime for $2,000$ consecutive primes starting with $N = 10^9 + 7$. The distributions are given in units of $s = t/\log N$. The smooth curves are the Poisson distributions $p(0; s) = e^{-s}$ and $p(1; s) = se^{-s}$.

distance to the following prime (in relation to $p^{(N)}(0; s)$) and the distance to the second biggest prime after that (in relation to $p^{(N)}(s; 1)$). We form a histogram, with the scale on the horizontal axis measured in units of $s = t/\log N$, where t is the actual spacing. The natural units for t are multiples of 2, and this provides a width for the bars of the histogram. We see from Figure 1 that the general trend of the histograms do indeed follow the respective Poisson distributions.

1.3 Empirical determination of spacing distributions for matrix ensembles

Wigner's interest was in the statistical properties of the eigenvalues of large real symmetric random matrices. More particularly, he sought the statistical properties of the eigenvalues in what may be termed the bulk of the spectrum (as opposed to the edge of the spectrum [9]). The eigenvalues in this region are characterized by having a uniform density, which after rescaling (referred to as 'unfolding') may be taken as unity (at the edge of the spectrum, which is the neighbourhood of the largest or smallest eigenvalue, the density does not have this property, as there is no scale for which the eigenvalues are evenly spaced). In distinction to the situation with the sequence of primes, for random matrices it is not necessary to study the statistical properties of a large sequence of (unfolded) eigenvalues from a single matrix. Rather the spacing distributions with respect to the middle eigenvalue (this is the eigenvalue most in the bulk) in multiple samples from the class of random matrices in question can be listed, and then this list used to create a histogram. Moreover, to approximate large matrix size behaviour, it is only necessary to consider quite small matrix sizes, say 13×13.

In Figure 2 we have plotted the empirical determination of $p(0; s)$ and $p(1; s)$ obtained from lists of eigenvalue spacings for realizations of the so called GOE (Gaussian orthogonal ensemble) eigenvalue distribution. The GOE consists of real symmetric random matrices, with each diagonal element chosen from the normal distribution $N[0, 1]$, and each (strictly) upper triangular element chosen from the normal distribution $N[0, 1/\sqrt{2}]$. For such matrices, it is well known that to leading order in the matrix rank N, the eigenvalue density is given by the Wigner semi-circle law

$$\rho_{(1)}(x) = \frac{\sqrt{2N}}{\pi}\sqrt{1 - \frac{x^2}{2N}}.$$

Multiplying the eigenvalues at point x by this factor allows us to unfold the sequence giving a mean eigenvalue spacing of unity.

A less well known, and much more recent result relating to GOE matrices is that their spectrum can be realized without having to diagonalize a matrix [4] (see also [13]). Thus one has that the roots of the random polynomial $P_N(\lambda)$, defined recursively by the stochastic three term recurrence

$$P_k(\lambda) = (\lambda - a_k)P_{k-1}(\lambda) - b_{k-1}^2 P_{k-2}(\lambda) \tag{1.6}$$

where

$$a_k \sim N[0, 1], \qquad b_k^2 \sim \text{Gamma}[k/2, 1],$$

have the same distribution as the eigenvalues of GOE matrices (the notation Gamma$[s, \sigma]$ denotes the gamma distribution with density proportional to $x^{s-1}e^{-x/\sigma}$). Generating such polynomials and finding their zeros then provides us with a sequence distributed as for GOE eigenvalues, from which we have determined $p(0; s)$ and $p(1; s)$.

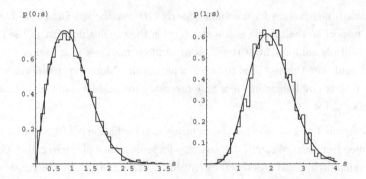

Figure 2: Plot of the distribution of the unfolded spacing between the 6th and 7th, and 7th and 8th eigenvalues (pooled together) for 2,000 samples from the 13×13 GUE eigenvalue distribution. The smooth curve is the Wigner surmise (2.12). The rightmost graph is the distribution between the 6th and 8th eigenvalues in the same setting, while in the smooth curve in this case is $(1/2)p_4(0; s/2)$ with p_4 given by (2.13).

2 Eigenvalue product formulas for gap probabilities

2.1 Theory relating to $p(n; s)$

Consider a point process consisting of a total of N points. Let the joint probability density function of the N points be denoted $p(x_1, \ldots, x_N)$. A quantity closely related to the spacing distribution $p(0; s)$ is the gap probability

$$E^{\mathrm{bulk}}(0; s) := \lim_{N \to \infty} a_N^N \int_{\bar{I}} dx_1 \cdots \int_{\bar{I}} dx_N \, p(a_N x_1, \ldots, a_N x_N) \qquad (2.1)$$

where $\bar{I} = (-\infty, \infty) - (-s/2, s/2)$ and a_N is the leading large N form of the local density at the origin (and thus the unfolding factor). Thus it is easy to see that

$$p(0; s) = \frac{d^2}{ds^2} E^{\mathrm{bulk}}(0; s). \qquad (2.2)$$

More generally we can define

$$E^{\mathrm{bulk}}(n; s) := \qquad (2.3)$$

$$\lim_{N \to \infty} \binom{N}{n} a_N^N \int_{-s/2}^{s/2} dx_1 \cdots \int_{-s/2}^{s/2} dx_n \int_{\bar{I}} dx_{n+1} \cdots \int_{\bar{I}} dx_N \, p(a_N x_1, \ldots, a_N x_N).$$

These quantities can be calculated from the generating function

$$E^{\text{bulk}}(s;\xi) := \tag{2.4}$$

$$\lim_{N\to\infty} a_N^N \int_{-\infty}^{\infty} dx_1 \cdots \int_{-\infty}^{\infty} dx_N \prod_{l=1}^{N}(1 - \xi\chi_{(-s/2,s/2)}^{(l)})p(a_N x_1, \ldots, a_N x_N),$$

where $\chi_J^{(l)} = 1$ for $x^{(l)} \in J$ and $\chi_J^{(l)} = 0$ otherwise, according to the formula

$$E^{\text{bulk}}(n;s) = \frac{(-1)^n}{n!} \frac{\partial^n}{\partial\xi^n} E^{\text{bulk}}(s;\xi)\Big|_{\xi=1}. \tag{2.5}$$

It follows from the definitions that

$$p(n;s) = \frac{d^2}{ds^2} E^{\text{bulk}}(n;s) + 2p(n-1;s) - p(n-2;s), \tag{2.6}$$

or equivalently

$$p(n;s) = \frac{d^2}{ds^2} \sum_{j=0}^{n}(n-j+1)E^{\text{bulk}}(j;s). \tag{2.7}$$

Hence knowledge of $\{E^{\text{bulk}}(j;s)\}_{j=0,\ldots,n}$ is sufficient for the calculation of $p(n;s)$.

It is possible to relate (2.4) to the k-point distribution functions.[2] In the finite system the latter are given by

$$\rho_{(k)}^{(N)}(x_1,\ldots,x_k) = \frac{N!}{(N-k)!} \int_{-\infty}^{\infty} dx_{k+1} \cdots \int_{-\infty}^{\infty} dx_N\, p(x_1,\ldots,x_N). \tag{2.8}$$

With

$$\rho_{(k)}^{\text{bulk}}(x_1,\ldots,x_k) := \lim_{N\to\infty} a_N^k \rho_{(k)}^{(N)}(a_N x_1,\ldots,a_N x_k),$$

by expanding (2.4) in a power series in ξ and making use of (2.8) we see that

$$E^{\text{bulk}}(s;\xi) = 1 + \sum_{k=1}^{\infty} \frac{(-\xi)^k}{k!} \int_{-s/2}^{s/2} dx_1 \cdots \int_{-s/2}^{s/2} dx_k\, \rho_{(k)}^{\text{bulk}}(x_1,\ldots,x_k). \tag{2.9}$$

For the limiting process to be rigorously justified, because $[-s/2, s/2]$ is a compact interval, it is sufficient that $\rho_{(k)}^{\text{bulk}}(x_1,\ldots,x_k)$ be bounded by M^k for some $M > 0$.

With these basic formulas established, we will now proceed to survey some of the main results relating to spacing distributions in the bulk of the various matrix ensembles (orthogonal, unitary and symplectic symmetry classes).

---

[2]Editors' comment: These are also called k-point correlation functions and are discussed in Sections 3 and 4 of the lectures of Y.V. Fyodorov, page 31.

2.2   Wigner surmise

For the Poisson process we have seen that $p(0; s) = e^{-s}$. Thus in this case
the spacing distribution is actually maximum at zero separation between the
points. The opposite feature is expected for $p(0; s)$ in relation to the eigenvalues
of random real symmetric matrices, as can be seen by examining the 2×2 case
of matrices of the form

$$A = \begin{bmatrix} a & b \\ b & c \end{bmatrix}.$$

This matrix is diagonalized by the decomposition $A = R\mathrm{diag}[\lambda_+, \lambda_-]R^T$ where

$$R = \begin{bmatrix} \cos\theta & -\sin\theta \\ \sin\theta & \cos\theta \end{bmatrix}.$$

Expressing a, b, c in terms of $\lambda_+, \lambda_-, \theta$ it is simple to show

$$da\,db\,dc = |\lambda_+ - \lambda_-|d\lambda_+d\lambda_-d\theta. \tag{2.10}$$

Thus for small separation $s := |\lambda_+ - \lambda_-|$ the probability density function
vanishes linearly.

Let $\mu(s)$ denote the small s behaviour of $p(0; s)$. We have seen that for
the Poisson process $\mu(s) = 1$, while for the bulk eigenvalues of real symmetric
matrices $\mu(s) \propto s$. Wigner hypothesized [38] that as with the Poisson process,
$p(0; s)$ for the bulk eigenvalues of random real symmetric matrices could be
deduced from the ansatz

$$p(0; s) = c_1\mu(s)\exp\left(-c_2\int_0^s \mu(t)\,dt\right) \tag{2.11}$$

where the constants c_1 and c_2 are determined by the normalization requirements

$$\int_0^\infty p(0; s)\,ds = 1, \qquad \int_0^\infty sp(0; s)\,ds = 1$$

(the second of these says that the mean spacing is unity). Thus one arrives at
the so called Wigner surmise

$$p(0; s) = \frac{\pi}{2}se^{-\pi s^2/4} \tag{2.12}$$

for the spacing distribution of the bulk eigenvalues of random real symmetric
matrices.

The ansatz (2.11) does not apply if instead of real symmetric matrices
one considers complex Hermitian matrices, or Hermitian matrices with real

quaternion elements. Examining the 2×2 case (see the introductory article by Porter in [31]) one sees that in the analogue of (2.10), the factor $|\lambda_+ - \lambda_-|$ should be replaced by $|\lambda_+ - \lambda_-|^\beta$ with $\beta = 2$ (complex elements) or $\beta = 4$ (real quaternion elements). Choosing the elements to be appropriate Gaussians, one can reclaim (2.12) and furthermore obtain

$$p_2(0; s) = \frac{32s^2}{\pi^2} e^{-4s^2/\pi}, \qquad p_4(0; s) = \frac{2^{18}s^4}{3^6\pi^3} e^{-64s^2/9\pi} \qquad (2.13)$$

as approximations to the spacing distributions in the cases $\beta = 2$ and $\beta = 4$ respectively.

2.3 Fredholm determinant evaluations

A unitary invariant matrix ensemble of $N \times N$ random complex Hermitian matrices has as its eigenvalue probability density function

$$\frac{1}{C} \prod_{l=1}^{N} w_2(x_l) \prod_{1 \le j < k \le N} (x_k - x_j)^2, \qquad (2.14)$$

which we will denote by $\mathrm{UE}_N(g)$. We know[3] that the k-point distribution function can be expressed in terms of the monic orthogonal polynomials $\{p_k(x)\}_{k=0,1,\ldots}$ associated with the weight function $w_2(x)$,

$$\int_{-\infty}^{\infty} w_2(x) p_j(x) p_k(x) \, dx = h_j \delta_{j,k}.$$

Thus with

$$\begin{aligned} K_N(x, y) &= (w_2(x)w_2(y))^{1/2} \sum_{k=0}^{N-1} \frac{p_k(x)p_k(y)}{h_k} \\ &= (w_2(x)w_2(y))^{1/2} \frac{p_N(x)p_{N-1}(y) - p_N(y)p_{N-1}(x)}{x - y} \end{aligned} \qquad (2.15)$$

we have

$$\rho_{(k)}^{(N)}(x_1, \ldots, x_k) = \det \left[K_N(x_j, x_l) \right]_{j,l=1,\ldots,k}. \qquad (2.16)$$

This structure is significant for the evaluation of the generating function

$$E_{N,2}(J; \xi; w_2) := \left\langle \prod_{l=1}^{N} (1 - \xi \chi_J^{(l)}) \right\rangle_{\mathrm{UE}_N(g)}. \qquad (2.17)$$

---

[3]Editors' comment: See the discussion of k-point correlation functions in the lectures of Y.V. Fyodorov, page 31, and in particular Section 4.

(the subscript 2 on $E_{N,2}$ indicates the exponent in (2.14)). Expanding (2.17) in a power series analogous to (2.9) we obtain

$$E_{N,2}(J;\xi;w_2) = 1 + \sum_{k=1}^{N} \frac{(-\xi)^k}{k!} \int_J dx_1 \cdots \int_J dx_k \det\left[K_N(x_j,x_l)\right]_{j,l=1,\ldots,k},$$

(2.18)

where use has been made of (2.16). The sum in (2.18) occurs in the theory of Fredholm integral equations [36], and is in fact an expansion of the determinant of an integral operator,

$$E_{N,2}(J;\xi;w_2) = \det(1 - \xi K_J)$$ (2.19)

where K_J is the integral operator on the interval J with kernel $K_N(x,y)$,

$$K_N[f](x) = \int_J K_N(x,y)f(y)\,dy.$$

It is well known that in the bulk scaling limit, independent of the precise functional form of $w_2(x)$,

$$\lim_{N\to\infty} a_N K_N(a_N x, a_N y) = \frac{\sin\pi(x-y)}{\pi(x-y)} =: K^{\text{bulk}}(x,y)$$ (2.20)

for a suitable scale factor a_N. Thus we have

$$E_2^{\text{bulk}}(J;\xi) = \det(1 - \xi K_J^{\text{bulk}})$$ (2.21)

where K_J^{bulk} is the integral operator on the interval J with kernel (2.20) (the so called sine kernel). This is a practical formula for the computation of E_2^{bulk} if we can compute the eigenvalues $\{\mu_j\}_{j=0,1,\ldots}$ of K_J^{bulk}, since we have

$$E_2^{\text{bulk}}(J;\xi) = \prod_{j=0}^{\infty}(1 - \xi\mu_j).$$ (2.22)

In fact for $J = (-s,s)$ the eigenvalues can be computed [18] by relating $K_{(-s,s)}^{\text{bulk}}$ to a differential operator which has the prolate spheroidal functions as its eigenfunctions, and using previously computed properties of this eigensystem.

Wigner's interest was not in complex Hermitian random matrices, but rather real symmetric random matrices. Orthogonally invariant ensembles of the latter have an eigenvalue probability density function of the form

$$\frac{1}{C}\prod_{l=1}^{N} w_1(x_l) \prod_{1\le j<k\le N} |x_k - x_j|,$$ (2.23)

to be denoted $\text{OE}_N(w_1)$. For such matrix ensembles, the k-point distribution function can be written as a quaternion determinant (or equivalently Pfaffian) with an underlying 2×2 matrix kernel (see e.g. [8, Ch. 5]). From this it is possible to show that

$$\left(E_1^{\text{bulk}}(J;\xi)\right)^2 = \det(1 - \xi K_{1,J}^{\text{bulk}}) \qquad (2.24)$$

where $K_{1,J}^{\text{bulk}}$ is the integral operator on J with matrix kernel

$$K_1^{\text{bulk}}(x,y) = \begin{bmatrix} \dfrac{\sin \pi(x-y)}{\pi(x-y)} & \dfrac{1}{\pi}\displaystyle\int_0^{\pi(x-y)} \dfrac{\sin t}{t}dt - \dfrac{1}{2}\text{sgn}(x-y) \\[2ex] \dfrac{\partial}{\partial x}\dfrac{\sin \pi(x-y)}{\pi(x-y)} & \dfrac{\sin \pi(x-y)}{\pi(x-y)} \end{bmatrix}. \qquad (2.25)$$

However, unlike the result (2.21), this form has not been put to any practical use.

Instead, as discovered by Mehta [26], a tractable formula results from the scaling limit of an inter-relationship between the generating function of an orthogonal symmetry gap probability and a unitary symmetry gap probability. The inter-relationship states

$$E_{2N,1}((-t,t);\xi;e^{-x^2/2})\Big|_{\xi=1} = E_{N,2}((0,t^2);\xi;y^{-1/2}e^{-y}\chi_{y>0})\Big|_{\xi=1}, \qquad (2.26)$$

and in the scaling limit leads to the result

$$E_1^{\text{bulk}}((-s,s);\xi)\Big|_{\xi=1} = \det(1 - K_{(-s,s)}^{\text{bulk}+}) \qquad (2.27)$$

where $K_{(-s,s)}^{\text{bulk}+}$ is the integral operator on $(-s,s)$ with kernel

$$\frac{1}{2}\left(\frac{\sin \pi(x-y)}{\pi(x-y)} + \frac{\sin \pi(x+y)}{\pi(x+y)}\right), \qquad (2.28)$$

which we recognize as the even part of the sine kernel (2.20). (For future reference we define $K_{(-s,s)}^{\text{bulk}-}$ analogously, except that the kernel consists of the difference of the two terms in (2.28), or equivalently the odd part of the sine kernel (2.20).) Because the eigenvalues μ_{2j} of the integral operator on $(-s,s)$ with kernel (2.20) correspond to even eigenfunctions, while the eigenvalues μ_{2j+1} correspond to odd eigenfunctions, we have that

$$E_1^{\text{bulk}}((-s,s);\xi)\Big|_{\xi=1} = \prod_{l=0}^{\infty}(1 - \mu_{2l}). \qquad (2.29)$$

Gaudin [18] used this formula, together with (2.2), to tabulate $p_1^{\text{bulk}}(0; s)$ and so test the accuracy of the Wigner surmise (2.12). In fact this confirmed the remarkable precision of the latter, with the discrepancy between it and the exact value no worse than a few percent.

The case of Hermitian matrices with real quaternion elements and having a symplectic symmetry remains. The eigenvalue p.d.f. of the independent eigenvalues (the spectrum is doubly degenerate) is then

$$\frac{1}{C} \prod_{l=1}^{N} w_4(x_l) \prod_{1 \le j < k \le N} (x_k - x_j)^4, \tag{2.30}$$

which we denote by $\text{SE}_N(w_4)$. The computation of the corresponding bulk gap probability relies on further inter-relationships between matrix ensembles with different underlying symmetries. These apply to the eigenvalue probability density function for Dyson's circular ensembles,

$$\frac{1}{C} \prod_{1 \le j < k \le N} |e^{i\theta_k} - e^{i\theta_j}|^\beta,$$

where $\beta = 1, 2$ or 4 according to the underlying symmetry being orthogonal, unitary or symplectic respectively. The corresponding matrix ensembles are referred to as the COE_N, CUE_N and CSE_N in order. In the $N \to \infty$ scaling limit these ensembles correspond with the bulk of the ensembles $\text{OE}_N(w_1)$, $\text{UE}_N(w_2)$ and $\text{SE}_N(w_4)$ respectively. (We repeat again that here the orthogonal, unitary or symplectic label refers to invariance properties of the ensembles. Apart from the CUE_N, which is equivalent to $U(N)$, the circular ensembles are not the same as the classical compact groups discussed in the lectures of J.B. Conrey, page 111.)

The first of the required inter-relationships was formulated by Dyson [5] and proved by Gunson [19]. It states that

$$\text{alt}(\text{COE}_N \cup \text{COE}_N) = \text{CUE}_N \tag{2.31}$$

where the operation $\text{COE}_N \cup \text{COE}_N$ refers to the superposition of two independent realizations of the COE_N and alt refers to the operation of observing only every second member of the sequence. The second of the required inter-relationships is due to Dyson and Mehta [7]. It states that

$$\text{alt}\,\text{COE}_{2N} = \text{CSE}_N. \tag{2.32}$$

(For generalizations of (2.31) and (2.32) to the ensembles $\text{OE}_N(w_1)$, $\text{UE}_N(w_2)$ and $\text{SE}_N(w_4)$ with particular w_1, w_2 and w_4 see [12].) Using (2.31) and (2.32) together one can deduce that in the scaled limit

$$E_4^{\text{bulk}}(0;(-s/2,s/2)) = \frac{1}{2}\Big(E_1^{\text{bulk}}(0;(-s,s)) + \frac{E_2^{\text{bulk}}(0;(-s,s))}{E_1^{\text{bulk}}(0;(-s,s))}\Big), \qquad (2.33)$$

which upon using (2.22) and (2.29) reads

$$E_4^{\text{bulk}}(0;(-s/2,s/2)) = \frac{1}{2}\Big(\prod_{l=0}^{\infty}(1-\lambda_{2l}) + \prod_{l=0}^{\infty}(1-\lambda_{2l+1})\Big). \qquad (2.34)$$

Another consequence of (2.32) is that

$$p_4(0;s) = 2p_1(1;2s). \qquad (2.35)$$

It is this relationship, used together with the approximation for $p_4(0;s)$ in (2.13), which is used to approximate $p(1;s)$ as a smooth curve in Figure 2.

In summary, as a consequence of the pioneering work of Mehta, Gaudin and Dyson, computable formula in terms of the eigenvalues of the integral operator on $(-s,s)$ with the sine kernel (2.20) were obtained for

$$E_2^{\text{bulk}}((-s,s);\xi), \qquad E_1^{\text{bulk}}(0;(-s,s)), \qquad E_4^{\text{bulk}}(0;(-s/2,s/2)).$$

3   Painlevé transcendent evaluations

3.1   The results of Jimbo et al.

An explicit connection between the multiple interval gap probability

$$E_2^{\text{bulk}}\Big(\cup_{j=1}^{p}(a_{2j-1},a_{2j});\xi\Big)$$

and integrable systems theory — specifically the theory of isomondromic deformations of linear differential equations — was made by Jimbo, Miwa, Môri and Sato in 1980. Here the endpoints $a_1,\ldots,a_{2p}$ of the gap free intervals become dynamical time like variables, inducing flows which turn out to be integrable.

As part of this study the quantity

$$E_2^{\text{bulk}}((-s,s);\xi) = \det(1-\xi K_{(-s,s)}^{\text{bulk}}) = \prod_{j=0}^{\infty}(1-\xi\mu_j) \qquad (3.1)$$

292 Peter J. Forrester

was expressed in terms of the solution of a nonlinear equation. In fact knowledge of (3.1) is sufficient to calculate the products appearing in (2.29) and
(2.34). Thus with

$$D_+(s;\xi) := \prod_{j=0}^{\infty}(1 - \xi\mu_{2j}), \qquad D_-(s;\xi) := \prod_{j=0}^{\infty}(1 - \xi\mu_{2j+1})$$

Gaudin (see [28]) has shown

$$\log D_\pm(s;\xi) = \frac{1}{2}\log E_2^{\text{bulk}}((-s,s);\xi) \pm \frac{1}{2}\int_0^s \sqrt{-\frac{d^2}{dx^2}\log E_2^{\text{bulk}}((-x,x);\xi)}\, dx.$$
(3.2)

The result of [23] is that

$$E_2^{\text{bulk}}((-s,s);\xi) = \exp\int_0^{\pi s}\frac{\sigma(u;\xi)}{u}\, du$$
(3.3)

where $\sigma(u;\xi)$ satisfies the nonlinear differential equation

$$(u\sigma'')^2 + 4(u\sigma' - \sigma)(u\sigma' - \sigma + (\sigma')^2) = 0$$
(3.4)

subject to the boundary condition

$$\sigma(u;\xi) \underset{u\to 0^+}{\sim} -\frac{\xi u}{\pi}.$$

In fact the equation (3.4) is an example of the so called σ-form of a Painlevé
V equation. In view of this it is appropriate to give some background into the
Painlevé theory, following [21]. First we remark that the Painlevé differential
equations are second order nonlinear equations isolated as part of the study
of Painlevé and his students into the moveable singularities of the solution of
such equations. Earlier Fuchs and Poincaré had studied first order differential
equations of the form

$$P(y', y, t) = 0$$
(3.5)

where P is a polynomial in y', y with coefficients meromorphic in t. In contrast
to linear differential equations, nonlinear equations have the property that the
position of the singularities of the solution will depend in general on the initial
condition. The singularities are then said to be moveable. For example

$$\frac{dy}{dt} = y^2$$
(3.6)

has the general solution $y = 1/(c-t)$, where c determines the initial condition,
and so exhibits a moveable first order pole. The nonlinear equation

$$y\frac{dy}{dt} = \frac{1}{2}$$

has the general solution $y = (t - c)^{1/2}$, which exhibits a moveable branch point (essential singularity). Fuchs and Poincaré sought to classify all equations of the form (3.5) which are free of moveable essential singularities. They were able to show that up to an analytic change of variables, or fractional linear transformation, the only such equations with this property were the differential equation of the Weierstrass $\mathcal{P}$-function,

$$\left(\frac{dy}{dt}\right)^2 = 4y^3 - g_2 y - g_3, \tag{3.7}$$

or the Riccati equation

$$\frac{dy}{dt} = a(t)y^2 + b(t)y + c(t) \tag{3.8}$$

where a, b, c are analytic in t (note that (3.6) is of the latter form).

Painlevé then took up the same problem as that addressed by Fuchs and Poincaré, but now with respect to second order differential equations of the form

$$y'' = R(y', y, t)$$

where R is a rational function in all arguments. It was found that the only equations of this form and with no moveable essential singularities were either reducible to (3.7) or (3.8), reducible to a linear differential equation, or were one of six new nonlinear differential equations, now known as the Painlevé equations. As an explicit example of the latter, we note the Painlevé V equation reads

$$y'' = \left(\frac{1}{2y} + \frac{1}{1-y}\right)(y')^2 - \frac{1}{x}y' + \frac{(y-1)^2}{x^2}\left(\alpha y + \frac{\beta}{y}\right) + \frac{\gamma y}{x} + \frac{\delta y(y+1)}{y-1} \tag{3.9}$$

where α, β, γ are parameters.

An immediate question is to how (3.9) relates to (3.4). For this one must develop a Hamiltonian theory of the Painlevé equations. The idea is to present a Hamiltonian $H = H(p, q, t; \vec{v})$, where the components of $\vec{v}$ are parameters, such that after eliminating p in the Hamilton equations

$$q' = \frac{\partial H}{\partial p}, \qquad p' = -\frac{\partial H}{\partial q}, \tag{3.10}$$

q' and p' denoting derivatives with respect to t, the equation in q is the appropriate Painlevé equation. Malmquist [25] was the first to present such Hamiltonians, although his motivation was not to further the development of the Painlevé theory itself. This was left to Okamoto in a later era, and it is aspects of his theory we will briefly present here.

The Hamiltonian for the PV equation as presented by Okamoto [29] is

$$
\begin{aligned}
tH_V &= q(q-1)^2 p^2 - \{(v_1 - v_2)(q-1)^2 - 2(v_1 + v_2)q(q-1) + tq\}p \\
&\quad + (v_3 - v_2)(v_4 - v_2)(q-1),
\end{aligned} \tag{3.11}
$$

where the parameters are constrained by $v_1 + v_2 + v_3 + v_4 = 0$ and are further related to those in (3.9) according to

$$
\alpha = \frac{1}{2}(v_3 - v_4)^2, \ \ \beta = -\frac{1}{2}(v_1 - v_2)^2, \ \gamma = v_1 + 2v_2 - 1, \ \ \delta = -\frac{1}{2}.
$$

It turns out that, as a consequence of the Hamilton equations (3.10), tH_V itself satisfies a nonlinear differential equation. It is this differential equation which relates to (3.4). Okamoto made use of this equation for the symmetry it exhibits in the parameters $v_1, \ldots, v_4$.

The equation in question, which is fairly straightforward to derive, is presented for the so called auxilary Hamiltonian

$$
h_V(t) = tH_V + (v_3 - v_2)(v_4 - v_2) - v_2 t - 2v_2^2.
$$

Okamoto showed

$$
(th_V'')^2 - (h_V - th_V' + 2(h_V')^2)^2 + 4\prod_{k=1}^{4}(h_V' + v_k) = 0.
$$

Setting

$$
\sigma_V(t) = h_V(t) + v_2 t + 2v_2^2, \qquad \nu_{j-1} = v_j - v_2 \ \ (j = 1, \ldots, 4)
$$

in this one obtains the so called Jimbo-Miwa-Okamoto σ-form of the Painlevé V equation

$$
\begin{aligned}
(t\sigma_V'')^2 &- \left(\sigma_V - t\sigma_V' + 2(\sigma_V')^2 + (\nu_0 + \nu_1 + \nu_2 + \nu_3)\sigma_V'\right)^2 \\
&+ 4(\nu_0 + \sigma_V')(\nu_1 + \sigma_V')(\nu_2 + \sigma_V')(\nu_3 + \sigma_V') = 0
\end{aligned} \tag{3.12}
$$

(Jimbo and Miwa [22] arrived at (3.12) in their study of isomonodromic deformations of linear differential equations). We note that (3.4) is an example of this equation with

$$
\nu_0 = \nu_1 = \nu_2 = \nu_3 = 0, \qquad t \mapsto -2iu.
$$

3.2 Unveiling more structure

The result of Jimbo et al. relates to the Fredholm determinant of the integral operator with the sine kernel. What is special about the sine kernel that relates it to integrable systems theory? This question was answered by Its, Izergin, Korepin and Slanov [20] who exhibited integrability features of all kernels of the Christoffel-Darboux type (recall (2.15) in relation to the latter terminology)

$$\xi K(x,y) = \frac{\phi(x)\psi(y) - \phi(y)\psi(x)}{x - y}, \tag{3.13}$$

the sine kernel begin the special case

$$\phi(x) = \sqrt{\xi}\sin x, \qquad \psi(y) = \sqrt{\xi}\cos y. \tag{3.14}$$

One of their key results related to the form of the kernel $R(x,y)$ for the so called resolvent operator

$$R_J := \xi K_J (1 - \xi K_J)^{-1}.$$

With

$$Q(x) := (1 - \xi K_J)^{-1}\phi, \qquad P(x) := (1 - \xi K_J)^{-1}\psi \tag{3.15}$$

they showed

$$R(x,y) = \frac{Q(x)P(y) - P(x)Q(y)}{x - y}. \tag{3.16}$$

The significance of the resolvent kernel is evident from the general formula

$$\frac{\partial}{\partial a_j} \log \det(1 - \xi K_{(a_1,a_2)}) = (-1)^{j-1} R(a_j, a_j) \quad (j = 1, 2). \tag{3.17}$$

To derive this formula, one notes that

$$\begin{aligned}
\log \det(1 - \xi K_{(a_1,a_2)}) &= \operatorname{Tr}\, \log(1 - \xi K_{(a_1,a_2)}) \\
&= \int_{-\infty}^{\infty} \log(1 - \xi K(x,x)\chi_{(a_1,a_2)}^{(x)})\, dx.
\end{aligned}$$

Thus

$$\frac{\partial}{\partial a_j} \log \det(1 - \xi K_{(a_1,a_2)}) = (-1)^{j-1}(1 - \xi K(a_j,a_j))^{-1}\xi K(a_j, a_j)$$

as required.

According to (3.16)

$$R(a_j, a_j) = -Q(x)P'(x) + P(x)Q'(x)\Big|_{x=a_j}, \tag{3.18}$$

so we see from (3.17) that the Fredholm determinant is determined by the quantities (3.15) and their derivatives evaluated at the endpoints of the interval. Indeed a close examination of the workings of [23], undertaken by Mehta [27], Dyson [6] and Tracy and Widom [32], revealed that the former study indeed proceeds via the equations (3.17) and (3.18), and in fact $\sigma(t)$ in (3.3) is related to the resolvent kernel evaluated at an endpoint by $\sigma(t) = -tR(t/2, t/2)$. Moreover it was realized that like (3.16) there are other equations contained in the working of [23] which apply to all kernels of the form (3.13). However it was also clear that other equations used in [23] were specific to the form of ϕ and ψ in (3.14).

Tracy and Widom were able to identify these latter properties, which are that ϕ and ψ are related by the coupled first order differential equations

$$
\begin{aligned}
m(x)\phi'(x) &= A(x)\phi(x) + B(x)\psi(x) \\
m(x)\psi'(x) &= -C(x)\phi(x) - A(x)\psi(x)
\end{aligned}
\tag{3.19}
$$

where m, A, B, C are polynomials. This structure allows the so called universal equations (independent of the specific form of (3.13)) such as (3.18) to be supplemented by a number of case specific equations. For some choices of ϕ and ψ in addition to that corresponding to sine kernel, the resulting system of equations closes. Examples relevant to spacing distributions at the soft and hard edge of matrix ensembles with unitary symmetry are

$$
\phi(x) = \sqrt{\xi}\mathrm{Ai}(x), \ \psi(x) = \phi'(x), \qquad \phi(x) = \sqrt{\xi}J_a(\sqrt{x}), \ \psi(x) = x\phi'(x).
$$

The soft edge refers to the neighbourhood of the largest eigenvalue (it is referred to as soft because the corresponding eigenvalue density is non-zero for all x values, even those greater than the mean value of the largest eigenvalue). The hard edge is the neighbourhood of the smallest eigenvalue for matrices with non-negative eigenvalues (it is referred to as hard because the eigenvalue density is strictly zero for $x < 0$). In both these cases it was possible to obtain an evaluation of the generating function for the corresponding gap probability in a form analogous to (3.3) [33, 34].

We will make note of the hard edge result because it, by virtue of Mehta's inter-relationship (2.26), relates to the gap probability in the bulk in the case of an underlying orthogonal symmetry. First, we define the hard edge gap probability in the case of an underlying unitary symmetry as the scaled limit of the ensemble (2.14) with $w_2(x) = x^a e^{-x}\chi_{x>0}$. Explicitly

$$
E_2^{\mathrm{hard}}((0, s); \xi) = \lim_{N \to \infty} E_2\Big((0, \frac{s}{4N}); \xi; x^a e^{-x}\chi_{x>0}\Big).
\tag{3.20}
$$

It was shown in [9] that

$$E_2^{\text{hard}}((0,s);\xi) = \det(1 - \xi K_{(0,s)}^{\text{hard}}) \qquad (3.21)$$

where $K_{(0,s)}^{\text{hard}}$ is the integral operator on $(0,s)$ with kernel

$$K^{\text{hard}}(x,y) = \frac{J_a(x^{1/2})y^{1/2}J_a'(y^{1/2}) - x^{1/2}J_a'(x^{1/2})J_a(y^{1/2})}{2(x-y)}.$$

As part of the study [34] the Fredholm determinant (3.21) was given the evaluation

$$E_2^{\text{hard}}((0,s);\xi) = \exp\int_0^s u(t;a;\xi)\frac{dt}{t} \qquad (3.22)$$

where u satisfies the differential equation

$$(tu'')^2 - a^2(u')^2 - u'(4u'+1)(u-tu') = 0 \qquad (3.23)$$

subject to the boundary condition

$$u(t;a;\xi) \underset{t\to 0^+}{\sim} -\xi t K^{\text{hard}}(t,t).$$

The equation (3.23) is a special case of the σ-form of the Painlevé III$'$ system [30].

It follows from (2.26), (3.20) and (3.22) that [10]

$$\begin{aligned}
E_1^{\text{bulk}}(0;(-s,s)) &= E_2^{\text{hard}}(0;(0,\pi^2 s^2))\Big|_{a=-1/2} \\
&= \exp\int_0^{(\pi s)^2} u(t;a;\xi)\frac{dt}{t}\Big|_{\substack{a=-1/2 \\ \xi=1}}. \qquad (3.24)
\end{aligned}$$

This is an alternative Painlevé transcendent evaluation to that implied by (2.28), (3.2) and (3.3). Similarly, by noting that

$$2\sqrt{xy}K^{\text{hard}}(x^2,y^2)\Big|_{a=1/2} = \frac{1}{2}\left(\frac{\sin(x-y)}{x-y} - \frac{\sin(x+y)}{x+y}\right)$$

we see from (2.34), (3.21) and (3.22) that [10]

$$E_4^{\text{bulk}}(0;(-s/2,s/2)) = \qquad (3.25)$$
$$\frac{1}{2}\left(\exp\int_0^{(\pi s)^2} u(t;a;\xi)\frac{dt}{t}\Big|_{\substack{a=-1/2 \\ \xi=1}} + \exp\int_0^{(\pi s)^2} u(t;a;\xi)\frac{dt}{t}\Big|_{\substack{a=1/2 \\ \xi=1}}\right).$$

In summary, the Fredholm determinants in the expressions for the bulk gap probabilities can each be written in terms of Painlevé transcendents. From a practical viewpoint these expressions are particularly well suited for generating power series expansions, and also allow for a numerical tabulation of each of $E_2^{\text{bulk}}(0;(-s,s))$, $E_1^{\text{bulk}}(0;(-s,s))$ and $E_4^{\text{bulk}}(0;(-s,s))$, as well as $E_2^{\text{bulk}}(n;(-s,s))$ for $n \geq 1$. For the latter quantity, according to (2.5) we must differentiate $E_2^{\text{bulk}}((-s,s);\xi)$ with respect to ξ then set $\xi = 1$. Doing this in (3.4) gives a coupled system of differential equations for $\partial^j \sigma(u;\xi)/\partial \xi^j|_{\xi=1}$ $(j = 0,\ldots,n)$ which is only numerically stable for small values of n.

3.3 Distribution of bulk right or left nearest neighbour spacings

The spacing distribution refers to the distribution of the distance between consecutive points as we move along the line left to right. Another simple to measure statistic of this type is the distribution of the smallest of the left neighbour spacing and right neighbour spacing for each point. Let us denote this by $p_\beta^{\text{n.n.}}(s)$ (the superscript n.n. stands for nearest neighbour, while the subscript β indicates the symmetry class). Let $E_\beta^{\text{n.n.}}(0;(-s,s))$ denote the probability that about a fixed eigenvalue at the origin, there is no eigenvalue at distance s either side. Analogous to (2.2) it is easy to see that

$$p_\beta^{\text{n.n.}}(s) = -\frac{d}{ds}E_\beta^{\text{n.n.}}(0;(-s,s)). \tag{3.26}$$

In the case $\beta = 2$ (unitary symmetry) the generating function $E_\beta^{\text{n.n.}}((-s,s);\xi)$ can be expressed as a Fredholm determinant

$$E_\beta^{\text{n.n.}}((-s,s);\xi) = \det(1 - \xi K_{(-s,s)}^{\text{n.n.}}) \tag{3.27}$$

where $K_{(-s,s)}^{\text{n.n.}}$ is the integral operator on $(-s,s)$ with kernel

$$K^{\text{n.n.}}(x,y) := (\pi x)^{1/2}(\pi y)^{1/2}\frac{\left(J_{a+1/2}(\pi x)J_{a-1/2}(\pi y) - J_{a+1/2}(\pi y)J_{a-1/2}(\pi x)\right)}{2(x-y)} \tag{3.28}$$

evaluated at $a = 1$. Following the strategy which leds to (3.22), the Fredholm determinant (3.27) for general $a \in \mathbb{Z}_{\geq 0}$ can be characterized as the solution of a nonlinear equation. Explicitly [11]

$$E_\beta^{\text{n.n.}}((-s,s);\xi) = \exp\left(\int_0^{2\pi s}\frac{\sigma_a(t;\xi)}{t}\,dt\right) \tag{3.29}$$

where σ_a satisfies the nonlinear equation

$$(s\sigma_a'')^2 + 4(-a^2 + s\sigma_a' - \sigma_a)\left((\sigma_a')^2 - \{a - (a^2 - s\sigma_a' + \sigma_a)^{1/2}\}^2\right) = 0 \quad (3.30)$$

subject to the boundary condition

$$\sigma_a(s;\xi) \underset{s\to 0^+}{\sim} -\xi\frac{2(s/4)^{2a+1}}{\Gamma(1/2+a)\Gamma(3/2+a)}.$$

In the case $a = 0$, (3.28) reduces to the sine kernel and the differential equation (3.30) reduces to (3.4). For general a the differential equation (3.30) is satisfied by an auxilary Hamiltonian for PIII (as distinct from PIII′) [39].

Substituting (3.29) in (3.26) gives

$$p_2^{\text{n.n.}}(s) = -\frac{\sigma_a(2\pi s;\xi)}{2\pi s}\exp\int_0^{2\pi s}\frac{\sigma_a(t;\xi)}{t}\,dt\bigg|_{a=\xi=1}. \quad (3.31)$$

An application of this result can be made to the study of the zeros of the Riemann zeta function on the critical line (Riemann zeros). We recall that the Montgomery-Odlyzko law states that the statistics of the large Riemann zeros coincide with the statistics of bulk eigenvalues of an ensemble of random matrices with unitary symmetry, where both the zeros and eigenvalues are assumed to be unfolded so as to have mean spacing unity. As a test of this law, in [11] the empirical determination of $p_2^{\text{n.n.}}(s)$ for large sequences of Riemann zeros, starting at different positions along the critical line, was compared with (3.31). The results, which are consistent with the Montgomery-Odlyzko law, are reproduced in Figure 3. A significant feature is that the empirical determination of $p_2^{\text{n.n.}}(s)$ for the Riemann zeros is so accurate that it is not possible to compare against an approximate form of $p_2^{\text{n.n.}}(s)$ for the random matrices. Thus the exact, readily computable, Painlevé evaluation (3.31) is of a practical importance.

4 Gap probabilities from the Okamoto τ-function theory

4.1 Other strategies

The method of Tracy and Widom may be described as being based on function theoretic properties of Fredholm determinants. Alternative methods which

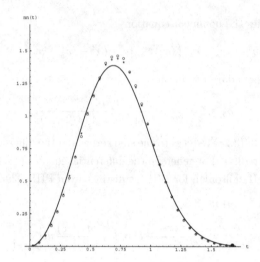

Figure 3: Comparison of $nn(t) := p_2^{\text{n.n.}}(s)$ for the matrix ensembles with unitary symmetry in the bulk (continuous curve) and for 10^6 consecutive Riemann zeros, starting near zero number 1 (open circles), 10^6 (asterisks) and 10^{20} (filled circles).

also lead to the characterization of gap probabilities in terms of the solution of nonlinear equations have been given by a number of authors. One alternative method is due to Adler and van Moerbeke [35], who base their strategy on the fact that for suitable underlying weight w_2, gap probabilities in the case of a unitary symmetry satisfy the KP hierarchy of partial differential equations known from soliton theory. The first member of this hierarchy is then used in conjunction with a set of equations referred to as Virasoro constraints, satisfied by the gap probabilities as a function of the endpoints of the gap free regions, to arrive at third order equations for some single interval gap probabilities. These third order equations are reduced to the σ-form of the Painlevé theory, making use of results of Cosgrove [3, 2]. Borodin and Deift [1] have given a method based on the Riemann-Hilbert formulation of the resolvent kernel (3.16) [24]. This makes direct contact with the Schlesinger equations from the theory of the isomonodromic deformation of linear differential equations, and is thus closely related to the work of Jimbo et al. [23]. The other approach to be mentioned is due to Forrester and Witte [14]. It is based on Okamoto's development of the Hamiltonian approach to Painlevé systems, and proceeds by inductively constructing sequences of multi-dimensional integral solutions of the σ-form of the Painlevé equations, and identifying these solutions with gap probabilities for certain random matrix ensembles with unitary symmetry.

For detailed accounts of all these methods, see [8, Ch. 6&7]. In the remainder of these lectures we will restrict ourselves to results from the work of Forrester and Witte which relate directly to gap probabilities in the bulk.

4.2 Direct calculation of spacing distributions

We have taken as our objective the exact evaluation of the bulk spacing distributions for the three symmetry classes of random matrices. So far exact evaluations have been presented not for the spacing distribution itself, but rather the corresponding gap probability, which is related to the spacing distribution by (2.2). It was realized by Forrester and Witte [15] that in all three cases one of the derivatives could be performed analytically by using theory relating to the σ-form of the Painlevé transcendents.

As an explicit example, consider the result (3.24). It was shown in [15] that

$$\frac{d}{ds} \exp \int_0^{(\pi s)^2} u(t; a; \xi) \frac{dt}{t} \bigg|_{\substack{a=-1/2 \\ \xi=1}} = -\exp\left(-\int_0^{(\pi s)^2} \tilde{u}(t) \frac{dt}{t}\right) \qquad (4.1)$$

where $\tilde{u}$ satisfies the nonlinear equation

$$s^2 (\tilde{u}'')^2 = (4(\tilde{u}')^2 - \tilde{u}')(s\tilde{u}' - \tilde{u}) + \frac{9}{4}(\tilde{u}')^2 - \frac{3}{2}\tilde{u}' + \frac{1}{4}$$

subject to the boundary condition

$$\tilde{u}(s) \underset{s\to 0^+}{\sim} \frac{s}{3} - \frac{s^2}{45} + \frac{8s^{5/2}}{135\pi}.$$

Recalling now (2.2) we see that

$$p_1^{\text{bulk}}(0; s) = \frac{2\tilde{u}((\pi s/2)^2)}{s} \exp\left(-\int_0^{(\pi s/2)^2} \frac{\tilde{u}(t)}{t} dt\right) \qquad (4.2)$$

(cf. (2.12)).

The identity (4.1) can be understood from the approach to gap probabilities of Forrester and Witte. The key advance from earlier studies is that the generating function (2.4), with p given by (2.14), can be generalized to the quantity

$$E^{\text{bulk}}(s; \mu; \xi) := \lim_{N\to\infty} a_N^N \int_{-\infty}^{\infty} dx_1 \cdots \int_{-\infty}^{\infty} dx_N \prod_{l=1}^{N}(1 - \xi\chi_{(-s/2,s/2)}^{(l)})$$
$$\times |s/2 - a_N x_l|^\mu p(a_N x_1, \ldots, a_N x_N) \qquad (4.3)$$

and still be characterized as the solution of a nonlinear equation. This is also true at the hard and soft edges, and in the neighbourhood of a spectrum singularity (before the generalization the latter is controlled by the kernel (3.28)).

It is the generalization in the case of the hard edge which leads to (4.1). The quantity of interest is defined by

$$E_2^{\text{hard}}((0,s);\mu;\xi) = \lim_{N\to\infty} \frac{I_N(a)}{I_N(a+\mu)} E_2\left((0,\frac{s}{4N});\xi;(x-\frac{s}{4N})^\mu x^a e^{-x}\chi_{x>0}\right)$$
(4.4)

where

$$I_N(a) := \int_0^\infty dx_1 \cdots \int_0^\infty dx_N \prod_{l=1}^N e^{-x_l} x_l^a \prod_{1\le j<k\le N} (x_k - x_j)^2$$

(the factor $I_N(a)/I_N(a+\mu)$, which is readily evaluated in terms of gamma functions, is chosen so that when $s = 0$, (4.4) is equal to unity). By using theory from the Okamoto τ function approach to the Painlevé systems PV and PIII' it is shown in [16] that

$$\tilde{E}_2^{\text{hard}}((0,s);\mu;\xi) = \exp\int_0^s u^h(t;a,\mu;\xi)\,\frac{dt}{t},$$

where u^h satisfies the differential equation

$$(tu'')^2 - (\mu+a)^2(u')^2 - u'(4u'+1)(u-tu') - \frac{\mu(\mu+a)}{2}u' - \frac{\mu^2}{4^2} = 0. \quad (4.5)$$

Thus we have

$$-\frac{1}{\xi}\frac{d}{ds}\exp\left(\int_0^s u^h(t;a,\mu;\xi)|_{\mu=0}\,\frac{dt}{t}\right)$$
(4.6)
$$= \frac{s^a}{2^{2a+2}\Gamma(a+1)\Gamma(a+2)}\exp\left(\int_0^s u^h(t;a,\mu;\xi)|_{\mu=2}\,\frac{dt}{t}\right),$$

which in the case $a = -1/2$ reduces to (4.1).

We also read off from (4.6) that

$$\frac{d}{ds}\exp\int_0^{(\pi s)^2} u(t;a;\xi)\frac{dt}{t}\Big|_{\substack{a=1/2\\ \xi=1}} = -\frac{2}{3}(\pi s)^2\exp\left(-\int_0^{(\pi s)^2}\tilde{v}(t)\frac{dt}{t}\right) \quad (4.7)$$

where $\tilde{v}(t) = -u^h(t;a,\mu;\xi)|_{a=1/2,\mu=2,\xi=1}$ and thus satisfies (4.5) appropriately specialized. The boundary condition consistent with (4.7) is

$$\tilde{v}(t) \underset{t\to 0^+}{\sim} \frac{t}{5}(1+O(t)) + \frac{8t^{7/2}}{3^3\cdot 5^3\cdot 7\pi}(1+O(t)). \quad (4.8)$$

Hence, according to (2.34) and (2.2),

$$p_4^{\text{bulk}}(0; s) = 2p_1^{\text{bulk}}(0; 2s) + \frac{2\pi^2 s}{3}\left(\tilde{v}((\pi s)^2) - 1\right)\exp\left(-\int_0^{(\pi s)^2}\tilde{v}(t)\frac{dt}{t}\right). \quad (4.9)$$

The Okamoto τ-function theory of PVI and PV allows (4.3) to be computed for general μ, and also its generalization in which there is a further factor $|-s/2 - a_N x_l|^a$ in the product over l in the integrand [17]. These results allow not only the first derivative with respect to s of (3.3) to be computed by an identity analogous to (4.1), but also the second derivative. In particular, it is found that

$$p_2^{\text{bulk}}(0; s) = \frac{\pi^2}{3}s^2 \exp\int_0^{2\pi s}v(t)\frac{dt}{t} \quad (4.10)$$

where v satisfies the nonlinear equation (which can be identified in terms of the σ-form of the PIII′ equation)

$$(sv'')^2 + (v - sv')\{v - sv' + 4 - 4(v')^2\} - 16(v')^2 = 0$$

subject to the boundary condition

$$v(s) \underset{s\to 0}{\sim} -\frac{1}{15}s^2.$$

The exact evaluations (4.2), (4.9) and (4.10) are perhaps the most compact Painlevé evaluations possible for the bulk spacing distributions. A striking feature of (4.2) and (4.10) is that they are of the functional form $a(s)\exp(-\int_0^s b(t)\,dt)$ and thus extend the Wigner surmise (2.12) and its $\beta = 2$ analogue in (2.13) to exact results.

Acknowledgement

It is a pleasure to thank the organisers for putting together such a stimulating workshop, and program in general. Also, the financial support of the Newton Insitute and the Australian Research Council is acknowledged.

References

[1] A. Borodin and P. Deift. Fredholm determinants, Jimbo-Miwa-Ueno tau-functions and representation theory. *Commun. Pur. Appl. Math.*, 55:1160–1230, 2002.

[2] C.M. Cosgrove. Chazy classes IX-XI of third-order differential equations. *Stud. in Appl. Math.*, 104:171–228, 2000.

[3] C.M. Cosgrove and G. Scoufis. Painlevé classification of a class of differential equations of the second order and second degree. *Stud. in Appl. Math.*, 88:25–87, 1993.

[4] I. Dumitriu and A. Edelman. Matrix models for beta ensembles. *J. Math. Phys.*, 43:5830–5847, 2002.

[5] F.J. Dyson. Statistical theory of energy levels of complex systems III. *J. Math. Phys.*, 3:166–175, 1962.

[6] F.J. Dyson. The Coulomb fluid and the fifth Painlevé transcendent. In S.-T. Yau, editor, *Chen Ning Yang*, page 131. International Press, Cambridge MA, 1995.

[7] F.J. Dyson and M.L. Mehta. Statistical theory of the energy levels of complex systems. IV. *J. Math. Phys.*, 4:701–712, 1963.

[8] P.J. Forrester. Log-gases and Random Matrices. www.ms.unimelb.edu.au/~matpjf/matpjf.html.

[9] P.J. Forrester. The spectrum edge of random matrix ensembles. *Nucl. Phys. B*, 402:709–728, 1993.

[10] P.J. Forrester. Inter-relationships between gap probabilities in random matrix theory. Preprint, 1999.

[11] P.J. Forrester and A.M. Odlyzko. Gaussian unitary ensemble eigenvalues and Riemann ζ function zeros: a non-linear equation for a new statistic. *Phys. Rev. E*, 54:R4493–R4495, 1996.

[12] P.J. Forrester and E.M. Rains. Inter-relationships between orthogonal, unitary and symplectic matrix ensembles. In P.M. Bleher and A.R. Its, editors, *Random matrix models and their applications*, volume 40 of *Mathematical Sciences Research Institute Publications*, pages 171–208. Cambridge University Press, United Kingdom, 2001.

[13] P.J. Forrester and E.M. Rains. Interpretations of some parameter dependent generalizations of classical matrix ensembles. *Prob. Th. Related Fields*, 131(1):1-61, 2005.

[14] P.J. Forrester and N.S. Witte. Application of the τ-function theory of Painlevé equations to random matrices: PIV, PII and the GUE. *Commun. Math. Phys.*, 219:357–398, 2000.

[15] P.J. Forrester and N.S. Witte. Exact Wigner surmise type evaluation of the spacing distribution in the bulk of the scaled random matrix ensembles. *Lett. Math. Phys.*, 53:195–200, 2000.

[16] P.J. Forrester and N.S. Witte. Application of the τ-function theory of Painlevé equations to random matrices: PV, PIII, the LUE, JUE and CUE. *Commun. Pure Appl. Math.*, 55:679–727, 2002.

[17] P.J. Forrester and N.S. Witte. Application of the τ-function theory of Painlevé equations to random matrices: PVI, the JUE,CyUE, cJUE and scaled limits. *Nagoya J. Math.*, 174:29–114, 2004.

[18] M. Gaudin. Sur la loi limite de l'espacement des valeurs propres d'une matrice aléatoire. *Nucl. Phys.*, 25:447–458, 1961.

[19] J. Gunson. Proof of a conjecture of Dyson in the statistical theory of energy levels. *J. Math. Phys.*, 4:752–753, 1962.

[20] A.R. Its, A.G. Izergin, V.E. Korepin, and N.A. Slavnov. Differential equations for quantum correlation functions. *Int. J. Mod. Phys B*, 4:1003–1037, 1990.

[21] K. Iwasaki, H. Kimura, S. Shimomura, and M. Yoshida. *From Gauss to Painlevé. A modern theory of special functions*. Vieweg Verlag, Braunschweig, 1991.

[22] M. Jimbo and T. Miwa. Monodromony preserving deformations of linear ordinary differential equations with rational coefficients II. *Physica*, 2D:407–448, 1981.

[23] M. Jimbo, T. Miwa, Y. Môri, and M. Sato. Density matrix of an impenetrable Bose gas and the fifth Painlevé transcendent. *Physica*, 1D:80–158, 1980.

[24] A.A. Kapaev and E. Hubert. A note on the Lax pairs for Painlevé equations. *J. Math. Phys.*, 32:8145–8156, 1999.

[25] J. Malmquist. Sur les équations différentialles du second ordre dont l'intégrale générale a ses points critiques fixes. *Arkiv Mat. Astron. Fys.*, 18:1–89, 1922.

[26] M.L. Mehta. On the statistical properties of the level-spacings in nuclear spectra. *Nucl. Phys. B*, 18:395–419, 1960.

[27] M.L. Mehta. A non-linear differential equation and a Fredholm determinant. *J. de Physique I (France)*, 2:1721–1729, 1991.

[28] M.L. Mehta. *Random Matrices*. Academic Press, New York, 2nd edition, 1991.

[29] K. Okamoto. Studies of the Painlevé equations. II. Fifth Painlevé equation P_V. *Japan J. Math.*, 13:47–76, 1987.

[30] K. Okamoto. Studies of the Painlevé equations. IV. Third Painlevé equation P_{III}. *Funkcialaj Ekvacioj*, 30:305–332, 1987.

[31] C.E. Porter. *Statistical theories of spectra: fluctuations*. Academic Press, New York, 1965.

[32] C.A. Tracy and H. Widom. Introduction to random matrices. In G.F. Helminck, editor, *Geometric and quantum aspects of integrable systems*, volume 424 of *Lecture notes in physics*, pages 407–424. Springer, New York, 1993.

[33] C.A. Tracy and H. Widom. Level-spacing distributions and the Airy kernel. *Commun. Math. Phys.*, 159:151–174, 1994.

[34] C.A. Tracy and H. Widom. Level-spacing distributions and the Bessel kernel. *Commun. Math. Phys.*, 161:289–309, 1994.

[35] P. van Moerbeke. Integrable lattices: random matrices and random permutations. In P.M. Bleher and A.R. Its, editors, *Random matrix models and their applications*, volume 40 of *Mathematical Sciences Research Institute Publications*, pages 321–406. Cambridge University Press, United Kingdom, 2001.

[36] E.T. Whittaker and G.N. Watson. *A course of modern analysis*. CUP, Cambridge, 2nd edition, 1965.

[37] E.P. Wigner. Characteristic vectors of bordered matrices with infinite dimensions. *Annals Math.*, 62:548–564, 1955.

[38] E.P. Wigner. Gatlinberg conference on neutron physics. Oak Ridge National Laboratory Report ORNL 2309:59, 1957.

[39] N.S. Witte. Gap probabilities for double intervals in Hermitian random matrix ensembles as τ-functions — spectrum singularity case. *Lett. Math. Phys.*,68(3):139–149, 2004.

Department of Mathematics and Statistics
University of Melbourne
Victoria 3010, Australia

308

Toeplitz Determinants, Fisher-Hartwig Symbols, and Random Matrices

Estelle L. Basor[*]

1   Introduction

These notes consist of three topics, the asymptotics of Toeplitz determinants for both smooth and singular symbols, the connection of the asymptotics to ensembles of random matrices, in particular the Circular Unitary Ensemble, and finally generalizations to other classes of ensembles and Fredholm determinants. We begin by describing finite Toeplitz matrices.

Consider a sequence of complex numbers $\{a_i\}_{i=-\infty}^{\infty}$ and the associated matrix

$$T_n = (a_{i-j})_{i,j=0}^{n-1}.$$

___
[*]Supported in part by NSF Grant 0200167 and the Isaac Newton Institute.

The matrix T_n is constant along its diagonals and has the following structure:

$$\begin{pmatrix} a_0 & a_{-1} & a_{-2} & \cdots & a_{-n+1} \\ a_1 & a_0 & a_{-1} & \cdots & a_{-n+2} \\ a_2 & a_1 & a_0 & \cdots & a_{-n+3} \\ \vdots & \vdots & \vdots & & \vdots \\ a_{n-1} & a_{n-2} & a_{n-3} & \cdots & a_0 \end{pmatrix}$$

The matrix is called a finite Toeplitz matrix and the basic problem is to determine what happens to $D_n = \det T_n$ as $n \to \infty$.

The finite matrix also looks like a "truncation" of the infinite one

$$\begin{pmatrix} a_0 & a_{-1} & a_{-2} & \\ a_1 & a_0 & a_{-1} & \ddots \\ a_2 & a_1 & a_0 & \ddots \\ & \ddots & \ddots & \ddots \end{pmatrix} \tag{1.1}$$

and since our finite matrix is growing in size it makes sense to ask if we can somehow get information about determinants from the infinite array. In order to do this, we will try to view the infinite array as an operator on a Hilbert space. We do this because the theory of Fredholm determinants defined for operators on Hilbert spaces is fairly well understood. If we imagine that we are multiplying the infinite matrix on the right by a column vector it is quite natural to choose the Hilbert space as an extension of n-dimensional complex space. We use the Hilbert space of unilateral sequences (usually denoted by l^2)

$$\left\{ \{f_k\}_{k=0}^{\infty} \mid \sum_{k=0}^{\infty} |f_k|^2 < \infty \right\},$$

which we identify with the Hardy space

$$H^2 = \{f \in L^2(S^1) \mid f_k = 0, k < 0\},$$

where

$$f_k = \frac{1}{2\pi} \int_{-\pi}^{\pi} f(e^{i\theta}) e^{-ik\theta} d\theta.$$

For this association we think of f_k as the kth Fourier coefficient of the function f defined on the circle. In other words, in the L^2 sense,

$$f(e^{i\theta}) = \sum_{k=0}^{\infty} f_k e^{ik\theta}.$$

We remind the reader that H^2 is a closed subspace of L^2, and that the inner product of two functions is given by

$$\langle f, g \rangle = \frac{1}{2\pi} \int_{-\pi}^{\pi} f(e^{i\theta}) \overline{g(e^{i\theta})} d\theta.$$

The two-norm of f, denoted by $\|f\|_2$, is given by $\langle f, f \rangle^{1/2}$ and also known to be equal to

$$\left(\sum_{k=0}^{\infty} |f_k|^2 \right)^{1/2}.$$

We denote the orthogonal projection of L^2 onto H^2 by P. The operator P simply takes the Fourier series for f and removes terms with negative index and satisfies $P^2 = P^* = P$. There is one more fact about H^2 that will be useful in what follows. Every function in H^2 has an analytic extension into the interior of the unit circle given by

$$f(z) = \sum_{k=0}^{\infty} f_k z^k.$$

When we work with Toeplitz determinants it is often convenient to mix up all these ideas. In other words, sometimes we think of an infinite sequence, sometimes a function on the circle and sometimes the analytic function in the interior of the circle. There is a large body of literature devoted to these topics. For additional information, we refer the reader to [12, 19, 20, 22].

Now let $\phi \in L^\infty(S^1)$ and define the operator

$$T(\phi) : H^2 \to H^2$$

by

$$T(\phi)f = P(\phi f).$$

The operator $T(\phi)$ is called a Toeplitz operator with symbol ϕ. Let $e_k = e^{ik\theta}$. The functions $\{e_k\}_{k=0}^{\infty}$ form a Hilbert space basis for H^2. To find the matrix representation of the operator $T(\phi)$ we compute

$$\langle T(\phi)e_k, e_j \rangle = \langle P(\phi e_k), e_j \rangle = \langle \phi e_k, P(e_j) \rangle$$

$$= \langle \phi e_k, e_j \rangle = \frac{1}{2\pi} \int_{-\pi}^{\pi} \phi(e^{i\theta}) e^{ik\theta} e^{-ij\theta} d\theta = \phi_{j-k}.$$

This shows that this operator has exactly the matrix representation of the infinite array given in (1.1).

It turns out that when we compose Toeplitz operators, as we will soon do, another operator will appear. It is the Hankel operator with symbol ϕ

$$H(\phi) : H^2 \to H^2$$

defined by

$$H(\phi)f = P(\phi J(f))$$

where

$$J(f)(e^{i\theta}) = e^{-i\theta}f(e^{-i\theta}).$$

The matrix representation of this operator has a different structure than that of a Toeplitz operator. It is found by computing

$$\langle H(\phi)e_k, e_j \rangle = \langle P(\phi J(e_k)), e_j \rangle = \langle \phi e_{-k-1}, P(e_j) \rangle$$

$$= \langle \phi e_{-k-1}, e_j \rangle = \frac{1}{2\pi} \int_{-\pi}^{\pi} \phi(e^{i\theta})e^{-i(k+j+1)\theta}d\theta = \phi_{j+k+1}.$$

Thus the matrix representation has constants running down the "opposite" diagonals and has the form

$$\begin{pmatrix} \phi_1 & \phi_2 & \phi_3 & \cdots \\ \phi_2 & \phi_3 & & \\ \phi_3 & & & \\ \vdots & & & \end{pmatrix} \tag{1.2}$$

Much is known about the invertibility of Toeplitz operators and their structure. However, we only need to recall a few things about Toeplitz operators before we begin to connect the infinite operators to the finite determinants. Here is a list of them.

Theorem 1.1.

(a) $T(\phi)$ is a bounded operator

(b) If $\psi_+ \in L^\infty \bigcap H^2$, then $T(\phi\psi_+) = T(\phi)T(\psi_+)$

(c) $T(\phi)^* = T(\bar\phi)$

(d) If $\overline{\psi_-} \in L^\infty \bigcap H^2$, then $T(\psi_-\phi) = T(\psi_-)T(\phi)$

(e) $T(\phi\psi) = T(\phi)T(\psi) + H(\phi)H(\tilde\psi)$, where $\tilde\psi(e^{i\theta}) = \psi(e^{-i\theta})$.

To prove (a) notice that

$$\|T(\phi)f\|_2 = \|P(\phi f)\|_2 \le \|\phi f\|_2 \le \|\phi\|_\infty \|f\|_2.$$

It is actually the case that the norm of $T(\phi)$ is equal to $\|\phi\|_\infty$, although this fact is not essential for our purposes. To prove (b) consider

$$T(\phi\psi_+)f = P(\phi(\psi_+ f)) = P(\phi P(\psi_+ f)) = T(\phi)T(\psi_+)f.$$

The middle inequality holds since both ψ and f are already in H^2. (Another way to say this is the product of two functions with only non-negative Fourier coefficients has only non-negative Fourier coefficients.) The proof of (c) follows straight from the definition and then using (b) and (c) property (d) holds as well. The proof of property (e) requires a little more work, but is reduced to algebra once we note that the kth Fourier coefficient of $\phi\psi$ is given by

$$\sum_{l=-\infty}^{\infty} \phi_l \psi_{k-l}.$$

Notice that these properties say that Toeplitz operators with symbols that have only non-negative coefficients can be factored on the right and Toeplitz operators with symbols having only non-positive coefficients can be factored on the left. These factorizations are used frequently.

Next, define $P_n : H^2 \to H^2$ by

$$P_n(f_0, f_1, f_2, \cdots) = (f_0, f_1, f_2, \cdots, f_{n-1}, 0, 0, \cdots).$$

This last definition allows us to identify $T_n(\phi)$, the finite Toeplitz matrix generated by the Fourier coefficients of ϕ as the truncation of $T(\phi)$ or as $P_n T(\phi) P_n$.

Now we have the framework to think of a finite matrix as a truncation of an infinite one, and so it seems reasonable to ask if we can also get information about finite determinants from an infinite determinant. To get a notion of how to define a determinant for an infinite array, recall that if M is a finite matrix then $\det M$ is the product $\prod_{i=1}^{n} \beta_i$ where the β_i are the eigenvalues of M. If we extend this to an infinite product $\prod_{i=1}^{\infty} \beta_i$ then we are guaranteed the product will converge if $\beta_i = 1 + \lambda_i$ with

$$\sum_{i=1}^{\infty} |\lambda_i| < \infty.$$

Hence we look for operators of the form $I + K$ where K has a discrete set of eigenvalues λ_i which satisfy $\sum_{i=1}^{\infty} |\lambda_i| < \infty$. A class of operators with exactly

this property is the set of **trace class** operators. We say that K is trace class
if

$$\|K\|_1 = \sum_{n=1}^{\infty} \langle (KK^*)^{1/2} e_n, e_n \rangle < \infty \tag{1.3}$$

for any set of orthonormal basis vectors $\{e_n\}$ for our Hilbert space. (We are
assuming here that our space is separable.) It can be shown that if the above
sum is finite with respect to some basis it is finite and independent for any
choice of basis. It is not always convenient to check whether or not an operator
is trace class from the definition. However, it is fairly straight forward to check
if an operator is **Hilbert-Schmidt** and it is well known that a product of two
Hilbert-Schmidt operators is trace class. The definition of the Hilbert-Schmidt
class of operators is the set of K such that the sum

$$\sum_{i,j} |\langle Ke_i, e_j \rangle|^2 < \infty$$

is finite for some choice of orthonormal basis. If it is, then the above sum is
independent of the choice of basis and its square root is called the Hilbert-
Schmidt norm of the operator. The basic facts about trace class and Hilbert-
Schmidt operators are contained in the following theorem. We state this theorem without proof, but refer the reader to [12] or [22] for general results. The
last part of this theorem first appeared in [24] which is really the first paper
where the idea of trace class operators was used to make the results about
Toeplitz operators seem natural. The reader is strongly advised to look at this
paper.

Theorem 1.2.

(a) Trace class operators form an ideal in the set of all bounded operators
    and are closed in the topology defined by the trace norm defined in (1.3).

(b) Hilbert-Schmidt operators form an ideal in the set of all bounded op-
    erators with respect to the Hilbert-Schmidt norm and are closed in the
    topology defined by the Hilbert-Schmidt norm.

(c) The product of two Hilbert-Schmidt operators is trace class.

(d) If K is trace class, then $\det P_n(I+K)P_n \to \det(I+K)$ as $n \to \infty$. Here
    P_n is the orthogonal projection on the linear span of the basis elements
    $\{e_0, e_1, \ldots, e_{n-1}\}$. (For the first determinant we think of $P_n(I+K)P_n$ as
    the finite rank operator defined on the image of P_n.)

(e) If $A_n \rightarrow A, B_n^* \rightarrow B^*$ strongly (pointwise in the Hilbert space) and if K is trace class, then $A_n K B_n \rightarrow AKB$ in the trace norm.

(f) The functions defined by $\operatorname{tr} K$ and $\det(I + K)$ are continuous on the set of trace class operators with respect to the trace norm.

(g) If $K_1 K_2$ and $K_2 K_1$ are trace class then $\operatorname{tr}(K_1 K_2) = \operatorname{tr}(K_2 K_1)$ and $\det(I + K_1 K_2) = \det(I + K_2 K_1)$.

Now that we have a way to define an infinite determinant for an operator of the form $I + K$ when K is trace class, the question still remains as to how one can compute the determinant in some concrete way. This will be crucial for our applications to random matrices. There are two basic formulas (among many others) that we will consider. The first is that if

$$I + K = e^A$$

where A is trace class, then

$$\det(I + K) = \det e^A = e^{\operatorname{tr} A}.$$

The second is that if operators K_1 and K_2 satisfy the condition that

$$K_1 K_2 - K_2 K_1$$

is trace class (note that neither K_1 or K_2 need be) then

$$\det(e^{K_1} e^{K_2} e^{-K_1} e^{-K_2}) = e^{\operatorname{tr}(K_1 K_2 - K_2 K_1)}.$$

The first formula follows from the definition of the exponential and properties (a), (d) and (e) above and the fact that the formula is true for finite matrices. To see this compare the expressions $(e^{P_n A P_n})$ and (e^A). The second we shall not prove but remark that it follows by using the Baker-Campbell-Hausdorf formula to expand products of exponentials on non-commuting operators. This is shown in [24].

At this point the reader may wonder if $T(\phi)$ is I plus trace class. If it were, then limits of finite Toeplitz matrices would be very easy to compute. This is not the case unless $\phi(e^{i\theta}) \equiv 1$, (just think about the diagonal) but something very close to this statement is true. If we recall Theorem 1.1 part (e) we see that if ϕ^{-1} is a bounded function, then

$$T(\phi)T(\phi^{-1}) = I - H(\phi)H(\tilde{\phi}^{-1}),$$

so the operator $T(\phi)T(\phi^{-1})$ will be $I+$ trace class if both $H(\phi)$ and $H(\widetilde{\phi^{-1}})$ are Hilbert-Schmidt. From the matrix representation of $H(\phi)$ it is very easy to see that $H(\phi)$ will be Hilbert-Schmidt if

$$\sum_{k=1}^{\infty} |k||\phi_k|^2 < \infty.$$

Putting this all together we have the following lemma.

Lemma 1.3. Suppose that ϕ and ϕ^{-1} are bounded functions satisfying

$$\sum_{k=1}^{\infty} |k||\phi_k|^2 < \infty \quad \text{and} \quad \sum_{k=1}^{\infty} |k||(\phi^{-1})_{-k}|^2 < \infty.$$

Then the operator $T(\phi)(T\phi^{-1})$ is $I + K$ where K is trace class.

How does knowing the above help us with the determinant of $P_n T(\phi) P_n$? Let us proceed informally for awhile. It is certainly the case that if $T(\phi)$ is upper or lower triangular, the determinants are easy to compute. This is precisely when ϕ or $\bar{\phi}$ belong to $L^{\infty} \bigcap H^2$. Of course, this is a special case, but it is known that for a fairly large class of functions ϕ there exist functions ϕ_+ and ϕ_- such that both $\phi_+, \bar{\phi}_- \in L^{\infty} \bigcap H^2$, and $\phi = \phi_- \phi_+$. Thus we have

$$P_n T(\phi) P_n = P_n T(\phi_-) T(\phi_+) P_n.$$

Here we have used Theorem 1.1, parts (b) and (d). It would be nice if we could move the P_n into the middle because then we would have the product of upper and lower triangular matrices. Unfortunately we cannot do this, but notice we could if the factors were reversed. This is because

$$P_n T(\phi_+) = P_n T(\phi_+) P_n$$

and

$$P_n T(\phi_-) P_n = T(\phi_-) P_n.$$

Using this motivation, we write

$$P_n T(\phi) P_n = P_n T(\phi_+) T(\phi_+^{-1}) T(\phi) T(\phi_-^{-1}) T(\phi_-) P_n$$

$$= P_n T(\phi_+) P_n T(\phi_+^{-1}) T(\phi) T(\phi_-^{-1}) P_n T(\phi_-) P_n.$$

Now the upper-left blocks of $P_n T(\phi_{\pm}) P_n$ are $T_n(\phi_{\pm})$, which are triangular matrices with diagonal entries $(\phi_{\pm})_0$. Therefore they have determinant $((\phi_-)_0)^n ((\phi_+)_0)^n$, so

$$D_n(\phi) = \det T_n(\phi)$$

equals this product times the determinant of the upper-left block of

$$P_n \, T(\phi_+^{-1}) \, T(\phi) \, T(\phi_-^{-1}) \, P_n.$$

To find the limit of this determinant, we notice by using Lemma 1.3 that if ϕ^{-1} and ϕ satisfy the conditions of the lemma then

$$T(\phi)T(\phi^{-1}) = I - H(\phi)H(\widetilde{\phi^{-1}})$$

and the product of the Hankels is trace class. If we multiply on the left of this equation by $T(\phi_+^{-1})$ and on the right by $T(\phi_+)$ then we have

$$T(\phi_+^{-1}) \, T(\phi)T(\phi^{-1}) \, T(\phi_+) = I + K$$

where K is trace class. This is immediate since trace class operators form an ideal. For the expression on the right side of this equality, we have by Thoerem 1.2 (b)

$$T(\phi_+^{-1}) \, T(\phi)T(\phi^{-1}) \, T(\phi_+) = T(\phi_+^{-1}) \, T(\phi)T(\phi_-^{-1}\phi_+^{-1}) \, T(\phi_+)$$

$$= T(\phi_+^{-1}) \, T(\phi) \, T(\phi_-^{-1})$$

and thus

$$T(\phi_+^{-1}) \, T(\phi) \, T(\phi_-^{-1}) = I + K.$$

Therefore using Theorem 1.2 (d) we have that

$$\lim_{n\to\infty} \det T_n(\phi)/((\phi_-)_0)^n((\phi_+)_0)^n = \lim_{n\to\infty} \det P_n \, T(\phi_+^{-1}) \, T(\phi) \, T(\phi_-^{-1}) \, P_n$$

$$= \det(T(\phi_+^{-1}) \, T(\phi) \, T(\phi_-^{-1})) = \det(T(\phi)T(\phi_-^{-1})T(\phi_+^{-1})) = \det(T(\phi)T(\phi^{-1})).$$

This last limit statement is almost the classical form of the Strong Szegö Limit Theorem. It will be in exactly that form once we rewrite the constants. To become completely rigorous we assume that the function $\phi = \phi_-\phi_+$ has a logarithm $\log \phi$ that is bounded such that

$$\log \phi = \log \phi_- + \log \phi_+$$

where $\overline{\log \phi_-}$ and $\log \phi_+$ are in $L^\infty \bigcap H^2$, satisfy

$$\sum_{k=1}^{\infty} |k| \, |(\log \phi_-)_{-k}|^2 < \infty$$

and

$$\sum_{k=1}^{\infty} |k| \, |(\log \phi_+)_k|^2 < \infty.$$

Conveniently the bounded functions f satisfying $\sum_{k=-\infty}^{\infty}|k|\,|f_k|^2 < \infty$ form a Banach algebra under a natural norm and for any such f the Hankel matrix (f_{i+j+1}) is the matrix of a Hilbert-Schmidt operator. If $\log\phi_{\pm}$ belong to this algebra so do ϕ_-, ϕ_+, $(\phi_+)^{-1}$, $(\phi_-)^{-1}$, ϕ and ϕ^{-1}, and it follows that all associated Hankel operators are Hilbert-Schmidt. With these assumptions, we consider

$$\det(T(\phi_+^{-1})\,T(\phi)\,T(\phi_-^{-1})) = \det(T(\phi_+^{-1})\,T(\phi_-)\,T(\phi_+)T(\phi_-^{-1})).$$

Now we know that $(T(\log\phi_\pm))^n = T((\log\phi_\pm)^n)$ by Theorem 1. (b) and (d), and thus we can write the above determinant as

$$\det e^{K_1}e^{K_2}e^{-K_1}e^{-K_2}$$

where

$$K_1 = -T(\log\phi_+), \quad K_2 = T(\log\phi_+).$$

Then using

$$\det(e^{K_1}e^{K_2}e^{-K_1}e^{-K_2}) = e^{\operatorname{tr}(K_1K_2-K_2K_1)}$$

yields

$$\det(T(\phi_+^{-1})\,T(\phi_-)\,T(\phi_+)T(\phi_-^{-1})) = \exp(\operatorname{tr}(K_1K_2 - K_2K_1).$$

The operator

$$K_1K_2 - K_2K_1 = -T(\log\phi_+)T(\log\phi_-) + T(\log\phi_-)T(\log\phi_+)$$

is equal to

$$-T(\log\phi_+)T(\log\phi_-) + T((\log\phi_-)(\log\phi_+))$$

by Theorem 1.2 (b) and is equal to

$$H(\log\phi_+)H(\widetilde{\log\phi_-})$$

by Theorem 1.2 (e). The trace of this product is easily seen to be

$$\sum_{k=1}^{\infty} k s_k s_{-k},$$

where

$$s_k = (\log\phi)_k.$$

(Note $(\log\phi)_k = ((\log\phi)_+)_k$ for k positive and $((\log\phi)_-)_k$ for k negative.) Finally, the term

$$((\phi_-)_0(\phi_+)_0)^n$$

is the same as
$$G(\phi) = \exp(s_0)$$

since
$$(\phi_\pm)_0 = \exp(\log(\phi_\pm))_0$$

and
$$\log \phi = \log \phi_- + \log \phi_+.$$

The constant $G(\phi)$ is called the geometric mean of ϕ. We collect the above results in this theorem.

Theorem 1.4. (Strong Szegö Limit Theorem) Suppose the functions $\log \phi_\pm$ belong to the algebra of bounded functions f satisfying $\sum_{k=-\infty}^{\infty} |k| |f_k|^2 < \infty$, and in addition suppose $\overline{\log \phi_-}, \log \phi_+ \in H^2$. Let $\phi = \phi_- \phi_+$. Then

$$\lim_{n \to \infty} D_n(\phi)/G(\phi)^n = \exp\left(\sum_{k=1}^{\infty} k s_k s_{-k}\right)$$

where $G(\phi)$ and s_k are as previously defined.

We end this section with an identity for Toeplitz determinants that actually implies the Strong Szegö Limit Theorem. The identity has an interesting history and useful consequences since it is not an asymptotic result, but an actual identity. It is called the Geronimo-Case-Borodin-Okounkov identity. It was first proved by Geronimo and Case in 1979 [18] but unfortunately unnoticed by most specialists in the area. It was independently rediscovered in 1999 by Borodin and Okounkov [11]. Our proof of the identity is an easy adaptation of the proof just presented [8].

Theorem 1.5. Suppose ϕ satisfies the conditions of the previous theorem. Then

$$D_n(\phi) = G(\phi)^n \exp\left(\sum_{k=1}^{\infty} k s_k s_{-k}\right) \det(I - K_n)$$

where
$$K_n = Q_n H(\phi_-/\phi_+) H(\widetilde{\phi_+/\phi_-}) Q_n$$

and $Q_n = I - P_n$.

To prove this let us return to the point in the previous proof where we have the equality

$$D_n(\phi)/((\phi_-)_0)^n((\phi_+)_0)^n = \det P_n T(\phi_+^{-1}) T(\phi) T(\phi_-^{-1}) P_n$$

or

$$D_n(\phi) = G(\phi)^n \det P_n \, T(\phi_+^{-1}) \, T(\phi) \, T(\phi_-^{-1}) \, P_n.$$

Let

$$A = T(\phi_+^{-1}) \, T(\phi) \, T(\phi_-^{-1}).$$

An identity for finite matrices due to Jacobi works equally well for any operator $A = I + K$ with K trace class. It says that

$$\det P_n A P_n = \det A \det Q_n A^{-1} Q_n.$$

The proof is completed by noting that, as we have already seen,

$$\det A = \exp\left(\sum_{k=1}^{\infty} k s_k s_{-k}\right)$$

and that

$$A^{-1} = T(\phi_-) \, (T(\phi))^{-1} \, T(\phi_+), \;\; T(\phi)^{-1} = T(\phi_+^{-1})T(\phi_-^{-1}).$$

This shows that

$$A^{-1} = T(\phi_-/\phi_+)T(\phi_+/\phi_-)$$

which equals

$$I - H(\phi_-/\phi_+)H(\widetilde{\phi_+/\phi_-}).$$

The final step in the proof is to notice that

$$\det(Q_n(I - H(\phi_-/\phi_+)H(\widetilde{\phi_+/\phi_-}))Q_n)$$

is the same as

$$\det(P_n + Q_n(I - H(\phi_-/\phi_+)H(\widetilde{\phi_+/\phi_-})Q_n))$$

or

$$\det(I - K_n).$$

Notice that since Q_n tends to zero strongly (pointwise), the identity together with Theorem 1.2 (e), yields Szegö's theorem as a corollary.

2 Fisher-Hartwig Symbols

If the conditions of the Strong Szegö Limit Theorem (Theorem 1.3) are not met, then the theorem may not hold and indeed the formula may not even make sense. In 1968, Fisher and Hartwig [21] considered a class of symbols of the form

$$\Psi(e^{i\theta}) = \phi(e^{i\theta}) \prod_{j=1}^{R} \phi_{\alpha_j, \beta_j}(e^{i(\theta - \theta_j)}) \tag{2.1}$$

where

$$\phi_{\alpha, \beta}(e^{i\theta}) = (2 - 2\cos\theta)^{\alpha} e^{i\beta(\theta - \pi)}, \ 0 < \theta < 2\pi$$

and $\mathrm{Re}\,\alpha > -\frac{1}{2}$. The function ϕ is assumed to be sufficiently smooth, continuous, non-zero, and have winding number zero. The factor

$$(2 - 2\cos\theta)^{\alpha}$$

may have a zero, be unbounded, or have a singularity of oscillatory type. The factor

$$e^{i\beta(\theta - \pi)}$$

has a jump if $\beta \notin \mathbb{Z}$. Note that for this last factor the Fourier coefficients of the logarithm have order $O(1/k)$ and hence the constant in Szegö's Theorem does not converge. However, Fisher and Hartwig conjectured that

$$D_n(\psi) \sim G(\phi)^n n^{\Omega} E^* \tag{2.2}$$

where

$$\Omega = \sum_{j=1}^{R}(\alpha_j^2 - \beta_j^2)$$

and E^* is some constant whose value they did not identify. The conjecture is now confirmed in many cases and the constant is given by

$$E^*(\psi) = E(\phi) \prod_{j=1}^{R} \phi_+(e^{i\theta_j})^{-\alpha_j + \beta_j} \phi_-(e^{i\theta_j})^{-\alpha_j - \beta_j}$$

$$\times \prod_{1 \le s \ne r \le R} (1 - e^{i(\theta_s - \theta_r)})^{-(\alpha_s + \beta_s)(\alpha_r - \beta_r)} \prod_{j=1}^{R} \frac{G(1 + \alpha_j + \beta_j)G(1 + \alpha_j - \beta_j)}{G(1 + 2\alpha_j)},$$

where $G(z)$ is the Barnes G-function satisfying

$$G(1 + z) = \Gamma(z)G(z)$$

and defined by

$$G(1 + z) = (2\pi)^{z/2} e^{-(z+1)z/2 - \gamma z^2/2} \prod_{k=1}^{\infty} (1 + \frac{z}{k})^k e^{-z + \frac{z^2}{2k}}.$$

The factors need an explanation. We first normalize ϕ so that the geometric mean is 1. Then we may assume that the factors ϕ_+, ϕ_- ($\phi_- \phi_+ = \phi$) are one at zero and infinity respectively and this defines the logarithms for the first product. (Note here we are thinking of the analytic extension of these factors.) The $E(\phi)$ term is the constant in Szegö's Theorem, and the argument of a term of the form $(1 - e^{i(\theta_s - \theta_r)})$ is taken between $-\pi/2$ and $\pi/2$. This conjecture is now known to be true in every case where it "should be true." We intend to make this last statement clearer by the end of this section. What we will do is sketch how to prove it in the cases where $\alpha_j = 0$. This will give the reader an indication of how it is proved in general, since the ideas are more or less the same, but due to technicalities the proof is more complicated. For more information about the general proof see [12, 13, 15, 16]. While not attempting to give a history of the conjecture we will mention that there is a nearly forty year history of the conjecture. Many authors have contributed to its proof including P. Bleher, A. Böttcher, T. Ehrhardt, B. Silbermann, H. Widom and the author. The most recent complete results are by Ehrhardt and are described in [16].

Before we start to give a proof of the conjecture, let us note that for the symbol $\phi_{0,\beta}$ the corresponding Toeplitz matrix is Cauchy and its determinant can be computed. This case was actually confirmed by Fisher and Hartwig in their orginal paper. We of course know what happens when there are no singularities, so it seems plausible to piece together answers by considering what happens to

$$\frac{D_n(\phi\psi)}{D_n(\phi)D_n(\psi)}$$

when ϕ and ψ have no common singularities. Recall that

$$T(\phi\psi) = T(\phi)T(\psi) + H(\phi)H(\widetilde{\psi}).$$

An analogue for finite matrices was noticed first by Widom [24],

$$T_n(\phi\psi) = T_n(\phi)T_n(\psi) + P_n H(\phi)H(\widetilde{\psi})P_n + W_n H(\widetilde{\phi})H(\psi)W_n, \qquad (2.3)$$

where

$$W_n(a_0, a_1, \ldots, a_{n-1}, a_n, \ldots) = (a_{n-1}, a_{n-2}, \ldots, a_1, a_0, 0, 0, \ldots).$$

The operator W_n has the property that

$$\langle W_n e_j, e_k \rangle = \delta_{n-1-j.k} = 0$$

if n is large enough. From this it follows that $W_n \to 0$ weakly or more precisely, that

$$\langle W_n f, g \rangle \to 0$$

for all f, g in H^2. Also, note that $W_n^2 = P_n$, $W_n^* = W_n$ and $W_n T(\phi) W_n = W(\tilde{\phi})$. Let us now suppose that our finite matrices are invertible. From the identity (2.3) we have

$$T_n(\phi\psi)T_n(\psi)^{-1}T_n(\phi)^{-1} = I_n + K_n + L_n$$

where

$$K_n = P_n H(\phi)H(\tilde{\psi})P_n T_n(\psi)^{-1}T_n(\phi)^{-1}$$
$$L_n = W_n H(\tilde{\phi})H(\psi)W_n T_n(\psi)^{-1}T_n(\phi)^{-1}.$$

The idea is now to see if

$$K_n + L_n$$

converges to something in the trace norm. If this were the case, we would indeed have a nice limit for

$$\frac{D_n(\phi\psi)}{D_n(\phi)D_n(\psi)}.$$

Once again this is not quite true. A modification is necessary but, fortunately, everything in the end works out. Let us begin by investigating the term K_n. It is not immediate that the middle term of this operator

$$H(\phi)H(\tilde{\psi})$$

is trace class as it was before. To see that this Hankel product is indeed trace class for disjoint singularities, we prove the following lemma.

Lemma 2.1. Suppose $\phi, \psi \in L^\infty$ and there exists a smooth partition of unity f, g such that $\phi f, \psi g$ satisfy the condition that $H(\phi f), H(\tilde{\psi}g)$ are both trace class operators. Then $H(\phi)H(\tilde{\psi})$ is trace class.

To prove this, notice that since $f + g = 1$,

$$H(\phi)H(\tilde{\psi}) = T(\phi\psi) - T(\phi)T(\psi) = T(\phi(f+g)\psi) - T(\phi)T(f+g)T(\psi)$$

$$= T(\phi f\psi) + T(\phi g\psi) - T(\phi)T(f)T(\psi) - T(\phi)T(g)T(\psi)$$

$$= T(\phi f)T(\psi) + H(\phi f)H(\tilde{\psi}) + T(\phi)T(g\psi) + H(\phi)H(\widetilde{g\psi})$$
$$-T(\phi f)T(\psi) + H(\phi)H(\tilde{f})T(\psi) - T(\phi)T(g\psi) + T(\phi)H(g)H(\tilde{\psi})$$
$$= H(\phi f)H(\tilde{\psi}) + H(\phi)H(\widetilde{g\psi}) + H(\phi)H(\tilde{f})T(\psi) + T(\phi)H(g)H(\tilde{\psi}).$$

By assumption

$$H(\phi f), \ H(\widetilde{g\psi})$$

are trace class, and since trace class operators form an ideal, the products

$$H(\phi f)H(\tilde{\psi}) \text{ and } H(\phi)H(\widetilde{g\psi})$$

are also trace class. For the remaining two terms notice that if h is any function with Fourier coefficients of order $O(k^{-2})$ then $H(h) = AB$ where

$$A_{jk} = h_{j+k+1}(k+1)^{3/4}, \ B_{jk} = \delta_{jk}(k+1)^{-3/4}.$$

Both of these are Hilbert-Schmidt and hence $H(h)$ is trace class. This shows that if our partition of unity functions f and g are smooth then both

$$H(\tilde{f}) \text{ and } H(g)$$

are trace class and the proof of the lemma is complete. It also shows that in the case of isolated singularities, if ϕ and ψ are smooth enough away from their singularities, then the conditions of the lemma are easily satisfied. In particular, this is the case for singularities in the products of the Fisher-Hartwig symbols. Let us return to the term

$$K_n = P_n H(\phi)H(\tilde{\psi})P_n T_n(\psi)^{-1}T_n(\phi)^{-1}.$$

Now that we know that the operator $H(\phi)H(\tilde{\psi})$ is trace class we would like to conclude that K_n converges to something in the trace norm. It is certainly true that P_n converges to the identity strongly. To be able to employ Theorem 1.2 (e) we need to know something about the convergence of

$$T_n(\phi)^{-1}.$$

Fortunately it is the case that if a piecewise continuous function ϕ has the property that the corresponding Toeplitz operator $T(\phi)$ is invertible, then for sufficiently large n the matrices $T_n(\phi)$ are invertible and converge strongly to $T(\phi)^{-1}$. The spectra of Toeplitz operators with piecewise symbols are also well understood. It turns out that for Fisher-Hartwig symbols given in (2.1), if $|\mathrm{Re}\,\beta_j| < \frac{1}{2}$ then the corresponding Toeplitz operators are invertible. (Recall if $\alpha_j = 0$ the symbols have only jump discontinuities.) The same is true for the adjoints of these operators since they correspond to the conjugates of the symbols. Results along these lines can be found in [12] and [13]. Putting this all together, we have that

$$K_n = P_n H(\phi) H(\widetilde{\psi}) P_n T_n(\psi)^{-1} T_n(\phi)^{-1}$$

converges to

$$H(\phi) H(\widetilde{\psi}) T(\psi)^{-1} T(\phi)^{-1}$$

in the trace norm. But what about our other term

$$L_n = W_n H(\widetilde{\phi}) H(\psi) W_n T_n(\psi)^{-1} T_n(\phi)^{-1}?$$

We have shown that W_n converges to zero only weakly, hence we cannot repeat the argument above. However, what we can do is to consider

$$I_n + K_n + L_n + K_n L_n - K_n L_n.$$

Since W_n converges to zero weakly and $H(\widetilde{\phi}) H(\psi)$ is trace class and hence compact, the term L_n converges to zero strongly. We can conclude that $K_n L_n$ converges to zero in the trace norm. In addition,

$$I_n + K_n + L_n + K_n L_n = (I_n + K_n)(I_n + L_n).$$

The operators $I_n + K_n$ are invertible for sufficiently large n and have uniformly bounded inverses with respect to the operator norm. (See [12] or [24] Prop. 2.1, Prop. 2.2.) For the term $I_n + L_n$ we note that

$$
\begin{aligned}
I_n + L_n &= I_n + W_n H(\widetilde{\phi}) H(\psi) W_n T_n(\psi)^{-1} T_n(\phi)^{-1} \\
&= W_n^2 + W_n H(\widetilde{\phi}) H(\psi) W_n T_n(\psi)^{-1} T_n(\phi)^{-1} \\
&= W_n (I_n + H(\widetilde{\phi}) H(\psi) W_n T_n(\psi)^{-1} T_n(\phi)^{-1} W_n) W_n \\
&= W_n (I_n + H(\widetilde{\phi}) H(\psi) T_n(\widetilde{\psi})^{-1} T_n(\widetilde{\phi})^{-1}) W_n
\end{aligned}
$$

Hence the inverses of $I_n + L_n$ are also uniformly bounded. We have now reduced the asymptotic computation of

$$\frac{D_n(\phi\psi)}{D_n(\phi) D_n(\psi)}$$

to

$$\det(I_n + K_n)(I_n + L_n).$$

We have

$$\det(I_n + K_n) = \det(I_n + P_n H(\phi) H(\widetilde{\psi}) P_n T_n(\psi)^{-1} T_n(\phi)^{-1})$$

converges to

$$\det(I + H(\phi) H(\widetilde{\psi}) T(\psi)^{-1} T(\phi)^{-1})$$
$$= \det(I + (T(\phi\psi) - T(\phi) T\psi)) T(\psi)^{-1} T(\phi)^{-1})$$
$$= \det T(\phi\psi) T(\phi)^{-1} T(\psi)^{-1}.$$

Also,

$$\det(I_n + L_n) = \det(I_n + W_n H(\tilde{\phi}) H(\psi) W_n T_n(\psi)^{-1} T_n(\phi)^{-1})$$

$$= \det(I_n + P_n H(\tilde{\phi}) H(\psi) W_n T_n(\psi)^{-1} T_n(\phi)^{-1} W_n)$$
$$= \det(I_n + P_n H(\tilde{\phi}) H(\psi) W_n T_n(\psi)^{-1} W_n^2 T_n(\phi)^{-1} W_n)$$
$$= \det(I_n + P_n H(\tilde{\phi}) H(\psi) T_n(\tilde{\psi})^{-1} T_n(\tilde{\phi})^{-1})$$

converges to

$$\det(I + H(\tilde{\phi}) H(\psi) T(\tilde{\psi})^{-1} T(\tilde{\phi})^{-1})$$
$$= \det(I + (T(\tilde{\phi}\tilde{\psi}) - T(\tilde{\phi}) T(\tilde{\psi})) T(\tilde{\psi})^{-1} T(\tilde{\phi})^{-1}) = \det T(\tilde{\phi}\tilde{\psi}) T(\tilde{\phi})^{-1} T(\tilde{\psi})^{-1}.$$

We summarize the above in the following theorem.

Theorem 2.2. Suppose ϕ and ψ are Fisher-Hartwig symbols with disjoint singularities such that $\alpha_j = 0$, $|Re\beta_j| < 1/2$. Then

$$\lim_{n\to\infty} \frac{D_n(\phi\psi)}{D_n(\phi) D_n(\psi)} = \det T(\phi\psi) T(\phi)^{-1} T(\psi)^{-1} \; \det T(\tilde{\phi}\tilde{\psi}) T(\tilde{\phi})^{-1} T(\tilde{\psi})^{-1}.$$

This theorem can be extended to the case where $|Re\,\alpha_j| < 1/2$, but since the corresponding symbols or their inverses may not be bounded, the Banach algebra approach that was used above must be modified. The reader is referred to [12, 13, 16] for details in the general case.

It may not be clear how this theorem yields the constant in the Fisher-Hartwig conjecture. Suppose we start with a function with one singularity $\phi_{\alpha,\beta}$. While this matrix is not Cauchy, it has a similar form and a direct computation can be made to compute the determinant asymptotically [8]. These asymptotics are of the form

$$D_n(\phi_{\alpha,\beta}) \sim n^{\alpha^2 - \beta^2} \frac{G(1 + \alpha + \beta) G(1 + \alpha + \beta)}{G(1 + 2\alpha)}.$$

Now if we translate the singularity the determinants do not change, since this amounts to changing the Fourier coefficients by $e^{i(k-j)\theta_r}$, which shows the corresponding Toeplitz matrix is unitarily equivalent to one with symbol $\phi_{\alpha,\beta}$. If we add one new factor to the product we get products of G-function terms, powers of n and the interaction terms from the previous theorem. It turns out these can be written as determinants of multiplicative commutators and the resulting evaluations yield the middle terms of the constant E^*. We add all the factors with singularities that are needed and then finally the smooth part. The resulting constants from this part are in the first product of E^*.

We end this section with some additional results about the Fisher-Hartwig conjecture. The following simple example shows that it is not always true! Let

$$\psi(e^{i\theta}) = \begin{cases} -1 & -\pi < \theta < 0 \\ 1 & 0 < \theta < \pi \end{cases}$$

The corresponding Toeplitz matrix is skew-symmetric and hence if n is odd the determinant is zero. If n is even, using some elementary row and column operations it is possible to put the matrix in block form and reduce the problem to one of computing determinants of Cauchy matrices. From this one can show

$$D_n(\psi) \sim (i)^n n^{-\frac{1}{2}} 2^{\frac{1}{2}} G(1/2)^2 G(3/2)^2$$

as $n \to \infty$ for n even. Thus the Fisher-Hartwig conjecture cannot possibly hold. Notice for this example,

$$\begin{aligned} \psi(e^{i\theta}) &= \phi_{0,1/2}(e^{i\theta})\phi_{0,-1/2}(e^{i(\theta-\pi)}) \\ &= \phi_{0,-1/2}(e^{i\theta})\phi_{0,1/2}(e^{i(\theta-\pi)}). \end{aligned}$$

This representation tells us that the conjecture does not even make sense since there is more than one way to describe ψ using the standard Fisher-Hartwig factors. In general, if we let

$$\phi_{\alpha,\beta}(e^{i(\theta-\gamma)}) = \phi_{\alpha,\beta,\gamma}$$

then if

$$\psi = \phi \prod_{j=1}^{R} \phi_{\alpha_j,\beta_j,\theta_j},$$

it is also the case that

$$\psi = \phi^* \prod_{j=1}^{R} \phi_{\alpha_j,\beta_j+n_j,\theta_j},$$

where

$$\sum_{j=1}^{R} n_j = 0, \text{ and } \phi^* = \phi \prod_{j=1}^{R} (-e^{i\theta_j})^{n_j}.$$

In the example, above $\beta_1 = 1/2, \beta_2 = -1/2, \theta_1 = 0, \theta_2 = \pi, n_1 = -1, n_2 = 1$. If we look at the result for our counterexample, combined with what is known for the case of integer values of α and β, one is led to the following generalized conjecture. Suppose

$$\psi(e^{i\theta}) = \phi^k \prod_{j=1}^{R} \phi_{\alpha_j^k, \beta_j^k, \theta_j},$$

for some set of indices k. Define $Q(k) = \sum_{j=1}^{R} (\alpha_j^k)^2 - (\beta_j^k)^2$. Let $Q = \max_k \operatorname{Re} Q(k)$ and

$$\mathcal{K} = \{k \mid \operatorname{Re} Q(k) = Q\}.$$

The generalized conjecture says that

$$D_n(\psi) = \sum_{k \in \mathcal{K}} G(\psi^k) n^{Q(k)} E_k + o(|G(\phi)|^n n^Q).$$

If may turn out that there is only one element in $\mathcal{K}$ and for these symbols there is a unique representation that yields the highest power in the exponent of the asymptotic expansion. These are the symbols for which the original Fisher-Hartwig conjecture should be true. It is now confirmed in these cases. For a full report of the status of the conjecture, please see [15].

3   Connections to Random Matrix Theory

In this section we present the connections of the two main results of the previous sections to random matrix theory. We begin by recalling two formulas. The first is a well-known Fourier transform identity. It reads

$$\frac{1}{\sigma\sqrt{2\pi}} \int_{-\infty}^{\infty} e^{-\frac{(t-\mu)^2}{2\sigma^2}} e^{ixt} \, dt = e^{ix\mu} e^{-x^2\sigma^2}. \tag{3.1}$$

Notice this says that for a Gaussian distribution, the mean and the variance can be easily identified from the Fourier transform. The second formula, first proved by Andréief in 1883 [1], says

$$\frac{1}{N!} \int \cdots \int \det(f_j(x_k)) \det(g_j(x_k)) dx_1 \cdots dx_N$$
$$= \det \left(\int f_j(x) g_k(x) dx \right)_{j,k=1,\cdots N}.$$

The last formula provides a very direct connection between the Circular Unitary Ensemble (CUE) (see the lectures of J.B. Conrey, page 111) in Random Matirx Theory (RMT) and Toeplitz determinants. By CUE we mean the group of $N \times N$ unitary matrices with probability measure being normalized Haar measure. The Haar measure induces a probability distribution on the space of eigenangles $(\theta_1, \dots, \theta_N)$ whose density is given by

$$\frac{1}{N!} \prod_{j<k} |e^{i\theta_j} - e^{i\theta_k}|^2.$$

Now suppose we have a random variable of the form

$$\sum_{j=1}^{N} f(e^{i\theta_j})$$

with f real-valued. This kind of random variable is called a linear statistic and has been studied extensively for different ensembles of random matrices. From probability theory we know that the Fourier transfom of the distribution function of the random variable is given by

$$g(\lambda) = \frac{1}{N!} \int_{-\pi}^{\pi} \dots \int_{-\pi}^{\pi} e^{i\lambda \sum_{j=1}^{N} f(e^{i\theta_j})} \prod_{j<k} |e^{i\theta_j} - e^{i\theta_k}|^2 d\theta_1 \dots d\theta_N,$$

or

$$g(\lambda) = \frac{1}{N!} \int_{-\pi}^{\pi} \dots \int_{-\pi}^{\pi} \prod_{j=1}^{N} e^{i\lambda f(e^{i\theta_j})} \prod_{j<k} |e^{i\theta_j} - e^{i\theta_k}|^2 d\theta_1 \dots d\theta_N.$$

Now consider

$$\prod_{j<k} |e^{i\theta_j} - e^{i\theta_k}|^2 = \prod_{j<k} (e^{i\theta_j} - e^{i\theta_k})(e^{-i\theta_j} - e^{-i\theta_k})$$

$$= \prod_{j<k} (e^{i\theta_j} - e^{i\theta_k}) \prod_{j<k} (e^{-i\theta_j} - e^{-i\theta_k}).$$

The product

$$\prod_{j<k} (e^{i\theta_j} - e^{i\theta_k})$$

is a Vandermonde determinant

$$\det((e^{i\theta_k})^{j-1})_{j,k=1}^{N}$$

as is

$$\prod_{j<k} (e^{-i\theta_j} - e^{-i\theta_k}) = \det((e^{-i\theta_k})^{j-1})_{j,k=1}^{N}.$$

Thus our formula for $g(\lambda)$ becomes

$$g(\lambda) = \frac{1}{N!} \int_{-\pi}^{\pi} \cdots \int_{-\pi}^{\pi} \prod_{j=1}^{N} e^{i\lambda f(e^{i\theta_j})} \det((e^{i\theta_k})^{j-1})_{j,k=1}^{N}$$
$$\times \det((e^{-i\theta_k})^{j-1})_{j,k=1}^{N} d\theta_1 \ldots d\theta_N.$$

We can incorporate a factor of the form $e^{i\lambda f(e^{i\theta_j})}$ into a column of either determinant in the last integral and apply Andréief's formula[1] to find

$$g(\lambda) = \det\left(\int_{-\pi}^{\pi} e^{i\lambda f(\theta)} e^{ik\theta} e^{-ij\theta} d\theta\right)_{j,k=0}^{N-1}.$$

This is exactly the formula for a finite Toeplitz determinant $D_N(\phi)$ with symbol $\phi = e^{i\lambda f(\theta)}$. In other words, the Fourier transform (or characteristic function) of the distribution function of a linear statistic for CUE is a Toeplitz determinant! To obtain asymptotic information about the linear statistic we apply the Strong Szegö Limit Theorem. This shows

$$g(\lambda) \sim G(\phi)^N \exp\left(\sum_{i}^{\infty} k s_k s_{-k}\right)$$

where

$$G(\phi)^N = \exp\left(i\lambda \frac{N}{2\pi} \int_{-\pi}^{\pi} f(e^{i\theta}) d\theta\right)$$

and

$$\exp\left(\sum_{i}^{\infty} k s_k s_{-k}\right) = \exp\left(-\lambda^2 \sum_{i}^{\infty} k f_k f_{-k}\right).$$

Now if we return to formula (3.1) we see that we can interpret the last formula as saying that asymptotically for a smooth function f the linear statistic

$$\sum_{j=1}^{N} f(e^{i\theta_j})$$

has a Gaussian distribution with mean

$$\mu = \frac{N}{2\pi} \int_{-\pi}^{\pi} f(e^{i\theta}) d\theta$$

and variance given by

$$\sigma^2 = \sum_{1}^{\infty} k f_k f_{-k} = \sum_{1}^{\infty} k|f_k|^2.$$

---

[1]Editors' comment: See Section 5 of the lectures of J.B. Conrey, page 111, for a discussion and proof of Andréief's identity.

(The last equality holds since f is real-valued.)

For functions f that are discontinuous, we generally are not able to apply the Strong Szegö Limit Theorem. However, there are some interesting linear statistics that are important in RMT that correspond to singular symbols. For example, consider

$$f(e^{i\theta}) = \chi_I(e^{i\theta}) = \begin{cases} 1 & \text{if } e^{i\theta} \in I \\ 0 & \text{otherwise} \end{cases}$$

This random variable counts the number of eigenvalues in an arc I on the circle. The corresponding symbol for the Toeplitz determinant representing the Fourier transform of the distribution function of the linear statistic is $\phi(e^{i\theta}) = e^{i\lambda\chi_I}$. This is a Fisher-Hartwig symbol. To compute the correct α, β parameters, we note this function only has two jump discontinuities so $\alpha_j = 0, j = 1, 2$. To compute the β parameters notice that for our standard factor $\phi_{0,\beta}$

$$\beta = \frac{1}{2\pi i} \log \left(\frac{\phi(1^-)}{\phi(1^+)} \right).$$

For

$$\phi(e^{i\theta}) = e^{i\lambda\chi_I}$$

with I equal to the arc from $e^{-i\gamma}$ to $e^{i\gamma}$ we have two jumps with

$$\beta_1 = \frac{1}{2\pi i} \log \left(\frac{1}{e^{i\lambda}} \right) = -\lambda/2\pi$$

and the point $e^{-i\gamma}$ and

$$\beta_2 = \frac{1}{2\pi i} \log \left(e^{i\lambda} \right) = \lambda/2\pi$$

at the point $e^{i\gamma}$. It is straightforward to check that

$$\phi(e^{i\theta}) = e^{i\lambda\chi_I} = e^{i\lambda\gamma/\pi} \phi_{0,\lambda/2\pi,\gamma} \; \phi_{0,-\lambda/2\pi,-\gamma}.$$

If we apply the Fisher-Hartwig results directly, we have that

$$D_N(\phi) \sim e^{i\lambda N\gamma/\pi} N^{-\frac{\lambda^2}{2\pi^2}} (2 - 2\cos 2\gamma)^{\frac{\lambda^2}{4\pi^2}} G\left(1 - \frac{\lambda}{2\pi} \right) G\left(1 + \frac{\lambda}{2\pi} \right)$$

This does not have the nice Gaussian form as before. Notice, though that the term

$$N^{-\frac{\lambda^2}{2\pi^2}} = e^{-\frac{\lambda^2}{2\pi^2} \log N}$$

so that σ^2 is on the order of $(1/2\pi^2)\log N$. Thus if we "re-scale" our random variable in a fairly natural way to be of the form

$$\frac{1}{\sqrt{\frac{\log N}{2\pi^2}}} \sum_{j=1}^{N} (\chi_I(e^{i\theta_j}) - \gamma/\pi)$$

then $g(\lambda)$ to tends to $e^{-\lambda^2}$. This is only valid however, if there is sufficient uniformity in the estimates from Section 2 used to prove the Fisher-Hartwig conjecture in the case of jump discontinuities. This is true in a certain range of λ, for $|\lambda/\sqrt{\frac{\log N}{2\pi^2}}| \le c < 1/2$. This can be checked by a careful analysis of the estimates and convergence tools used in proving the conjecture.

It is also the case that the statistics of a characteristic polynomial of a random unitary matrix from CUE can be described using Toeplitz determinants with Fisher-Hartwig symbols. We will not give these results here, but refer the reader to [23].

We end this section with results about other classes of operators that are important in RMT and which have asymptotic expansions that yield information about the corresponding linear statistics. We will not give proofs, but just state results. To understand how these results are analogous to what we have described so far, think of the finite Toeplitz matrix as the operator

$$P_n \mathcal{T}^{-1} M_\phi \mathcal{T} P_n$$

where M_ϕ is multiplication by the function ϕ and $\mathcal{T}$ is the discrete Fourier transform, associating the Fourier coefficients to the corresponding function in H^2.

It turns out for the Gaussian Unitary Ensemble (see the lectures of Y.V. Fyodorov, page 31), for the study of linear statistics it is necessary to consider a finite Wiener-Hopf operator, $W_\alpha(\phi)$, defined on $L^2(0,\alpha)$ by

$$P_\alpha \mathcal{F}^{-1} M_\phi \mathcal{F} P_\alpha$$

where P_α is multiplication by the characteristic function of $(0,\alpha)$, and $\mathcal{F}$ is the Fourier transform. For linear statistics the important quantity is

$$\det(I + W_\alpha(\phi))$$

where $\phi = e^{i\lambda f} - 1$. (The above determinant is well-defined for sufficiently nice ϕ.) The analogue of the Strong Szegö Limit Theorem says that if $\phi = e^b - 1$

then as $\alpha \to \infty$

$$\det(I + W_\alpha(\phi)) \sim \exp\left(\frac{\alpha}{2\pi}\int_{-\infty}^{\infty} b(x)dx + \int_0^{\infty} x\hat{b}(x)\hat{b}(-x)dx\right),$$

where $\hat{b}(x)$ is the Fourier transform of b. This formula again implies that for smooth f the linear statistics are asymptotically Gaussian. Analogues of this theorem have also recently been proved for Fisher-Hartwig type symbols [10, 13].

In RMT Laguerre ensembles are defined on the space of positive Hermitian matrices and for these ensembles, the study of linear statistics lead to the study of Bessel operators $B_\alpha(\phi)$ defined on $L^2(0,\alpha)$ by

$$P_\alpha \mathcal{H}_\nu M_\phi \mathcal{H}_\nu P_\alpha$$

where $\mathcal{H}_\nu$ is the Hankel transform of order ν given by

$$H_\nu(f)(x) = \int_0^{\infty} \sqrt{tx}\, J_\nu(tx)f(t)\, dt,$$

and J_ν is the Bessel function of order ν. If $\phi = e^b - 1$, then

$$\det(I + B_\alpha(\phi)) \sim \exp\left(\frac{\alpha}{2\pi}\int_{-\infty}^{\infty} b(x)dx - \frac{\nu}{2} + \frac{1}{2}\int_0^{\infty} x(\hat{b}(x))^2 dx\right).$$

Some results are known in this case for singular symbols as well, but only in the case of $\nu = \pm\frac{1}{2}$. For results about these ensembles and a proof of the above result see [3, 4, 7].

If we scale in GUE ensemble on the edge of the spectrum, linear statistics problems reduce to the study of the Airy operators, integral operators on $L^2(0,\alpha)$ with kernel

$$A_\alpha(f)(x,y) = f(x/\alpha)\int_0^{\infty} A(x+z)A(z+y)dz$$

where $A(x)$ is the Airy function. (This operator also has an equivalent definition in terms of multiplication and a "Airy" transform, see [9].) The asymptotic formula reads

$$\det(I + A_\alpha(f)) \sim \exp\left(c_1\alpha^{3/2} + c_2\right)$$

where

$$c_1 = \frac{1}{\pi}\int_0^{\infty} \sqrt{x}\log(1 + f(-x))\, dx$$

and

$$c_2 = \frac{1}{2} \int_0^\infty x G(x)^2 dx$$

$$G(x) = \frac{1}{2\pi} \int_{-\infty}^\infty e^{ixy} \log(1 + f(-y^2)) dy.$$

Results for discontinuous symbols are not yet known in the Airy case.

In all of the above cases, CUE, GUE, Laguerre, and Airy there is always a Szegö type limit theorem for a particular class of operators which implies that after scaling, linear statistics have a Gaussian or normal distribution in the limit, at least for smooth symbols. Hence there seems to exist a universality or central limit theorem for such quantities.

We end these notes with one last example. There are also Fisher-Hartwig analogues for finite Toeplitz plus Hankel matrices

$$M_n(\phi) = T_n(\phi) + H_n(\phi)$$

$$= (\phi_{i-j} + \phi_{i+j+1})_{i,j=0}^{N-1}.$$

The asymptotic formula is similar to what was given before

$$\det M_n(\phi) \sim G(\phi)^n n^\Omega E^{**}(\phi)$$

where Ω and E^{**} are certain constants which we will not describe here. We remark here that the constant Ω is not the same as given in the Fisher-Hartwig conjecture, but depends on the location of the singularities. For applications of these asymptotics to RMT see [17] and for proofs and a survey of known results see [4, 5].

References

[1] C. Andréief. Note sur une relation les intégrales définies des produits des fonctions, *Mém. de la Soc. Sci. Bordeaux* 2 (1883) 1-14.

[2] E. W. Barnes. The theory of the G-function, *Quart. J. Pure and Appl. Math.* 31 (1900) 264-313.

[3] E. L. Basor. Distribution Functions for Random Variables for Ensembles of Positive Hermitian Matrices, *Comm. Math. Phys.* 188 (1997), 327-350.

[4] E. L. Basor, T. Ehrhardt. Asymptotics of determinants of Bessel operators, *Comm. Math. Physics* 234 (2003), 491-516.

[5] E. L. Basor, T. Ehrhardt. Asymptotic formulas for determinants of a sum of finite Toeplitz plus Hankel matrices, *Math. Nach.* 122 (2001) 5-45.

[6] E. L. Basor, T. Ehrhardt. Determinants of symmetric Toeplitz plus Hankel matrices, *Operator Th: Advances and App.* 135 (2002) 61-90.

[7] E. L. Basor, T. Ehrhardt, H. Widom. On the determinant of a certain Wiener-Hopf + Hankel operator, *Integral Equations and Operator Theory* 47 (2003) 257-288.

[8] E. L. Basor, H Widom. On a Toeplitz Identity of Borodin and Okounkov, *Integral Equations and Operator Theory* 37 (2002) 397-401.

[9] E. L. Basor, H Widom. Harold Determinants of Airy operators and applications to random matrices. *J. Statist. Phys.* 96 (1999) no. 1-2, 1-20.

[10] E. L. Basor, H. Widom. Wiener-Hopf determinants with Fisher-Hartwig symbols, *Operator Th: Advances and App.* 147 (2004) 131-149.

[11] A. Borodin, A. Okounkov. A Fredholm determinant formula for Toeplitz determinants *Integral Equations and Operator Theory* 37 (2000) 386-396.

[12] A. Böttcher, B. Silbermann. *Introduction to Large Truncated Toeplitz Matrices*, Springer-Verlag, Berlin, 1998.

[13] A. Böttcher, B. Silbermann. *Analysis of Toeplitz Operators*, Akademie-Verlag, Berlin, 1990.

[14] A. Böttcher, H. Widom. Two elementary derivations of the pure Fisher-Hartwig determinant, *Integral Equations and Operator Theory* 53(4) (2005) 593-596.

[15] T. Ehrhardt. A status report on the asymptotic behavior of Toeplitz determinants with Fisher-Hartwig singularities, *Operator Th: Advances and App.* 124 (2001), 217-241.

[16] T. Ehrhardt, B. Silbermann. Toeplitz determinants with one Fisher-Hartwig singularity, *J. Func. Anal.* 148 (1997), 229-256.

[17] P. J. Forrester, N. E. Frankel. Applications and generalizations of Fisher-Hartwig asymptotics, *J. Math. Phys.* 45(5) (2004) 2003-2028.

[18] J. Geronimo, K. Case. Scattering theory and polynomials orthogonal on the unit circle, *J. Math. Phys.* 20 (1979), no. 2, 299-310.

[19] Ronald G. Douglas. *Banach Algebra Techniques in Operator Theory*, Academic Press, New York, 1972.

[20] Peter L. Duren. *Theory of H^p Spaces*, Academic Press, New York, 1970.

[21] M. E. Fisher, R. E. Hartwig. Toeplitz determinants: some applications, theorems, and conjecture, *Adv. Chem. Phys.* 15 (1968) 333–353.

[22] I.C. Gohberg, M.G. Krein. *Introduction to the theory of linear nonselfadjoint operators* Vol. 18, Translations of Mathematical Monographs, Amer. Math. Soc., Rhode Island, 1969.

[23] C. Hughes, J. P. Keating, N. O'Connell. On the characteristic polynomial of a random unitary matrix, *Commun. Math. Phys.* 220 (2001) 429-451.

[24] H. Widom. Asymptotic behavior of block Toeplitz matrices and determinants. II, *Adv. in Math.* **21**:1 (1976) 1-29.

Department of Mathematics
California Polytechnic State University
San Luis Obispo, CA 93407, USA

Mock-Gaussian Behaviour

C.P. Hughes

Abstract

We show that the moments of the smooth counting function of a
set of points encodes the same information as the n-level density of
these points. If the points are the eigenvalues of matrices taken from
the classical compact groups with Haar measure, then we show that
the first few moments of the smooth counting function are Gaussian,
while the distribution is not. The same phenomenon occurs for smooth
counting functions of the zeros of L–functions, and we give examples
relating to each classical compact group. The advantage of calculating
moments of the counting function is that combinatorially, they are far
easier to handle than the n-level densities.

One of the connections between random matrix theory and number theory
is that the correlations and densities between eigenangles of matrices chosen at
random from the classical compact groups appear to be the same as correlations
and densities between zeros of L–functions taken from certain families. This

connection was first suggested by Katz and Sarnak [7], and is discussed in more detail in the article by J. B. Conrey on page 225.

Rather than study densities of zeros, the purpose of this note is to argue that the same results can be obtained more easily by calculating the moments of smooth counting functions. The mock-Gaussian behaviour of the title refers to the fact that in all the cases examined here, the first few moments are Gaussian, while the overall distribution is not.

In the first section we will explicitly demonstrate the connections between the moments of smooth counting functions and the n-level densities. In the middle section we will sketch the proofs of mock-Gaussian behaviour for the classical compact groups. At the end of the paper we will give examples from number theory where mock-Gaussian behaviour holds, and therefore re-prove, in a manner that does not require a lot of combinatorial sieving, that the n-level densities of these examples agree with what one obtains from random matrix theory.

1 General connections between moments and n-level densities

Let $(x_1, \ldots, x_N)$ be chosen from some probability distribution on $\mathbb{R}^N$. The n-level density function for this distribution is

$$D_n(g_1, \ldots, g_n) = \mathbb{E}\left[\sum_{1 \le j_1, \ldots, j_n \le N}' g_1(x_{j_1}) \ldots g_n(x_{j_n}) \right]. \quad (1.1)$$

Here $\sum'$ denotes the sum over *distinct* indices, that is $j_i \ne j_l$ for $i \ne l$, and $\mathbb{E}$ denotes expectation with respect to the density function.

Remark. Sometimes the numbers x_j have a symmetry condition. An example would be if for each j there exists an j' such that $x_j = -x_{j'}$. In that case sometimes the n-level density is defined with the further condition that $j_i \ne j_{l'}$ imposed, as well as the current distinctness condition that $j_i \ne j_l$. We assume the x_j are desymmetrized.

Another statistic is the moments of the smooth counting function,

$$M_n(f) = \mathbb{E}\left[\left(\sum_{j=1}^N f(x_j) \right)^n \right] = \mathbb{E}\left[\sum_{1 \le j_1, \ldots, j_n \le N} f(x_{j_1}) \cdots f(x_{j_n}) \right] \quad (1.2)$$

which is also just the nth moment of the one level density.

We will show that the two statistics provide the same information. This might seem a little surprising since the n-level density appears to have a more general test function, being a product of n different functions.

1.1 The n-level density implies the moments of the counting function

Note that the sums in (1.2) range unrestrictedly over all variables (they include both diagonal and off-diagonal terms), whereas the sum in (1.1) is over distinct variables (off-diagonals). This problem can be overcome by summing over the diagonals separately.

Definition 1. We say σ is a set partition of n elements into r non-empty blocks if

$$\sigma : \{1, \ldots, n\} \longrightarrow \{1, \ldots, r\} \tag{1.3}$$

satisfies

1. For every $q \in \{1, \ldots, r\}$ there exists at least one j such that $\sigma(j) = q$ (this is the non-emptiness of the blocks).

2. For all j, either $\sigma(j) = 1$ or there exists a $k < j$ such that $\sigma(j) = \sigma(k) + 1$.

The collection of all set partitions of n elements into r blocks is denoted $P(n, r)$.

Roughly speaking, if we think of $\{1, \ldots, r\}$ as denoting ordered pigeonholes, then $\sigma(j)$ either goes into a non-empty pigeonhole, or into the next empty hole.

Remark. The number of $\sigma \in P(n, r)$ is equal to $S(n, r)$, a Stirling number of the second kind. The number of set partitions of n elements into any number of non-empty blocks is $\sum_{r=1}^{n} S(n, r) = B_n$, a Bell number.

Therefore,

$$\sum_{1 \leq j_1, \ldots, j_n \leq N} g_1(x_{j_1}) \ldots g_n(x_{j_n}) = \sum_{r=1}^{n} \sum_{\sigma \in P(n,r)} \underset{\substack{1 \leq i_1, \ldots, i_r \leq N \\ i_j \text{ all distinct}}}{\sideset{}{'}\sum} g_1(x_{i_{\sigma(1)}}) \ldots g_n(x_{i_{\sigma(n)}})$$

$$\tag{1.4}$$

(think of this as summing over the diagonals separately).

From this we may conclude that

$$M_n(f) = \mathbb{E}\left[\sum_{1 \le j_1,\dots,j_n \le N} f(x_{j_1})\dots f(x_{j_n})\right] = \sum_{r=1}^{n}\sum_{\sigma \in P(n,r)} D_r(f^{\lambda_1},\dots,f^{\lambda_r})$$
(1.5)

where $\lambda_q = \#\{j : 1 \le j \le n , \sigma(j) = q\}$.

Therefore, from knowing the r-level densities for $1 \le r \le n$, one can immediately deduce the moments of the smooth counting function.

1.2 The moments of the counting function imply the n-level densities

Let us create an inductive hypothesis that for all $1 \le r < n$, $D_r(g_1,\dots,g_r)$ can be written in terms of $M_m(f)$ for various f with $1 \le m \le r$.

An inclusion / exclusion type formula gives

$$\sum_{S \subseteq \{1,\dots,n\}} (-1)^{n-|S|}\left(\sum_{i \in S} a_i\right)^n = n! a_1 \dots a_n$$
(1.6)

where the sum is over *all* subsets of $\{1,\dots,n\}$, and so removing the subset $\{1,\dots,n\}$ we get

$$M_n(g_1 + \dots + g_n) = \mathbb{E}\left[\left(\sum_{j=1}^{N} g_1(x_j) + \dots + \sum_{j=1}^{N} g_n(x_j)\right)^n\right]$$
$$= n!\,\mathbb{E}\left[\left(\sum_{j=1}^{N} g_1(x_j)\right)\dots\left(\sum_{j=1}^{N} g_n(x_j)\right)\right] - \sum_{S \subsetneq \{1,\dots,n\}} (-1)^{n-|S|} M_n\left(\sum_{i \in S} g_i\right)$$
(1.7)

where the sum is over all *proper* subsets of $\{1,\dots,n\}$.

By (1.4) the term

$$n!\,\mathbb{E}\left[\left(\sum_{j=1}^{N} g_1(x_j)\right)\dots\left(\sum_{j=1}^{N} g_n(x_j)\right)\right]$$
(1.8)

equals $n! D_n(g_1,\dots,g_n)$ plus terms involving D_r for $r < n$. By the inductive hypothesis those terms can be written in terms of the moments of the counting function, and we are done.

This is more easily seen in terms of an example. Consider the 2-level density. We have

$$M_2(g_1 + g_2) = \mathbb{E}\left[\left(\sum_{j=1}^{N} g_1(x_j) + \sum_{j=1}^{N} g_2(x_j)\right)^2\right] \tag{1.9}$$

$$= 2\,\mathbb{E}\left[\left(\sum_{j=1}^{N} g_1(x_j)\right)\left(\sum_{j=1}^{N} g_2(x_j)\right)\right] \tag{1.10}$$

$$+ \mathbb{E}\left[\left(\sum_{j=1}^{N} g_1(x_j)\right)^2\right] + \mathbb{E}\left[\left(\sum_{j=1}^{N} g_2(x_j)\right)^2\right] \tag{1.11}$$

$$= 2\,\mathbb{E}\left[\left(\sum_{j=1}^{N} g_1(x_j)\right)\left(\sum_{j=1}^{N} g_2(x_j)\right)\right] + M_2(g_1) + M_2(g_2). \tag{1.12}$$

Now,

$$\mathbb{E}\left[\left(\sum_{j=1}^{N} g_1(x_j)\right)\left(\sum_{j=1}^{N} g_2(x_j)\right)\right] \tag{1.13}$$

$$= \mathbb{E}\left[\sideset{}{'}\sum_{1\leq j_1,j_2\leq N} g_1(x_{j_1})g_2(x_{j_2})\right] + \mathbb{E}\left[\sum_{j=1}^{N} g_1(x_j)g_2(x_j)\right] \tag{1.14}$$

$$= D_2(g_1, g_2) + M_1(g_1 g_2) \tag{1.15}$$

and so we see that from knowing $M_2(f)$ and $M_1(f)$, we have recovered $D_2(g_1, g_2)$, since

$$D_2(g_1, g_2) = \tfrac{1}{2}M_2(g_1 + g_2) - M_1(g_1 g_2) - \tfrac{1}{2}M_2(g_1) - \tfrac{1}{2}M_2(g_2). \tag{1.16}$$

1.3 Restricted range

Often in number theory, it is only possible to prove the n-level density or the moments of the counting function for test functions whose Fourier transforms are supported in a restricted range. However, the above arguments go through without change, and if we know $M_m(f)$ for all f with supp $\widehat{f} \in [-\alpha/m, \alpha/m]$, for all $1 \leq m \leq n$, then we know $D_n(g_1, \ldots, g_n)$ for all g_i with supp $\widehat{g}_i \in [-\alpha/n, \alpha/n]$. We should remark that this is a little bit weaker than what is often proved within number theory, where the support restriction is frequently of the form $\sum_{j=1}^{n} \alpha_i = \alpha$, where supp $\widehat{g}_i \in [-\alpha_i, \alpha_i]$. Clearly the result above, where $\alpha_i = \alpha/n$, fits this.

2   Mock-Gaussian behaviour in the classical compact groups

The classical compact groups are:

- U(N), the group of all $N \times N$ unitary matrices.

- SO($2N$), the subgroup of U($2N$) containing the even orthogonal matrices with determinant one. If $e^{i\theta}$ is an eigenvalue, then so is $e^{-i\theta}$.

- SO($2N + 1$), the subgroup of U($2N + 1$) containing the odd orthogonal matrices with determinant one. If $e^{i\theta}$ is an eigenvalue, then so is $e^{-i\theta}$, and there is an additional eigenvalue at 1.

- USp($2N$), the subgroup of U($2N$) containing the symplectic unitary matrices. That is, $UU^\dagger = I_{2N}$ and $U^t J U = J$ where $J = \begin{pmatrix} 0 & I_N \\ -I_N & 0 \end{pmatrix}$. Again, if $e^{i\theta}$ is an eigenvalue then $e^{-i\theta}$ is also an eigenvalue.

Averages with respect to Haar measure over all of these compact classical groups can be written in the form

$$\mathbb{E}_{G(N)}\left[\prod_{n=1}^{N} f(\theta_n)\right] = \frac{1}{N!} \int_{\mathbb{T}^N} \det_{N \times N}\{Q^{G(N)}(x_i, x_j)\} \prod_{n=1}^{N} f(x_n)\, \mathrm{d}x_n \qquad (2.1)$$

where G(N) denotes one of U(N), SO($2N$), SO($2N + 1$) or USp($2N$), (N being the number of independent eigenvalues). We call $Q^{G(N)}$ the kernel of the group, and $\mathbb{T}$ the range. Let

$$S_N(z) = \frac{1}{2\pi} \frac{\sin(Nz/2)}{\sin(z/2)}. \qquad (2.2)$$

then, the kernels and ranges are given by

Group G(N)	Kernel $Q^{G(N)}(x,y)$	Range $\mathbb{T}$
U(N)	$S_N(x - y)$	$(-\pi, \pi]$
SO($2N$)	$S_{2N-1}(x - y) + S_{2N-1}(x + y)$	$[0, \pi]$
SO($2N + 1$)	$S_{2N}(x - y) - S_{2N}(x + y)$	$[0, \pi]$
USp($2N$)	$S_{2N+1}(x - y) - S_{2N+1}(x + y)$	$[0, \pi]$

Choose a function ϕ from the set of all real functions whose Fourier transform,

$$\widehat{\phi}(u) := \int_{-\infty}^{\infty} \phi(x)e^{-2\pi i x u} \, dx, \tag{2.3}$$

is smooth and compactly supported. Note that for any $A > 1$, this means[1] $\phi(x) \ll (1 + |x|)^{-A}$ for all $x \in \mathbb{R}$. From such a ϕ we create a 2π-periodic function

$$F_M(\theta) := \sum_{n=-\infty}^{\infty} \phi \left(\frac{M}{2\pi}(\theta + 2\pi n) \right). \tag{2.4}$$

Given an $M \times M$ unitary matrix U with eigenangles θ_n, the smooth counting function, or one-level density, or linear statistic, of the eigenangles of the matrix U is

$$Z_\phi(U) := \operatorname{Tr} F_M(U) := \sum_{n=1}^{M} F_M(\theta_n). \tag{2.5}$$

Remark. Note that the matrix is chosen to be size $M \times M$, and it has N independent eigenangles. The counting function sums over all M of the eigenangles, but Haar measure integrates only over the N independent terms.

Note that due to the rapid decay on ϕ, $Z_\phi(U)$ has the largest contribution from the eigenvalues of U close to 1. We will study moments of $Z_\phi(U)$ when U is averaged over one of the classical compact groups, and show that the first few moments are Gaussian, but the higher ones are not.

Theorem 1 (Hughes and Rudnick [5, 6]). If ϕ is chosen so that $\widehat{\phi}$ is smooth and has compact support, then:

i) If $\operatorname{supp} \widehat{\phi} \subseteq [-2/m, 2/m]$ then the first m moments of $Z_\phi(U)$ over the unitary group $\mathrm{U}(N)$ converge as $N \to \infty$ to the moments of a Gaussian random variable with mean

$$\mu_\phi^{\mathrm{U}} = \int_{-\infty}^{\infty} \phi(x) \, dx \tag{2.6}$$

and variance

$$(\sigma_\phi^{\mathrm{U}})^2 = \int_{-1}^{1} |u||\widehat{\phi}(u)|^2 \, du. \tag{2.7}$$

---

[1]Editors' comment: See the Appendix of the lectures of D.W. Farmer, page 185, for a discussion of the $\ll$ notation.

ii) If ϕ is even, and supp $\widehat{\phi} \subseteq [-1/m, 1/m]$, then the first m moments of $Z_\phi(U)$ when averaged over the symplectic group $\mathrm{USp}(2N)$ converge to the moments of a Gaussian with mean

$$\mu_\phi^{\mathrm{USp}} = \int_{-\infty}^{\infty} \phi(x)\,\mathrm{d}x - \int_0^1 \widehat{\phi}(u)\,\mathrm{d}u \qquad (2.8)$$

and variance

$$(\sigma_\phi^{\mathrm{USp}})^2 = 2 \int_{-1/2}^{1/2} |u| \widehat{\phi}(u)^2 \,\mathrm{d}u. \qquad (2.9)$$

iii) If ϕ is even, and supp $\widehat{\phi} \subseteq [-1/m, 1/m]$, then the first m moments of $Z_\phi(U)$ when averaged over either $\mathrm{SO}(2N)$ or $\mathrm{SO}(2N+1)$ converge to the moments of a Gaussian with mean

$$\mu_\phi^{\mathrm{SO}} = \int_{-\infty}^{\infty} \phi(x)\,\mathrm{d}x + \int_0^1 \widehat{\phi}(u)\,\mathrm{d}u \qquad (2.10)$$

and variance

$$(\sigma_\phi^{\mathrm{SO}})^2 = 2 \int_{-1/2}^{1/2} |u| \widehat{\phi}(u)^2 \,\mathrm{d}u. \qquad (2.11)$$

To re-phrase this theorem, part (i) says that if supp $\widehat{\phi} \subseteq [-\frac{2}{m}, \frac{2}{m}]$, then

$$\lim_{N \to \infty} \mathbb{E}_{\mathrm{U}(N)} \left[\left(Z_\phi(U) - \mu_\phi^{\mathrm{U}} \right)^m \right] = \begin{cases} \frac{(2k)!}{2^k k!} (\sigma_\phi^{\mathrm{U}})^{2k} & \text{if } m = 2k \text{ is even} \\ 0 & \text{otherwise} \end{cases} \qquad (2.12)$$

Remark. This theorem is sharp, in the sense that if the support of $\widehat{\phi}$ was increased beyond $[-1/m, 1/m]$ (or $[-2/m, 2/m]$ in the unitary case), then the mth moment ceases to be Gaussian for $m \geq 3$.

This theorem can be proven via a study of the cumulants (though in [6] a different approach is taken), since if one knows the first ℓ cumulants then one knows the first ℓ moments. If $\theta_1, \ldots, \theta_N$ are the independent eigenangles of a matrix $U \in G(N)$, then for a 2π-periodic function g, the cumulants of $\sum_{n=1}^{N} g(\theta_n)$ are defined as

$$\log \mathbb{E}_{G(N)} \left[\exp \left(t \sum_{n=1}^{N} g(\theta_n) \right) \right] = \sum_{\ell=1}^{\infty} \frac{t^\ell}{\ell!} C_\ell^{G(N)}(g) \qquad (2.13)$$

and for the classical compact groups they can be written in terms of the kernel as follows (this is non-obvious: See, for example, [9])

$$C_\ell^{G(N)}(g) = \sum_{m=1}^{\ell} \sum_{\sigma \in P(\ell,m)} (-1)^{m+1} (m-1)! \int_{\mathbb{T}^m} \prod_{j=1}^{m} g^{\lambda_j}(x_j) Q^{G(N)}(x_j, x_{j+1}) \,\mathrm{d}x_j$$

$$(2.14)$$

where we identify x_{m+1} with x_1. Here $P(\ell, m)$ is the set of all partitions of ℓ objects into m non-empty blocks, as in Definition 1, where the jth block has $\lambda_j = \lambda_j(\sigma)$ elements.

Put

$$C_{\ell,N}^{\text{even}}(g) = \frac{1}{2} \sum_{m=1}^{\ell} \sum_{\sigma \in P(\ell,m)} (-1)^{m+1}(m-1)! \int_{[-\pi,\pi]^m} \prod_{j=1}^{m} g^{\lambda_j}(x_j) S_N(x_j - x_{j+1}) \, dx_j \tag{2.15}$$

and

$$C_{\ell,N}^{\text{odd}}(g) = \frac{1}{2} \sum_{m=1}^{\ell} \sum_{\sigma \in P(\ell,m)} (-1)^{m+1}(m-1)! \int_{[-\pi,\pi]^m} \prod_{j=1}^{m} g^{\lambda_j}(x_j) S_N(x_j - \epsilon_j x_{j+1}) \, dx_j \tag{2.16}$$

where $\epsilon_j = +1$ for $j = 1, \ldots, m-1$ and $\epsilon_m = -1$, and where $S_N(z)$ is defined in (2.2).

Expanding out the kernels $Q^{G(N)}(x,y)$ for the various groups, we find

$$C_\ell^{U(N)}(g) = 2C_{\ell,N}^{\text{even}}(g), \tag{2.17}$$

$$C_\ell^{USp(2N)}(g) = C_{\ell,2N+1}^{\text{even}}(g) - C_{\ell,2N+1}^{\text{odd}}(g), \tag{2.18}$$

$$C_\ell^{SO(2N)}(g) = C_{\ell,2N-1}^{\text{even}}(g) + C_{\ell,2N-1}^{\text{odd}}(g), \tag{2.19}$$

$$C_\ell^{SO(2N+1)}(g) = C_{\ell,2N}^{\text{even}}(g) - C_{\ell,2N}^{\text{odd}}(g). \tag{2.20}$$

Extending the combinatorics introduced by Soshnikov [9], we deduced in [5] that if $\ell \geq 2$,

$$\left| C_{\ell,L}^{\text{odd}}(g) \right| \leq \text{const}_\ell \sum_{\substack{\mathbf{k} \in \mathbb{Z}^\ell \\ |k_1| + \cdots + |k_\ell| > L}} |g_{k_1}| \ldots |g_{k_\ell}| \tag{2.21}$$

and if $\ell \geq 3$

$$\left| C_{\ell,L}^{\text{even}}(g) \right| \leq \text{const}_\ell \sum_{\substack{k_1 + \cdots + k_\ell = 0 \\ |k_1| + \cdots + |k_\ell| > 2L}} |g_{k_1}| \ldots |g_{k_\ell}| \tag{2.22}$$

where g_k is the kth Fourier coefficient of g, so $g(\theta) = \sum_{k=-\infty}^{\infty} g_k e^{i\theta}$.

In order to prove Theorem 1 we must show that for $\ell \geq 3$, the ℓth cumulant of

$$Z_\phi(U) := \sum_{n=1}^{M} F_M(\theta_n) \tag{2.23}$$

tends to zero when averaged over the unitary group if $\operatorname{supp}\widehat{\phi} \subseteq [-2/\ell, 2/\ell]$, and tends to zero when averaged over the symplectic or orthogonal groups if $\operatorname{supp}\widehat{\phi} \subseteq [-1/\ell, 1/\ell]$. Recall that M is the total number of eigenangles of the matrix U, while N is the number of independent ones. Therefore, we choose g as follows:

- If $G(N) = U(N)$, we choose $g(\theta) = F_N(\theta)$.

- If $G(N) = USp(2N)$, ϕ must be even, and we choose $g(\theta) = 2F_{2N}(\theta)$.

- If $G(N) = SO(2N)$, ϕ must be even, and we choose $g(\theta) = 2F_{2N}(\theta)$.

- If $G(N) = SO(2N+1)$, ϕ must be even, and we choose $g(\theta) = 2F_{2N+1}(\theta) + F_{2N+1}(0)$.

Note that from the definition of $F_M(\theta)$, (2.4), the Fourier coefficients of g can be computed, since

$$\frac{1}{2\pi}\int_{-\pi}^{\pi} F_M(\theta)e^{ik\theta}\,d\theta = \frac{1}{2\pi}\int_{-\pi}^{\pi}\sum_{j=-\infty}^{\infty}\phi\left(\frac{M}{2\pi}(\theta - 2\pi j)\right)d\theta \qquad (2.24)$$

$$= \frac{1}{M}\widehat{\phi}\left(\frac{k}{M}\right). \qquad (2.25)$$

Therefore part (i) of Theorem 1 follows from (2.17) and (2.22), since if $\operatorname{supp}\widehat{\phi} \in (-2/\ell, 2/\ell)$ the ℓth cumulant of $Z_\phi(U)$ is zero (for $\ell \geq 3$). The mean and variance of $Z_\phi(U)$ equals $C_1^{U(N)}(g)$ and $C_2^{U(N)}(g)$ can be calculated from (2.15).

Similarly, from (2.19)–(2.18) and (2.21)–(2.22), we have that if $\operatorname{supp}\widehat{\phi} \in [-1/\ell, 1/\ell]$ the first ℓ cumulants of $Z_\phi(U)$ are Gaussian for $USp(2N)$, $SO(2N)$, and $SO(2N+1)$, and this proves parts (ii) and (iii) of Theorem 1.

2.1 Connections to moments of traces of matrices

From the cumulants one can obtain moments of traces of powers of U, first investigated by Diaconis and Shahshahani [1].

Expanding $Z_\phi(U)$ out as a Fourier series, we obtain

$$\mathbb{E}_{G(N)}\{(Z_\phi)^m\}$$
$$= \frac{1}{N^m} \sum_{n_1=-\infty}^{\infty} \cdots \sum_{n_m-\infty}^{\infty} \hat{\phi}\left(\frac{n_1}{N}\right) \cdots \hat{\phi}\left(\frac{n_m}{N}\right) \mathbb{E}_{G(N)}\{\operatorname{Tr} U^{n_1} \ldots \operatorname{Tr} U^{n_m}\}. \quad (2.26)$$

Writing the moments in terms of cumulants, using (2.21) and (2.22), and comparing Fourier coefficients, we find that

Corollary 1.1. Let Z_j be independent standard normal random variables, and let

$$\eta_j = \begin{cases} 1 & \text{if } j \text{ is even} \\ 0 & \text{if } j \text{ is odd} \end{cases} \quad (2.27)$$

Let $a_j \in \{0, 1, 2, \ldots\}$ for $j = 1, 2, \ldots$.

- If $\sum j a_j \leq M - 1$, then

$$\mathbb{E}_{SO(M)}\left\{\prod (\operatorname{Tr} U^j)^{a_j}\right\} = E\left\{\prod (\sqrt{j} Z_j + \eta_j)^{a_j}\right\}. \quad (2.28)$$

- If $\sum j a_j \leq M + 1$, where M is even, then

$$\mathbb{E}_{USp(M)}\left\{\prod (\operatorname{Tr} U^j)^{a_j}\right\} = E\left\{\prod (\sqrt{j} Z_j - \eta_j)^{a_j}\right\}. \quad (2.29)$$

These results were found by Diaconis and Shahshahani [1], though only for half the range of the parameters (and they dealt with the full orthogonal group, not the special orthogonal group). Recently Michael Stolz [10] has provided a further proof of this theorem, for the full range, using invariant theory (though again, he does not deal with the special orthogonal group).

Analogously, for the unitary group, one obtains

Corollary 1.2. For $a_j, b_j \in \{0, 1, 2, \ldots\}$, if

$$\max\left(\sum_{j\geq 1} j a_j, \sum_{j\geq 1} j b_j\right) \leq N \quad (2.30)$$

then

$$\mathbb{E}_{U(N)}\left\{\prod_{j\geq 1} (\operatorname{Tr} U^j)^{a_j} (\operatorname{Tr} U^{-j})^{b_j}\right\} = \delta_{a,b} \prod_{j\geq 1} j^{a_j} a_j! \quad (2.31)$$

where $\delta_{a,b} = 1$ if $a_j = b_j$ for all j, and zero otherwise.

This is exactly the result of Diaconis and Shahshahani, [1]. Indeed, in [6] the mock-Gaussian result for the unitary case was proved via this result, rather than evaluating the cumulants, as this was a more direct approach.

3   Number theory examples

3.1   The Riemann zeta function: A unitary example

Consider the non-trivial zeros of the Riemann zeta function, $\frac{1}{2} + i\gamma$. The Riemann Hypothesis (which we do not assume) is the statement that $\gamma \in \mathbb{R}$ for all γ.

The counting function of Riemann zeros is[2]

$$N(T) = \#\{\gamma \ : \ 0 \le \mathfrak{Re}(\gamma) \le T\} \tag{3.1}$$
$$= \overline{N}(T) + S(T), \tag{3.2}$$

where

$$\overline{N}(T) = 1 + \frac{1}{\pi}\mathfrak{Im}\log\left(\pi^{-iT/2}\Gamma(\tfrac{1}{4}+\tfrac{1}{2}iT)\right) \tag{3.3}$$
$$= \frac{T}{2\pi}\log\frac{T}{2\pi e} + \frac{7}{8} + \mathcal{O}(\frac{1}{T}), \tag{3.4}$$

and the error term is

$$S(T) = \frac{1}{\pi}\mathfrak{Im}\log\zeta(\tfrac{1}{2}+iT) \tag{3.5}$$
$$= \mathcal{O}(\log T). \tag{3.6}$$

To motivate the study of the smooth counting function, we ask the question: What is the distribution of the number of zeros lying in an interval of size h around height T? That is: What is the distribution of $N(t+h)-N(t)$ averaged over $T \le t \le 2T$?

Clearly, the mean is

$$\frac{1}{T}\int_T^{2T} \overline{N}(t+h) - \overline{N}(t)\ dt \sim \frac{h}{2\pi}\log T \tag{3.7}$$

and Fujii [2] (among others) has shown that the centered moments are

$$\frac{1}{T}\int_T^{2T}\left(\frac{S(t+h)-S(t)}{\sigma}\right)^{2k} dt = \frac{(2k)!}{2^k k!} + \mathcal{O}(\frac{1}{\sigma}), \tag{3.8}$$

where

$$\sigma^2 = \begin{cases} \frac{1}{\pi^2}\int_0^{h\log T}\frac{1-\cos t}{t}\ dt & 0 < h \ll 1 \\ \frac{1}{\pi^2}(\log\log T - \log|\zeta(1+ih)|) & 1 \ll h \ll T \end{cases} \tag{3.9}$$

---

[2]Editors' comment: See also Section 7 of the contribution by D.R. Heath-Brown starting on page 1.

Thus if $h \log T \to \infty$, the moments converge to the Gaussian moments, and so the distribution is normal.

However, when $h \log T = \mathcal{O}(1)$, the main term, $\frac{(2k)!}{2^k k!}$, is of the same order as the error term $\mathcal{O}(1/\sigma)$. Therefore, we cannot conclude from (3.8) that the distribution is normal. In fact, the distribution is not normal, as it is discrete.

This motivates the study of the smooth counting function when the zeros are critically scaled,

$$N_\phi(t) = \sum_\gamma \phi\left(\frac{\log T}{2\pi}(\gamma - t)\right). \tag{3.10}$$

In [4] the moments of $N_\phi(t)$ were calculated, and the first few were found to be Gaussian.

For technical reasons we change the average. Instead of integrating over $t \in [T, 2T]$ we define the average to be

$$\langle N_\phi \rangle_T = \int_{-\infty}^{\infty} N_\phi(t) \omega(\frac{t-T}{T}) \frac{\mathrm{d}t}{T} \tag{3.11}$$

where $\int_{-\infty}^{\infty} \omega(x)\, \mathrm{d}x = 1$ and $\widehat{\omega}$ is compactly supported. The previous average would come from setting ω to be the indicator function of the interval $[0, 1]$, but this is not allowed.

Theorem 2 (Hughes and Rudnick [4]). If $\operatorname{supp} \widehat{\phi} \subset (-2/m, 2/m)$ then the first m moments of N_ϕ converge as $T \to \infty$ to those of a Gaussian random variable with mean

$$\int_{-\infty}^{\infty} \phi(x)\, \mathrm{d}x \tag{3.12}$$

and variance

$$\sigma_\phi^2 = \int_{-\infty}^{\infty} \min(|u|, 1)\widehat{\phi}(u)^2\, \mathrm{d}u. \tag{3.13}$$

Sketch of proof. From a smooth version of Riemann's explicit formula we have that $N_\phi(\tau) = \overline{N_\phi}(\tau) + S_\phi(\tau)$, where

$$\overline{N_\phi}(\tau) = \frac{1}{2\pi} \int_{-\infty}^{\infty} \phi\left(\frac{\log T}{2\pi}(r - \tau)\right) \Omega(r)\mathrm{d}r$$
$$+ \phi\left(\frac{\log T}{2\pi}(\frac{i}{2} - \tau)\right) + \phi\left(\frac{\log T}{2\pi}(-\frac{i}{2} - \tau)\right), \tag{3.14}$$

with

$$\Omega(r) = \frac{1}{2}\Psi(\frac{1}{4} + \frac{1}{2}ir) + \frac{1}{2}\Psi(\frac{1}{4} - \frac{1}{2}ir) - \log \pi, \tag{3.15}$$

and where

$$S_\phi(\tau) = -\frac{1}{\log T} \sum_{n \geq 2} \frac{\Lambda(n)}{\sqrt{n}} \widehat{\phi}(\frac{\log n}{\log T}) 2 \cos(\tau \log n). \tag{3.16}$$

(For examples of explicit formulae, see the contribution of D.R. Heath-Brown (page 1) Section 10, and that of D.A. Goldston (page 79) Section 3)

Asymptotic analysis gives that if $\widehat{\phi} \in C_c^\infty(\mathbb{R})$, then the mean value of N_ϕ is given by

$$\langle N_\phi \rangle_T = \langle \overline{N_\phi} \rangle_T \tag{3.17}$$

$$= \int_{-\infty}^{\infty} \phi(x) \mathrm{d}x + \mathcal{O}(\frac{1}{\log T}), \qquad T \to \infty. \tag{3.18}$$

Since

$$\lim_{T \to \infty} \langle (N_\phi - \langle N_\phi \rangle_T)^m \rangle_T = \lim_{T \to \infty} \langle (S_\phi)^m \rangle_T \tag{3.19}$$

it is therefore sufficient to show that the mth moment of S_ϕ is the same as that of a centered normal random variable with variance σ_ϕ^2.

Multiplying out and integrating

$$\langle (S_\phi)^m \rangle_T = (\frac{-1}{\log T})^m \sum_{\epsilon_1,\dots,\epsilon_m = \pm 1} \sum_{n_1,\dots,n_m} \prod_{j=1}^{m} \frac{\Lambda(n_j)}{\sqrt{n_j}} \widehat{\phi}(\frac{\log n_j}{\log T})$$

$$\times \widehat{w}(\frac{T}{2\pi} \sum_{j=1}^{m} \epsilon_j \log n_j) e^{-\mathrm{i}T \sum_{j=1}^{m} \epsilon_j \log n_j}. \tag{3.20}$$

Note $n_j \leq T^{2/m-\epsilon}$ since supp $\widehat{\phi} \in (-2/m, 2/m)$.

Since $\widehat{w}$ has compact support, in order to get a nonzero contribution we need

$$|\sum_{j=1}^{m} \epsilon_j \log n_j| \ll \frac{1}{T} \tag{3.21}$$

and thus $\sum \epsilon_j \log n_j = 0$.

Thus for $T \gg 1$, we find (using $\widehat{w}(0) = \int_{-\infty}^{\infty} w(x) \mathrm{d}x = 1$)

$$\langle (S_\phi)^m \rangle_T = (\frac{-1}{\log T})^m \sum_{\substack{n_1,\dots,n_m \geq 2 \\ \epsilon_1,\dots,\epsilon_m = \pm 1 \\ \sum_{j=1}^{m} \epsilon_j \log n_j = 0}} \prod_{j=1}^{m} \frac{\Lambda(n_j)}{\sqrt{n_j}} \widehat{\phi}(\frac{\log n_j}{\log T}). \tag{3.22}$$

The only terms which do not vanish as $T \to \infty$ are those where $m = 2k$ is even, and there is a partition $\{1,\dots,2k\} = S \cup S'$ into disjoint subsets and a bijection $\sigma : S \to S'$ such that $n_j = n_{\sigma(j)}$ and $\epsilon_j = -\epsilon_{\sigma(j)}$. There are $k!\binom{2k}{k}$ such terms, and so

$$\langle (S_\phi)^{2k} \rangle_T \to \frac{(2k)!}{k!} \left(\frac{1}{\log^2 T} \sum_n \frac{\Lambda(n)^2}{n} \widehat{\phi}(\frac{\log n}{\log T})^2 \right)^k \tag{3.23}$$

$$\to \frac{(2k)!}{k!} \left(\int_0^\infty u \widehat{\phi}(u)^2 \mathrm{d}u \right)^k \tag{3.24}$$

by the Prime Number Theorem (see Theorem 3 of the contribution of D.R. Heath-Brown (starting on page 1)). $\qquad\Box$

This theorem compares perfectly with part (i) of Theorem 1.

3.2 Real Dirichlet L–functions : A symplectic family

Consider the zeros of quadratic L-functions, that is of L-functions of the form

$$L(s, \chi_d) = \sum_{n=1}^{\infty} \frac{\chi_d(n)}{n^s} \tag{3.25}$$

where $\chi_d(n) = \left(\frac{d}{n} \right)$ is the Kronecker symbol. These real Dirichlet characters are discussed in more depth in the lectures of J.B. Conrey, page 225, Section 2.3.

Rather than averaging over t, we will average the low-lying zeros of the L-function over characters, that is over $d \in D(X) := \{d : |d| \leq X , \chi_d \text{ primitive}\}$.

From an explicit formula we can show that the smooth counting function equals

$$N_\phi(\chi_d) := \sum_{\gamma_d} \phi \left(\frac{\log X}{2\pi} \gamma_d \right) \tag{3.26}$$

$$= \int_{-\infty}^{\infty} \phi(x) \, \mathrm{d}x - \int_0^{\infty} \widehat{\phi}(u) \, \mathrm{d}u - \frac{2}{\log X} \sum_{\substack{n=1 \\ n \neq \square}}^{\infty} \frac{\Lambda(n)}{n^{1/2}} \chi_d(n) \widehat{\phi} \left(\frac{\log n}{\log X} \right).$$

$$\tag{3.27}$$

The mean of $N_\phi(\chi_d)$ is $\mu_\phi^{\mathrm{USp}} := \int_{-\infty}^{\infty} \phi(x) \, \mathrm{d}x - \int_0^1 \widehat{\phi}(u) \, \mathrm{d}u$, and so the centered moments are

$$\frac{1}{|D(X)|} \sum_{d \in D(X)} \left(N_\phi(\chi_d) - \mu_\phi^{\mathrm{USp}} \right)^m$$

$$= \frac{1}{|D(X)|} \sum_{d \in D(X)} \left(-\frac{2}{\log X} \sum_{\substack{n=1 \\ n \neq \square}}^\infty \frac{\Lambda(n)}{n^{1/2}} \chi_d(n) \widehat{\phi} \left(\frac{\log n}{\log X} \right) \right)^m . \quad (3.28)$$

Expanding out the bracket, one can show that if $\operatorname{supp} \widehat{\phi} \in (-1/m, 1/m)$, then the only contribution comes from the terms where $n_1 \ldots n_m = \square$ (in which case $\chi_d(n_1 \ldots n_m) = 1$ for all d). That is, we have the following theorem:

Theorem 3. Let $D(X)$ be the set of primitive quadratic characters χ_d with $|d| \leq X$. If $\operatorname{supp} \widehat{\phi} \in [-1/m, 1/m]$ then

$$\frac{1}{|D(X)|} \sum_{d \in D(X)} \left(N_\phi(\chi_d) - \mu_\phi^{\mathrm{USp}} \right)^m$$

$$\rightarrow \begin{cases} \frac{(2k)!}{2^k k!} \left(4 \int_0^{1/2} u \widehat{\phi}(u)^2 \, \mathrm{d}u \right)^k & \text{if } m = 2k \text{ is even} \\ 0 & \text{if } m \text{ is odd} \end{cases} \quad (3.29)$$

where

$$\mu_\phi^{\mathrm{USp}} = \int_{-\infty}^\infty \phi(x) \, \mathrm{d}x - \int_0^1 \widehat{\phi}(u) \, \mathrm{d}u. \quad (3.30)$$

This theorem agrees perfectly with part (ii) of Theorem 1. By the work in Section 1, this theorem implies the n-level densities of the zeros $L(s, \chi_d)$ are the same as the n-level densities of the symplectic group (within a restricted range), a result first derived by Mike Rubinstein [8], though this approach avoids the combinatorial sieving necessary there. Indeed, also by the work in Section 1, one can derive this theorem immediately from Rubinstein's result.

3.3 L–functions arising from cuspidal newforms: An orthogonal example

Let $H_k^\star(N)$ be the set of all holomorphic cusp forms which are newforms of weight k and level N.

Let the Fourier coefficients of an $f \in H_k^\star(N)$ be $a_f(n)$, and let $\lambda_f(n) = a_f(n) n^{-(k-1)/2}$. The L-function associated with f is

$$L(s, f) = \sum_{n=1}^\infty \lambda_f(n) n^{-s}. \quad (3.31)$$

It satisfies the functional equation mapping $s \longrightarrow 1 - s$, and the sign in the functional equation is $\epsilon_f = \pm 1$.

Therefore $H_k^\star(N)$ splits into two disjoint subsets, $H_k^+(N) = \{f \in H_k^\star(N) : \epsilon_f = +1\}$ and $H_k^-(N) = \{f \in H_k^\star(N) : \epsilon_f = -1\}$.

For $\widehat{\phi} \in C_c^\infty(\mathbb{R})$, define the smooth counting function

$$N_\phi(f) = \sum_{\gamma_f} \phi\left(\frac{\log(k^2 N)}{2\pi}\gamma_f\right). \tag{3.32}$$

Here, γ_f runs through the non-trivial zeros of $L(s, f)$. We rescale the zeros by $\log(k^2 N)$ as this is the order of the number of zeros with imaginary part less than a large absolute constant.

We define the average over either $H_k^+(N)$ or $H_k^-(N)$ by

$$\langle N_\phi(f) \rangle_\pm := \frac{1}{|H_k^\pm(N)|} \sum_{f \in H_k^\pm(N)} N_\phi(f). \tag{3.33}$$

We let $N \to \infty$ through the primes, with k held fixed.

Theorem 4 (Hughes and Miller [3]). If $\operatorname{supp}\widehat{\phi} \subseteq (-\frac{1}{m}, \frac{1}{m})$ then the mth moment of $N_\phi(f)$, when averaged over the elements of either $H_k^+(N)$ or $H_k^-(N)$, converges to the mth moment of a normal distribution with mean

$$\mu = \widehat{\phi}(0) + \frac{1}{2}\int_{-1}^{1} \widehat{\phi}(y)\,\mathrm{d}y \tag{3.34}$$

and variance

$$\sigma^2 = 2\int_{-1/2}^{1/2} |u|\widehat{\phi}(y)\,\mathrm{d}y. \tag{3.35}$$

This result is in complete agreement with part (iii) of Theorem 1, and also shows that the n-level densities do, as expected, agree with the n-level densities for the special orthogonal group. However, in this case we are able to go beyond the diagonal, and show that Gaussian behaviour ceases at the point predicted by random matrix theory.

Theorem 5 (Hughes and Miller [3]). Let $\mu_\pm = \langle N_\phi(f) \rangle_\pm$, and let $S(x) = \frac{\sin \pi x}{\pi x}$. For $n \geq 2$, let $\operatorname{supp}(\widehat{\phi}) \subset (-\frac{2}{2n-1}, \frac{2}{2n-1})$. Then as $N \to \infty$ through the primes, if $n = 2m$ is an even integer,

$$\lim_{N \to \infty} \left\langle (N_\phi(f) - \mu_\pm)^{2m} \right\rangle_\pm =$$

$$\frac{(2m)!}{2^m m!} \left(2\int_{-1}^{1} \widehat{\phi}(x)^2|x|\,\mathrm{d}x\right)^m \mp 2^{2m-1}\left[\int_{-\infty}^{\infty} \phi(x)^{2m} S(2x)\,\mathrm{d}x - \frac{1}{2}\phi(0)^{2m}\right], \tag{3.36}$$

and if $n = 2m + 1$ is an odd integer, then

$$\lim_{N \to \infty} \left\langle (N_\phi(f) - \mu_\pm)^{2m+1} \right\rangle_\pm \qquad (3.37)$$

$$= \pm 2^{2m} \left[\int_{-\infty}^\infty \phi(x)^{2m+1} S(2x) \mathrm{d}x - \frac{1}{2}\phi(0)^{2m+1} \right].$$

In particular, as the Fourier Transform of $S(2x)$ is $\frac{1}{2}\mathbb{1}_{\{|x| \le 1\}}$, the third centered moment is zero if $\operatorname{supp} \widehat{\phi} \subset (1/3, 1/3)$, but non-zero if the support exceeds this interval. These non-Gaussian results still agree with the random matrix results for $Z_\phi(U)$.

Acknowledgements

Apart from the work with Steven Miller at the end, all the work presented in this talk is joint with Zeév Rudnick, and was carried out in 2001–2002 while funded by the EC TMR network "Mathematical Aspects of Quantum Chaos", EC-contract no HPRN-CT-2000-00103. This talk was given on 8th April 2004 at the *School on Recent Perspectives in Random Matrix Theory and Number Theory*, held at the Isaac Newton Institute, Cambridge, where my stay was funded by EPSRC Grant N09176

References

[1] P. Diaconis and M. Shahshahani, "On the eigenvalues of random matrices", *J. Appl. Probab.* **31A** (1994) 49–62.

[2] A. Fujii, "Explicit formulas and oscillations", *Emerging Applications of Number Theory*, D.A. Hejhal, J. Friedman, M.C. Gutzwiller, A.M. Odlyzko, eds. (Springer, 1999) 219–267.

[3] C.P. Hughes and S.J. Miller, "Low-lying zeros of L–functions with orthogonal symmetry" *Duke Math. J.* **136**(1) (2007) 115–172.

[4] C.P. Hughes and Z. Rudnick, "Linear statistics for zeros of Riemann's zeta function", *C. R. Acad. Sci. Paris* Ser I **335** (2002) 667–670.

[5] C.P. Hughes and Z. Rudnick, "Mock Gaussian behaviour for linear statistics of classical compact groups", *J. Phys. A.* **36** (2003) 2919–2932.

[6] C.P. Hughes and Z. Rudnick, "Linear statistics for low-lying zeros of L–functions", *Quart. J. Math. Oxford* **54** (2003) 309–333.

[7] N.M. Katz and P. Sarnak, *Random Matrices, Frobenius Eigenvalues, and Monodromy*, (AMS Colloquium Publications, 1999).

[8] M. Rubinstein, *Low-lying zeros of L–functions and random matrix theory*, Duke Math. J. **109** (2001) 147–181.

[9] A. Soshnikov, *Central limit theorem for local linear statistics in classical compact groups and related combinatorial identities*, Ann. Probab. **28** (2000) 1353–1370.

[10] M. Stolz, "On the Diaconis-Shahshahani method in random matrix theory", PhD Thesis, (University of Tuebingen, 2004) .

hughes@aimath.org

Some specimens of L-functions

Philippe Michel [*]

Contents

1 Introduction   360

2 Preamble: first contact with L-functions   361

  2.1 What to expect from an L-function ? 361

  2.2 L-functions of modular forms 368

    2.2.1 Theta-series 368

    2.2.2 Modular forms and Hecke L-functions 370

    2.2.3 Automorphic forms 372

[*]P. M. was supported by the Institut Universitaire de France and by the ACI
"Arithmétique des fonctions L".

3　Artin *L*-functions　　　　　　　　　　　　　　　　　　**373**

　3.1　Algebraic number theory background 374

　　3.1.1　The ring of integers 374

　　3.1.2　Ideals 374

　　3.1.3　Decomposition of primes 375

　3.2　Galois representations 376

　　3.2.1　The frobenius 377

　　3.2.2　A relative version of the frobenius 378

　　3.2.3　Example: cyclotomic fields 378

　　3.2.4　Extending a representation over the absolute Galois group 379

　3.3　Chebotareff density theorem 379

　　3.3.1　Sketch of the proof of Chebotareff density theorem . . . 381

　3.4　Artin *L*-functions 381

　　3.4.1　Continuation of the proof of Chebotareff's theorem . . . 383

　　3.4.2　Properties of Artin *L*-functions 383

4　*L*-functions of elliptic curves　　　　　　　　　　　　　　　**386**

　4.1　Background on elliptic curves 386

　　4.1.1　Elliptic curves as algebraic curves 386

　　4.1.2　The discriminant 386

　　4.1.3　Projective elliptic curves 387

　　4.1.4　The group law on an elliptic curve 388

　　4.1.5　Isogenies 389

　　4.1.6　Elliptic curves over the complex numbers 391

　　4.1.7　Elliptic curves over finite fields 393

　　4.1.8　Hasse's bound and character sums 394

　　4.1.9　Proof of Hasse's Theorem 395

　　4.1.10　Elliptic curves over the rationals 396

　4.2　The *L*-function of an elliptic curve 398

4.2.1   *L*-functions of elliptic curves over finite fields 398

4.2.2   *L*-functions of elliptic curves over the rationals 399

4.2.3   Reduction of an elliptic curve 399

4.3   The Galois representation attached to an elliptic curve 401

4.3.1   The Tate module attached to the roots of unity 401

4.3.2   Action of the ℓ-adic integers 402

4.3.3   Action of the Galois group 403

4.3.4   The Tate module of an elliptic curve 404

4.3.5   The Galois action 405

4.3.6   The Weil pairing 405

4.3.7   Comparison with the Tate module associated to roots of
        unity 406

4.3.8   *L*-function of an elliptic curve as an Artin *L*-function . . 406

4.4   The Sato-Tate conjecture and a Theorem of Serre 407

5   *L*-functions over function fields   412

5.1   Function fields 413

5.2   Elliptic curves over function fields 415

5.3 ℓ-adic sheaves and their *L*-functions 417

5.4   Sato/Tate laws and Random Matrices 422

5.4.1   Deligne's equidistribution Theorem 422

5.4.2   Connection with RMT 422

1   Introduction

These notes are an expanded version of a series of lectures given at the Isaac Newton Institute for Mathematical Sciences, Cambridge, on the occasion of the Spring School "Recent Perspectives in Random Matrix Theory and Number Theory" in April 2004. The main aim of these lectures was to give a general overview of L-functions to a mixed audience of participants working in various branches of mathematics and theoretical physics.

In particular, these lectures, which are addressed to non-specialists in number theory, do not explain the many methods and techniques that are involved in the study of the analytical properties of L-functions and of their zeros. Instead our objective is to present in some detail the basic theoretical material for the construction of several specimens of L-functions. Note that our presentation is slightly unusual as we have to chosen to present things from the perspective of Artin L-functions (that is L-functions associated to – possibly ℓ-adic, rather than complex – representations of some Galois group): our choice had been dictated by our belief that an important part of the connections between Random Matrices and L-function is rooted into the so-called *Sato/Tate laws* and that such laws are more easily motivated in the framework of Galois representations. Such belief is strongly supported by the work of Katz/Sarnak relating unconditionally Random Matrices to zeros of L-functions over function fields.

The lectures are organized as follows: the first lecture is a presentation of some of the most basic, yet important, analytic properties that are to be expected from a generic L-function. There we "only" reach the level of the Generalised Riemann Hypothesis and do not touch with the most precise conjectures available today; in particular we never discuss the many existing conjectures which predict the equality between limits of several natural statistics formed out of L-functions with their counterparts coming from Random Matrix theory (say, the pair or higher correlation conjecture for zeros, the conjectures on moments of critical values etc...). In the next lecture, we describe the construction of Artin L-functions (constructed out of a Galois representation). In particular, we give some rough algebraic number theoretic background and explain the structure of the proof of the *Chebotareff density theorem* which is one of the simplest instances of a *Sato/Tate law*. The third lecture is devoted to the presentation of the L-function associated to an elliptic curve (defined over $\mathbf{Q}$). After presenting again some general material, we explain that the associated L-function, $L(E, s)$, is in fact the Artin L-function of some Galois (ℓ-adic) representation; on this occasion we have tried to make accessible to non-specialists the notion of Tate module and of its associated Galois representation. This connection with Artin L-functions is illustrated by an important application (due to Serre) of the Chebotareff density theorem to bound the frequency of vanishing for the p-th coefficient of the $L(E, s)$: in this application, the Sato/Tate laws for complex Galois representations (that is Chebotareff's theorem) are used to establish unconditionally a consequence of the Sato/Tate conjecture for elliptic curves. The last chapter is concerned with the function field case where, thanks to the works of Deligne and Katz, several instances of Sato/Tate laws for ℓ-adic Galois representations can be obtained (through Deligne's equidistribution theorem). The material presented in these lectures is necessarily sketchy but the interested reader will find at the end of each lecture some more detailed references.

As said above, these lectures are directed essentially towards people working in domains like mathematical physic or probability theory; in particular, number theorists will learn little by reading these notes. My hope is that these lectures will help people become more familiar with several objects and concepts which are quite familiar in number theory (like the frobenius, elliptic curves, ℓ-adic representations and their associated L-functions) so that they can effectively combine them with methods imported from their own field of research. I would like to conclude this introduction by addressing my biggest thanks to the organizers of the school for their kind invitation, to the lively audience for its attention and to Nina Snaith and the referee for their careful reading of earlier versions of this text.

2 Preamble: first contact with L-functions

2.1 What to expect from an L-function ?

In this short section we give a presentation of what a "generic" L-function should be and which standard analytical properties are expected from it.

L-functions form an important class of analytic invariants which can be associated to many arithmetical objects. These invariants are Dirichlet series of a complex variable s, say

$$L(s) = \sum_{n \geq 1} \frac{\lambda_\pi(n)}{n^s},$$

and, moreover, they are built as products of local factors over the primes (this is this crucial feature which qualifies them for the term of L-function). More precisely, given π an adequate arithmetical object, one associates to π a collection of local objects indexed by the prime numbers

$$\{\pi_p\}_{p \in \mathcal{P}}$$

(note that these local data are often sufficient to characterize π completely). There is also an integer $d \geq 1$ such that to each local object π_p, one can associate a d-uple of complex numbers

$$\{(\alpha_{\pi,1}(p), \ldots, \alpha_{\pi,d}(p))\}$$

(the local numerical parameters of π at p) and a *local Euler factor*

$$L(\pi_p, s) = \prod_{i=1}^{d} (1 - \frac{\alpha_{\pi,i}(p)}{p^s})^{-1}.$$

The global L-function of π is then the Euler product

$$L(\pi, s) = \prod_p L(\pi_p, s),$$

and d is the degree of the L-function . Expanding formally the above Euler product and using the fundamental theorem of arithmetic one obtains the formal expression of $L(\pi, s)$ as a Dirichlet series

$$L(\pi, s) = \prod_p L(\pi_p, s) = \sum_{n \geq 1} \frac{\lambda_\pi(n)}{n^s}, \tag{2.1}$$

where $n \to \lambda_\pi(n)$ is a multiplicative function (ie. $\lambda_\pi(mn) = \lambda_\pi(m)\lambda_\pi(n)$ whenever $(m, n) = 1$) such that for p a prime

$$\lambda_\pi(p) = \alpha_{\pi,1}(p) + \cdots + \alpha_{\pi,d}(p).$$

In all cases I know of, the numerical parameters are bounded polynomially in p so that both the Euler product and the Dirichlet series (2.1) converge absolutely in some right half-plane $\Re es \geq A$ where it defines an analytic function.

This definition is essentially formal and so far represents only a convenient way to "pack" together the collection of local numerical data attached to π.

The reason why L-functions are interesting comes from the fact that they have (or are expected to have) nice extra analytic properties from which one can deduce, amongst other things, important information about the average value of the Dirichlet coefficients $\lambda_\pi(n)$ as n ranges over the full set of integers or over the set of prime numbers (if $n = p$ is a prime,

$$\lambda_\pi(p) = \sum_{i=1}^{d} \alpha_{\pi,i}(p)$$

represents the "trace" of the local data at p). These extra analytic properties, which we are about to describe, arise from the fact that the sequence of the d-uples

$$\{(\alpha_{\pi,1}(p), \ldots, \alpha_{\pi,d}(p))\}_{p \in \mathcal{P}}$$

is not just a set of random data, but comes from a global object. In these lectures we will not discuss how such properties are established but we refer to the bibliography given the end of this Chapter.

The first manifestation of this principle is that L-functions have an extension beyond their trivial region of analyticity: in order to perform analytic continuation it is first necessary to complete the collection of local factors at the various primes $p \in \mathcal{P}$ by a local factor "at infinity" $L(\pi_\infty, s)$ given by a product of Gamma functions

$$L(\pi_\infty, s) = \prod_{i=1}^{d} \Gamma_{\mathbf{R}}(s + \mu_{\pi,i}(\infty)), \text{ with } \Gamma_{\mathbf{R}}(s) = \pi^{-s/2}\Gamma(\frac{s}{2}).$$

With this complementary local factor, one expects the following:

Meromorphic continuation and Functional equation (FCT). *The L-function $L(\pi, s)$ has meromorphic continuation to the whole complex plane and satisfies a functional equation of the following form: there exists some integer $w = w_\pi \geqslant 0$, some complex number ω_π of modulus one and some integer $Q_\pi \geqslant 1$ such that, setting*

$$\Lambda(\pi, s) = Q_\pi^{\frac{s}{2}} L_\infty(\pi, s) L(\pi, s)$$

one has

$$\Lambda(\pi, s) = \omega_\pi \overline{\Lambda(\pi, \overline{1 + w_\pi - s})}.$$

Moreover, the poles of $\Lambda(\pi, s)$ are finite in number ($\leqslant 2d$) and are all located on the two lines

$$\mathfrak{Res} = \frac{1 + w_\pi \pm 1}{2},$$

and away from such poles $\Lambda(\pi, s)$ is of order 1.

Definition. *The integer w_π is called the* weight *of π, ω_π the* root number *and Q_π the* conductor *of π. Following Iwaniec/Sarnak, the* analytic conductor *of $L(\pi, s)$ is the function of $\mathbf{R}$ given as the product*

$$Q(\pi, t) = Q_\pi \prod_{i=1\ldots d} (1 + |\mu_{\pi,i}(\infty) + it|).$$

Remark 1. The weight w_π has the following natural interpretation: in general one expects (and one can prove on some occasions) that the $\alpha_{\pi,i}(p)$ satisfy the bound

$$|\alpha_{\pi,i}(p)| \leqslant p^{w_\pi/2}. \tag{2.2}$$

Such a bound is called a *Ramanujan/Petersson type bound* ; moreover this is expected to be optimal, in the sense that for almost all p the inequality above is an equality. However, it is more often the case that, instead of the pointwise bound above, an averaged version is available unconditionally: for any $\varepsilon > 0$ one has[1]

$$(Q(\pi, 0)X)^{-\varepsilon} X^{1+w_\pi} \ll_{\varepsilon,d} \sum_{n \leqslant X} |\lambda_\pi(n)|^2 \ll_{\varepsilon,d} (Q(\pi, 0)X)^\varepsilon X^{1+w_\pi}.$$

In any case this implies that $1 + \frac{w_\pi}{2}$ is the abscissa of absolute convergence of the Dirichlet series $L(\pi, s)$.

Remark 2. At that level imposing the weight to be an integer seems pretty arbitrary, for if $L(\pi, s)$ is an L-function as above then for any $w \in \mathbf{R}$, the shifted L-function $L(\pi, s + w/2) =: L(\pi_{w,s})$ is an L-function with genuine functional equation and weight $w_\pi - w$. In fact it is quite common in analytic number theory to normalize L-functions using such a shift as to have weight 0. On the other hand, in arithmetic geometry, L-functions often arise with an integral weight; this weight is natural in that the coefficients $\lambda_\pi(p)$ satisfy nice integrality properties which of course do not persist after a shift.

To get some feeling on how delicate FCT should be in general, consider the following example: given $q \geqslant 1$ a prime, let $\chi : (\mathbf{Z}/q\mathbf{Z})^\times \to \{1, -1\}$ be the

---

[1] See the Appendix of the lectures of D.W. Farmer, page 185, for a discussion of the $\ll$ notation.

unique non-trivial (Legendre) character of order 2; the L-function associated to χ

$$L(\chi, s) = \prod_{p \nmid q}(1 - \frac{\chi(p)}{p^s})^{-1}.$$

Then FCT is known in that case (see below): $L(\chi, s)$ has holomorphic continuation to $\mathbf{C}$ and satisfies a functional equation of the above shape with weight 0 and conductor q. On the other hand, consider a sequence of independent random variables, one for each prime $p \nmid q$,

$$X_p : \Omega \to \{1, -1\},$$

each taking value $+1$ with probability $1/2$; then, the probability that the random Euler product

$$L(\omega, s) = \prod_{p \nmid q}(1 - \frac{X_p(\omega)}{p^s})^{-1}$$

admits analytic continuation to $\mathbf{C}$ or satisfies a functional equation of the above shape, is zero.

The next very important properties expected from such L-functions is a precise location of their zeros: in many occasions (but not always) one is capable of proving that (at least on average) the $\lambda_\pi(p)$ are bounded in absolute value by $dp^{w_\pi/2}$ which immediately implies that $L(\pi, s)$ does not vanish in the half-plane $\Re es > 1 + \frac{w_\pi}{2}$. Likewise, by FCT, this implies the non-vanishing of $\Lambda(\pi, s)$ in the union of the half-planes $\Re es > 1 + \frac{w_\pi}{2}$ and $\Re es < \frac{w_\pi}{2}$. Hence the zeros of $\Lambda(\pi, s)$ are all located in the critical strip

$$\frac{w_\pi}{2} \leqslant \Re es \leqslant 1 + \frac{w_\pi}{2}.$$

One can also often prove that $\Lambda(\pi, s)$ is a function of order one: it follows that $\Lambda(\pi, s)$ has infinitely many zeros and that one has the Hadamard factorization formula

$$\Lambda(\pi, s) = A(\pi)e^{B(\pi)s} \prod_{\rho'}(s - \rho')^{-1} \prod_{\rho}(1 - \frac{s}{\rho})e^{s/\rho},$$

where ρ (resp. ρ') ranges over the zeros (resp. the poles) of $\Lambda(\pi, s)$ counted with multiplicity. Taking the logarithmic derivative on both sides gives the identity

$$-\frac{L'}{L}(\pi, s) = \sum_{\rho'} \frac{1}{s - \rho'} - \sum_{\rho}(\frac{1}{s - \rho} + \frac{1}{\rho}) + \frac{L'}{L}(\pi_\infty, s) + \frac{1}{2}\log Q_\pi - B(\pi).$$

One the other hand for s in the zone of absolute convergence of $L(\pi, s)$ (usually for $\Re e s > 1 + \frac{w_\pi}{2}$) one can use the expression of $L(\pi, s)$ as an Euler product (2.1) to obtain

$$-\frac{L'}{L}(\pi, s) = \sum_p \frac{\lambda_\pi(p)\log(p)}{p^s} + \sum_{k \geq 2}\sum_p \frac{\Lambda_\pi(p^k)}{p^{ks}}$$

$$= \sum_{\rho'} \frac{1}{s - \rho'} - \sum_\rho (\frac{1}{s - \rho} + \frac{1}{\rho}) + \frac{L'}{L}(\pi_\infty, s) + \frac{1}{2}\log Q_\pi - B(\pi), \quad (2.3)$$

where

$$\Lambda_\pi(p^k) = (\sum_{i=1\ldots d} \alpha_{\pi,i}(p)^k)\log p$$

is the Von-Mangolt type function. One can give a sense to this formula even for s in the critical strip as an identity between two distributions: the outcome is the so-called *Weil explicit formulae*. Note that the Ramanujan/Petersson bound (2.2) (or a weaker version of it) implies that the $\sum_{k \geq 2}$-sum is very well controlled, so that (2.3) provides a direct link between

1. The distribution of the coefficients $\lambda_\pi(p)$ as p ranges over the primes,

2. The localization of the zeros of $L(\pi, s)$ within the critical strip.

Concerning point 2. the highest expectation is the *Generalised Riemann Hypothesis*

Generalised Riemann Hypothesis (GRH). *The zeros of $\Lambda(\pi, s)$ are all located on the critical line $\Re e s = \frac{1+w_\pi}{2}$.*

Consequently one has

$$\sum_{p \leq X} \lambda_\pi(p)\log p = \sum_{\rho'} X^{\rho'} + O_\pi(\log^2(X)X^{\frac{1+w_\pi}{2}})$$

where ρ' ranges over the (finite, possibly empty) set of poles of $\Lambda(\pi, s)$ located on the line $\Re e s = 1 + \frac{w_\pi}{2}$ and counted with multiplicity.

One is probably very far from proving GRH anytime soon, but fortunately, at least some non-trivial zero free region is available unconditionally: in many cases one is capable of proving the

Generalised Prime Number Theorem (GPNT). *$L(\pi, s)$ does not vanish on the line $\Re e s = 1 + \frac{w_\pi}{2}$, and consequently, one has for $X \to +\infty$*

$$\sum_{p \leq X} \lambda_\pi(p)\log p = \sum_{\rho'} X^{\rho'} + o_\pi(X^{1+\frac{w_\pi}{2}})$$

where ρ ranges over the (finite in number) poles of $\Lambda(\pi, s)$ located on $\Re es = 1 + \frac{w_\pi}{2}$ and counted with multiplicity.

In most known instances of the GPNT, a larger zero-free region is available.[2]

Hadamard/de la Vallée-Poussin zero free region (HdVP). *There is a positive constant c_d depending on d only, such that for $t \in \mathbf{R}$, $L(\pi, s)$ does not vanish for s on the segment*

$$\Im ms = t, \quad \Re es \in [1 + \frac{w_\pi}{2} - \frac{c_d}{\log Q(\pi, t)}, 1 + \frac{w_\pi}{2}].$$

Besides the above expected analytical properties of a typical L-function, some extra information of a more arithmetical nature is often expected from the local behavior of $L(\pi, s)$ in the neighborhood of some special[3] points (such points usually depend on the parity of w_π and on the numerical parameters at ∞): indeed the order of vanishing or the (regularized) values of $L(\pi, s)$ at such point is expected to be related to deep global arithmetic invariants of π. Typical examples are given by the Dirichlet class number formula or by the Birch/Swinnerton-Dyer conjecture which are now to be understood as special cases of wider conjectures of Beilinson.

In view of these connections, it becomes useful to have good quantitative information about the behavior of such L-values, in particular upper (and whenever possible) lower bounds. In many interesting situations the special point is located precisely in the critical strip and even on the critical line and the localization of zeros have an influence on the size of such an L-value. In particular one can often show the GRH implies the following

Generalised Lindelöf Hypothesis (GLH). *For $\Re es = \frac{1+w_\pi}{2}$ and any $\varepsilon > 0$, one has the bound*

$$L(\pi, s) \ll_{d,\varepsilon} Q(\pi, \Im ms)^\varepsilon.$$

If only FCT is available, one can often prove the following weaker statement

Convexity Bound (CB). *For $\Re es = \frac{1+w_\pi}{2}$ and any $\varepsilon > 0$,*

$$L(\pi, s) \ll_{d,\varepsilon} Q(\pi, \Im ms)^{1/4+\varepsilon}$$

---

[2] Here we have deliberately ignored the issue regarding a possible Landau/Siegel zero.

[3] The usual terminology is "critical points" but in the present context this might be misleading as such points may well be on the border or even out of the critical strip.

and formulate the following problem which is some cases can be proven without recourse to GRH

Subconvexity Problem (ScP). *Find $\delta > 0$ such that for $\Re es = \frac{1+w_\pi}{2}$ and any $\varepsilon > 0$,*

$$L(\pi, s) \ll_{d,\varepsilon} Q(\pi, \Im ms)^{1/4-\delta+\varepsilon}.$$

2.2   *L*-functions of modular forms

All the examples L-functions I know, for which some of the expected analytical properties described above are known, are in one way or another connected with L-functions of automorphic forms. Automorphic forms are functions living on some symmetric spaces which satisfy some invariance properties with respect to the action of a discrete group of isometries (periodic functions of period one on **R** are a prototypical example). The associated L-functions are related to automorphic forms via some integral transform and in particular their analytical properties (like FCT) are obtained via harmonic analysis and by exploiting the invariance properties of the form. For the rest of this lecture, we describe a few cases of L-functions obtained along this principle.

2.2.1   Theta-series

Riemann's zeta function

$$\zeta(s) = \sum_{n \geqslant 1} \frac{1}{n^s}$$

does not make exception to this rule, indeed in order to establish FCT for $\zeta(s)$, Riemann introduced the theta-series

$$\theta(z) = \sum_{n \in \mathbf{Z}} e(n^2 z), \ e(z) = \exp(2\pi i z)$$

which is holomorphic in the upper-half plane

$$\mathbf{H} = \{z \in \mathbf{C}, \ \Im mz > 0\}.$$

Indeed, one has

$$\Gamma_{\mathbf{R}}(s)\zeta(s) = \int_0^\infty \frac{1}{2}(\theta(iy/2) - 1)y^{s/2}\frac{dy}{y}.$$

It follows from the Poisson summation formula that θ satisfies a functional equation

$$\theta(-1/2iy) = y^{1/2}\theta(iy/2). \tag{2.4}$$

Splitting the y-integral into two ranges $[0, 1]$ and $[1, \infty]$, and using (2.4) on the first portion and making the change of variable $y \to 1/y$ yields the analytic continuation of

$$\xi(s) := \frac{s(s-1)}{2}\Gamma_{\mathbf{R}}(s)\zeta(s)$$

to $\mathbf{C}$; moreover one has the functional equation

$$\xi(s) = \xi(1-s).$$

The same kind of result holds for Dirichlet character L-functions: for $\chi :$ $(\mathbf{Z}/q\mathbf{Z})^{\times} \to \mathbf{C}^{\times}$ a primitive character of modulus $q > 1$, which we extend to a multiplicative function on $\mathbf{Z}$, periodic of period q, by setting

$$\chi(n) = \begin{cases} \chi(n(\mathrm{mod} q)) & for\ (n, q) = 1 \\ 0 & else, \end{cases}$$

the L-function associated with χ is the L-series

$$L(\chi, s) := \prod_p (1 - \frac{\chi(p)}{p^s})^{-1} = \sum_{n \geqslant 1} \frac{\chi(n)}{n^s}.$$

This time one considers the theta series

$$\theta_\chi(z) = \sum_{n \in \mathbf{Z}} \chi(n)n^\nu e(nz)$$

with $\nu = (1 - \chi(-1))/2$. Let $G(\chi) = \sum_{x(\mathrm{mod} q)} \chi(x)e(x/q)$ denote the Gauss sum. Then for χ a primitive character, one has for any n

$$\chi(n) = \frac{1}{G(\overline{\chi})} \sum_{x(q)} \overline{\chi}(x)e(nx/q).$$

Using this formula and the Poisson summation formula, one can infer, for $y > 0$, the functional equation

$$\theta_\chi(-1/2qiy) = i^{-\nu}\frac{G(\chi)}{\sqrt{q}}y^{1/2+\nu}\theta_{\overline{\chi}}(iy/2q),$$

which by analytic continuation extends to $z \in \mathbf{H}$

$$\theta_\chi(-1/2qz) = i^{-\nu}\frac{G(\chi)}{\sqrt{q}}(-iz)^{1/2+\nu}\theta_{\overline{\chi}}(z/2q).$$

Using this functional equation and the identity

$$\Gamma_{\mathbf{R}}(s + \nu)L(\chi, s) = \frac{1}{2}\int_0^\infty \theta_\chi(iy/2q)(\frac{y}{q})^{s/2}\frac{dy}{y}$$

one infers the analytic continuation of $L(\chi, s)$ to the complex plane and the FCT:

$$q^{s/2}\Gamma_{\mathbf{R}}(s + \nu)L(\chi, s) = i^{-\nu}\frac{G(\chi)}{\sqrt{q}}q^{(1-s)/2}\Gamma_{\mathbf{R}}(1 - s + \nu)L(\overline{\chi}, 1 - s).$$

2.2.2   Modular forms and Hecke L-functions

The theta series presented above are special cases of modular forms on the upper half-plane: let us recall that the group $SL(2, \mathbf{R})$ acts on $\mathbf{H}$ by fractional linear transformations

$$\gamma.z = \frac{az + b}{cz + d}, \gamma = \begin{pmatrix} a & b \\ c & d \end{pmatrix} \in GL(2, \mathbf{R}), \ \det \gamma > 0$$

and this action induces various actions for various discrete subgroups Γ of $SL(2, \mathbf{R})$ on the spaces of functions on $\mathbf{H}$. A class of subgroups important in arithmetic, are the Hecke congruence subgroups of $SL(2, \mathbf{Z})$ of level q

$$\Gamma_0(q) = \{\begin{pmatrix} a & b \\ c & d \end{pmatrix} \in SL(2, \mathbf{Z}), \ c \equiv 0(q)\}.$$

For instance the theta-series θ_χ can be shown to satisfy the automorphy relations: for any $\gamma = \begin{pmatrix} a & b \\ c & d \end{pmatrix} \in \Gamma_0(4q^2)$

$$\theta_\chi(\gamma z) = \chi(d)\overline{\varepsilon}_d(\frac{c}{d})(cz + d)^{1/2}\theta_\chi(z),$$

where $(\frac{c}{d})$ denote the (extended) Legendre-Jacobi symbol and (as d is odd) $\varepsilon_d = 1$ or i according whether $d \equiv 1$ or $3 \bmod 4$. These are examples of modular forms of weight $1/2 + \nu$ and level $4q^2$.

Another important class of modular forms (which are not unrelated to the previous ones) are modular forms of integral weight k: for $k \geqslant 1$, $q \geqslant 1$ and χ a Dirichlet character of modulus q, a modular form of weight k, level q and nebentypus χ is an holomorphic function $f(z)$ on $\mathbf{H}$ which satisfies the following automorphy relations: for any $\gamma = \begin{pmatrix} a & b \\ c & d \end{pmatrix} \in \Gamma_0(q)$

$$f(\gamma z) = \chi(d)(cz + d)^k f(z),$$

and which satisfies some growth condition as z approaches the rationals. The space of such forms is denoted by $M_k(q, \chi)$ is finite dimensional and contains the space of cusp forms $S_k(q, \chi)$ which are the square integrable modular forms with respect to the Petersson inner product

$$\langle f, g \rangle = \int_{\Gamma_0(q)\backslash \mathbf{H}} \overline{f}(z) g(z) y^k \frac{dxdy}{y^2}, \quad (with \ z = x + iy).$$

For f a cusp form, using the periodicity relation $f(z+1) = f(z)$, one sees that f admits a Fourier expansion

$$f(z) = \sum_{n \geqslant 1} a_n(f) e(nz).$$

The Dirichlet series associated to f is

$$L(f, s) = \sum_{n \geqslant 1} \frac{a_n(f)}{n^s}.$$

It is absolutely convergent in some right half-plane and is related to $f(z)$ via the formula

$$\Lambda(f, s) := (2\pi)^{-s} \Gamma(s) L(f, s) = \int_0^\infty f(iy) y^s \frac{dy}{y}.$$

The analytic continuation of $\Lambda(f, s)$ and its functional equation follow from the properties of f and of the group action. Indeed the matrix $w_q := \begin{pmatrix} 0 & -1 \\ q & 0 \end{pmatrix}$ normalizes $\Gamma_0(q)$ and this implies that the function $g(z)$ given by

$$g(z) = q^{k/2}(qz)^{-k} f(w_q z)$$

belongs to $S_k(q, \overline{\chi})$. This implies that $\Lambda(f, s)$ has analytic continuation to $\mathbf{C}$ and satisfies

$$\Lambda(f, s) = i^k \Lambda(g, k - s).$$

This however does not qualify $L(f, s)$ as an L-function for there is so far no Euler product. This can be accomplished but only for a very special subset of cusp forms, namely the *primitive* forms; note that the restriction to such forms is not an essential issue since in an appropriate sense, primitive forms generate the whole space of cusp forms. The key fact for the existence of Euler products is the existence of a rich (and explicit) commutative algebra (the Hecke algebra) of linear operators (the Hecke operators) acting on $S_k(q, \chi)$,

$$\mathbf{T} = \{T_n, \ n \geqslant 1\}.$$

These operators satisfy a muliplicativity relation, and the Hecke operators of prime power index satisfy a two term recurrence relation:

$$T_{mn} = T_m T_n \ for \ (m,n) = 1, \ T_{p^{n+1}} = T_p T_{p^n} - \chi(p) p^{k-1} T_{p^{n-1}};$$

in particular one has the formal identity

$$\sum_{n \geqslant 1} \frac{T_n}{n^s} = \prod_p (1 - \frac{T_p}{p^s} + \frac{\chi(p) p^{k-1}}{p^{2s}})^{-1}. \qquad (2.5)$$

Primitive forms are eigenforms of all Hecke operators and given such a form f, if we denote by $\lambda_f(n)$ the corresponding eigenvalue for the Hecke operator one has the relation

$$a_n(f) = \lambda_f(n) a_1(f),$$

and in particular

$$L(f,s) = a_1(f) \prod_p (1 - \frac{\lambda_f(p)}{p^s} + \frac{\chi(p) p^{k-1}}{p^{2s}})^{-1}.$$

2.2.3   Automorphic forms

The notion of modular form can be widely generalized by considering functions living on an hermitian symmetric space which are eigenforms under some action of a discrete subgroup of isometries and which are eigenforms of a commutative ring of elliptic differential operators which commute with the group action (this condition comes in replacement of the holomorphicity condition). For instance, on **H**, Maass was the first to replace the condition of being holomorphic by the (weaker) condition of being an eigenform of the hyperbolic Laplacian

$$\Delta = -y^2 (\frac{\partial^2}{\partial^2 x} + \frac{\partial^2}{\partial^2 y}),$$

and this gave rise to the notion of Maass forms. For many of these generalized modular forms, it is possible to associate one or several Dirichlet series, but the construction would involve many non-canonical choices and more importantly would be very unlikely to have an Euler product.

On the other hand, if the subgroup of isometries is *arithmetic* in some appropriate sense, Hecke operators exist and they may be used to "select" the analog of primitive forms from which one can construct a genuine L-function. Quite remarkably, by using harmonic analysis on these higher rank spaces and the

Euler product structure, it is then possible to establish, for these *L*-functions, many of the analytic properties mentioned above: FCT, GPNT, HdVP, CB. So a whole supply of new *L*-functions of any degree become available to play with. More importantly, it is believed that all "reasonable" *L*-functions should be automorphic (that is, should arise in that way). An example of this is the *L*-functions presented in the next lectures, which carry a strong arithmetical content.

References

[1] Harold Davenport, *Multiplicative number theory*, 3rd ed., Graduate Texts in Mathematics, **74**, Springer-Verlag, New York, 2000. Revised and with a preface by Hugh L. Montgomery.

[2] Henryk Iwaniec, *Topics in classical automorphic forms*,Graduate Studies in Mathematics, **17**,American Mathematical Society,Providence, RI,1997.

[3] Henryk Iwaniec and Emmanuel Kowalski, *Analytic number theory*, American Mathematical Society Colloquium Publications, **53**, American Mathematical Society, Providence, RI, 2004.

3   Artin *L*-functions

Let $\mathbf{Q} \subset K \subset \mathbf{C}$ be a finite algebraic extension of $\mathbf{Q}$ contained in $\mathbf{C}$ of degree n say: K is a subfield of the field of complex numbers which are roots of a rational polynomial, and moreover as a $\mathbf{Q}$ vector space $\dim_{\mathbf{Q}} K = n$. The Galois group of K over $\mathbf{Q}$, $\mathrm{Gal}(K/\mathbf{Q})$ say, is by definition the group of ($\mathbf{Q}$-linear) field-homomorphisms of K into itself. The Galois group is a subset of $\mathrm{Hom}_{\mathbf{Q}}(K, \mathbf{C})$, the set of ($\mathbf{Q}$-linear) field homomorphisms of K into $\mathbf{C}$ which has n elements and one says that K is *Galois* if the inclusion is an equality.

Given K a Galois extension and $\rho : \mathrm{Gal}(K/\mathbf{Q}) \rightarrow GL(V)$ a(n irreducible) representation of $\mathrm{Gal}(K/\mathbf{Q})$ on a finite dimensional complex vector space V, one associates an Artin *L*-function $L(\rho, K/\mathbf{Q}, s)$. In this lecture, we present the construction of this *L*-function and its basic analytic properties.

3.1   Algebraic number theory background

We consider K a finite (not necessarily Galois) algebraic extension of $\mathbf{Q}$ of dimension n. A fundamental tool in the study of the algebraic and arithmetic properties of K is the action of K onto itself by multiplication. For any x in K one defines the endomorphism (of $\mathbf{Q}$-vector space) of multiplication by $[x]$:

$$[x] \; : \; \begin{array}{ccc} K & \to & K \\ y & \to & xy \end{array}.$$

Obviously $x \to [x]$ defines an injection of $K^{\times}$ into $\mathrm{Aut}_{\mathbf{Q}}(K)$; then the *trace*, $\mathrm{tr}_{K/\mathbf{Q}}(x)$, and the *norm*, $\mathrm{N}_{K/\mathbf{Q}}(x)$, of x are defined, respectively, as the trace and the determinant of $[x]$ and it is obvious from the definition that these functions are $\mathbf{Q}$ valued. In particular for $x \in \mathbf{Q}$, $\mathrm{tr}_{K/\mathbf{Q}}(x) = nx$ and $\mathrm{N}_{K/\mathbf{Q}}(x) = x^n$.

3.1.1   The ring of integers

The *ring of integers* of K, $\mathcal{O}_K \subset K$, is the set of elements which are annihilated by a monic polynomial with integral coefficients

$$\mathcal{O}_K = \{ x \in K, \; \exists P \in \mathbf{Z}[X], \; monic \; , \; s.t. \; P(x) = 0 \}.$$

That this set forms a ring is a nice exercise in elementary linear algebra using the endomorphism $[x]$. Moreover one can show that (as an abelian group under addition) $\mathcal{O}_K$ is a free $\mathbf{Z}$-module of rank n (ie. is isomorphic to $\mathbf{Z}^n$). This latter property comes from the fact that the bilinear form

$$\langle x, y \rangle = \mathrm{tr}_{K/\mathbf{Q}}(xy)$$

is non-degenerate. In particular, the *discriminant* of K is the determinant of the $n \times n$ matrix

$$\Delta_K = \det(\langle x_i, x_j \rangle_{i,j}) \neq 0.$$

where $\{x_1, \ldots, x_n\}$ is any $\mathbf{Z}$-basis of $\mathcal{O}_K$.

3.1.2   Ideals

Let us denote by $\mathrm{Id}(\mathcal{O}_K)$, the set of all non zero ideals of $\mathcal{O}_K$ (the $\mathcal{O}_K$-submodules of $\mathcal{O}_K$). Any non-zero ideal, I, is a free $\mathbf{Z}$-module of rank n. In particular $\mathcal{O}_K/I$ is finite and the *norm* of I, $N_{K/\mathbf{Q}}(I)$, is by definition $|\mathcal{O}_K/I|$. In particular, for $x \in \mathcal{O}_K^\times$ and $I = x.\mathcal{O}_K$ the ideal generated by x, one has $N_{K/\mathbf{Q}}(I) = |N_{K/\mathbf{Q}}(\mathcal{O}_K)|$.

The set $\mathrm{Id}(\mathcal{O}_K)$ forms a monoid (with $\mathcal{O}_K$ as the identity element) under the product of ideals

$$\mathfrak{a}.\mathfrak{b} := \text{Ideal generated by the products a.b, } a \in \mathfrak{a}, b \in \mathfrak{b},$$

and the norm is multiplicative with respect to this product:

$$N_{K/\mathbf{Q}}(\mathfrak{a}.\mathfrak{b}) = N_{K/\mathbf{Q}}(\mathfrak{a})N_{K/\mathbf{Q}}(\mathfrak{b}).$$

Recall that a non-zero ideal $\mathfrak{p} \in \mathrm{Id}(\mathcal{O}_K)$ is *prime* iff. $\mathcal{O}_K/\mathfrak{p}$ is integral: this is equivalent to $\mathfrak{p}$ being *maximal* since $\mathcal{O}_K/\mathfrak{p}$ is finite and a finite integral domain is a field.

We are now in a position to state the following fundamental result of Dedekind which is the analog for rings of integers of the fundamental theorem of arithmetic for $\mathbf{Z}$:

Theorem. *(Dedekind)* $\mathrm{Id}(\mathcal{O}_K)$ *is freely generated by the set of all prime ideals, or in other words every* $\mathfrak{a} \in \mathrm{Id}(\mathcal{O}_K)$ *admits a factorization of the form*

$$\mathfrak{a} = \prod_{\mathfrak{p}\ prime} \mathfrak{p}^{v_\mathfrak{p}(\mathfrak{a})}$$

where the $v_\mathfrak{p}(\mathfrak{a})$ are, almost always zero, integers; moreover the above decomposition is unique up to permutation.

In particular, the notion of a divisibility relation is available ($\mathfrak{a}|\mathfrak{b}$ iff. $v_\mathfrak{p}(\mathfrak{a}) \leqslant v_\mathfrak{p}(\mathfrak{b})$ for all $\mathfrak{p}$) which turns out to be equivalent to inclusion ($\mathfrak{a}|\mathfrak{b}$ iff. $\mathfrak{b} \subset \mathfrak{a}$).

3.1.3 Decomposition of primes

The above theorem reduces much of the study of integral ideals to that of the primes: one basic question is then "where to find the prime ideals ?".

If $\mathfrak{p}$ is a prime ideal then, as noted before, the quotient ring $\mathcal{O}_K/\mathfrak{p}$ is a finite field with $N(\mathfrak{p})$ elements; this field is called the *residue field at* $\mathfrak{p}$ and will be noted by $k_\mathfrak{p}$. In particular if p is the characteristic of the field $k_\mathfrak{p}$, then $p = 0$ in $k_\mathfrak{p} = \mathcal{O}_K/\mathfrak{p}$ hence $p \subset \mathcal{O}_K$, which means that $(p) = p.\mathcal{O}_K$ (the ideal generated by p) is divisible by $\mathfrak{p}$. So the prime ideals of $\mathcal{O}_K$ are exactly the primes appearing in the decomposition

$$p.\mathcal{O}_K = \prod_{\mathfrak{p} \; prime} \mathfrak{p}^{v_\mathfrak{p}(p\mathcal{O}_K)}$$

when p ranges over the prime of $\mathbf{Z}$.

To each prime are attached two important invariants to describe how a prime ideal occurs in the decomposition of a prime number: the ramification index and the residual degree.

If $v_\mathfrak{p}(p\mathcal{O}_K)$ is non zero, one says that $\mathfrak{p}$ divides p or lies above p, and $v_\mathfrak{p}(p\mathcal{O}_K)$ is called the *ramification index* of $\mathfrak{p}$ and will be noted $e_\mathfrak{p}$. Note that each prime ideal in $\mathcal{O}_K$ lies above exactly one prime of $\mathbf{Z}$ so there is no ambiguity in not specifying p in $e_\mathfrak{p}$. By definition a prime $\mathfrak{p}$ is *ramified* iff. $e_\mathfrak{p} > 1$. One can show that $\mathfrak{p}$ is ramified iff. $\mathfrak{p}$ divides the discriminant Δ_K, hence there are only finitely many ramified primes.

If $\mathfrak{p}|p$ then one has a natural injection $\mathbf{Z}/p\mathbf{Z} \rightarrow \mathcal{O}_K/\mathfrak{p}$ given by the inclusion $\mathbf{Z} \subset \mathcal{O}_K$ and $p\mathbf{Z} \subset \mathfrak{p}$. In particular $k_\mathfrak{p}$ is a finite extension of the finite field $\mathbf{F}_p$ and the degree of this extension is called the *residual degree* and is noted $f_\mathfrak{p}$. One has then the following mass formula:

$$n = \sum_{\mathfrak{p}|p} e_\mathfrak{p} f_\mathfrak{p}.$$

3.2   Galois representations

Given that K is a Galois extension of $\mathbf{Q}$ with Galois group $\mathrm{Gal}(K/\mathbf{Q}) = G$, many problems in arithmetic boil down to understanding the action of G on various arithmetically interesting sets. When the set, S say, is finite, one can linearize the problem by considering $\mathbf{C}^S$ the complex $|S|$-dimensional vector space with basis indexed by the elements of S; G then acts linearly on $\mathbf{C}^S$ by permutations on the basis.

More generally, a complex Galois representation (of G) is an homomorphism

$$\rho : \; G \; \rightarrow \; Aut_\mathbf{C}(V)$$

of G into the group of automorphisms of a finite dimensional complex vector space V. Of course this is simply a complex representation of the finite group G, a classical topic (see [5] for instance), but what one really wants is to understand this representation in terms of the Galois data (rather than in terms of the abstract group G). As we shall see below the Galois structure can be captured in a large part via an action of special elements of G (in fact conjugacy classes in G): the *frobenius* .

3.2.1 The frobenius

There is a natural action of G (by automorphisms) on the set of ideals $\mathrm{Id}(\mathcal{O}_K)$ simply given by

$$\sigma.\mathfrak{a} = \{\sigma(x),\ x \in \mathfrak{a}\}$$

which of course preserves the primes. Given p a prime, one can show that G acts transitively on the set of primes in $\mathcal{O}_K$ dividing p. This implies (since $\sigma(\mathfrak{p}) \simeq \mathfrak{p}$) that the ramification indices and the residual degrees for each of these primes are equal so we denote their common values by e_p, f_p; in particular $e_p = 1$ iff. $p|\Delta_K$ and one then says that p is ramified .

Taking any $\mathfrak{p}$ dividing p, the *decomposition subgroup* at $\mathfrak{p}$ is the stabilizer of $\mathfrak{p}$ under the G-action

$$D_{\mathfrak{p}} := \{\sigma \in G,\ \sigma.\mathfrak{p} = \mathfrak{p}\}$$

and there is a canonical map $r_{\mathfrak{p}} : D_{\mathfrak{p}} \to \mathrm{Gal}(k_{\mathfrak{p}}/\mathbf{F}_p)$ which to $\sigma \in D_{\mathfrak{p}}$ associates the Galois automorphism

$$r_{\mathfrak{p}}(\sigma) :\ x\mathfrak{p} = x\ (mod\ \mathfrak{p}) \to \sigma(x\mathfrak{p}) = \sigma(x)\mathfrak{p} = \sigma(x)(mod\ \mathfrak{p}).$$

The map $r_{\mathfrak{p}}$ turns out to be surjective and its kernel, $I_{\mathfrak{p}}$, is called the inertia subgroup at $\mathfrak{p}$: one has

$$1 \to I_{\mathfrak{p}} \to D_{\mathfrak{p}} \to \mathrm{Gal}(k_{\mathfrak{p}}/\mathbf{F}_p) \to 1.$$

One can show that $I_{\mathfrak{p}}$ is trivial ($r\mathfrak{p}$ is injective) iff. $\mathfrak{p}$ is unramified.

We recall that the Galois group of a finite extension k of the finite field $\mathbf{F}_p$ is cyclic and has a canonical generator, the *frobenius*, given by $\mathrm{Frob}_p(x) = x^p$ for $x \in k$. Returning to our situation, and assuming that p is unramified, the *frobenius at* $\mathfrak{p}$, $\sigma_{\mathfrak{p}}$, is by definition the inverse image under $r_{\mathfrak{p}}$ of the frobenius Frob_p of the extension $k_{\mathfrak{p}}/\mathbf{F}_p$. Let $\mathfrak{p}'$ be another prime above p, and let $\sigma \in G$ such that $\mathfrak{p}' = \sigma.\mathfrak{p}$, then one computes that

$$\sigma_{\mathfrak{p}'} = \sigma\sigma_{\mathfrak{p}}\sigma^{-1}.$$

Hence one defines

Definition. *Let p be unramified, then the frobenius at p, σ_p, is the conjugacy class in G generated by any $\sigma_{\mathfrak{p}}$ for $\mathfrak{p}|p$.*

3.2.2 A relative version of the frobenius

Let $\mathbf{Q} \subset K \subset L \subset \mathbf{C}$ be a tower of finite algebraic extension such that $K \subset L$ is Galois of degree n (i.e. L is a K vector space of degree n and every K linear automorphism of L into $\mathbf{C}$ is an automorphism of L) of Galois group $\mathrm{Gal}(L/K)$. Then the results described above essentially carry over to the case of the extension $K \subset L$. The (relative) discriminant of L/K, $\Delta_{L/K}$, is the ideal of $\mathcal{O}_K$ generated by the elements

$$\{(\mathrm{tr}_{L/K}(x_i x_j))_{i,j}, \text{ for } (x_1, x_2, \ldots, x_n) \text{ ranging over the } n\text{-uples in } \mathcal{O}_K\}.$$

Then if $\mathfrak{p} \nmid \Delta_{L/K}$, ($\mathfrak{p}$ is unramified in $\mathcal{O}_K$) and $\mathfrak{P}|\mathfrak{p}$, the frobenius at $\mathfrak{P}$ is the inverse image under the natural map $D_{\mathfrak{P}} \to \mathrm{Gal}(k_{\mathfrak{P}}/k_{\mathfrak{p}})$ of the frobenius element $\mathrm{Frob}_{\mathfrak{p}} : x \to x^{N(\mathfrak{p})}$. Finally the frobenius at $\mathfrak{p}$ is the conjugacy class in $\mathrm{Gal}(L/K)$ generated by the $\sigma_{\mathfrak{P}}$, $\mathfrak{P}|\mathfrak{p}$; to emphasize the relative nature of the construction, the frobenius is denoted by $[\mathfrak{p}, L/K]$.

3.2.3 Example: cyclotomic fields

An important example of the above is the case of cyclotomic extension: for $q \geqslant 3$ let ξ_q be a primitive q-th root of unity, that is $\xi_q^q = 1$ and $\xi_q^d \neq 1$ for any d dividing q (for example take $\xi_q = \exp(\frac{2\pi i}{q})$) or in other words ξ_q is a generator of the cyclic group μ_q of q-th roots of 1. Let $K_q = \mathbf{Q}(\xi_q)$ be the extension generated by ξ_q. Then K_q is called the q-th cyclotomic extension, and it is Galois of degree $\varphi(q) = |(\mathbf{Z}/q\mathbf{Z})^\times|$. Moreover the Galois group G is isomorphic (non-canonically) to the multiplicative group $(\mathbf{Z}/q\mathbf{Z})^\times$ via the isomorphism (which depends on a choice of ξ_q)

$$(\mathbf{Z}/q\mathbf{Z})^\times \quad \to \quad G$$
$$\bar{a} = a \ (\mathrm{mod} \ q) \quad \to \quad \sigma_{\bar{a}} : \xi_q \to \xi_q^a.$$

One can further show that $\mathcal{O}_K = \mathbf{Z}[\xi_q]$ and that the ramified primes are precisely the prime divisors of q; moreover for $p \nmid q$, the frobenius σ_p (which is reduced to a single element since G is commutative) is the map which sends ξ_q to ξ_q^p, or in other words $\sigma_p = p \ (\mathrm{mod} \ q)$.

3.2.4   Extending a representation over the absolute Galois group

So far we have considered only finite Galois extensions of $\mathbf{Q}$, their associated (finite) Galois group and the action of it on some sets. However in several situations, one needs to work simultaneously with many Galois extensions, for example with an infinite increasing tower of Galois extensions, each Galois group acting in a "compatible" manner on an increasing sequence of sets. A convenient way to achieve this is to consider extending the action of some Galois group $\mathrm{Gal}(K/\mathbf{Q})$ to the absolute Galois group over $\mathbf{Q}$. Let $\overline{\mathbf{Q}} \subset \mathbf{C}$ be the (algebraically closed) field of algebraic complex numbers; we denote by $\mathrm{Gal}(\overline{\mathbf{Q}}/\mathbf{Q})$ the group of $\mathbf{Q}$-linear automorphisms of $\overline{\mathbf{Q}}$. Then by Galois theory, the Galois group of a finite extension $\mathrm{Gal}(K/\mathbf{Q})$ is a finite index quotient of $\mathrm{Gal}(\overline{\mathbf{Q}}/\mathbf{Q})$: more precisely one has

$$1 \to \mathrm{Gal}(\overline{\mathbf{Q}}/\mathbf{Q})^K \to \mathrm{Gal}(\overline{\mathbf{Q}}/\mathbf{Q}) \to \mathrm{Gal}(K/\mathbf{Q}) \to 1,$$

where $\mathrm{Gal}(\overline{\mathbf{Q}}/\mathbf{Q})^K$ denotes the subgroup of the absolute Galois group fixing all the elements of K. Suppose we are given $\mathrm{Gal}(K/\mathbf{Q})$ with an action on some set S, then the projection above extends this to an action of $\mathrm{Gal}(\overline{\mathbf{Q}}/\mathbf{Q})$ on S. Moreover to each prime p, one can associate a frobenius conjugacy class $\mathrm{Frob}_p \subset \mathrm{Gal}(\overline{\mathbf{Q}}/\mathbf{Q})$. This reformulation allows us to consider more general actions (or representations) of the absolute Galois group $\mathrm{Gal}(\overline{\mathbf{Q}}/\mathbf{Q})$ (at the associated frobenius) which do not necessarily factor through a finite quotient.

Remark 3. Note however that a (continuous- for the natural Krull topology-) finite dimensional complex representation of $\mathrm{Gal}(\overline{\mathbf{Q}}/\mathbf{Q})$ does necessarily factor through a finite quotient: that is, is given as a representation of a finite Galois group $\mathrm{Gal}(K/\mathbf{Q})$.

3.3   Chebotareff density theorem

We return to the situation of a finite Galois extension $K/\mathbf{Q}$ and denote it by $G = \mathrm{Gal}(K/\mathbf{Q})$.

As we have seen above, as p ranges over the unramified primes, one obtains a collection of conjugacy classes in G ($G^\natural$ say) and one may ask whether this collection of conjugacy classes fills entirely the (finite) set $G^\natural$. That is indeed the case (and in a very strong form indeed) is the content of Chebotareff's density theorem:

Theorem. *(Chebotareff) Given K a finite Galois extension of the rationals, then the Frobenius map from the set of unramified primes to $G^\natural$ which associates to each p the frobenius σ_p is surjective. More precisely, for any $C \in G^\natural$, and $X \geqslant 2$ set*

$$\pi(X; K) = |\{p \leqslant X, \ unramified\}|,$$
$$\pi(X; K, C) = |\{p \leqslant X, \ unramified, \ s.t. \ \sigma_p = C\}|.$$

Then as $X \to +\infty$ one has

$$\pi(X; K, C) = \frac{|C|}{|G|}\pi(X; K)(1 + o(1)) = \frac{|C|}{|G|}\frac{X}{\log X}(1 + o(1)).$$

In other words not only is the frobenius map surjective on $G^\natural$ but the proportion of primes such that the frobenius falls in a given conjugacy class C is proportional to the size of G. Here is a fancy (complicated) way of restating this simple statement (which is more apt to the generalization explained below):

For G any compact group, there is on G a unique left-(and right-)translation invariant probability measure, namely the *Haar* measure μ_{Haar}. Let $G^\natural$ be the space of conjugacy classes of G and $\pi : G \to G^\natural$ be the canonical projection; by definition the *Sato/Tate* measure μ_{ST} on $G^\natural$ is the direct image under π of μ_{Haar}.

In the present case, G being equipped with the discrete topology, the Haar measure is simply the uniform probability measure: for any $C \subset G$

$$\mu_{Haar}(C) = \frac{|C|}{|G|},$$

and the Sato/Tate measure on $G^\natural$ is given for any set of conjugacy classes $E \subset G^\natural$ by

$$\mu_{ST}(E) = \sum_{C \in E} \mu_{Haar}(C) = \sum_{C \in E} \frac{|C|}{|G|}.$$

The Chebotareff's density theorem states that for any function f on $G^\natural$, one has

$$\frac{1}{\pi(X; K)} \sum_{p \in \pi(X; K)} f(\sigma_p) \to \int_{G^\natural} f(g^\natural)\mu_{ST}(g^\natural) \tag{3.1}$$

or in fancy words that the sequence of averaged Dirac measures

$$\frac{1}{\pi(X; K)} \sum_{p \in \pi(X; K)} \delta_{\sigma_p}$$

weak-* converges to the Sato/Tate measure μ_{ST} as $X \to +\infty$ (one also says that the set of conjugacy classes $\{\sigma_p, \; p \leqslant X, \; unramified\}$ becomes equidistributed w.r.t. μ_{ST} as $X \to +\infty$).

Example. Consider again the case of the q-th cyclotomic extension. Then by the discussion above, $G = G^\natural \simeq (\mathbf{Z}/q\mathbf{Z})^\times$ and σ_p is in the conjugacy class given by $\bar{a} = a + q\mathbf{Z}$ iff $p \equiv a(\mathrm{mod}\ q)$. Thus in this case Chebotareff's theorem is equivalent to (a stronger form of) Dirichlet's theorem for primes in arithmetic progressions: let a be coprime with q and $\pi(X; q, a)$ be the number of primes less than X congruent to a modulo q then

$$\pi(X; q, a) = \frac{1}{\varphi(q)}\pi(X)(1 + o(1)) = \frac{1}{\varphi(q)}\frac{X}{\log X}(1 + o(1)).$$

3.3.1 Sketch of the proof of Chebotareff density theorem

For the proof we consider the alternative formulation (3.1), and remark that it is sufficient to test this equality for f ranging over a set of functions generating $\mathbf{C}^{|G^\natural|}$ the space of (continuous) functions on $G^\natural$: this is the so-called *Weyl equidistribution criterion*. It is well know from representation theory (see [5] for example) that $\mathbf{C}^{|G^\natural|}$ is generated by the characters of G, that is the function $g \to \mathrm{tr}\rho(g)$ where ρ ranges over the irreducible representations of G. When ρ is the trivial representation then (3.1) is trivially true, so it is sufficient to prove that for ρ a non-trivial irreducible representation of G, one has

$$\frac{1}{\pi(X; K)} \sum_{p \in \pi(X; K)} \mathrm{tr}\rho(\sigma_p) \to 0 \tag{3.2}$$

since when ρ is irreducible and non-trivial

$$\int_{G^\natural} \mathrm{tr}\rho(g^\natural)\mu_{ST}(g^\natural) = \frac{1}{|G|}\sum_{g \in G} \mathrm{tr}\rho(g) = 0.$$

It is via the introduction of L-functions (first done by Dirichlet) that (3.2) is proved.

This will be continued in section 3.4.1.

3.4 Artin L-functions

Let K be a Galois extension of $\mathbf{Q}$ with Galois group G and let $\rho : G \to \mathrm{Aut}(V)$ be an irreducible representation of G on some complex (finite dimensional) vector space V. For any prime p unramified, one defines the local factor at p

$$L_p(\rho, s) = \det(Id - \frac{1}{p^s}\rho(\sigma_{\mathfrak{p}}))^{-1}$$

where $\mathfrak{p}$ denotes any ideal above p: note that this definition does not depend on the choice of $\mathfrak{p}$ since all the frobenius are conjugate with each other. When p is ramified, the definition is slightly more complicated: indeed the map $r_{\mathfrak{p}}$ is not injective anymore but induces an isomorphism $D_{\mathfrak{p}}/I_{\mathfrak{p}} \simeq \mathrm{Gal}(k_{\mathfrak{p}}/\mathbf{F}_p)$ thus one can still lift the frobenius of $\mathrm{Gal}(k_{\mathfrak{p}}/\mathbf{F}_p)$ to some element $\sigma_{\mathfrak{p}} \in D_{\mathfrak{p}}$ but the latter is only well defined modulo the inertia subgroup $I_{\mathfrak{p}}$, and consequently the local L-factor is defined by

$$L_p(\rho, s) = \det(Id - \frac{1}{p^s}\rho(\sigma_{\mathfrak{p}})|V^{I_{\mathfrak{p}}})^{-1},$$

where the determinant is taken relative to the subspace space of the $I_{\mathfrak{p}}$-invariants of V. This is well defined and does not depend on the choice of $\mathfrak{p}$ above p.

Definition. *The Artin L-function of ρ is the Euler product*

$$L(\rho, s) = \prod_p L_p(\rho, s) = \prod_p \det(Id - \frac{1}{p^s}\rho(\sigma_{\mathfrak{p}})|V^{I_{\mathfrak{p}}})^{-1}.$$

Since G is finite, the eigenvalues of any $\rho(\sigma_{\mathfrak{p}})$ have modulus 1, hence $L(\rho, s)$ is absolutely convergent and non-vanishing in the half-plane $\Re es > 1$. When ρ is the trivial representation, $L(1, s) = \zeta(s)$ is Riemann's zeta and hence has analytic continuation to $\mathbf{C} - \{1\}$ with a simple pole at $s = 1$. The famous Artin conjecture predicts that for ρ irreducible and non-trivial, $L(\rho, s)$ has analytic continuation to the whole complex plane: in fact one has the more precise

Conjecture. *(Artin) Let $G = \mathrm{Gal}(K/\mathbf{Q})$ and ρ an irreducible non-trivial representation of G, then $L(\rho, s)$ has analytic continuation to $\mathbf{C}$ and satisfies a functional equation of the form:*

$$\Lambda(\rho, s) = \varepsilon(\rho)\overline{\Lambda(\rho, 1 - \bar{s})},$$

where $|\varepsilon(\rho)| = 1$ and

$$\Lambda(\rho, s) = q_\rho^{s/2} L_\infty(\rho, s) L(\rho, s), \ \ with \ L_\infty(\rho, s) = \Gamma_{\mathbf{R}}(s)^{n_+}\Gamma_{\mathbf{R}}(s + 1)^{n_-};$$

here q_ρ is an explicit integer (the Artin conductor) and n_+ (resp n_-) denote the multiplicity of the eigenvalue $+1$ (resp. -1) of the restriction to K of the complex conjugation automorphism.

Example. For K_q the q-th cyclotomic field, the irreducible representations of $G \simeq (\mathbf{Z}/q\mathbf{Z})^\times$ correspond to the (Dirichlet) characters of $(\mathbf{Z}/q\mathbf{Z})^\times$. For such a character $\chi : (\mathbf{Z}/q\mathbf{Z})^\times \to \mathbf{C}^\times$, one defines $\chi(n) = \chi(n(\mathrm{mod}\ q))$ if $(n, q) = 1$ and 0 otherwise. Then $q_\chi = q$ (unless χ is trivial), $n_+ = (1 + \chi(-1))/2$ and

$$L(\chi, s) = \prod_p (1 - \frac{1}{p^s}\chi(p))^{-1} = \sum_{n \geqslant 1} \frac{\chi(n)}{n^s},$$

in particular (cf. the first lecture) Artin's conjecture is true for cyclotomic fields.

3.4.1 Continuation of the proof of Chebotareff's theorem

For ρ irreducible, recall that one has to establish (3.2). Such an estimate is closely related to the analytic properties of the converging series (for res > 1)

$$\sum_p \frac{\mathrm{tr}\rho(\sigma_p)}{p^s}.$$

The latter equals

$$-\frac{L'}{L}(\rho, s) = \sum_{k \geqslant 1} \sum_p \frac{\mathrm{tr}\rho(\sigma_p^k)}{p^{ks}}$$

up to a function holomorphic in the half-plane $\Re es > 1/2$ and one can show that (3.2) follows from the

Theorem 1. *(Artin+Brauer+Hecke) For ρ irreducible, $L(\rho, s)$ admit meromorphic continuation to $\mathbf{C}$ and is holomorphic and non-vanishing for $\Re es \geqslant 1$.*

3.4.2 Properties of Artin L-functions

In this last section we discuss briefly the proof of Theorem 1. First of all we note that an Artin L-function can be defined for a Galois extension L of a general number field K. Namely, if ρ is a representation of $\mathrm{Gal}(L/K)$, the associated Artin L-function is given by

$$L_K(\rho, s) = \prod_{\mathfrak{p}} \det(Id - \frac{1}{\mathrm{N}_{K/\mathbf{Q}}(\mathfrak{p})^s}\rho([\mathfrak{p}, L/K]))^{-1}.$$

Artin L-functions have the following formal properties:

1. For ρ the trivial representation of the Galois group of some number field K,

$$L_K(\rho, s) = \prod_{\mathfrak{p}} (1 - \frac{1}{N_{K/\mathbf{Q}}(\mathfrak{p})^s})^{-1} = \sum_{\mathfrak{a} \subset \mathcal{O}_K} \frac{1}{N_{K/\mathbf{Q}}(\mathfrak{a})^s}$$

is the Dedekind zeta function .

2. (Additivity) $L_K(\rho_1 \oplus \rho_2, s) = L_K(\rho_1, s) L_K(\rho_2, s)$

3. (Invariance under induction) For $H \subset G$ a subgroup, set $K' = L^H$ (then $H = \mathrm{Gal}(L/K')$). For ρ' a representation of H, let $\rho = Ind_H^G \rho'$ be the induced representation to G, then

$$L_{K'}(\rho', s) = L_K(\rho, s).$$

The Artin/Brauer/Hecke theorem is proven by a conjunction of three main ingredients.

– The first ingredient is due to Hecke; generalizing work of Dedekind on zeta functions of number fields, he established that L-functions associated to characters (ie. one dimensional continuous representations) of a locally compact group naturally associated to a given general number field (the *Idèle group*[4] $\mathbf{I}_K$), have analytic continuation to the whole complex plane and satisfy a functional equation. Due to lack of space, we will merely state Hecke's theorem without defining the terms and use it as a black box in the sequel.

Theorem. *(Hecke) Let K be a number field and $\chi : \mathbf{I}_K/K^\times \to \mathbf{S}^1 := \{z \in \mathbf{C}, |z| = 1\}$ be a continuous unitary representation of the idèle group of K, trivial on $K^\times$ (ie. a Hecke character). Then χ has an associated L-function $L(\chi, s)$ (this is an Euler product of degree $[K : \mathbf{Q}]$) which has analytic continuation – with possibly at most one (well identified) pole – to $\mathbf{C}$. Moreover when χ is primitive, $L(\chi, s)$ satisfies a functional equation on the (usual) form*

$$\Lambda(\chi, s) = \varepsilon(\chi)\Lambda(\overline{\chi}, 1 - s)$$

where $|\varepsilon(\chi)| = 1$, $\overline{\chi}$ denotes the complex conjugate of χ and

$$\Lambda(\chi, s) = q_\chi^{s/2} L_\infty(\chi, s) L(\chi, s),$$

with q an integer and $L_\infty(\chi, s)$ a product of Gamma factors. Moreover, $L(\chi, s)$ does not vanish on the line $\Re es = 1$.

---

[4]In fact Hecke did not formulate his result in terms of idèles but rather in terms of ideals

We shall not enter into the definition of what is an Hecke character or its L-function but simply note that Hecke characters are automorphic in nature and can be handled by methods of harmonic analysis over number fields.

– The second key fact is of a more arithmetic nature, and is due to Artin: it has as a consequence that Artin's conjecture is true for one dimensional Galois representation of number fields:

Theorem. *(Artin) Let $K \subset L$ be a Galois extension of number fields and ρ a non-trivial one dimensional representation of* $\mathrm{Gal}(L/K)$. *Then there exists a Hecke character χ of* $\mathbf{I}_K$ *such that*

$$L(\rho, s) = L(\chi, s).$$

Consequently $L(\rho, s)$ has analytic continuation of $\mathbf{C}$, *satisfies a functional equation and does not vanish on the line* $\Re es = 1$.

– The last ingredient is purely group theoretic and due to Brauer,

Theorem. *(Brauer) Let G be a finite group and ρ be an irreducible representation of G, then there exists a finite set of finite subgroups of G, $\{G_i\}_i$, one dimensional representations of G_i, χ_i and integers $n_i \in \mathbf{Z}$ such that one has*

$$\rho = \sum_i n_i Ind^G_{G_i} \chi_i$$

as an identity of virtual representation.

In particular by additivity and by invariance under induction one has for $\rho : G = \mathrm{Gal}(L/K) \to Aut(V)$ a general complex irreducible representation, one has

$$L_K(\rho, s) = \prod_i L_K(Ind^G_{G_i} \chi_i, s)^{n_i} = \prod_i L_{K_i}(\chi_i, s)^{n_i},$$

and the proof follows from Artin's and Hecke's theorems above and the observation that the $\{\chi_i\}$ does not contain the trivial representation.

In fact we want to stress that, for all the few cases where the analytic continuation of Artin L-functions (or more generally of L-functions attached to Galois representations) has been proved, it has always been by identifying the Artin L-function with an L-function associated to an automorphic form, for which analytic continuation is easier to establish. In the case of L-functions of degree 1, the identification is provided by Class Field Theory [3].

References

[1] James Milne, *Galois Theory*, available at http://www.jmilne.org/math/.

[2] James Milne, *Algebraic Number Theory*, available at http://www.jmilne.org/math/.

[3] James Milne, *Class Field Theory*, available at http://www.jmilne.org/math/.

[4] Pierre Samuel, *Algebraic theory of numbers*, Translated from the French by Allan J. Silberger, Houghton Mifflin Co., Boston, Mass., 1970.

[5] Jean-Pierre Serre, *Représentations linéaires des groupes finis*, Hermann, Paris, 1971 (French). Deuxième édition, refondue.

[6] Jean-Pierre Serre, *Harvard lectures in analytic number theory*.

4 *L*-functions of elliptic curves

4.1 Background on elliptic curves

4.1.1 Elliptic curves as algebraic curves

Let k be a field. An (affine plane) elliptic curve, E^a (say), over k is the algebraic curve given by an equation of the following form: let $a_1, a_2, a_3, a_4, a_6 \in k$ be given, and denote by (E) the equation

$$(E) = (E_{a_1,a_2,a_3,a_4,a_6}) : \quad y^2 + a_1 xy + a_3 y = x^3 + a_2 x^2 + a_4 x + a_6;$$

for K a field containing k (or more generally for K a k-algebra) the set of solutions

$$E^a(K) := \{(x,y) \in K^2 \text{ s.t. } y^2 + a_1 xy + a_3 y = x^3 + a_2 x + a_4 x + a_6\}$$

is called the *set of K-rational points of E*.

4.1.2 The discriminant

However in order to properly qualify for the name of elliptic curve, one requires E^a to be non-singular: setting

$$f_E(x,y) := y^2 + a_1 xy + a_3 y - (x^3 + a_2 x^2 + a_4 x + a_6),$$

non-singular means that the the system of equations

$$f_E(x,y) = \frac{\partial}{\partial x} f_E(x,y) = \frac{\partial}{\partial y} f_E(x,y) = 0$$

has no solutions in an(y) algebraic closure of k ($\overline{k}$ say). It can be checked that the latter is equivalent to the non-vanishing of some

$$\Delta_E = \Delta(a_1, a_2, a_3, a_4, a_6)$$

where Δ is some universal polynomial in five variables with coefficients in $\mathbf{Z}$. For example, if $a_1 = a_3 = 0$ then Δ_E is (up to a power of 2 factor) the discriminant of the polynomial $P(x) = x^3 + a_2 x^2 + a_4 x + a_6$. The quantity Δ_E is called the *discriminant* of E. A curve E as above with zero discriminant is called a degenerate (or singular) elliptic curve. From now on, an elliptic curve will always be non-singular.

An elliptic curve is *defined over some field k* if it is given by a non-singular equation as above with coefficients a_1, a_2, a_3, a_4, a_6 in k.

4.1.3 Projective elliptic curves

It is both fundamental and natural for the theory, to "compactify" the affine curve E^a by adding to it "a point at infinity". For this one consider the *projective plane* $\mathbf{P}_k^2$ which is the algebraic variety defined over k, whose set of K rational points ($\mathbf{P}_k^2(K)$ say) is the set of all K-rational directions of lines in the affine 3-dimensional space

$$\mathbf{A}_k^3(K) = \{(x,y,z) \in K^3\};$$

in other words $\mathbf{P}_k^2(K)$ is the quotient set

$$\mathbf{P}_k^2(K) = \{(x,y,z) \in K^3 - \{(0,0,0)\}\}/\sim$$

where $\sim$ stand for the equivalence relation induced by the diagonal action of $K^\times$ on $K^3 - \{(0,0,0)\}$: two points (x,y,z), $(x',y',z') \in K^3 - \{(0,0,0)\}$ are identified iff. $(x',y',z') = (\lambda x, \lambda y, \lambda z)$ for some $\lambda \in K^\times$. In particular the projective plane $\mathbf{P}_k^2(K)$ contains the affine plane $\mathbf{A}_k^2(K) = K^2$ as the set of classes of triples of the form $(x,y,1)$, $(x,y) \in K^2$, the complementary subset being a "projective line at infinity" $\mathbf{P}_{\infty,k}^1(K)$ consisting the set of classes of points of the form $\{(x,y,0), \ x,y \in K\}/\sim$ and the latter line decomposes as an affine line $\{(x, 1, 0),\ x \in K\} \simeq \mathbf{A}^1_k(K)$ and a point which we denote by ∞ corresponding to the class of $(0, 1, 0)$. In particular $E^a(K)$ embeds into $P^2_k(K)$, via $(x, y) \to (x, y, 1)$.

The *compactified* elliptic curve E corresponding to equation (E) is defined as the set of solutions, in P^2_k, of the homogeneous version of equation (E):

$$zy^2 + a_1 xyz + a_3 yz^2 = x^3 + a_2 x^2 z + a_4 xz^2 + a_6 z^3. \qquad (4.1)$$

In other words, for $k \subset K$ a field, the K-rational points of E are given by

$$E(K) = \left\{ (x, y, z) \in \mathbf{P}^2_k(K) \text{ s.t. } zy^2 + a_1 xyz + a_3 yz^2 = x^3 + a_2 x^2 z + a_4 xz^2 + a_6 z^3 \right\},$$

indeed the above equation is homogeneous (of degree 3) so if $(x, y, z) \in K^3$ satisfies (E) then $(\lambda x, \lambda y, \lambda z)$ satisfies (E) for any $\lambda \in K^\times$. As noted before, $E(K)$ contains the affine points $E^a(K)$ as the set of points with third coordinate z equal to 1. The set of points on E at infinity (ie. the points contained in $\mathbf{P}^1_{\infty,k}$ or in other words, the point for which the third coordinate is $z = 0$) consists of the single point $\infty = (0, 1, 0)$. It is called *the point at infinity of E* and is denoted O_E.

4.1.4 The group law on an elliptic curve

A fundamental fact in the theory of elliptic curves is the existence of the structure of a group on the set of its points. In this section, we recall the definition of the group law.

Let E be a (non-singular) elliptic curve. As we have seen above, E always contains a k-rational point, namely the point at infinity $(0, 1, 0)$: we choose this point as the neutral element for the group law and denote it by O_E.

Given $P_1, P_2 \in E(K)$, we consider the (projective) line $L_{P_1 P_2}$ passing through both points (the point on such line are the solutions of a homogeneous equation of degree 1

$$ax + by + cz = 0$$

and if $P_1 = P_2$, $L_{P_1 P_1}$ is the tangent line to E passing through P_1) then $L_{P_1 P_2}$ intersects E at a third point P_3 and then $P_1 + P_2$ is defined to be the intersection of E with the line joining P_3 and O_E. One can verify that this process defines a commutative group law on E, with O_E as the neutral element and such that $P_3 = -(P_1 + P_2)$. Moreover one can show that the coordinates of $P_1 + P_2$ are given by rational functions in the coordinates of P_1 and P_2 with coefficients in k (one says that the group law is k-rational): in other words this law defines a group law on each set of K-rational points $E(K)$.

Example. Given an elliptic curve, many formal computational systems allows for the computation of the group law of an elliptic curve, for example the PARI-GP system. Consider the curve E over $\mathbf{Q}$ given by the equation

$$y^2 + y = x^3 - x \iff y(y+1) = x(x-1)(x+1).$$

this curve has a discriminant 37 and $E(\mathbf{Q})$ contains 7 trivial $\mathbf{Q}$-rational points, namely $O_E = \infty$ and the affine points with (x, y) coordinates (and $z = 1$)

$$(0,0), \ (1,0), \ (-1,0), \ (0,-1), \ (1,-1), \ (-1,-1).$$

set $P = (0,0)$ then

$$-P = (0,-1), \ -2P = (1,-1), \ 2P = (1,0), \ -3P = (-1,0), \ 3P = (-1,-1),$$

which exhaust all the trivial points of $E(\mathbf{Q})$ but there are many more

$$4P = (2,-3), \ 5P = (\frac{1}{4}, -\frac{5}{8}), \ 6P = (6,-14),$$

$$7P = (-\frac{5}{9}, -\frac{8}{27}), \ 8P = (\frac{21}{25}, -\frac{69}{125}), \ 9P = (-\frac{20}{49}, -\frac{435}{343}),$$

$$10P = (\frac{161}{16}, -\frac{2065}{64}), \ 12P = (\frac{1357}{841}, \frac{28888}{24389}).$$

In fact, one can show (see below) that the group $E(\mathbf{Q}) = \langle P \rangle$ is the infinite cyclic group generated by P.

4.1.5   Isogenies

By definition, the *function field* of E, $k(E)$, is the field of fractions of the integral domain $k[x, y]/(f_E(x, y))$. An *isogeny* between two elliptic curves E, E' defined over k is a non-constant k-rational morphism $\varphi : E \to E'$ (that is a k-homomorphism $k(E') \hookrightarrow k(E)$) such that $\varphi(O_E) = O_{E'}$. We recall that a non-constant morphism between curves is *separable* if the extension of fields over the algebraic closure of k, $\overline{k}$ say, $\overline{k}(E') \hookrightarrow \overline{k}(E)$ is. One denotes by $\deg \varphi$ the degree of this extension. In the case of isogenies, one can also show that φ is automatically a group morphism:

$$\varphi(P_1 +_E P_2) = \varphi(P_1) +_{E'} \varphi(P_2);$$

In particular, $\varphi^{-1}(O_{E'}) = \ker \varphi$ is a subgroup of E defined over k and moreover if φ is separable, then $|\ker \varphi(\overline{k})| = \deg \varphi$.

We denote by $Hom(E, E')$ the set of isogenies from E onto E' which is viewed as a **Z**-module under addition. We denote by $End(E)$ the set of isogenies of E onto itself. This is a (non-necessarily commutative) ring under addition and composition: the endomorphism ring of E.

Example. An example of isogeny is given for $n \in \mathbf{Z}$ by $[n]_E \in End(E)$ the multiplication by n in E; one can show that $\deg[n]_E = n^2$ and that if n is coprime with the characteristic of k, $[n]_E$ is separable. The kernel of this isogeny $\ker[n]_E$ is also denoted $E(k)[n]$: this is the set of n-torsion points on $E(k)$. If n is coprime with the characteristic of k, one has

$$E(\overline{k})[n] \simeq (\mathbf{Z}/n\mathbf{Z}) \times (\mathbf{Z}/n\mathbf{Z}).$$

Note that the map $n \to [n]_E$ gives an embedding of **Z** into $End(E)$; in many cases this embedding is also surjective and in general one has the following structural result

Theorem. *Given E an elliptic curve defined over k, then the endomorphism ring of E is either*

$$End(E) \simeq \begin{cases} \mathbf{Z} \text{ or} \\ \text{an order of an imaginary quadratic field or} \\ \text{a maximal order in a quaternion algebra} \end{cases}$$

and $Aut(E) = End(E)^{\times}$ is either one of $\mathbf{Z}/2\mathbf{Z}$ or $\mathbf{Z}/4\mathbf{Z}$ or $\mathbf{Z}/6\mathbf{Z}$.

If $End(E) \neq \mathbf{Z}$, one says that E has complex multiplication *(or E is CM) and if moreover $End(E)$ is isomorphic to maximal order in a quaternion algebra, one says that E has quaternionic multiplication.*

If k has characteristic zero, quaternionic multiplication never occur.

If k has positive characteristic, then E has always complex multiplication (quaternionic or not); if E has quaternionic multiplication one also says that E is supersingular.

To conclude our presentation of isogenies, we mention that to every isogeny $\varphi : E \to E'$ is associated a *dual isogeny* $\hat{\varphi} : E' \to E$ satisfying

$$\hat{\hat{\varphi}} = \varphi, \quad \widehat{\varphi +_E \varphi'} = \hat{\varphi} +_{E'} \hat{\varphi}',$$

$$\widehat{\varphi \circ \varphi'} = \hat{\varphi} \circ \hat{\varphi}', \quad \widehat{[n]_E} = [n]_E, \quad \hat{\varphi} \circ \varphi = [\deg \varphi]_E, \quad \varphi \circ \hat{\varphi} = [\deg \hat{\varphi}]_{E'} = [\deg \varphi]_{E'}.$$

Thus the passage to dual is an involution on $End(E)$ which plays a key role in the determination of the structure of $End(E)$.

4.1.6   Elliptic curves over the complex numbers

In this section we recall several facts concerning the structure of elliptic curves over the complex numbers by using the complex uniformization of an elliptic curve provided by the Weierstrass $\mathfrak{P}$-function.

A *lattice* $\Lambda \subset \mathbf{C}$ is a free subgroup of rank 2 of $\mathbf{C}$ which generates $\mathbf{C}$ as a real vector space: $\Lambda = \mathbf{Z}\omega_1 + \mathbf{Z}\omega_2$ with $\omega_1/\omega_2 \notin \mathbf{R}$. The Weierstrass function associated to Λ is the series of a complex variable

$$\mathfrak{P}_\Lambda(s) = \frac{1}{s^2} + \sum_{\lambda \in \Lambda - \{0\}} \frac{1}{(s-\lambda)^2} - \frac{1}{\lambda^2}.$$

$\mathfrak{P}_\Lambda(s)$ defines a meromorphic function on $\mathbf{C}$, periodic of period Λ (ie. $\mathfrak{P}_\Lambda(s + \lambda) = \mathfrak{P}_\Lambda(s)$ for any $\lambda \in \Lambda$) , is holomorphic on $\mathbf{C} - \Lambda$ and has a pole of order 2 at every point of Λ. Moreover one can show that it satisfies the differential equation

$$(\mathfrak{P}'_\Lambda)^2(s) = 4\mathfrak{P}^3_\Lambda(s) - g_2(\Lambda)\mathfrak{P}_\Lambda(s) - g_3(\Lambda),$$

with

$$g_2(\Lambda) = 60 \sum_{\lambda \in \Lambda - \{0\}} \frac{1}{\lambda^4}, \quad g_3(\Lambda) = 140 \sum_{\lambda \in \Lambda - \{0\}} \frac{1}{\lambda^6},$$

and also that

$$\Delta(\Lambda) = g_2^3(\Lambda) - 27g_3^2(\Lambda) \neq 0. \tag{4.2}$$

The differential equation above gives a map

$$(\mathbf{C} - \Lambda)/\Lambda \quad \rightarrow \quad E_\Lambda^a : y^2 = 4x^3 - g_2(\Lambda)x - g_3(\Lambda)$$
$$s \quad \rightarrow \quad (\mathfrak{P}_\Lambda(s), \mathfrak{P}'_\Lambda(s)),$$

from the torus $\mathbf{C}/\Lambda$ minus the origin in the affine elliptic curve with equation $y^2 = 4x^3 - g_2(\Lambda)x - g_3(\Lambda)$ (we note that this curve is non-singular because of (4.2)). This map can be extended to a map from the complete torus $\mathbf{C}/\Lambda$ to the corresponding projective elliptic curve E_Λ, by sending the origin of the torus to the point at infinity O_{E_Λ}, noting in particular that as $s \rightarrow 0$, $\mathfrak{P}_\Lambda(s) \simeq s^{-2}$, $\mathfrak{P}'_\Lambda(s) \simeq -2s^{-3}$. One can then show that this map is bijective and in fact is a group isomorphism from the commutative torus $(\mathbf{C}/\Lambda, +)$ to the elliptic curve $E_\Lambda(\mathbf{C})$ endowed with its group law.

There is a converse to this result: firstly (because we are in characteristic $\neq 2,3$) the equation defining an elliptic curve over $\mathbf{C}$ can always be rewritten up to linear change of variables (completing squares and cubes) in the form

$$y^2 = 4x^3 - g_2 x - g_3.$$

One can then show that there exists a lattice $\Lambda \subset \mathbf{C}$ such that $g_2(\Lambda) = g_2$, $g_3(\Lambda) = g_3$. In particular the map

$$\Lambda \to \mathbf{C}/\Lambda \to E_\Lambda$$

induces a one to one correspondence between the set of isomorphism classes of elliptic curves over $\mathbf{C}$ and the set of lattices of $\mathbf{C}$ up to homothety (two lattices Λ, Λ' are homothetic iff. there exists $\alpha \in \mathbf{C}^\times$ such that $\alpha.\Lambda = \Lambda'$).

From this description one can recover several results concerning isogenies between elliptic curves over $\mathbf{C}$. For instance, one has

$$Hom(E_\Lambda, E_{\Lambda'}) \simeq \{\alpha \in \mathbf{C}, \ s.t. \ \alpha.\Lambda \subset \Lambda'\}$$

and in particular

$$End(E_\Lambda) \simeq \{\alpha \in \mathbf{C}, \ s.t. \ \alpha.\Lambda \subset \Lambda\}.$$

From this one shows that $End(E_\Lambda) = \mathbf{Z}$ unless ω_1/ω_2 generates an imaginary quadratic field K in which case $End(E_\Lambda)$ is an order of K (and one says that E has CM by K).

In the same vein, the multiplication by n-isogeny, $[n]_E$, is simply the map on $\mathbf{C}/\Lambda$ induced by multiplication by n: in particular its kernel is given by

$$E(\mathbf{C})[n] \simeq \frac{1}{n}.\Lambda/\Lambda = (\mathbf{Z}\omega_1/n + \mathbf{Z}\omega_2/n)/(\mathbf{Z}\omega_1 + \mathbf{Z}\omega_2) \simeq (\mathbf{Z}/n\mathbf{Z})^2. \qquad (4.3)$$

This statement is a result of transcendence type but can be used in a more arithmetic context: namely when E is defined over $\mathbf{Q}$ (or more generally over any number field). Indeed since the multiplication by n map, $[n]$, is given by polynomials with coefficients in $\mathbf{Q}$ evaluated at the coordinate of the points, it turn out that the n-torsion points have coordinates in $\overline{\mathbf{Q}}$, hence we have from (4.3)

$$E(\overline{\mathbf{Q}})[n] = E(\mathbf{C})[n] \simeq (\mathbf{Z}/n\mathbf{Z})^2. \qquad (4.4)$$

4.1.7 Elliptic curves over finite fields

We suppose now that $k = \mathbf{F}_q$ is a finite field (a finite extension of some field $\mathbf{F}_p \simeq \mathbf{Z}/p\mathbf{Z}$ for some prime p; then $q = p^n$ where n is the degree of the extension). If E is an elliptic curve defined over k then $End(E)$ is always strictly bigger than $\mathbf{Z}$. Indeed, one has the frobenius morphism Frob_q on P_k^2 which is induced by the polynomial map

$$\mathrm{Frob}_q : (x, y, z) \rightarrow (x^q, y^q, z^q).$$

Indeed one verifies that for any k-extension K the frobenius maps the set of K-rational points of E to itself: namely for $(x, y, z) \in K^3$ satisfying

$$zy^2 + a_1 xyz + a_3 yz^2 - (x^3 + a_2 x^2 z + a_4 xz^2 + a_6 z^3) = 0$$

one has

$$z^q(y^q)^2 + a_1 x^q y^q z^q + a_3 y^q (z^q)^2 - ((x^q)^3 + a_2 (x^q)^2 z^q + a_4 x^q (z^q)^2 + a_6 (z^q)^3)$$
$$= z^q (y^q)^2 + (a_1)^q x^q y^q z^q + (a_3)^q y^q (z^q)^2$$
$$- ((x^q)^3 + (a_2)^q (x^q)^2 z^q + (a_4)^q x^q (z^q)^2 + (a_6)^q (z^q)^3)$$

since $a^q = a$ for any $a \in k$. Since elevating to the q-th power is a k-linear field homomorphism, this equals

$$(zy^2 + a_1 xyz + a_3 yz^2 - (x^3 + a_2 x^2 z + a_4 xz^2 + a_6 z^3))^q = 0.$$

It is also easily checked that $\mathrm{Frob}_q(O_E) = O_E$ thus Frob_q defines an isogeny of E. Moreover one can show that that $\deg \mathrm{Frob}_q = q$.

The frobenius map is of fundamental importance in the study of algebraic varieties over fields of finite characteristic. In particular it helps counting points on such varieties; in the case of elliptic curves one has the following

Theorem. *(Hasse) Given E/k an elliptic curve defined over the finite field $k = \mathbf{F}_q$, the number of its k-rational points satisfies*

$$|E(k)| = q - a_q(E) + 1,$$

where $a_q(E)$ satisfies the (Hasse) bound

$$|a_q(E)| \leqslant 2\sqrt{q}. \tag{4.5}$$

Remark 4. Hasse's theorem states that when the cardinality, q, of k is large, the number of k-rational points of an elliptic curve defined over k is asymptotic to q (which is also asymptotic to the number of points ($= q + 1$) on the projective line $\mathbf{P}^1_k(k)$). A. Weil generalized Hasse's bound to any non-singular projective algebraic curve defined over a finite field: for C defined over $\mathbf{F}_q$

$$\big| |C(\mathbf{F}_q)| - q - 1 \big| \leqslant 2g_C\sqrt{q}$$

(g_C is the genus of C).

4.1.8   Hasse's bound and character sums

Before displaying the proof of Hasse's bound, we would like to interpret it as a non-trivial cancellation property for some character sums. Suppose (for simplicity) that $k = \mathbf{F}_p$ and that the equation defining E is of the form

$$y^2 = x^3 + ax + b$$

and that the discriminant $\Delta = -4a^3 + 27b^2 \neq 0$ so that the curve is non-singular. Then the number of points in the affine curve $E^a(\mathbf{F}_p)$, which by definition is $p - a_p(E)$, equals the sum over all x in $\mathbf{F}_p$ of the number of solutions of the equation of degree 2, $y^2 = x^3 + ax + b$. To detect analytically this number we introduce the following function on $\mathbf{F}_p^\times$ (the *Legendre symbol*)

$$\left(\frac{x}{p}\right) = \begin{cases} 1 \text{ if } x \in (\mathbf{F}_p^\times)^2 \text{ (ie. is a square)}, \\ -1 \text{ if } x \notin (\mathbf{F}_p^\times)^2 \text{ (ie. is not a square)}, \end{cases}$$

and it is easy to see that $x \to \left(\frac{x}{p}\right) \in \{\pm 1\}$ is a character of the group $\mathbf{F}_p^\times$, since $\left(\frac{xy}{p}\right) = \left(\frac{x}{p}\right)\left(\frac{y}{p}\right)$. If $p > 2$ this is the unique real character of $\mathbf{F}_p^\times$. We extend this function to $\mathbf{F}_p$ by setting $\left(\frac{0}{p}\right) = 0$ and with this definition one has for any $x \in \mathbf{F}_p$

The number of solutions of the equation $y^2 = x$ is equal to $1 + \left(\dfrac{x}{p}\right)$.

In particular

$$|E^a(\mathbf{F}_p)| = \sum_{x \in \mathbf{F}_p} 1 + \left(\frac{x^3 + ax + b}{p}\right)$$

and in particular

$$a_p(E) = -\sum_{x \in \mathbf{F}_p} \left(\frac{x^3 + ax + b}{p}\right).$$

Observe now that since $|(\frac{x^3+ax+b}{p})| \leqslant 1$, one has the trivial bound

$$| \sum_{x\in F_p} (\frac{x^3 + ax + b}{p})| \leqslant p$$

while the Hasse bound yields

$$| \sum_{x\in F_p} (\frac{x^3 + ax + b}{p})| \leqslant 2\sqrt{p}.$$

In particular, when p is large Hasse's bound shows that there is considerable cancellation in the sum due to the sign changes of the function $x \to (\frac{x^3+ax+b}{p})$.

4.1.9 Proof of Hasse's Theorem

The proof of this theorem makes crucial use of the frobenius endomorphism $\mathrm{Frob}_q \in End(E)$: namely one remarks that

$$E(k) = \ker(Id_E - \mathrm{Frob}_q)$$

(that is, a point on E has all its coordinates in k iff. it is fixed under the frobenius map – this is easily verified). It follows from the fact that Frob_q is purely inseparable, that $Id_E - \mathrm{Frob}_q$ is separable, hence one has

$$|E(k)| = |\ker(Id_E - \mathrm{Frob}_q)| = \deg(Id_E - \mathrm{Frob}_q) = (Id_E - \mathrm{Frob}_q).(\widehat{Id_E - \mathrm{Frob}_q}).$$

Expanding this last expression we obtain the following identity in the ring $End(E)$

$$|E(k)| = 1 - (\mathrm{Frob}_q + \widehat{\mathrm{Frob}_q}) + \deg \mathrm{Frob}_q = 1 - (\mathrm{Frob}_q + \widehat{\mathrm{Frob}_q}) + q,$$

hence in $End(E)$ one has

$$a_q(E) = \mathrm{Frob}_q + \widehat{\mathrm{Frob}_q}.$$

Now for any $m, n \in \mathbf{Z}$, one has

$$0 \leqslant \deg(m + n.\mathrm{Frob}_q) = (m + n.\mathrm{Frob}_q).(m + \widehat{n.\mathrm{Frob}_q}) = m^2 + mna_q(E) + n^2 q$$

hence the quadratic form $m^2 + mna_q(E) + n^2 q$ is positive which implies that its discriminant $a_q(E)^2 - 4q$ is negative.

Q.E.D.

Remark 5. From the proof above, we see that Frob_q is annihilated by the polynomial with integral coefficients $X^2 - a_q(E)X + q$. In particular if q is an odd power of a prime, this polynomial has no roots since its discriminant is non-positive and the polynomial cannot be a square. In that case, we see that $End(E)$ contains the imaginary quadratic order $\mathbf{Z}[\mathrm{Frob}_q]$.

This kind of method can be used to prove other interesting statements: for instance

Theorem 2. *Given E, E' two elliptic curves defined over some finite field k. If $Hom(E, E') \neq \emptyset$, then $a_q(E) = a_q(E')$. In other words two isogenous elliptic curves defined over a finite field have the same number of points in that field.*

Indeed, let $\varphi : E \to E'$ be an isogeny, then, denoting by $\mathrm{Frob}_{q,E}$ (resp. $\mathrm{Frob}_{q,E'}$) the frobenius isogeny on E (resp. E'), one has

$$
\begin{aligned}
\deg\varphi \times |E(k)| &= \deg\varphi \times \deg(Id - \mathrm{Frob}_q) \\
&= \deg(\varphi.(Id - \mathrm{Frob}_q)) = \deg(\varphi - \varphi\mathrm{Frob}_q);
\end{aligned}
$$

now one can check from the definition of Frob_q that it commutes with any isogeny defined over k (since an isogeny is defined by polynomials with coefficients in k)

$$
\varphi.\mathrm{Frob}_{q,E} = \mathrm{Frob}_{q,E'}.\varphi
$$

and one has

$$
\deg\varphi \times |E(k)| = \deg(\varphi - \mathrm{Frob}_{q,E'}\varphi) = \deg(1 - \mathrm{Frob}_{q,E'})\deg\varphi = |E'(k)|\deg\varphi
$$

and the proof follows since $\deg\varphi \geqslant 1 \neq 0$.

4.1.10   Elliptic curves over the rationals

Let E be an elliptic curve defined over the field of rationals $\mathbf{Q}$. One of the main goals of the theory (which maybe goes back to Diophantus) is to have a good understanding of the group $E(\mathbf{Q})$ of its $\mathbf{Q}$-rational points. A fundamental step in this direction is the finiteness theorem of Mordell

Theorem. *(Mordell) The abelian group $E(\mathbf{Q})$ is finitely generated: i.e.*

$$
E(\mathbf{Q}) \simeq \mathbf{Z}^r \times E(\mathbf{Q})^{tors}
$$

where $r \geqslant 0$ is an integer (the rank *of $E(\mathbf{Q})$) and $E(\mathbf{Q})^{tors}$ is a finite group: the torsion subgroup of $E(\mathbf{Q})$ (the group of rational $P \in E(\mathbf{Q})$ such that $n_P.P = O_E$ for some integer $n_P \neq 0$).*

Remark 6. The same result is valid when $\mathbf{Q}$ is replaced by an arbitrary number field (due to A. Weil).

Remark 7. Mordell's theorem has the following geometric interpretation. Consider the following elementary operation on an elliptic curve over $\mathbf{Q}$: given two points P_1, $P_2 \in E(\mathbf{Q})$, form P_3 from the intersection of E with the line joining P_1, P_2 (possibly the tangent to E at P_1 if $P_1 = P_2$). Then Mordell's theorem says that there exists a finite set of points on $E(\mathbf{Q})$ (containing O_E) such that any point of $E(\mathbf{Q})$ can be obtained from the original set of point by performing a finite number of times the above elementary operation.

Thus Mordell's (or the Mordell/Weil) theorem reduces the study of $E(\mathbf{Q})$ (or of $E(K)$ for K a number field) to the determination of its rank and of a basis of the corresponding free group, and of its torsion subgroup $E(\mathbf{Q})^{tors}$. the torsion subgroup of an elliptic curve a is rather well understood and can be computed in any explicit example (Nagell/Lutz); moreover its general structure is known

Theorem. *(Mazur)*

$$E(\mathbf{Q})^{tors} = \begin{cases} \mathbf{Z}/n\mathbf{Z} \ for \ n = 1, 2, \ldots, 10 \ or \ 12 \ or \\ (\mathbf{Z}/2\mathbf{Z}) \times (\mathbf{Z}/2n\mathbf{Z}) \ for \ n = 1, 2, 3 \ or \ 4. \end{cases}$$

For a general number field the precise structure of the torsion point is not so well known but at least one has the following

Theorem. *(Merel) Let K be a number field. Then there exists a constant C_K (depending only on K) such that for any elliptic curve, E, defined over K one has*

$$|E(K)^{tors}| \leqslant C_K.$$

Concerning the rank, unfortunately, much less is known: for example it is not known yet if there exist elliptic curves defined over $\mathbf{Q}$ with arbitrarily large rank.[5] A possibility to study this kind of question may be the use of the Birch/Swinnerton-Dyer conjecture (see below), in order to relate the rank of E to the analytic properties of the L-function attached to E.

---

[5] If one replaces $\mathbf{Q}$ by a function field over a finite field, for example $K = \mathbf{F}_p(T)$, the field of fractions in one variable with coefficients in $\mathbf{F}_p$, then one can construct elliptic curves over K with arbitrarily large rank [14].

4.2 The *L*-function of an elliptic curve

In this section we describe the *L*-functions attached to elliptic curves.

4.2.1 *L*-functions of elliptic curves over finite fields

For simplicity we suppose that E is defined over $k = \mathbf{F}_p$. The *L*-function of E is *by definition* the formal series

$$L(E, s) = \exp\left(\sum_{n \geqslant 1} \frac{|E(\mathbf{F}_{p^n})|}{n} p^{-ns}\right),$$

(remember that $|E(\mathbf{F}_{p^n})|$ is the number of points on E with coordinates in $\mathbf{F}_{p^n}$). By Hasse's theorem this series is convergent for $\Re e s > 1$

Remark 8. It is not a priori clear why this is a natural definition: some motivation for this definition is to be found in the last lecture.

Theorem 3. *One has the identity (valid for $\Re e s > 1$)*

$$L(E, s) = \frac{(1 - a_p(E)p^{-s} + p^{1-2s})}{(1 - p^{-s})(1 - p^{1-s})}.$$

Remark 9. Observe that this identity provides a meromorphic continuation of $L_p(E, s)$ to the complex plane, with the functional equation

$$L(E, s) = L(E, 1 - s)$$

and with (infinitely many) poles on $\Re e s = 0, 1$ and zeros (by Hasse's theorem) on the line $\Re e s = 1/2$. Thus Hasse's theorem can be considered as an analog of Riemann's Hypothesis for (*L*-functions of) elliptic curves over finite fields.

We prove now Theorem 3: by the proof of Hasse's theorem, one has for any $n \geqslant 1$

$$
\begin{aligned}
|E(\mathbf{F}_{p^n})| &= \deg(1 - \mathrm{Frob}_{p^n}) = \deg(1 - \mathrm{Frob}_p^n) \\
&= (1 - \widehat{\mathrm{Frob}_p}^n).(1 - \mathrm{Frob}_p^n) = 1 - (\widehat{\mathrm{Frob}_p}^n + \mathrm{Frob}_p^n) + p^n
\end{aligned}
$$

since $\deg \mathrm{Frob}_p^n = (\deg \mathrm{Frob}_p)^n = p^n$. Hence we have (as formal power series with coefficients in $End(E)$)

$$
\begin{aligned}
\exp\left(\sum_{n \geqslant 1} \frac{|E(\mathbf{F}_{p^n})|}{n} T^n\right) &= \exp\left(\sum_{n \geqslant 1} \frac{1 - (\widehat{\mathrm{Frob}_p}^n + \mathrm{Frob}_p^n) + p^n}{n} T^n\right) \\
&= \frac{(1 - \widehat{\mathrm{Frob}_p}T)(1 - \mathrm{Frob}_p T)}{(1 - T)(1 - pT)} = \frac{(1 - a_p(E)T + pT^2)}{(1 - T)(1 - pT)}.
\end{aligned}
$$

4.2.2 *L*-functions of elliptic curves over the rationals

We consider an elliptic curve E defined over $\mathbf{Q}$. For simplicity we assume that its equation is given by

$$E : y^2 = x^3 + ax + b$$

where a, b are integers and $\Delta_E = -16(4a^3 + 27b^2) \neq 0$.

4.2.3 Reduction of an elliptic curve

If p is any prime, we can consider the above equation modulo p: setting $\bar{a}, \bar{b}$ to be $a(\operatorname{mod} p)$, $b(\operatorname{mod} p)$ respectively, the equation

$$E_p : y^2 = x^3 + \bar{a}x + \bar{b}$$

then defines an elliptic curve over $\mathbf{F}_p$ granted that $p \nmid \Delta_E$. Thus in this situation, given E as above, we have attached to each prime not dividing Δ_E an elliptic curve E_p defined over $\mathbf{F}_p$. The curve E_p is called the *reduction of E modulo p*. For a general prime p (possibly dividing Δ_E), a more general notion of reduction of E modulo p is available (and matches the above one) but defining it requires more sophisticated tools (in particular the concept of minimal model which we will not describe in detail here); in any case, one can associate to each prime p, the reduction modulo p of E, E_p say, which most of the time is an elliptic curve but which may also be a conic or even a singular curve. In the former case, we say that E has *good* reduction at p and *bad* reduction at p in the latter. In particular, the primes not dividing Δ_E are all primes of good reduction.

To each prime p, one associates a local L-factor at p, $L_p(E, s)$ which is given by

$$L_p(E, s) = (1 - \frac{a_p(E_p)}{p^s} + \frac{p}{p^{2s}})^{-1}$$

if E has good reduction at p (this is the inverse of the numerator of the L-function of E_p, $L(E_p, s)$) and

$$L_p(E, s) = (1 - \frac{a_p(E_p)}{p^s})^{-1}$$

with $a_p(E_p) \in \{1, -1\}$ if E has bad reduction; then the *global L*-function attached to E is

$$L(E, s) = \prod_p L_p(E, s).$$

By Hasse's bound, $L(E, s)$ is analytic and non-vanishing in the half-plane $\Re e s > 3/2$; what is more remarkable is that in fact $L(E, s)$ has analytic continuation to the whole complex plane and satisfies a functional equation. This was proved by Deuring for CM elliptic curves,

Theorem. *Let $E/\mathbf{Q}$ be an elliptic curve with complex multiplication with underlying imaginary quadratic field K. Then there exists a Hecke character χ_E of $\mathbf{I}_K$ such that*

$$L(E, s) = L(\chi_E, s);$$

consequently $L(E, s)$ has analytic continuation to $\mathbf{C}$ and satisfies the functional equation given below.

In the non-CM case this is a former conjecture of Shimura/Tanyama and a major recent achievement of Wiles, Taylor/Wiles and Breuil/Conrad/ Diamond/Taylor ([15, 13, 1])

Theorem 4. *(Wiles, Taylor/Wiles, Breuil/Conrad/Diamond/Taylor) Given E an elliptic curve defined over $\mathbf{Q}$, its associated L-function has analytic continuation to $\mathbf{C}$ and satisfies the functional equation*

$$\Lambda(E, s) = \varepsilon(E)\Lambda(E, 2 - s)$$

where $\varepsilon(E) = \pm 1$ and

$$\Lambda(E, s) = N_E^{s/2} L_\infty(E, s) L(E, s)$$

with N_E (the conductor of E) is an integer (> 1) dividing Δ_E and

$$L_\infty(E, s) = \Gamma_{\mathbf{R}}(s)\Gamma_{\mathbf{R}}(s + 1).$$

Remark 10. As we have said before the fact that this L-function admits analytic continuation to the complex plane is a strong indication that the reductions of E at the various primes are not uncorrelated, but rather conspire together to create harmony. But we will see later that these reductions however show nevertheless some random behavior.

Remark 11. In fact the proof of the theorem above gives more, namely that $L(E, s)$ is the L-function associated to a modular form of weight 2 (see Lecture 2); in particular, the GNPT and even HdVP are known for $L(E, s)$.

Beyond analytic continuation there is another striking conjecture of Birch/Swinnerton-Dyer which shows that this L-function, even if it is built out of local data attached to the curve (ie. the curves the E_p), can nevertheless "capture" some of the global arithmetical structure of it:

Conjecture. *(Birch/Swinnerton-Dyer) The order of vanishing of $L(E,s)$ at $s = 1$ equals the rank of E.*

This conjecture is now known when $\mathrm{ord}_{s=1} L(E,s) \leqslant 1$, as a combination of the works of Gross/Zagier and Kolyvagin. In fact, Birch/Swinnerton-Dyer's conjecture is more precise as it predicts the exact value of the regularized value of $L(E,s)$ at $s = 1$, $\lim_{s\to 1} L(E,s)/(s-1)^{\mathrm{rank}(E)}$ in term of other arithmetical and transcendental invariants of the curve.

4.3 The Galois representation attached to an elliptic curve

Given E/k an elliptic curve defined over some field k, we have seen already in the case of function fields, that in order to get information about the k-rational points of E, $E(k)$, it may be useful to consider more generally the set $E(\overline{k})$ of $\overline{k}$-rational points of E over $\overline{k}$ some algebraic closure of the original field k: indeed the absolute Galois group $\mathrm{Gal}(\overline{k}/k)$ has a natural action on the set $E(\overline{k})$ (given by acting on the coordinate of the point) and by Galois theory the k-rational points $E(k)$ are exactly the set of fixed points under this action.

This rather simple principle, is one of the main motivations for trying to attach to E a finite dimensional representation of the absolute Galois group $\mathrm{Gal}(\overline{k}/k)$. We shall present this when $k = \mathbf{Q}$. Before doing so we will explain a simple but fundamental example of this type in the simpler setting of cyclotomy.

4.3.1 The Tate module attached to the roots of unity

For ℓ a prime number, and $n \geqslant 1$ an integer, we have seen in the first lecture that the Galois group G_{ℓ^n} of the cyclotomic extension $\mathbf{Q}(\xi_{\ell^n})$ acts on the set μ_{ℓ^n} of ℓ^n-th roots of unity. By Galois theory, this finite group is a quotient (of finite index) of the full Galois group $G := \mathrm{Gal}(\overline{\mathbf{Q}}/\mathbf{Q})$ of $\mathbf{Q}$-linear automorphisms of the field of algebraic numbers $\overline{\mathbf{Q}} \subset \mathbf{C}$:

$$1 \to \mathrm{Gal}(\overline{\mathbf{Q}}/K_{\ell^n}) \to \mathrm{Gal}(\overline{\mathbf{Q}}/\mathbf{Q}) \to \mathrm{Gal}(K_{\ell^n}/\mathbf{Q}) \to 1.$$

Hence the projection map defines an action of $\mathrm{Gal}(\overline{\mathbf{Q}}/\mathbf{Q})$ on μ_{ℓ^n}. Hence we have a natural action of G on the sequence of ℓ-th power roots of unity

$$\{1\} \subset \mu_\ell \subset \mu_{\ell^2} \subset \cdots \subset \mu_{\ell^n} \subset \ldots$$

and it is convenient to transfer this action on an infinite sequence of spaces into an action on a single space. One forms the *projective limit* of the $\{\mu_{\ell^n}\}_{n \geqslant 0}$, which is denoted by $\varprojlim \mu_{\ell^n}$; the latter is associated with the following descending sequence of maps

$$\{1\} \leftarrow \mu_\ell \leftarrow \mu_{\ell^2} \leftarrow \cdots \leftarrow \mu_{\ell^n} \leftarrow \ldots$$

where each arrow is the ℓ-power map

$$[\ell] : \begin{matrix} \mu_{\ell^{n+1}} & \to & \mu_{\ell^n} \\ \xi & \to & \xi^\ell \end{matrix}$$

and is defined as the set of sequences

$$(\xi_1, \xi_2, \ldots, \xi_n, \ldots) \text{ satisfying for any } n \geqslant 1, \ \xi_n \in \mu_{\ell_n} \text{ and } \xi_{n+1}^\ell = \xi_n.$$

This space is a subset of the product of finite sets $\prod_{n \geqslant 1} \mu_{\ell^n}$ and, by Tychonoff, has a structure of a topological compact space. For reasons which will be clear later we denote this space by $\mathbf{T}_\ell(\mathbf{G}_m)$ and call it the *Tate module* of the multiplicative group $\mathbf{G}_m$

This space is also endowed with two natural actions.

4.3.2 Action of the ℓ-adic integers

The first action is given as follows: for each n, there is a natural action of the additive group $\mathbf{Z}/\ell^n\mathbf{Z}$ on μ_{ℓ^n} via the power map

$$\begin{matrix} \mathbf{Z}/\ell^n\mathbf{Z} & \times & \mu_{\ell^n} & \to & \mu_{\ell^n} \\ (a(\mathrm{mod}\,\ell^n) & , & \xi_n) & \to & a.\xi_n = \xi_n^a, \end{matrix}$$

and one can see easily that the various $\mathbf{Z}/\ell^n\mathbf{Z}$ are *compatible*, in the following sense: for $\xi_{n+1} \in \mu_{\ell_{n+1}}$, and $a(\mathrm{mod}\,\ell^{n+1}) \in \mathbf{Z}/\ell^{n+1}\mathbf{Z}$ one has, setting $\xi_n :=$ $[\ell]x_{n+1} = x_{n+1}^\ell$,

$$[\ell](a.\xi_{n+1}) = (a\ell).\xi_{n+1} = a(\mathrm{mod}\,\ell^n).\xi_n.$$

As a consequence, one can combine the various $\mathbf{Z}/\ell^n\mathbf{Z}$ actions altogether by forming the following projective limit $\varprojlim \mathbf{Z}/\ell^n\mathbf{Z}$ associated to the descending sequence of maps

$$\{0\} \leftarrow \mathbf{Z}/\ell\mathbf{Z} \leftarrow \mathbf{Z}/\ell^2\mathbf{Z} \leftarrow \cdots \leftarrow \mathbf{Z}/\ell^n\mathbf{Z} \leftarrow \ldots$$

where each map $\mathbf{Z}/\ell^n\mathbf{Z} \leftarrow \mathbf{Z}/\ell^{n+1}\mathbf{Z}$ is the reduction modulo ℓ^n-map. The projective limit $\varprojlim \mathbf{Z}/\ell^n\mathbf{Z}$ is noted $\mathbf{Z}_\ell$, this is a topologically compact totally disconnected integral domain which contains $\mathbf{Z}$ as a dense subring (via the diagonal embedding

$$a \in \mathbf{Z} \to (a(\mathrm{mod}\,\ell), a(\mathrm{mod}\,\ell^2), \ldots, a(\mathrm{mod}\,\ell^n), \ldots)$$

and is called the *ring of ℓ-adic integers*; an alternate definition for $\mathbf{Z}_\ell$ is the completion of $\mathbf{Z}$ with respect to the ℓ-adic distance $|a - b|_\ell = \ell^{-v_\ell(a-b)}$, where $v_\ell(a) = \max\{\beta,\ \ell^\beta | a\}$ is the ℓ-adic valuation. We will also denote $\mathbf{Q}_\ell = \mathbf{Q} \otimes \mathbf{Z}_\ell$ its field of fractions ie. this is the *field of ℓ-adic numbers*.

The above $\mathbf{Z}/\ell^n\mathbf{Z}$-action on each μ_{ℓ^n} then induces a natural continuous action of the ring $\mathbf{Z}_\ell$ on $\varprojlim \mu_{\ell^n}$ given by componentwise action:

$$(a_1, a_2, \ldots, a_n, \ldots).(\xi_1, \xi_2, \ldots, \xi_n, \ldots) = (\xi_1^{a_1}, \xi_2^{a_2}, \ldots, \xi_n^{a_n}, \ldots).$$

Moreover, let $\xi = (\xi_1, \xi_2, \ldots, \xi_n, \ldots) \in \varprojlim \mu_{\ell^n}$ be chosen such that each ξ_n is a *primitive* ℓ^n-th root of unity (ie. is of order ℓ^n exactly). Then one can show that

$$\mathbf{Z}_\ell.\xi = \varprojlim \mu_{\ell^n} = \mathbf{T}_\ell(\mathbf{G}_m),$$

making of $\mathbf{T}_\ell(\mathbf{G}_m)$ a free $\mathbf{Z}_\ell$-module of rank one.

4.3.3 Action of the Galois group

Since the Galois action of G commutes with the ℓ-power map, the G-action on each μ_{ℓ^n} defines an action of G on $\varprojlim \mu_{\ell^n}$ by componentwise action:

$$\sigma.(\xi_1, \xi_2, \ldots, \xi_n, \ldots) = (\sigma(\xi_1), \sigma(\xi_2), \ldots, \sigma(\xi_n), \ldots);$$

we recall that G is equipped with a natural (Krull) topology (for which G is compact) and one can check that the action on the compact space $\varprojlim \mu_{\ell^n}$ is continuous, and commutes with the $\mathbf{Z}_\ell$-action. Thus we have defined a $\mathbf{Z}_\ell$-linear representation of G on the rank one $\mathbf{Z}_\ell$-module $\mathbf{T}_\ell(\mathbf{G}_m)$: $\chi_{cyc} : G \to$

$Aut_{\mathbf{Z}_\ell}(\mathbf{T}_\ell(\mathbf{G}_m))$. For this we can form the tensor product of $\mathbf{T}_\ell(\mathbf{G}_m)$ with $\mathbf{Q}_\ell$ (endowed with the trivial G-action),

$$\mathbf{T}_\ell(\mathbf{G}_m) \otimes_{\mathbf{Z}_\ell} \mathbf{Q}_\ell := \mathbf{V}_\ell(\mathbf{G}_m),$$

so as to obtain a genuine (non-trivial) representation of G on a 1-dimensional $\mathbf{Q}_\ell$-vector space

$$\chi_{cyc}: \ \mathrm{Gal}(\overline{\mathbf{Q}}/\mathbf{Q}) \ \rightarrow \ Aut_{\mathbf{Q}_\ell}(\mathbf{V}_\ell(\mathbf{G}_m)).$$

This representation is our first example of an ℓ-adic Galois representation: this is *the cyclotomic character*. For each prime $p \neq \ell$, one has a well defined Frobenius conjugacy class $\mathrm{Frob}_p \subset G$ and by Section 3.2.3 of the second lecture one has

$$\chi_{cyc}(\mathrm{Frob}_p) = p. \tag{4.6}$$

Remark 12. It is not difficult to show that a continuous representation of G on a complex vector space, always factors through a finite quotient and in particular has finite image. This is definitely not the case for the cyclotomic character as well as for "most" ℓ-adic representations: for example one can show (use (4.6)) that

$$\chi_{cyc}(G) = \mathbf{Q}_\ell^\times.$$

4.3.4   The Tate module of an elliptic curve

Let $E/\mathbf{Q}$ be an elliptic curve defined over $\mathbf{Q}$. In this section, we define a natural ℓ-adic Galois representation attached to E. Let ℓ be a fixed prime at which E has good reduction; for each $n \geqslant 1$, we consider the subgroup of ℓ^n-torsion points on E, $E(\overline{\mathbf{Q}})[\ell^n] \subset E(\overline{\mathbf{Q}})$, which by (4.4) is isomorphic as a group to $(\mathbf{Z}/\ell^n\mathbf{Z})^2$. As in the case of ℓ-power roots of unity it will prove natural to form the projective limit

$$\mathbf{T}_\ell(E) := \varprojlim E(\overline{\mathbf{Q}})[\ell^n]$$

formed out of the decreasing sequence of maps

$$\{O_E\} \leftarrow E(\overline{\mathbf{Q}})[\ell] \leftarrow E(\overline{\mathbf{Q}})[\ell^2] \leftarrow \cdots \leftarrow E(\overline{\mathbf{Q}})[\ell^n] \leftarrow \ldots$$

where each arrow is the ℓ-multiplication map

$$[\ell]: \begin{array}{ccc} E(\overline{\mathbf{Q}})[\ell^{n+1}] & \rightarrow & E(\overline{\mathbf{Q}})[\ell^n] \\ P_{n+1} & \rightarrow & P_n = \ell.P_{n+1} \end{array}$$

which is defined as the set of sequences

$(P_1, P_2, \ldots, P_n, \ldots)$ satisfying for any $n \geqslant 1$, $P_n \in E(\overline{\mathbf{Q}})[\ell^n]$ and $\ell.P_{n+1} = P_n$;

this is the *Tate module* of E at ℓ.

For each n, $\mathbf{Z}/\ell^n\mathbf{Z}$ acts on $E(\overline{\mathbf{Q}})[\ell^n]$ by multiplication and this defines an action of $\mathbf{Z}_\ell$ on $\mathbf{T}_\ell(E)$ and one can show using (4.4) that $\mathbf{T}_\ell(E) \simeq \mathbf{Z}_\ell \times \mathbf{Z}_\ell$ is a free $\mathbf{Z}_\ell$-module of rank 2.

4.3.5   The Galois action

The absolute Galois group $\mathrm{Gal}(\overline{\mathbf{Q}}/\mathbf{Q})$ acts on $E(\overline{\mathbf{Q}})$ and this induces an action on $E(\overline{\mathbf{Q}})[\ell^n]$: indeed the coordinates of the ℓ^n-torsion points are the roots of polynomials with rational coefficients. Since the roots of these polynomials are permuted by the action of $\mathrm{Gal}(\overline{\mathbf{Q}}/\mathbf{Q})$, the ℓ^n-torsion points are also permuted. Moreover, one can see that the various actions of $\mathrm{Gal}(\overline{\mathbf{Q}}/\mathbf{Q})$ on the $E(\overline{\mathbf{Q}})[\ell^n]$ are compatible (ie. commute with the multiplication by ℓ map) so that one has in fact a $\mathbf{Z}_\ell$-linear representation of $\mathrm{Gal}(\overline{\mathbf{Q}}/\mathbf{Q})$ on the $\mathbf{Z}_\ell$-module of rank 2 $\mathbf{T}_\ell(E)$. Tensoring with $\mathbf{Q}_\ell$, we form $\mathbf{V}_\ell(E) := \mathbf{T}_\ell(E) \otimes_{\mathbf{Z}_\ell} \mathbf{Q}_\ell$ and obtain a two dimensional ℓ-adic representation

$$\rho_{\ell,E} : \mathrm{Gal}(\overline{\mathbf{Q}}/\mathbf{Q}) \rightarrow Aut_{\mathbf{Q}_\ell}(\mathbf{V}_\ell(E)) \simeq GL_2(\mathbf{Q}_\ell).$$

Again, this representation does not factor through a finite quotient.

4.3.6   The Weil pairing

The theory of the Weil pairing on elliptic curves gives for each $n \geqslant 1$, the existence of a non-degenerate $\mathbf{Z}/\ell^n\mathbf{Z}$-bilinear alternating map

$$\langle\, , \,\rangle_{\ell^n} : E(\overline{\mathbf{Q}})[\ell^n] \times E(\overline{\mathbf{Q}})[\ell^n] \rightarrow \mu_{\ell^n},$$

hence passing to the inverse limit and tensoring with $\mathbf{Q}_\ell$, we obtain a non-degenerate $\mathbf{Q}_\ell$-bilinear alternating map

$$\langle\, , \,\rangle_\ell : \mathbf{V}_\ell(E) \times \mathbf{V}_\ell(E) \rightarrow \mathbf{V}_\ell(\mathbf{G}_m).$$

Since this pairing is compatible with the Galois action, one deduce that

$$\det \rho_{\ell,E} = \chi_{cyc}. \tag{4.7}$$

4.3.7 Comparison with the Tate module associated to roots of unity

In this section we justify the choice of the notation $\mathbf{T}_\ell(\mathbf{G}_m)$ for the Tate module $\varprojlim \mu_{\ell^n}$. Namely, for k a field, let $\mathbf{G}_m$ be the affine plane curve defined over k by the equation

$$(\mathbf{G}_m) \qquad xy = 1.$$

Then for any field extension K of k, the set of K-rational points, $\mathbf{G}_m(K)$, is identified with *the multiplicative group* $K^\times$ via the map

$$x \in K^\times \;\rightarrow\; (x, 1/x) \in \mathbf{G}_m(K) \subset K^2,$$

and in particular $\mathbf{G}_m(K)$ has the structure of a commutative group given by

$$\begin{aligned}
\mathbf{G}_m(K) \times \mathbf{G}_m(K) &\;\rightarrow\; \mathbf{G}_m(K) \\
(x,y), (x',y') &\;\rightarrow\; (xx', yy').
\end{aligned}$$

In particular for any m one has the "multiplication by m" map:

$$[m]: \; (x,y) \rightarrow [m].(x,y) = (x^m, y^m),$$

so under this identification μ_{ℓ^n} corresponds to the set $\mathbf{G}_m(\overline{\mathbf{Q}})[\ell^n]$ of ℓ^n-torsion points in $\mathbf{G}_m(\overline{\mathbf{Q}})$. Thus the Tate module associated to the roots of unity is in fact the Tate module associated to the sets of ℓ^n torsion points on the curve $\mathbf{G}_m$.

4.3.8 L-function of an elliptic curve as an Artin L-function

Let $p \neq \ell$ be a prime. One can associate to it a decomposition subgroup and an inertia subgroup $I_p \subset D_p \subset \mathrm{Gal}(\overline{\mathbf{Q}}/\mathbf{Q})$ which are defined up to conjugacy and such that one has the following exact sequence

$$1 \rightarrow I_p \rightarrow D_p \rightarrow \mathrm{Gal}(\overline{\mathbf{F}_p}/\mathbf{F}_p) = \langle \mathrm{Frob}_p \rangle \rightarrow 1.$$

In particular, one can lift to D_p the frobenius automorphism of $\mathrm{Gal}(\overline{\mathbf{F}_p}/\mathbf{F}_p)$, which is well defined modulo I_p. In particular, one has a well defined action of the frobenius $\rho_{\ell,E}(\mathrm{Frob}_p)$ on the $\rho_{\ell,E}(I_p)$-invariant subspace $\mathbf{V}_\ell(E)^{I_p}$ and one can then form the local Artin L-factor at p:

$$L_p^{Artin}(E,s) = L_p(\rho_{\ell,E}, s) = \det\left(Id - \frac{1}{p^s}\rho_{\ell,E}(\mathrm{Frob}_p)|\mathbf{V}_\ell(E)^{I_p}\right)^{-1},$$

which is well defined and only depends on p and not on the choice of the decomposition subgroup D_p. The next theorem shows that in fact the previously defined local L-factor matches the Artin L-factor.

Theorem 5. *Let ℓ be a prime of good reduction for E, then for any $p \nmid \ell$ one has the equality*

$$L_p(E, s) = L_p^{Artin}(E, s).$$

We write out several simple but important consequences of this result:

- From the original definition of the local factors, one knows that $L_p(E, s)^{-1}$ is a polynomial of degree 2 in p^{-s} iff. E has good reduction at p. Hence for $\ell \neq p$, E has good reduction at p iff. $\mathbf{V}_\ell(E)^{I_p} = \{0\}$ (ie. the inertia subgroup acts trivially on the Tate module), that is iff. $\rho_{\ell,E}$ is unramified at p: this is the *Neron/Ogg/Shafarevitch criterion*.

- For p a prime of good reduction (for example not dividing Δ_E), one has

$$\mathrm{tr}(\rho_{\ell,E}(\mathrm{Frob}_p)) = a_p(E) = p + 1 - |E_p(\mathbf{F}_p)|,$$
$$\det(\rho_{\ell,E}(\mathrm{Frob}_p)) = p = \chi_{cyc}(\mathrm{Frob}_p).$$

- For $p \neq \ell$, the Artin local factor does not depend on the choice of the auxiliary prime ℓ, thus justifying the notation $L_p^{Artin}(E, s)$. This enables us to define the Artin local factor at ℓ by

$$L_\ell^{Artin}(E, s) = L_\ell(\rho_{\ell',E}, s)$$

for any $\ell' \neq \ell$.

- The global L-function $L(E, s)$ is an Artin type L-function: for any pairs of primes $\ell \neq \ell'$,

$$L(E, s) = \prod_p L^{Artin}(E, s) = L_\ell(\rho_{\ell',E}, s) \prod_{p \neq \ell} L_p(\rho_{\ell,E}, s)$$

We cannot discuss in great detail the proof of the important Theorem 5

4.4 The Sato-Tate conjecture and a Theorem of Serre

Given E an elliptic curve, by Hasse's bound (4.5), one can write by any prime p

$$a_p(E) = 2\sqrt{p}\cos(\theta_{E,p}), \quad \theta_{E,p} \in [0, \pi];$$

the Sato/Tate conjecture predicts the distribution law of the angles $\theta_{E,p}$ as p ranges through the primes:

Sato/Tate Conjecture. *Let E be a non-CM elliptic curve defined over $\mathbf{Q}$. Then the angles $\{\theta_{E,p}, \ p \in \mathcal{P}\}$ are equidistributed on $[0, \pi]$ relatively to the Sato/Tate measure $d\mu_{ST} = \frac{2}{\pi} \sin^2 \theta d\theta$ or in other words for any $0 \leqslant \alpha < \beta \leqslant \pi$*

$$\lim_{X \to +\infty} \frac{1}{\pi(X)} \sum_{p \leqslant X} 1_{[\alpha,\beta]}(\theta_{E,p}) = \int_\alpha^\beta \frac{2}{\pi} \sin^2 \theta d\theta = \int_{[0,\pi]} 1_{[\alpha,\beta]}(\theta) d\mu_{ST}(\theta). \quad (4.8)$$

The case of CM-elliptic curve is different and settled by the works of Deuring and Hecke:

Theorem. *Let E be a CM-elliptic curve. Then the angles $\{\theta_{E,p}, \ p \in \mathcal{P}\}$ are equidistributed on $[0, \pi]$ relatively to the measure $\frac{1}{2}\delta_{\pi/2} + \frac{1}{2\pi}d\theta$ or in other words for any $0 \leqslant \alpha < \beta \leqslant \pi$*

$$\lim_{X \to +\infty} \frac{1}{\pi(X)} \sum_{p \leqslant X} 1_{[\alpha,\beta]}(\theta_{E,p}) = \frac{1}{2}\delta_{\pi/2 \in [\alpha,\beta]} + \frac{\beta - \alpha}{2\pi}.$$

The Sato/Tate conjecture (in the non-CM case) can be given the more conceptual interpretation: by Hasse's bound, the normalized coefficient $2\cos(\theta_{E,p})$ $= a_p(E)/\sqrt{p}$ is interpreted as the trace of the matrix of determinant one

$$\begin{pmatrix} e^{\theta_{E,p}} & 0 \\ 0 & e^{-\theta_{E,p}} \end{pmatrix},$$

or more precisely as the trace of any element of the conjugacy class in $SU(2, \mathbf{C})$ defined by the matrix above (this is certainly something natural as a Frobenius is defined only up to conjugacy). It is well known that the map (let $\natural$ denote the passage to the conjugacy class)

$$T := \begin{array}{ccc} \mathbf{R}/\pi\mathbf{Z} & \to & SU(2, \mathbf{C})^\natural \\[2mm] \theta & \to & \begin{pmatrix} e^\theta & 0 \\ 0 & e^{-\theta} \end{pmatrix}^\natural \end{array}$$

from $\mathbf{R}(\mathrm{mod}\pi\mathbf{Z})$ to the space of conjugacy classes of the compact subgroup $SU(2, \mathbf{C})$ is an homeomorphism. Recall that in the first chapter, any compact group G is endowed with its Haar measure and that the space of conjugacy classes of G is endowed with its Sato/Tate measure. In the present case, $G = SU(2, \mathbf{C})$, the inverse image under T of the Sato/Tate measure on $SU(2, \mathbf{C})^\natural$ is precisely the Sato/Tate measure $\mu_{ST}(\theta)$ (this is a special case of Weyl's integration formula). Hence the Sato/Tate conjecture is naturally interpreted by saying that the $a_p(E)/\sqrt{p}$ define via the map T a set of conjugacy classes in $SU(2\mathbf{C})$, which fill uniformly $SU(2,\mathbf{C})^{\natural}$ w.r.t. μ_{ST}. One should note the strong parallel between this conjecture and the Chebotareff density theorem and in fact the current strategy for proof follows the same lines.

One first notes that it is sufficient to replace in (4.8) the characteristic function of any interval by any continuous function on $G^{\natural}$. Next by the Peter/Weyl theorem, the space of continuous functions on $G^{\natural}$ is densely generated (for the L^{∞}-norm) by the characters of the irreducible representations of G (and the irreducible characters form an orthonormal basis of $L^2(G^{\natural}, \mu_{ST,G^{\natural}})$). Thus in order to prove the Sato/Tate conjecture it is sufficient (Weyl's equidistribution criterion) to prove that for any irreducible non-trivial representation ρ of G

$$\lim_{X \to +\infty} \frac{1}{\pi(X)} \sum_{p \leqslant X} tr(\rho(T(\theta_{E,p}))) = \int_{G^{\natural}} tr(\rho(g^{\natural})) d\mu_{ST}(g^{\natural}) = 0.$$

For $G = SU(2, \mathbf{C})$ the non-trivial irreducible representations are the k-th symmetric powers (for $k \geqslant 1$) of the standard (2-dimensional) representation. Recall that the k-th symmetric power representation, Sym^k say, is defined as follows: the associated complex vector space is $\mathbf{C}[X, Y]_k$, the space of homogeneous polynomials in two variables of degree k, and with $SU(2, \mathbf{C})$ acting on $P(X, Y) \in \mathbf{C}[X, Y]_k$ by

$$\mathrm{Sym}^k \begin{pmatrix} a & b \\ c & d \end{pmatrix} P(X, Y) = P(aX + cY, bX + cZ).$$

This representation is irreducible and for $D = \begin{pmatrix} \alpha_1 & 0 \\ 0 & \alpha_2 \end{pmatrix}$ diagonal, one has

$$tr(\mathrm{Sym}^k D) = \alpha_1^k + \alpha_1^{k-1}\alpha_2 + \cdots + \alpha_1\alpha_{k-1} + \alpha_2^k;$$

in particular one has for any $k \geqslant 1$

$$tr(\mathrm{Sym}^k(T(\theta))) = \frac{\sin((k+1)\theta)}{\sin(\theta)} =: \mathrm{sym}_k(\theta),$$

say. Hence, the Sato/Tate conjecture would follow from estimates for sums over the primes: for any $k \geqslant 1$

$$\lim_{X \to +\infty} \frac{1}{\pi(X)} \sum_{p \leqslant X} \mathrm{sym}_k(\theta_{E,p}) = 0. \tag{4.9}$$

Such sums are naturally associated to L-functions. For $k \geqslant 1$, let Sym^k : $GL(2) \to GL(k+1)$ denote the k-th symmetric power of the standard representation and for $\ell \geqslant 2$, let $\mathrm{Sym}^k_{\ell,E}$ denote the ℓ-adic Galois representation obtained by composition of $\rho_{\ell,E}$ with the k-th symmetric power:

$$\mathrm{Sym}_{\ell,E}^k : \mathrm{Gal}(\overline{\mathbf{Q}}/\mathbf{Q}) \to \mathrm{Aut}(T_\ell(E)) \simeq GL(2,\mathbf{Q}_\ell) \to GL(k+1,\mathbf{Q}_\ell).$$

$L(\mathrm{Sym}_{\ell,E}^k, s)$ is its associated Artin L-function. By construction, one has for $p \nmid \ell \Delta_E$

$$\lambda_{\mathrm{Sym}^k E}(p) = tr(\mathrm{Frob}_p|\mathrm{Sym}_{\ell,E}^k) = p^{k/2}\mathrm{sym}_k(\theta_{E,p}),$$

hence the Sato/Tate conjecture would follow from the following (weak form of a) conjecture of Langlands

Conjecture. *(Langlands) For any $k \geqslant 1$, the completed (by an explicit local factor at infinity) L-function, $\Lambda(\mathrm{Sym}_{E,\ell}^k, s)$ is automorphic; in particular, it satisfies FCT (and is holomorphic over $\mathbf{C}$) and GPNT.*

The case $k = 1$ of this conjecture is just Theorem 4 above. The case $k = 2$ follows from Theorem 4 – more precisely the fact that $L(E,s)$ is modular cf. Remark 11 – and earlier work of Gelbart/Jacquet [3]. The cases $k = 3, 4$ are due to Kim/Shahidi and Kim [4, 5, 6] and by the Rankin/Selberg theory of Jacquet/Piatetsky-Shapiro/Shalika, their results also imply that (4.9) is valid for for $k = 5, 6, 7, 8$ so that by now (4.9) holds unconditionally for $k \leqslant 8$.

We will now describe an application of Chebotareff's theorem to the Sato/Tate conjecture, due to Serre [10]. It concerns the proportion of primes for which $a_p(E) = 0$; note that the condition $a_p(E) = 0$ implies that E_p is a supersingular elliptic curve. As explained above if E is CM, the proportion of primes such that $a_p(E) = 0$ ($\theta_{E,p} = \pi/2$) is 1/2. On the other hand, when E has no CM, the Sato/Tate conjecture predicts that the proportion of such primes is 0.

Theorem. *(Serre) Let E be a non-CM elliptic curve, then for $X \to +\infty$*

$$|\{p \leqslant X, a_p(E) = 0\}| = o(\pi(X)).$$

For ℓ a prime, one considers the Galois representation of $\mathrm{Gal}(\overline{\mathbf{Q}}/\mathbf{Q})$ acting on the set of ℓ-torsion points of E, $E(\overline{Q})[\ell] \simeq (\mathbf{Z}/\ell\mathbf{Z}) \times (\mathbf{Z}/\ell\mathbf{Z}) = \mathbf{F}_\ell^2$. During the proof of Theorem 5 one obtains the following congruence relation: for $p \nmid \ell \Delta_E$ one has

$$tr(\mathrm{Frob}_p|E(\overline{Q})[\ell]) = a_p(E) \ (\mathrm{mod}\,\ell), \quad \det(\mathrm{Frob}_p|E(\overline{Q})[\ell]) = p \ (\mathrm{mod}\,\ell).$$

on the other hand, Serre proved the deep and remarkable fact that the image of this representation in $\mathrm{Aut}(E(\overline{Q})[\ell]) \simeq GL(2,\mathbf{F}_\ell)$ is most of the time as big as possible [9].

Theorem. *(Serre) Let E be a non-CM elliptic curve, for all but finitely many ℓ the image of* $\mathrm{Gal}(\overline{\mathbf{Q}}/\mathbf{Q})$ *in* $\mathrm{Aut}(E(\overline{Q})[\ell]) \simeq GL(2, \mathbf{F}_\ell)$ *is* $GL(2, \mathbf{F}_\ell)$.

Picking ℓ as above, by the Chebotareff density theorem, one sees that for $X \to +\infty$

$$\frac{1}{\pi(X)} |\{p \leqslant X,\ tr(\mathrm{Frob}_p | E(\overline{Q})[\ell]) = 0(\mathrm{mod}\,\ell)\}|$$

$$\to \frac{|\{g \in GL(2, \mathbf{F}_\ell),\ tr(g) = 0\}|}{|GL(2, \mathbf{F}_\ell)|} \ll \frac{\ell^3}{\ell^4} \ll \frac{1}{\ell};$$

hence by the congruence relation above we obtain that for any ℓ large enough

$$\limsup_{X \to +\infty} \frac{1}{\pi(X)} |\{p \leqslant X,\ a_p(E) = 0\}|$$

$$\leqslant \lim_{X \to +\infty} \frac{1}{\pi(X)} |\{p \leqslant X,\ a_p(E) \equiv 0\ (\mathrm{mod}\,\ell)\}| \ll \frac{1}{\ell}.$$

References

[1] Christophe Breuil, Brian Conrad, Fred Diamond and Richard Taylor, *On the modularity of elliptic curves over* **Q**: *wild 3-adic exercises*, J. Amer. Math. Soc., **14** (2001), no. 4, 843– 939 (electronic).

[2] Gary Cornell, Joseph H. Silverman and Glenn Stevens (eds.), *Modular forms and Fermat's last theorem*, Springer-Verlag, New York, 1997. Papers from the Instructional Conference on Number Theory and Arithmetic Geometry held at Boston University, Boston, MA, August 9–18, 1995.

[3] Stephen Gelbart, Hervé Jacquet, *A relation between automorphic representations of* GL(2) *and* GL(3), Ann. Sci. École Norm. Sup. (4), **11** (1978), no. 4, 471– 542.

[4] Henry H. Kim, *Functoriality for the exterior square of* GL$_4$ *and the symmetric fourth of* GL$_2$, J. Amer. Math. Soc., **16** (2003), no. 1, 139– 183 (electronic). With appendix 1 by Dinakar Ramakrishnan and appendix 2 by Kim and Peter Sarnak.

[5] Henry H. Kim and Freydoon Shahidi, *Functorial products for* GL$_2$ × GL$_3$ *and the symmetric cube for* GL$_2$, Ann. of Math. (2), **155** (2002), no. 3, 837– 893. With an appendix by Colin J. Bushnell and Guy Henniart.

[6] Henry H. Kim and Freydoon Shahidi, *Cuspidality of symmetric powers with applications*, Duke Math. J., **112** (2002), no. 1, 177– 197.

[7] B. Mazur, *Rational isogenies of prime degree (with an appendix by D. Goldfeld)*, Invent. Math., **44** (1978), no. 2, 129– 162.

[8] Loïc Merel, *Bornes pour la torsion des courbes elliptiques sur les corps de nombres*, Invent. Math., **124** (1996), no. 1-3, 437– 449 (French).

[9] Jean-Pierre Serre, *Propriétés galoisiennes des points d'ordre fini des courbes elliptiques*, Invent. Math., **15** (1972), no. 4, 259– 331 (French).

[10] Jean-Pierre Serre, *Abelian l-adic representations and elliptic curves*, McGill University lecture notes written with the collaboration of Willem Kuyk and John Labute, W. A. Benjamin, Inc., New York-Amsterdam, 1968.

[11] Jean-Pierre Serre, *Quelques applications du théorème de densité de Chebotarev*, Inst. Hautes Études Sci. Publ. Math. (1981), no. 54, 323– 401 (French).

[12] Joseph H. Silverman, *The arithmetic of elliptic curves*, Graduate Texts in Mathematics, **106**, Springer-Verlag, New York, 1994. Corrected reprint of the 1986 original.

[13] Richard Taylor and Andrew Wiles, *Ring-theoretic properties of certain Hecke algebras*, Ann. of Math. (2), **141** (1995), no. 3, 553– 572.

[14] Douglas Ulmer, *Elliptic curves with large rank over function fields*, Ann. of Math. (2), **155** (2002), no. 1, 295– 315.

[15] Andrew Wiles, *Modular elliptic curves and Fermat's last theorem*, Ann. of Math. (2), **141** (1995), no. 3, 443– 551.

5   *L*-functions over function fields

In this last lecture, we consider some analog in positive characteristic of the Artin *L*-functions considered before. For this we need to introduce some analog of the rationals or of number fields. this analog is to be found amongst the field of functions of a non-singular curve defined over a finite field.

5.1 Function fields

Let $k = \mathbf{F}_q$ be a finite field (q is a prime power) and $\overline{k}$ be an algebraic closure; an algebraic affine plane curve C^a is given by an equation

$$F(x, y) = 0 \qquad (5.1)$$

where $F(x, y) \in k[x, y]$ is a polynomial in two variables with coefficients in k which is irreducible as a polynomial in $\overline{k}$. In particular, the elliptic curves encountered in the previous lectures form a special class of curves. We will require the curve to be non-singular by which we mean that the system of equations

$$F(x, y) = \frac{\partial}{\partial x} F(x, y) = \frac{\partial}{\partial y} F(x, y) = 0$$

has no solutions in $\overline{k}^2$. If K is any field extension of k, we denote by $C^a(K)$ the set of K-rational points of C^a, that is the set of solutions in K^2 of the equation (5.1):

$$C^a(K) = \{(x, y) \in K^2, \ F(x, y) = 0\}.$$

Note that for a finite extension $k \subset k' \subset \overline{k}$, the Galois group $\mathrm{Gal}(\overline{k}/k)$ acts on the set of k'-rational point by acting on the coordinates: for $\sigma \in \mathrm{Gal}(\overline{k}/k)$ and $P = (x, y) \in C^a(K)$, $\sigma.P = P' = (\sigma(x), \sigma(y)) \in C^a(K)$. The *ring of (algebraic) functions* on C^a is by definition the quotient of the ring $k[x, y]$ by the ideal generated by $F(x, y)$

$$\mathcal{O}_{C^a} := k[x, y]/(F(x, y)).$$

This is an integral domain and the *function field* of C^a is the field of fractions of $\mathcal{O}_{C^a}$:

$$k(C) := Frac(\mathcal{O}_{C^a}).$$

As in the case of number fields, the ideals of $\mathcal{O}_{C^a}$ factor uniquely as a product of prime ideals. For $\mathfrak{p}$ a prime ideal, the quotient ring

$$k_{\mathfrak{p}} = \mathcal{O}_{C^a}/\mathfrak{p}.\mathcal{O}_{C^a}$$

is a finite extension of k and is called the residue field. Its degree is by definition the degree of $\mathfrak{p}$ and is written $\deg \mathfrak{p}$; the *norm* of $\mathfrak{p}$, $N(\mathfrak{p})$, is the cardinality of $k_{\mathfrak{p}}$. A prime ideal of C^a is also called a *closed point*: the reason is that there is a one-to-one correspondence between the prime ideals of $\mathcal{O}_{C^a}$ and the set of $\mathrm{Gal}(\overline{k}/k)$-orbits of geometric (ie. $\overline{k}$-rational) points $P \in C^a(\overline{k})$. This correspondence preserves the degree in the sense that if $\mathfrak{p}$ corresponds to the orbit, $\mathrm{Gal}(\overline{k}/k).P := \{P^\sigma,\ \sigma \in \mathrm{Gal}(\overline{k}/k)\}$, of $P = (x_P, y_P)$ then $k_\mathfrak{p} = k(x_P, y_P)$ is the field generated by the coordinates of P and $|\mathrm{Gal}(\overline{k}/k).P| = \deg \mathfrak{p}$.

Example. The most typical example is that of the affine line $\mathbf{A}_k^1$ defined by the equation $Y = 0$. In that case $\mathcal{O}_{C^a} = k[X]$ is a principal domain and the prime ideals are generated by the irreducible monic polynomials in $k[X]$. Given an irreducible polynomial $p(X)$ and $\mathfrak{p} = p.k[X] \subset k[X]$, one has $\deg \mathfrak{p} = \deg p$ and the orbit of $k_\mathfrak{p}$-rational points corresponding to $\mathfrak{p}$ is just the set of roots of $p(X)$ in $\overline{k}$.

The L-function attached to the affine curve C^a is given by the product

$$L(C^a, s) := \prod_\mathfrak{p} (1 - \frac{1}{N(\mathfrak{p})^s})^{-1} = \sum_\mathfrak{a} \frac{1}{N(\mathfrak{a})^s}.$$

Here $\mathfrak{a}$ ranges over the ideals of $\mathcal{O}_{C^a}$.

Lemma 1. *One has the following alternative expression*

$$L(C^a, s) = \exp(\sum_{n \geqslant 1} \frac{|C^a(\mathbf{F}_{q^n})|}{n} q^{-ns}).$$

Proof. The proof is obtained by decomposing each $C^a(\mathbf{F}_{q^n})$ into a disjoint union of $\mathrm{Gal}(\overline{k}/k)$-orbits:

$$|C^a(\mathbf{F}_{q^n})| = \sum_{d|n} d|\{Gal(\overline{k}/k)-\text{orbits of length } d\}| = \sum_{d|n} d|\{\mathfrak{p}, \deg \mathfrak{p} = d\}|;$$

then setting $T = q^{-s}$, one has

$$\sum_{n \geqslant 1} \frac{|C^a(\mathbf{F}_{q^n})|}{n} q^{-ns} = \sum_{n \geqslant 1} \frac{T^n}{n} \sum_{d|n} d|\{\mathfrak{p}, \deg \mathfrak{p} = d\}| = \sum_\mathfrak{p} \deg \mathfrak{p} \sum_{\substack{n \geqslant 1 \\ n \equiv 0 (\deg \mathfrak{p})}} \frac{T^n}{n}$$

$$= \sum_\mathfrak{p} \sum_{n \geqslant 1} \frac{T^{dn}}{n} = \sum_\mathfrak{p} \log(1 - T^{\deg \mathfrak{p}})^{-1}$$

$\square$

It is not difficult to see that $L(C^a, s)$ converges absolutely for $\mathfrak{Re} s > 1$; moreover it was proven by A. Weil that $L(C^a, s)$ has meromorphic continuation to $\mathbf{C}$ and has all its zeros satisfying $\mathfrak{Re} s \leqslant 1/2$ (this is the Riemann hypothesis for curves over finite fields). However the L-function of C^a, as it is defined, does not satisfy a nice functional equation and does not have all its zeros on the critical line $\Re es = 1/2$ in general: the reason it that (much as in the case of the standard L-function) it has to be completed with local factors corresponding to "points at infinity". To see this consider the two simple examples of $C^a = \mathbf{A}^1_k$ the affine line or $C^a = E^a$ is a affine elliptic curve. In the first case, $\mathbf{A}^1_k$ needs to be embedded into the projective line $\mathbf{P}^1_k$ and one needs to add an extra k-rational point at infinity so that

$$L(\mathbf{A}^1_k, s) = \frac{1}{(1 - q^{1-s})} \quad , \quad L(\mathbf{P}^1_k, s) = \frac{1}{(1 - q^{-s})(1 - q^{1-s})},$$
$$L(\mathbf{P}^1_k, 1 - s) = q^{1-2s} L(\mathbf{P}^1_k, s).$$

In the second case, again the point at infinity O_E was missing and (recall that $|a_q(E)| \leqslant 2\sqrt{q}$)

$$L(E^a, s) = \frac{1 - a_q(E)q^{-s} + q^{1-2s}}{(1 - q^{1-s})},$$
$$L(E, s) = \frac{1 - a_q(E)q^{-s} + q^{1-2s}}{(1 - q^{-s})(1 - q^{1-s})} = L(E, 1 - s).$$

In both cases, one see that the completed L-function satisfies a functional equation.

5.2   Elliptic curves over function fields

We have seen that there is a strong analogy between number fields and function fields of curves. One may then consider elliptic curves defined over a function field $k(C)$. From now on the function field considered will be the simplest one: the field associated to the affine line $\mathbf{A}^1_k$, $k(\mathbf{A}^1_k) = k(T)$. An elliptic curve over K is the curve defined by the equation

$$E_T : \quad y^2 = x^3 + a(T)x + b(T) \tag{5.2}$$

with $a(T), b(T) \in k[T]$ two polynomials in the indeterminate T such that $\Delta(a, b) = \Delta(T) = 4a^3(T) + 27b^2(T) \neq 0$. Moreover, we will assume that $(a(T), b(T)) = 1$.

Let $\mathfrak{p} \subset k[T]$ be the prime ideal generated by some monic irreducible polynomial $p(T) \in k[T]$. One can then consider the equation (5.2) and reduce its coefficients modulo p, this gives

$$E_\mathfrak{p} : \quad y^2 = x^3 + \overline{a(T)}x + \overline{b(T)},$$

(with $\overline{a(T)}$, $\overline{b(T)}$ the reductions of $a(T)$, $b(T)$ modulo p) and we obtain an elliptic curve over $k_\mathfrak{p} = k[T]/(p) \simeq \mathbf{F}_{q^{\deg p}}$; note moreover that $E_\mathfrak{p}$ is non-singular if $p(T) \nmid \Delta(T)$.

Remark 13. If $p(T)$ has degree 1, $p(T) = T - t$ with $t \in k$, then reducing modulo $p(T)$ amounts to replacing the indeterminate T in (5.2) by the scalar t, thus one obtain an elliptic curve over k. In that way, one sees E_T as a one parameter *family* of elliptic curves (defined over finite fields) indexed by the affine line.

For each prime $\mathfrak{p}$ such that $p \nmid \Delta(T)$, one can define as usual $a_\mathfrak{p}(E)$ by

$$|E_\mathfrak{p}(k_\mathfrak{p})| = N(\mathfrak{p}) + 1 - a_\mathfrak{p}(E)$$

and by Hasse's bound

$$|a_\mathfrak{p}(E)| \leqslant 2\sqrt{N(\mathfrak{p})} = 2q^{\deg p/2}.$$

One can then form the global (incomplete) L-function

$$L^{(\Delta)}(E, s) = \prod_{\mathfrak{p} \nmid \Delta(T)} (1 - \frac{a_\mathfrak{p}(E)}{N(\mathfrak{p})^s} + \frac{N(\mathfrak{p})}{N(\mathfrak{p})^{2s}})^{-1}.$$

This series is absolutely convergent for $\mathfrak{Re}\, s > 3/2$ and it is a consequence of deep results of Grothendieck, Deligne and others that

1. $L^{(\Delta)}(E, s)$ admits analytic continuation to $\mathbf{C}$ (in fact $L^{(\Delta)}(E, s)$ is a polynomial in q^{-s} with coefficients in $\mathbf{Q}$),

2. The zeros of $L^{(\Delta)}(E, s)$ satisfy $\mathfrak{Re}\, s \leqslant 1$ (this is the analog of GRH).

In fact the analog of the Sato/Tate conjecture hold in this case

Theorem 6. *(Deligne) Let E be an elliptic curve over $k(T)$ defined by the equation (5.2) and with $a(T), b(T)$ coprime; for $\mathfrak{p} \nmid \Delta(T)$ define $\theta_{E,\mathfrak{p}} \in [0, \pi]$ by*

$$\cos(\theta_{E,\mathfrak{p}}) = a_\mathfrak{p}(E)/2\sqrt{N(\mathfrak{p})},$$

then as $d \to +\infty$, the set $\{\theta_{E,\mathfrak{p}}\}_{\deg \mathfrak{p}=d}$ becomes equidistributed on $[0, \pi]$ relatively to the Sato/Tate measure $d\mu_{ST} = \frac{2}{\pi}\sin^2\theta d\theta$. In other words, for any $0 \leqslant \alpha < \beta \leqslant \pi$, one has, for $d \to +\infty$,

$$\frac{1}{|\{\mathfrak{p}, \deg \mathfrak{p} = d\}|} \sum_{\deg \mathfrak{p}=d} 1_{[\alpha,\beta]}(\theta_{E,\mathfrak{p}}) \to \int_\alpha^\beta \frac{2}{\pi}\sin^2\theta d\theta.$$

In the next section, we will explain the structure of the proof of that theorem.

5.3 ℓ-adic sheaves and their L-functions

Let $K = k(T)$. We denote by $\overline{k}$ an algebraic closure of k, by $\overline{K}$ an algebraic closure of K and $G_K = \mathrm{Gal}(\overline{K}/K)$ its Galois group. We also denote by $K^{sep} \subset \overline{K}$
the *separable closure* of K (ie. the set of elements of $\overline{K}$ whose minimal polynomial –over K– has no multiple root). The Galois group of K^{sep} over K will be
called the *arithmetic Galois group* and will be noted $G_K^{arith} := \mathrm{Gal}(K^{sep}/K)$.
In contrast with the number field case K^{sep} has another important subfield
namely $\overline{k}.K = \overline{k}(T)$, whose Galois group $\mathrm{Gal}(K^{sep}/\overline{k}.K)$ is called the *geometric* Galois group and will be noted G_K^{geom}. Hence we have the exact sequence

$$1 \to G_K^{geom} \to G_K^{arith} \to \mathrm{Gal}(\overline{k}.K/K) \to 1 \tag{5.3}$$

with $\mathrm{Gal}(\overline{k}.K/K) \simeq \mathrm{Gal}(\overline{k}/k) = \langle \mathrm{Frob}_q \rangle$; the isomorphism is obtained by
restriction to $\overline{k}$.

To $\mathfrak{p}$ a prime ideal of $k[T]$, one can associate a *decomposition subgroup*
$D_{\mathfrak{p}} \subset G_K^{arith}$ well defined up to conjugacy, together with a surjective *reduction
map*

$$D_{\mathfrak{p}} \mapsto \mathrm{Gal}(\overline{k}/k_{\mathfrak{p}}),$$

whose kernel is by definition the inertia subgroup at $\mathfrak{p}$ and is denoted $I_{\mathfrak{p}}$:

$$1 \to I_{\mathfrak{p}} \to D_{\mathfrak{p}} \to \mathrm{Gal}(\overline{k}/k_{\mathfrak{p}}) = \langle \mathrm{Frob}_{\mathfrak{p}} \rangle \to 1.$$

Here and for the rest of this lecture, we make the convention that $\mathrm{Frob}_{\mathfrak{p}}$ is not
the usual frobenius (which sends x to x^q) but its inverse: the latter is called
the *geometric frobenius* .

Let $\ell \nmid q$ be a prime invertible in k, and let $\mathbf{Q}_\ell$ be the field of ℓ-adic numbers.
We suppose there has been chosen once and for all an algebraic closure $\overline{\mathbf{Q}}_\ell$
contained in $\mathbf{C}$.

Definition. *Let $S = \{\mathfrak{p}\}$ be a finite set of places and let $\ell \nmid q$ be a prime. An
$\overline{\mathbf{Q}}_\ell$-sheaf on $\mathbf{A}_k^1$, $\mathcal{F} = (\rho_{\mathcal{F}}, V_{\mathcal{F}})$ say, lisse outside S, is a continuous representation of G_K^{arith} on a finite dimensional $\overline{\mathbf{Q}}_\ell$-vector space $V_{\mathcal{F}}$*

$$\rho_{\mathcal{F}} : G_K^{arith} \to Aut_{\overline{\mathbf{Q}}_\ell}(V_{\mathcal{F}}),$$

*such that for any $\mathfrak{p} \notin S$ and for some choice of a decomposition subgroup $D_{\mathfrak{p}}$,
the inertia subgroup $I_{\mathfrak{p}}$ acts trivially on $V_{\mathcal{F}}$ (this implies that the action of the
inertia of any decomposition subgroup at $\mathfrak{p}$ is trivial).*

- *By definition, the* rank *of $\mathcal{F}$ is the dimension of $V_{\mathcal{F}}$ and will be noted $r_{\mathcal{F}}$.*

- *By definition, one has, choosing a basis for $V_{\mathcal{F}}$, the inclusions*

$$\rho_{\mathcal{F}}(G_K^{geom}) \subset \rho_{\mathcal{F}}(G_K^{arith}) \subset Aut(V_{\mathcal{F}}) \simeq GL(r_{\mathcal{F}}, \overline{\mathbf{Q}}_\ell) \subset GL(r_{\mathcal{F}}, \mathbf{C}).$$

  The Zariski closure of $\rho_{\mathcal{F}}(G_K^{geom})$ (resp. $\rho_{\mathcal{F}}(G_K^{arith})$) inside $GL(r_{\mathcal{F}}, \mathbf{C})$ is called the geometric (resp. arithmetic) monodromy group and is noted $G_{\mathcal{F}}^{geom}$ (resp. $G_{\mathcal{F}}^{arith}$).

- *If G_K^{arith} (resp G_K^{geom}) acts irreducibly on $V_{\mathcal{F}}$ the sheaf $\mathcal{F}$ is called arithmetically (resp. geometrically) irreducible.*

Remark 14. Since $G_K^{geom} \subset G_K^{arith}$, $G_{\mathcal{F}}^{geom} \subset G_{\mathcal{F}}^{arith}$ and geometrical irreducibility is stronger than arithmetical irreducibility.

Remark 15. There is another, more geometrical description of a ℓ-adic sheaf as a local system over the base curve $\mathbf{A}_k^1$. Such a description is useful in the proof the two main theorems given below.

Because of (5.3) one can associate to any $\mathfrak{p} \notin S$ a well defined Frobenius conjugacy class in $Aut(V)$ which we denote by $\rho_{\mathcal{F}}(\mathrm{Frob}_{\mathfrak{p}})$ or $\mathrm{Frob}_{\mathfrak{p}}|\mathcal{F}$.

Definition. *For w an integer, the sheaf $\mathcal{F}$ is pure (resp. mixed) of weight w if for any $\mathfrak{p} \notin S$, and any eigenvalue α of $\mathrm{Frob}_{\mathfrak{p}}|\mathcal{F}$ (viewed as a complex number via the embedding chosen above), one has $|\alpha| = N(\mathfrak{p})^{w/2}$. (resp. $|\alpha| \leqslant N(\mathfrak{p})^{w/2}$).*

By a fundamental result of Deligne, the geometric monodromy group of a pure sheaf of some weight is reductive and even semi-simple.

The (incomplete) L-function associated to an ℓ-adic sheaf $\mathcal{F}$ is the Euler product

$$L(\mathcal{F}, s) = \prod_{\mathfrak{p} \notin S} \det(Id - \frac{1}{N(\mathfrak{p})^s} \mathrm{Frob}_{\mathfrak{p}}|\mathcal{F})^{-1}.$$

If $\mathcal{F}$ is mixed of weight w, then $L(\mathcal{F}, s)$ is absolutely convergent for $Res > 1 + w/2$.

A fundamental fact due to Grothendieck is that $L(\mathcal{F}, s)$ admits meromorphic continuation to $\mathbf{C}$: this is the content of the

Theorem. *(Grothendieck/Lefshetz trace formula) Let $\mathcal{F}$ be an $\overline{\mathbf{Q}}_\ell$-sheaf lisse on $U = \mathbf{A}_k^1 - S$ and pure of some weight w. There exist three finite dimensional $\overline{\mathbf{Q}}_\ell$-vector spaces, $H_c^0(U, \mathcal{F})$, $H_c^1(U, \mathcal{F})$, $H_c^2(U, \mathcal{F})$ say, on which $\mathrm{Gal}(\overline{k}/k)$ acts and such that for any $n \geqslant 1$ one has*

$$\sum_{\substack{\deg \mathfrak{p} \mid n \\ \mathfrak{p} \notin S}} \deg \mathfrak{p}.tr(\mathrm{Frob}_\mathfrak{p}^{n/\deg \mathfrak{p}} | \mathcal{F}) = \sum_{i=0}^2 (-1)^i tr(\mathrm{Frob}_q^n | H_c^i(U, \mathcal{F})). \qquad (5.4)$$

In particular (by using (5.4) and a computation similar to the proof of Lemma 1), one has the identity

$$L(\mathcal{F}, s) = \frac{\det(Id - q^{-s}\mathrm{Frob}_q | H_c^1(U, \mathcal{F}))}{\det(Id - q^{-s}\mathrm{Frob}_q | H_c^0(U, \mathcal{F})) \det(Id - q^{-s}\mathrm{Frob}_q | H_c^2(U, \mathcal{F}))}$$

which shows the meromorphic continuation of $L(\mathcal{F}, s)$ to the whole complex plane.

Moreover (as U is affine) $H_c^0(U, \mathcal{F})$ is always $\{0\}$ and, if $\mathcal{F}$ is geometrically irreducible,

$$H_c^2(U, \mathcal{F}) = \{0\},$$

so that in that case, $L(\mathcal{F}, s)$ extends to an holomorphic function over $\mathbf{C}$. In general, $\dim H_c^2(U, \mathcal{F})$ equals the dimension of the coinvariants of V under the action of G_K^{geom}.

The next fundamental result, due to Deligne [2], is a sharp bound for the eigenvalues of Frob_q acting on the $H_c^i(U, \mathcal{F})$ in terms of the weight of $\mathcal{F}$; this is the exact analog to GRH:

Theorem. *(Deligne) For $\mathcal{F}$ an ℓ-adic sheaf pure of weight w as above, for $i = 0, 1, 2$, the eigenvalues of Frob_q acting on $H_c^i(U, \mathcal{F})$ satisfy $|\alpha| \leqslant q^{(i+w)/2}$. In particular, the poles of $L(\mathcal{F}, s)$ are located in the half plane $\mathfrak{Re}s \leqslant 1 + \frac{w}{2}$ (and there are no such poles if $\mathcal{F}$ is geometrically irreducible) and the zeros of $L(\mathcal{F}, s)$ are located in the half plane $\mathfrak{Re}s \leqslant \frac{1+w}{2}$.*

Armed with these two powerful black boxes, we can at least describe the proof of the Sato/Tate conjectures in the function field case. By Weyl's equidistribution criterion, it is sufficient to prove that for any $k \geqslant 1$ (see Lecture 4)

$$\lim_{d \to +\infty} \frac{1}{|\{\mathfrak{p}, \ \deg \mathfrak{p} = d\}|} \sum_{\mathfrak{p}, \ \deg \mathfrak{p} = d} \mathrm{sym}_k(\theta_{E, \mathfrak{p}}) = 0.$$

The fundamental theoretical point is to realize the $\mathrm{sym}_k(\theta_{E,\mathfrak{p}})$ as traces of frobenius of an ℓ-adic sheaf. It turn out that such a sheaf exists: more precisely (see [2])

Theorem 7. *Let E_T be the elliptic curve over K defined by the equation (5.2) with $\Delta(T) \neq 0$. There exists a $\mathbf{Q}_\ell$-sheaf $\mathcal{E}$, of rank 2, lisse on $A^1_k - S$, where $S = \{\mathfrak{p}, \ \mathfrak{p}|\Delta(T)\}$, and of weight 1 such that*

- *For $\mathfrak{p} \notin S$, $\mathrm{tr}(\mathrm{Frob}_\mathfrak{p}|\mathcal{E}) = a_\mathfrak{p}(E)$, $\det(\mathrm{Frob}_\mathfrak{p}|\mathcal{E}) = N(\mathfrak{p})$.*

- *The rank one sheaf $\det(\mathcal{E})$ is geometrically trivial (ie. the representation $\det(\rho_\mathcal{E})$ restricted to G^{geom}_K is trivial); in particular $G^{geom}_\mathcal{E} \subset SL(2, \mathbf{C})$.*

- *When $a(T)$, $b(T)$ are coprimes, the geometric monodromy group is s big as possible:*
$$G^{geom}_\mathcal{E} = SL(2, \mathbf{C}). \tag{5.5}$$

In rough terms, the sheaf $\mathcal{E}$ is obtained as the local system formed by the family of ℓ-adic Tate modules $V_\ell(E_t)$ for t varying over $\mathbf{A}^1_k(\overline{k})$, but there exists a more formal and conceptual construction.

The key geometrical ingredient is the fact that for $(a(T), b(T)) = 1$ the geometric monodromy group is big: the coprimality condition is a simple criterion to insure that the family E_T is not geometrically constant. One can see this maximality statement as the analog in the function field case of the Theorem of Serre (see Lecture 4) asserting that for a non-CM elliptic curve E the image of the Galois representation on the ℓ-torsion points of E is almost always as big as possible.

Since $\rho_\mathcal{E}(G^{arith}_K) \subset Aut_{\mathbf{Q}_\ell}(V) \simeq GL(2, \mathbf{Q}_\ell)$ one can compose this representation with any representation ρ of $GL(2)$. Considering for $k \geqslant 1$ the k-th symmetric power of the standard representation of $GL(2)$, one obtains a new ℓ-adic sheaf of rank $k + 1$

$$\mathrm{Sym}^k\mathcal{E} : G^{arith}_K \to Aut_{\mathbf{Q}_\ell}(V) \to Aut_{\mathbf{Q}_\ell}(\mathrm{Sym}^k V) \simeq GL(k + 1, \mathbf{Q}_\ell).$$

For $\mathfrak{p} \notin S$, let $\{\alpha_{1,\mathfrak{p}}(E), \ \alpha_{2,\mathfrak{p}}(E)\}$ be the eigenvalues of $\mathrm{Frob}_\mathfrak{p}|\mathcal{F}$

$$\alpha_{1,\mathfrak{p}}(E) + \alpha_{2,\mathfrak{p}}(E) = a_\mathfrak{p}(E) = 2\sqrt{N(\mathfrak{p})}\cos(\theta_{E,\mathfrak{p}}), \ \alpha_{1,\mathfrak{p}}(E)\alpha_{2,\mathfrak{p}}(E) = N(\mathfrak{p}).$$

Then the eigenvalues of the frobenius at $\mathfrak{p}$ of $\mathrm{Sym}^k\mathcal{E}$ are given by

$$\{\alpha^k_{1,\mathfrak{p}}(E), \alpha^{k-1}_{1,\mathfrak{p}}(E)\alpha^1_{2,\mathfrak{p}}(E) \ldots, \alpha^k_{2,\mathfrak{p}}(E)\}.$$

Hence

$$tr(\text{Frob}_{\mathfrak{p}}|\text{Sym}^k\mathcal{E}) = \alpha^k_{1,\mathfrak{p}}(E) + \alpha^{k-1}_{1,\mathfrak{p}}(E)\alpha^1_{2,\mathfrak{p}}(E) + \cdots + \alpha^k_{2,\mathfrak{p}}(E)$$
$$= N(\mathfrak{p})^{k/2}\text{sym}_k(\theta_{E,\mathfrak{p}}).$$

It follows that $\text{Sym}^k\mathcal{E}$ is pure of weight k. Moreover since the representation Sym^k, $k \geqslant 1$, exhausts all the irreducible representations of $SL(2, \mathbf{Q}_\ell)$, it follows from (5.5) that $\text{Sym}^k\mathcal{E}$ is geometrically irreducible. Hence by the Grothendieck/Lefshetz trace formula (using that $H^2_c(\ldots) = 0$) and by Deligne's theorem on the weight, one has for $d \geqslant 1$,

$$\sum_{d'|d} d'q^{d'k/2} \sum_{\substack{\mathfrak{p}\nmid\Delta(T) \\ \deg \mathfrak{p}=d'}} \text{sym}_k(\theta_{E,\mathfrak{p}}) = O(\dim H^1_c(U, \text{Sym}^k\mathcal{E})q^{d\frac{k+1}{2}}), \qquad (5.6)$$

where the constant implied in $O(\ldots)$ is at most 1. We decompose the left-hand side as follows:

$$\sum_{\substack{\mathfrak{p}\nmid\Delta(T) \\ \deg \mathfrak{p}=d}} dq^{dk/2}\text{sym}_k(\theta_{E,\mathfrak{p}}) + \sum_{d'|d,\ d'<d} d'q^{d'k/2} \sum_{\substack{\mathfrak{p}\nmid\Delta(T) \\ \deg \mathfrak{p}=d'}} \text{sym}_k(\theta_{E,\mathfrak{p}}). \qquad (5.7)$$

Using Möbius inversion on the identity

$$q^n = \sum_{d|n} d|\{\mathfrak{p},\ \deg \mathfrak{p} = d\}|,$$

we obtain that

$$d|\{\mathfrak{p},\ \deg \mathfrak{p} = d\}| = \sum_{d'|d} \mu(d')q^{d/d'} = q^d + \sum_{\substack{d'|d \\ d'\geqslant 2}} \mu(d')q^{d/d'}.$$

Hence

$$|\{\mathfrak{p},\ \deg \mathfrak{p} = d\}| = \frac{q^d}{d} + O(q^{d/2}), \qquad (5.8)$$

where the constant implied is absolute (this is a function field version of the PNT). We use (5.8) and a trivial estimate on the right-hand side of (5.7) (if $d'|d$, $d' < d$ then $d' \leqslant d/2$) and after dividing by $q^{dk/2}$ we obtain

$$d \sum_{\substack{\mathfrak{p}\nmid\Delta(T) \\ \deg \mathfrak{p}=d}} \text{sym}_k(\theta_{E,\mathfrak{p}}) \ll (\dim H^1_c(U, \text{Sym}^k\mathcal{E}) + (k+1))q^{\frac{d}{2}}.$$

Hence

$$\frac{1}{|\{\mathfrak{p} \nmid \Delta(T),\ \deg \mathfrak{p} = d\}|} \sum_{\substack{\mathfrak{p}\nmid\Delta(T) \\ \deg \mathfrak{p}=d}} \text{sym}_k(\theta_{E,\mathfrak{p}}) \ll_{k,E_T} q^{-d/2}.$$

This concludes the proof of the Sato/Tate law for the function field case.

5.4   Sato/Tate laws and Random Matrices

5.4.1   Deligne's equidistribution Theorem

The above situation can be generalised as follows: suppose we are given an ℓ-adic sheaf $\mathcal{F}$ of rank $r \geqslant 1$, lisse on $\mathbf{A}_k^1 - S$ which is

1. pure of weight 0,

2. Geometrically irreducible,

3. the arithmetic and geometric monodromy groups coincide : $G_{\mathcal{F}}^{arith} = G_{\mathcal{F}}^{geom} := G \subset GL(r, \mathbf{C})$. Recall that $G_{\mathcal{F}}^{geom}$ is semisimple.

Let K be a fixed maximal compact subgroup of G. For $\mathfrak{p} \notin S$, using that the eigenvalues of the frobenius $\mathrm{Frob}_\mathfrak{p}|\mathcal{F}$ have absolute value 1 and that all maximal compact subgroups are conjugate, one can show that $\mathrm{Frob}_\mathfrak{p}|\mathcal{F}$ gives rise to a well defined K-conjugacy class in $K^\natural$ which we still denote by $\mathrm{Frob}_\mathfrak{p}|\mathcal{F}$. The Sato/Tate law explained above is a special case of the general

Theorem. *(Deligne equidistribution theorem) Under the above assumptions, the set of conjugacy classes*

$$\{(\mathrm{Frob}_\mathfrak{p}|\mathcal{F})\}_{\substack{\mathfrak{p} \notin S \\ \deg \mathfrak{p}=d}} \subset K^\natural$$

becomes equidistributed as $d \to +\infty$ w.r.t. the Sato/Tate measure $\mu_{ST,K^\natural}$ on $K^\natural$. Hence for any continuous function f on $K^\natural$, one has for $d \to +\infty$

$$\frac{1}{|\{\mathfrak{p} \notin S, \ \deg \mathfrak{p} = d\}|} \sum_{\mathfrak{p} \notin S, \ \deg \mathfrak{p}=d} f(\mathrm{Frob}_\mathfrak{p}|\mathcal{F}) \to \int_{K^\natural} f(g^\natural) d\mu_{ST,K}(g^\natural).$$

The proof of Theorem 6 is a specialization of Deligne's equidistribution theorem and the general proof is similar.

5.4.2   Connection with RMT

Suppose we are given a random matrix ensemble $\{K(r)\}_{r \geqslant 1}$ made of compact linear groups of rank increasing with r, each one equipped with its Haar measure. In very rough terms, Random Matrix Theory is concerned with the various statistics associated to the eigenvalues or the characteristic polynomial of typical elements g of $K(r)$ (for instance: for k fixed, the k-th power of the characteristic polynomial of such g evaluated at 1 or the pair correlation distribution) and aims at determining the possible limit as $r \to +\infty$ of the expectation or the variance of such statistics. In many occasions, RMT show that the limit exists and can be computed rather explicitly.

Note that these statistics depend solely on the eigenvalues and (at least all the ones I know) can be expressed in terms of central functions on $K(r)$ – possibly with value in a space of distributions. Thus after some reductions, the evaluation of these statistics via RMT amounts to evaluate the expectation of some function, f say, on $K(r)^\natural$.

Now suppose that for each r one has a corresponding ensemble of ℓ-adic sheaves $\{\mathcal{F}_r\}_{r \geqslant 1}$ such that for each r, $\mathcal{F}_r$ satisfy the hypotheses of Deligne's equidistribution theorem and such that the geometric monodromy group $G_{\mathcal{F}_r}^{geom}$ contains $K(r)$ as a maximal compact subgroup. Then the frobenius conjugacy classes $\{\mathrm{Frob}_{\mathfrak{p}} | \mathcal{F}_r\}_{\mathfrak{p}}$ give rise to well defined conjugacy classes in $K(r)$ and Deligne's equidistribution theorem says precisely that the expectation

$$\int_{K(r)^\natural} f(g^\natural) d\mu_{ST,K(r)}(g^\natural)$$

can be computed as the limiting averaged value of f evaluated over the frobenius conjugacy classes of large degree d: for $d \to +\infty$,

$$\frac{1}{|\{\mathfrak{p} \notin S, \ \deg \mathfrak{p} = d\}|} \sum_{\mathfrak{p} \notin S, \ \deg \mathfrak{p}=d} f(\mathrm{Frob}_{\mathfrak{p}} | \mathcal{F}) \to \int_{K(r)^\natural} f(g^\natural) d\mu_{ST,K(r)}(g^\natural).$$

Combining these two ingredients one usually deduces, that in the large r limit the eigenvalues of a typical frobenius conjugacy class $\mathrm{Frob}_{\mathfrak{p}} | \mathcal{F}_r$ of degree d sufficiently large $(d \geqslant d(r))$ follow approximately the statistics of a typical random matrix of $K(r)$. As the large r-limits of such statistics can be computed explicitly via RMT, one obtains quite precise (unconditional) information on the limiting behavior of typical frobenius eigenvalues or sheaves of large rank.

This concludes our short description of the possible links between random matrices and L-functions over function fields. For those interested in digging further in this direction as well as for a more complete description of the ℓ-adic methods presented, I recommend the reading the seminal book of Katz/Sarnak [7] as well as other books of Katz devoted more specifically to the ℓ-adic methods sketched here [3, 4, 5]. In particular [5] contains some striking applications (of diophantine and arithmetic nature) of the interactions between RMT and L-functions over function fields.

424   *Philippe Michel*

References

[1] Pierre Deligne, *La conjecture de Weil. I*, Inst. Hautes Études Sci. Publ. Math. (1974), no. 43, 273– 307 (French).

[2] Pierre Deligne, *La conjecture de Weil. II*, Inst. Hautes Études Sci. Publ. Math., (1980), no. 52, 137– 252 (French).

[3] Nicholas M. Katz, *Gauss sums, Kloosterman sums, and monodromy groups*, Annals of Mathematics Studies, **116**, Princeton University Press, Princeton, NJ, 1988.

[4] Nicholas M. Katz, *Exponential sums and differential equations*, Annals of Mathematics Studies, **124**, Princeton University Press, Princeton, NJ, 1990.

[5] Nicholas M. Katz, *Twisted L-functions and monodromy*, Annals of Mathematics Studies, **150**, Princeton University Press, Princeton, NJ, 2002.

[6] Nicholas M. Katz, *L-functions and monodromy: four lectures on Weil II*, Adv. Math., **160** (2001), no. 1, 81– 132.

[7] Nicholas M. Katz and Peter Sarnak, *Random matrices, Frobenius eigenvalues, and monodromy*, American Mathematical Society Colloquium Publications, **45**, American Mathematical Society, Providence, RI, 1999.

Computational methods and experiments in analytic number theory

Michael Rubinstein

1 Introduction

We cover some useful techniques in computational aspects of analytic number theory, with specific emphasis on ideas relevant to the evaluation of L-functions. These techniques overlap considerably with basic methods from analytic number theory. On the elementary side, summation by parts, Euler-Maclaurin summation, and Möbius inversion play a prominent role. In the slightly less elementary sphere, we find tools from analysis, such as Poisson summation, generating function methods, Cauchy's residue theorem, asymptotic methods, and the fast Fourier transform. We then describe conjectures and experiments that connect number theory and random matrix theory.

2   Basic methods

2.1   Summation by parts

Summation by parts can be viewed as a discrete form of integration by parts. Let f be a function from $\mathbb{Z}^+$ to $\mathbb{R}$ or $\mathbb{C}$, and g a real or complex valued function of a real variable. Then

$$\sum_{1\leqslant n\leqslant x} f(n)g(n) = \left(\sum_{1\leqslant n\leqslant x} f(n)\right) g(x) - \int_1^x \left(\sum_{1\leqslant n\leqslant t} f(n)\right) g'(t)dt. \qquad (2.1)$$

Here we are assuming that g' exists and is continous on $[1, x]$. One verifies this identity by writing the integral as $\int_1^2 + \int_2^3 + \ldots + \int_{\lfloor x\rfloor}^x$, noticing that the sum in each integral is constant on each open interval, integrating, and telescoping. Although our integral begins at $t = 1$, it is sometimes convenient to start earlier, for example at $t = 0$. This doesn't change the value of the integral, the sum in the integrand being empty if $t < 1$. Formula (2.1) can also be interpreted in terms of the Stieltjes integral.

A slightly more general form of partial summation is over a set $\{\lambda_1, \lambda_2, \ldots\}$ of increasing real numbers:

$$\sum_{\lambda_n\leqslant x} f(n)g(\lambda_n) = \left(\sum_{\lambda_n\leqslant x} f(n)\right) g(x) - \int_{\lambda_1}^x \left(\sum_{\lambda_n\leqslant t} f(n)\right) g'(t)dt.$$

As an application, let

$$\pi(x) = \sum_{p\leqslant x} 1$$

denote the number of primes less than or equal to x, and

$$\theta(x) = \sum_{p\leqslant x} \log p$$

denote the number primes up to x with each prime weighted by its logarithm. The famous equivalence between $\pi(x) \sim x/\log x$ and $\theta(x) \sim x$ can be verified using partial summation. Write

$$\pi(x) = \sum_{p\leqslant x} \log p\frac{1}{\log p} = \theta(x)\frac{1}{\log x} + \int_2^x \theta(t)\frac{dt}{t(\log t)^2},$$

from which it follows that if $\theta(x) \sim x$ then $\pi(x) \sim x/\log x$. The converse follows from

$$\theta(x) = \sum_{p\leqslant x} 1\cdot\log p = \pi(x)\log x - \int_2^x \pi(t)\frac{dt}{t}.$$

2.2 Euler-Maclaurin summation

A powerful application of partial summation occurs when the function $f(n)$ is identically equal to 1 and the function $g(t)$ is many times differentiable. In that case, summation by parts specializes to the Euler Maclaurin formula which involves one summation by parts with $f(n) = 1$ followed by repeated integration by parts. For $a, b \in \mathbb{Z}, a < b$, partial summation gives

$$\sum_{a<n\leqslant b} g(n) = (b-a)g(b) - \int_a^b (\lfloor t \rfloor - a)g'(t)dt = bg(b) - ag(a) - \int_a^b \lfloor t \rfloor g'(t)dt.$$

Here, we have chosen to start the integral at $t = a$, rather than at $t = a+1$. Writing $\lfloor t \rfloor = t - \{t\}$, with $\{t\}$ the fractional of t we get

$$\sum_{a<n\leqslant b} g(n) = \int_a^b g(t)dt + \int_a^b \{t\}g'(t)dt.$$

The second term on the r.h.s. should be viewed as the necessary correction that arises from replacing the sum on the left with an integral.

The next step is to write $\{t\} = 1/2 + (\{t\} - 1/2)$, the latter term having nicer properties than $\{t\}$, for example being odd and also having zero constant term in its Fourier expansion. So

$$\sum_{a<n\leqslant b} g(n) = \int_a^b g(t)dt + \frac{1}{2}(g(b) - g(a)) + \int_a^b (\{t\} - 1/2)g'(t)dt. \qquad (2.2)$$

Integrating the second integral repeatedly by parts leads naturally to the introduction of Bernoulli polynomials, named after Jacob Bernoulli (1654-1705), who discovered them in connection to the problem of studying sums of positive integer powers of consecutive integers. During the 1730's Euler (1707-1783), who studied mathematics from Jacob's brother Johann (1667-1748), developed the summation formula being described in connection with computing reciprocals of powers and Euler's constant.

2.2.1 Bernoulli Polynomials

The Bernoulli polynomials are defined recursively by the following relations

$$
\begin{aligned}
B_0(t) &= 1 \\
B_k'(t) &= kB_{k-1}(t), \quad k \geqslant 1 \\
\int_0^1 B_k(t)dt &= 0, \quad k \geqslant 1.
\end{aligned}
$$

The second equation determines $B_k(t)$ recursively up to the constant term, and the third equation fixes the constant. The first few Bernoulli polynomials are listed in Table 1.

k	$B_k(t)$
0	1
1	$t - 1/2$
2	$t^2 - t + 1/6$
3	$t^3 - 3/2t^2 + 1/2t$
4	$t^4 - 2t^3 + t^2 - 1/30$
5	$t^5 - 5/2t^4 + 5/3t^3 - 1/6t$

Table 1: The first few Bernoulli polynomials

Let $B_k = B_k(0)$ denote the constant term of $B_k(t)$. B_k is called the k-th Bernoulli number. We state basic properties of the Bernoulli polynomials. Expansion in terms of Bernoulli numbers:

$$B_k(t) = \sum_0^k \binom{k}{m} B_{k-m} t^m, \quad k \geq 0$$

Generating function:

$$\frac{ze^{zt}}{e^z - 1} = \sum_0^\infty B_k(t) z^k / k!, \quad |z| < 2\pi$$

Fourier series:

$$B_1(\{t\}) = -\frac{1}{\pi} \sum_1^\infty \frac{\sin(2\pi m t)}{m}, \quad t \notin \mathbb{Z} \tag{2.3}$$

$$B_k(\{t\}) = -k! \sum_{m \neq 0} \frac{e^{2\pi i m t}}{(2\pi i m)^k}, \quad k \geq 2. \tag{2.4}$$

Functional equation:

$$B_k(t) = (-1)^k B_k(1 - t), \quad k \geq 0$$

Difference equation:

$$\frac{B_{k+1}(t+1) - B_{k+1}(t)}{k+1} = t^k, \quad k \geq 0 \tag{2.5}$$

Special values:

$$B_k(1) = \begin{cases} (-1)^k B_k(0), & k \geqslant 0 \\ 0, & k \text{ odd}, k \geqslant 3 \\ 1/2, & k = 1 \end{cases}$$

i.e.

$$B_k(1) = B_k(0), \quad \text{unless } k = 1 \tag{2.6}$$

Recursion:

$$\sum_{m=0}^{k-1} \binom{k}{m} B_m = 0, \quad k \geqslant 2$$

Equation (2.3) can be obtained directly. The other formulae can be verified using the defining relations and induction.

Property (2.4) can be used to obtain a formula for $\zeta(2m)$. Let

$$\zeta(s) = \sum_1^\infty n^{-s}, \quad \Re s > 1.$$

Taking $t = 0$, $k = 2m$, even, in the Fourier expansion of $B_k(\{t\})$ gives

$$B_{2m} = \frac{(-1)^{m+1}(2m)!}{(2\pi)^{2m}} 2\zeta(2m)$$

so that

$$\zeta(2m) = \frac{(-1)^{m+1}(2\pi)^{2m}}{2(2m)!} B_{2m},$$

a formula discovered by Euler. Because $\zeta(2m) \to 1$ as $m \to \infty$, we have

$$B_{2m} \sim \frac{(-1)^{m+1} 2(2m)!}{(2\pi)^{2m}}$$

as $m \to \infty$.

2.2.2 Euler-Maclaurin continued

Returning to (2.2), we write

$$\int_a^b (\{t\} - 1/2) g'(t) dt = \int_a^b B_1(\{t\}) g'(t) dt.$$

Breaking up the integral $\int_a^b = \int_a^{a+1} + \int_{a+1}^{a+2} + \ldots \int_{b-1}^b$, integrating by parts, and noting that $B_2(1) = B_2(0)$, we get, assuming that $g^{(2)}$ exists and is continous on $[a, b]$,

$$\frac{B_2}{2}(g'(b) - g'(a)) - \int_a^b \frac{B_2(\{t\})}{2} g^{(2)}(t) dt.$$

Repeating, using $B_k(1) = B_k(0)$ if $k \geqslant 2$, leads to the Euler-Maclaurin summation formula. Let K be a positive integer. Assume that $g^{(K)}$ exists and is continous on $[a,b]$. Then

$$\sum_{a<n\leqslant b} g(n) = \int_a^b g(t)dt + \sum_{k=1}^K \frac{(-1)^k B_k}{k!}(g^{(k-1)}(b) - g^{(k-1)}(a))$$
$$+\frac{(-1)^{K+1}}{K!}\int_a^b B_K(\{t\})g^{(K)}(t)dt.$$

2.2.3 Application: Sums of consecutive powers

We apply Euler-Macluarin summation to obtain Bernoulli's formula for sums of powers of consecutive integers. Let $r \geqslant 0$ be an integer. Then

$$\sum_{n=1}^N n^r = \frac{B_{r+1}(N+1) - B_{r+1}(1)}{r+1}.$$

We can verify this directly using property (2.5), substituting $n = 1, 2, \ldots, N$, and telescoping. However, it is instructive to apply the Euler-Maclaurin formula, which, once begun, carries through in an automatic fashion. In this example, we have $g(t) = t^r$. Notice that $g^{(r+1)}(t) = 0$, and that

$$g^{(m)}(N) - g^{(m)}(0) = \begin{cases} r(r-1)\ldots(r-m+1)N^{r-m}, & m \leqslant r-1 \\ 0, & m \geqslant r. \end{cases}$$

If $m = 0$ we set $r(r-1)\ldots(r-m+1) = 1$. Then

$$\begin{aligned}\sum_{n=1}^N n^r &= \int_0^N t^r dt + \sum_{k=1}^r \frac{(-1)^k B_k}{k!}r(r-1)\ldots(r-k+2)N^{r-k+1}\\ &= \int_0^N t^r dt + \sum_{k=1}^r \frac{(-1)^k B_k}{r-k+1}\binom{r}{k}N^{r-k+1}\\ &= \int_0^N \sum_{k=0}^r (-1)^k B_k \binom{r}{k}t^{r-k}dt\\ &= \int_0^N (-1)^r B_r(-t)dt = \int_0^N B_r(t+1)dt\\ &= (B_{r+1}(N+1) - B_{r+1}(1))/(r+1).\end{aligned}$$

If $r \geqslant 1$, the last line simplifies according to (2.6) and equals

$$\frac{B_{r+1}(N+1) - B_{r+1}}{r+1}.$$

2.2.4   Application: $\zeta(s)$

The Euler-Maclaurin formula can be used to obtain the analytic continuation of $\zeta(s)$ and also provides a useful expansion for its numeric evaluation. Consider

$$\sum_1^N n^{-s} = 1 + \sum_2^N n^{-s}$$

with $\Re e s > 1$. We have started the sum at $n = 2$ rather than $n = 1$ to avoid difficulties near $t = 0$ below. Applying Euler-Maclaurin summation, with $g(t) = t^{-s}$, $g^{(m)}(t) = (-1)^m s(s+1)\ldots(s+m-1)t^{-s-m}$, we get

$$\sum_1^N n^{-s} = 1 + \int_1^N t^{-s} dt \quad - \quad \sum_{k=1}^K \frac{B_k}{k}\binom{s+k-2}{k-1}(N^{-s-k+1} - 1)$$

$$- \binom{s+K-1}{K}\int_1^N B_K(\{t\})t^{-s-K} dt.$$

Evaluating the first integral, taking the limit as $N \to \infty$, with $\Re e s > 1$, we get

$$\zeta(s) = \frac{1}{s-1} + \frac{1}{2} + \sum_2^K \binom{s+k-2}{k-1}\frac{B_k}{k} - \binom{s+K-1}{K}\int_1^\infty B_K(\{t\})t^{-s-K} dt.$$

$$(2.7)$$

While we started with $\Re e s > 1$, the r.h.s. is meromorphic for $\Re e s > -K+1$, so gives the meromorphic continuation of $\zeta(s)$ in this region, with the only pole being the simple pole at $s = 1$.

Taking $s = 2 - K$, $K \geqslant 2$,

$$\zeta(2-K) = 1 - \frac{1}{K-1}\sum_{k=0}^{K-1}(-1)^k\binom{K-1}{k}B_k = \frac{(-1)^K B_{K-1}}{K-1}.$$

Thus,

$$\begin{aligned}
\zeta(1-2m) &= -B_{2m}/(2m), \quad m = 1,2,3,\ldots \\
\zeta(-2m) &= 0, \quad m = 1,2,3,\ldots \\
\zeta(0) &= -1/2.
\end{aligned}$$

Applying the functional equation for ζ,[1]

$$\pi^{-s/2}\Gamma(s/2)\zeta(s) = \pi^{-(1-s)/2}\Gamma((1-s)/2)\zeta(1-s)$$

---

[1] Editors' comment: See for example D.R. Heath-Brown's lectures, page 1, Theorem 8.

and

$$\Gamma(1/2) = \pi^{1/2} = \frac{-1}{2}\frac{-3}{2}\frac{-5}{2}\cdots\frac{-(2m-1)}{2}\Gamma(1/2-m)$$

gives another proof of Euler's identity

$$\zeta(2m) = \frac{(-1)^{m+1}(2\pi)^{2m}}{2(2m)!}B_{2m}, \quad m \geqslant 1.$$

2.2.5 Computing $\zeta(s)$ using Euler-Maclaurin summation

Next we describe how to adapt the above to obtain a practical method for numerically evaluating $\zeta(s)$. From a computational perspective, the following works better than using (2.7). Let N be a large positive integer, proportional in size to $|s|$. We will make this more explicit shortly. For $\Re s > 1$, write

$$\zeta(s) = \sum_1^\infty n^{-s} = \sum_1^N n^{-s} + \sum_{N+1}^\infty n^{-s}. \tag{2.8}$$

The first sum on the r.h.s. is evaluated term by term, while the second sum is evaluated using Euler-Maclaurin summation

$$\sum_{N+1}^\infty n^{-s} = \frac{N^{1-s}}{s-1} + \sum_1^K \binom{s+k-2}{k-1}\frac{B_k}{k}N^{-s-k+1} \tag{2.9}$$
$$-\binom{s+K-1}{K}\int_N^\infty B_K(\{t\})t^{-s-K}\,dt.$$

As before, the r.h.s. above gives the meromorphic continuation of the l.h.s. to $\Re s > -K+1$. Breaking up the sum over n in this fashion allows us to throw away the integral on the r.h.s., and obtain sharp estimates for its neglected contribution. First, from property (2.4),

$$|B_K(\{t\})| \leqslant \frac{K!}{(2\pi)^K}2\zeta(K).$$

It is convenient to take $K = 2K_0$, even, in which case we have from (2.8)

$$|B_{2K_0}(\{t\})| \leqslant B_{2K_0}.$$

Therefore, for $s = \sigma + i\tau, \sigma > -2K_0 + 1$,

$$\left| \binom{s + 2K_0 - 1}{2K_0} \int_N^\infty B_{2K_0}(\{t\}) t^{-s-2K_0} dt \right|$$

$$\leqslant \left| \binom{s + 2K_0 - 1}{2K_0} B_{2K_0} \right| \frac{N^{-\sigma - 2K_0 + 1}}{\sigma + 2K_0 - 1}$$

$$= \frac{|s + 2K_0 - 1|}{\sigma + 2K_0 - 1} \left| \binom{s + 2K_0 - 2}{2K_0 - 1} \frac{B_{2K_0}}{2K_0} \right| N^{-\sigma - 2K_0 + 1}$$

$$= \frac{|s + 2K_0 - 1|}{\sigma + 2K_0 - 1} |\text{last term taken}|.$$

A more precise estimate follows by comparison of B_{2K_0} with $\zeta(2K_0)$, and we have that the remainder is

$$\leqslant \frac{\zeta(2K_0)}{\pi N^\sigma} \frac{|s + 2K_0 - 1|}{\sigma + 2K_0 - 1} \prod_{j=0}^{2K_0 - 2} \frac{|s + j|}{2\pi N}.$$

We start to win when $2\pi N$ is bigger than $|s|, |s+1|, \ldots, |s + 2K_0 - 2|$. There are two parameters which we need to choose: K_0 and N, and we also need to specify the number of digits accuracy, Digits, we desire. For example, with $\sigma \geqslant 1/2$, taking

$$2\pi N \geqslant 10|s + 2K_0 - 2|$$

with

$$2K_0 - 1 > \text{Digits} + \frac{1}{2} \log_{10}(|s + 2K_0 - 1|)$$

achieves the desired accuracy. The main work involves the computation of the sum $\sum_1^N n^{-s}$ consisting of $O(|s|)$ terms. Later we will examine the Riemann-Siegel formula and its smoothed variants which, for $\zeta(s)$, involves a main sum of $O(|s|^{1/2})$ terms. However, for high precision evaluation of $\zeta(s)$, especially with s closer to the real axis, the Euler-Maclaurin formula remains an ideal method allowing for sharp and rigorous error estimates and reasonable efficiency.

In fact, we can turn the above scheme into a computation involving $O(|s|^{1/2})$ operations but requiring $O((\text{Digits} + \log |s|) \log |s|)$ precision due to cancellation that occurs. In (2.8) choose $N \sim |10s/(2\pi)|^{1/2}$, and assume that $\Re s \geqslant 1/2$. Expand $B_K(\{t\})$ into its Fourier series (2.4). We only need $M = O(|s|^{1/2})$ terms of the Fourier expansion to assure a contribution from the neglected terms smaller than the desired precision. Each term contributes

$$K! \binom{s + K - 1}{K} \frac{1}{(2\pi i m)^K} \int_N^\infty e^{2\pi i m t} t^{-s-K} dt, \tag{2.10}$$

so the neglected terms contribute altogether less than

$$\frac{1}{N^\sigma}\frac{|s+K-1|}{\sigma+K-1}\left(\prod_{j=0}^{K-2}\frac{|s+j|}{2\pi N}\right)\left(\sum_{M+1}^{\infty}\frac{2}{m^K}\right).$$

Here we have combined the $\pm m$ terms together. Comparing to an integral, the sum above is $< 2/((K-1)M^{K-1})$ and so the neglected terms contribute less than

$$\frac{2}{(K-1)N^\sigma}\frac{|s+K-1|}{\sigma+K-1}\prod_{j=0}^{K-2}\frac{|s+j|}{2\pi MN}.$$

We start to win when $2\pi MN$ exceeds $|s|,\dots,|s+K-2|$. For $\sigma \geqslant 1/2$, choose $K > \text{Digits} + \log_{10}(|s+K-1|) + 1$ and $M = N$ with

$$2\pi MN \geqslant 10|s+K-2|.$$

Asymptotically, we can improve the above choices so as to achieve $M = N \sim |s|^{1/2}/(2\pi)$, the same as in the Riemann-Siegel formula. The only drawback is that extra precision as described above is needed. The individual terms summed in (2.9) are somewhat large in comparison to the final result, this coming form the binomial coefficients which have numerator $(s+k-2)\dots(s+1)s$, and this leads to cancellation.

Finally to compute the contribution to the Fourier expansion from the terms with $|m| \leqslant M$, we assume that $4|K$ so that the terms $\pm m$ together involve in (2.10) the integral

$$\int_N^\infty \cos(2\pi mt)t^{-s-K}dt = (2\pi m)^{s+K-1}\int_{2\pi mN}^\infty \cos(u)u^{-s-K}du.$$

This can be expressed in terms of the incomplete Γ function

$$\int_w^\infty \cos(u)u^{z-1}du = \frac{1}{2}\left(e^{-\pi iz/2}\Gamma(z,iw) + e^{\pi iz/2}\Gamma(z,-iw)\right).$$

See Section 3 which describes properties of the incomplete Γ function and methods for its evaluation.

The Euler-Maclaurin formula can also be used to evaluate Dirichlet L-functions. It works in that case due to the periodic nature of the corresponding Dirichlet coefficients. For general L-functions, there are smoothed Riemann-Siegel type formulae. These are described later.

2.3 Möbius inversion with an application to sums and products over primes

Computations in analytic number theory often involve evaluating sums or products over primes. For example, let $\pi_2(x)$ denote the number of twin primes $(p, p+2)$, with p and $p+2$ both prime and less than or equal to x. The famous conjecture of Hardy and Littlewood predicts that

$$\pi_2(x) \sim 2 \prod_{p>2} \frac{p(p-2)}{(p-1)^2} \frac{x}{(\log x)^2}.$$

Generally, it is easier to deal with a sum rather than a product, so we turn this product over primes into a sum by expressing it as

$$\exp\left(\sum_{p>2} \log(1-2/p) - 2\log(1-1/p)\right).$$

Letting $f(p) = \log(1-2/p) - 2\log(1-1/p)$, we have

$$f(p) = -\sum_{m=1}^{\infty} \frac{2^m-2}{mp^m}$$

hence

$$\sum_{p>2} f(p) = -\sum_{m=1}^{\infty} \frac{2^m-2}{m}(h(m) - 1/2^m) \tag{2.11}$$

with

$$h(s) = \sum_{p} p^{-s}, \quad \mathfrak{Re}\, s > 1.$$

We therefore need an efficient method for computing $h(m)$. This will be dealt with below. Notice that $h(m) - 1/2^m \sim 1/3^m$ so the sum on the r.h.s. of (2.11) converges exponentially fast. We can achieve faster convergence by writing

$$\sum_{p>2} f(p) = \sum_{2<p\leqslant P} f(p) + \sum_{p>P} f(p),$$

summing the terms in the first sum, and expressing the second sum as

$$-\sum_{m=1}^{\infty} \frac{2^m-2}{m}(h(m) - 1/2^m - \ldots - 1/P^m).$$

A second example involves the computation of constants that arise in conjectures for moments of $\zeta(s)$. The Keating-Snaith conjecture [KeS] asserts that

$$M_k(T) := \frac{1}{T}\int_0^T |\zeta(1/2+it)|^{2k}dt \sim \frac{a_k g_k}{k^2!}(\log T)^{k^2} \tag{2.12}$$

$$a_k = \prod_p \left(1 - \frac{1}{p}\right)^{k^2} \sum_{m=0}^{\infty} \binom{m+k-1}{m}^2 p^{-m} \tag{2.13}$$

$$= \prod_p \left(1 - \frac{1}{p}\right)^{(k-1)^2} \sum_{j=0}^{k-1} \binom{k-1}{j}^2 p^{-j}$$

and

$$g_k = k^2! \prod_{j=0}^{k-1} \frac{j!}{(k+j)!}.$$

The placement of $k^2!$ is to ensure that g_k is an integer [CF]. Keating and Snaith also provide a conjecture for complex values, $\Re e\, k > -1/2$, of which the above is a special case. Keating and Snaith used random matrix theory to identify the factor g_k. The form of (2.12), without identifying g_k, was conjectured by Conrey and Ghosh [CG].

The above conjecture gives the leading term for the asymptotics for the moments of $|\zeta(1/2+it)|$. In [CFKRS] a conjecture is given for the full asymptotics of $M_k(T)$:

$$M_k(T) \sim \sum_{r=0}^{k^2} c_r(k)(\log T)^{k^2-r}$$

where $c_0(k) = a_k g_k / k^2!$ coincides with the Keating-Snaith leading term and where the degree k^2 polynomial is given implicitly as an elaborate multiple residue. Explicit expressions for $c_r(k)$ are worked out in [CFKRS3] and are given as $c_0(k)$ times complicated rational functions in k, generalized Euler constants, and sums over primes involving $\log(p)$, $_2F_1(k,k,1;p^{-1})$ and its derivatives. One method for computing the $c_r(k)$'s involves as part of a single step the computation of sums of the form

$$\sum_p \frac{(\log p)^r}{p^m}, \quad m = 2,3,4,\dots \quad r = 0,1,2,\dots. \tag{2.14}$$

We now describe how to efficiently compute $h(s) = \sum_p p^{-s}$ and the sums in (2.14). Take the logarithm of

$$\zeta(s) = \prod_p (1 - p^{-s})^{-1}, \quad \Re e\, s > 1$$

and apply the Taylor series for $\log(1-x)$ to get

$$\log \zeta(s) = \sum_{m=1}^{\infty} \frac{1}{m} h(ms), \quad \Re e\, s > 1. \tag{2.15}$$

Let $\mu(n)$, the Möbius μ function, denote the Dirichlet coefficients of $1/\zeta(s)$:

$$1/\zeta(s) = \prod_p (1 - p^{-s}) = \sum_1^\infty \mu(n) n^{-s}.$$

We have

$$\mu(n) = \begin{cases} 0 & \text{if } n \text{ is divisible by the square of an integer} > 1 \\ (-1)^{\text{number of prime factors of } n} & \text{if } n \text{ is squarefree} \end{cases}$$

and

$$\sum_{n|r} \mu(n) = \begin{cases} 1 & \text{if } r = 1 \\ 0 & \text{otherwise.} \end{cases}$$

The last property can be proven by writing the sum of the left as $\prod_{p|r}(1-1)$, and it allows us to invert equation (2.15)

$$\begin{aligned}
\sum_{m=1}^\infty \frac{\mu(m)}{m} \log \zeta(ms) &= \sum_{m=1}^\infty \frac{\mu(m)}{m} \sum_{n=1}^\infty \frac{h(mns)}{n} \\
&= \sum_{r=1}^\infty \frac{h(rs)}{r} \sum_{m|r} \mu(m) = h(s),
\end{aligned}$$

i.e.

$$\sum_p p^{-s} = \sum_{m=1}^\infty \frac{\mu(m)}{m} \log \zeta(ms). \tag{2.16}$$

This is an example of Möbius inversion, and expresses $h(s)$ as a sum involving ζ. Möbius inversion can be interpreted as a form of the sieve of Eratosthenes.

Notice that $\zeta(ms) = 1 + 2^{-ms} + 3^{-ms} + \dots$ tends to 1, and hence $\log \zeta(ms)$ tends to 0, exponentially fast as $m \to \infty$. Therefore, the number of terms needed on the r.h.s. of (2.16) is proportional to the desired precision.

To compute the series appearing in (2.14) we can differentiate $h(s)$ r times, obtaining

$$\sum_p \frac{(\log p)^r}{p^s} = (-1)^r \sum_{m=1}^\infty \frac{\mu(m)}{m} (\log \zeta(ms))^{(r)}. \tag{2.17}$$

In both (2.16) and (2.17), we can use Euler-Maclaurin summation to compute ζ and its derivatives. The paper of Henri Cohen [C] is a good reference for computations involving sums or products of primes.

2.4 Poisson summation as a tool for numerical integration

Let $f \in L^1(\mathbb{R})$ and let

$$\hat{f}(y) = \int_{-\infty}^{\infty} f(t)e^{-2\pi i y t} dt.$$

denote its Fourier transform. The Poisson summation formula asserts, for $f, \hat{f} \in L^1(\mathbb{R})$ and of bounded variation, that

$$\sum_{n=-\infty}^{\infty} f(n) = \sum_{n=-\infty}^{\infty} \hat{f}(n).$$

We often encounter the Poisson summation formula as a potent theoretical tool in analytic number theory. For example, the functional equations of the Riemann ζ function and of the Dedekind η function can be derived by exploiting Poisson summation. However, Poisson summation is often overlooked in the setting of numerical integration where it provides justification for carrying out certain numerical integrals in a very naive way.

Let $\Delta > 0$. By a change of variable

$$\Delta \sum_{n=-\infty}^{\infty} f(n\Delta) = \sum_{n=-\infty}^{\infty} \hat{f}(n/\Delta) = \hat{f}(0) + \sum_{n\neq 0} \hat{f}(n/\Delta)$$

so that

$$\int_{-\infty}^{\infty} f(t)dt - \Delta \sum_{n=-\infty}^{\infty} f(n\Delta) = -\sum_{n\neq 0} \hat{f}(n/\Delta)$$

tells us how closely the Riemann sum $\Delta \sum_{n=-\infty}^{\infty} f(n\Delta)$ approximates the integral $\int_{-\infty}^{\infty} f(t)dt$.

The main point is that if $\hat{f}$ is rapidly decreasing then we get enormous accuracy from the Riemann sum, even with Δ not too small. For example, with $\Delta = 1/10$, the first contribution comes from $\hat{f}(\pm 10)$ which can be extremely small if $\hat{f}$ decreases sufficiently fast.

As a simple application, let $f(t) = \exp(-t^2/2)$. Then $\hat{f}(y) = \sqrt{2\pi}\exp(-2\pi^2 y^2)$, and so

$$\sum_{n\neq 0} \hat{f}(n/\Delta) = O(\exp(-2\pi^2/\Delta^2)).$$

Therefore

$$\int_{-\infty}^{\infty} \exp(-t^2/2)dt - \Delta \sum_{n=-\infty}^{\infty} \exp(-(n\Delta)^2/2) = O(\exp(-2\pi^2/\Delta^2)).$$

As everyone knows, the integral on the l.h.s. equals $\sqrt{2\pi}$. Taking $\Delta = 1/10$, we therefore get

$$\Delta \sum_{n=-\infty}^{\infty} \exp(-(n\Delta)^2/2) = \sqrt{2\pi} + \epsilon$$

with $\epsilon \approx 10^{-857}$. We can truncate the sum over n roughly when

$$\frac{(n\Delta)^2}{2} > \frac{2\pi^2}{\Delta^2},$$

i.e. when $n > 2\pi/\Delta^2$. So only 628 terms (combine $\pm n$) are needed to evaluate $\sqrt{2\pi}$ to about 857 decimal place accuracy!

This method can be applied to the problem of computing certain probability distributions that arise in random matrix theory. Let U be an $N \times N$ unitary matrix, with eigenvalues $\exp(i\theta_1), \ldots \exp(i\theta_N)$, and characteristic polynomial

$$Z(U, \theta) = \prod_{1}^{N}(\exp(i\theta) - \exp(i\theta_n))$$

evaluated on the unit circle at the point $\exp(i\theta)$. In making their conjecture for the moments of $|\zeta(1/2 + it)|$, Keating and Snaith [KeS] studied the analogous random matrix theory problem of evaluating the moments of $|Z(U, \theta)|$, averaged according to Haar measure on $U(N)$. The characteristic polynomial of a matrix is a class function that only depends on the eigenvalues of the matrix. For class functions, the Weyl integration formula gives Haar measure in terms of the eigenangles, the invariant probability measure on $U(N)$ being

$$\frac{1}{(2\pi)^N N!} \prod_{1 \leqslant j < m \leqslant N} |e^{i\theta_j} - e^{i\theta_m}|^2 d\theta_1 \ldots d\theta_N.$$

Therefore, $M_N(r)$, the rth moment of $|Z(U, \theta)|$, is given by

$$M_N(r) = \frac{1}{(2\pi)^N N!} \int_0^{2\pi} \cdots \int_0^{2\pi} \prod_{1 \leqslant j < m \leqslant N} |e^{i\theta_j} - e^{i\theta_m}|^2 |Z(U, \theta)|^r d\theta_1 \ldots d\theta_N,$$

for $\Re r > -1$. This integral happens to be a special case of Selberg's integral, and Keating and Snaith consequently determined that

$$M_N(r) = \prod_{j=1}^{N} \frac{\Gamma(j)\Gamma(j+r)}{\Gamma(j+r/2)^2}.$$

Notice that this does not depend on θ.

Say we are interested in computing the probability distribution of $|Z(U,\theta)|$. One can recover the probability density function from the moments as follows. We can express the moments of $|Z(U,\theta)|$ in terms of its probability density function. Let

$$\text{prob}(0 \leqslant a \leqslant |Z(U,\theta)| \leqslant b) = \int_a^b p_N(t)\,dt.$$

Then

$$M_N(r) = \int_0^\infty p_N(t)t^r\,dt \tag{2.18}$$

is a Mellin transform, and taking the inverse Mellin transform we get

$$p_N(t) = \frac{1}{2\pi i t}\int_{\nu-i\infty}^{\nu+i\infty}\prod_{j=1}^N \frac{\Gamma(j)\Gamma(j+r)}{\Gamma(j+r/2)^2}t^{-r}\,dr \tag{2.19}$$

with ν to the right of the poles of $M_N(r)$, $\nu > -1$. There is an extra $1/t$ in front of the integral since the Mellin transform (2.18) is evaluated at r rather than at $r-1$.

To compute $p_N(t)$ we could shift the line integral to the left picking up residues at the poles of $M_N(r)$, but as N grows this becomes burdensome. Instead, we can compute the inverse Mellin transform (2.19) as a simple Riemann sum.

Changing variables we have

$$p_N(t) = \frac{1}{2\pi t}\int_{-\infty}^{\infty}\prod_{j=1}^N \frac{\Gamma(j)\Gamma(j+\nu+iy)}{\Gamma(j+(\nu+iy)/2)^2}t^{-\nu-iy}\,dy.$$

Let

$$f_t(y) = \frac{1}{2\pi}\prod_{j=1}^N \frac{\Gamma(j)\Gamma(j+\nu+iy)}{\Gamma(j+(\nu+iy)/2)^2}t^{-\nu-1-iy}.$$

This function also depends on ν and N, but we do not include them explicitly on the l.h.s. so as to simplify our notation. The above integral equals

$$p_N(t) = \int_{-\infty}^{\infty} f_t(y)\,dy. \tag{2.20}$$

To estimate the error in computing this integral as a Riemann sum using increments of size Δ, we need bounds on the Fourier transform

$$\hat{f}_t(u) = \int_{-\infty}^{\infty} f_t(y)e^{-2\pi iuy}\,dy. \tag{2.21}$$

However,
$$f_t(y)e^{-2\pi i u y} = f_{te^{2\pi u}}(y)e^{2\pi u(\nu+1)}$$

and so
$$\hat{f}_t(u) = e^{2\pi u(\nu+1)}p_N(te^{2\pi u}).$$

Now, $p_N(t)$ is supported in $[0, 2^N]$, because $0 \leqslant |Z(U,\theta)| \leqslant 2^N$. Hence if $u > (N \log 2 - \log t)/(2\pi)$ then $\hat{f}_t(u) = 0$. Thus, for $0 < t < 2^N$, if we evaluate (2.20) as a Riemann sum with step size $\Delta < 2\pi/(N \log 2 - \log t)$ the error is
$$\sum_{n \neq 0} \hat{f}_t(n/\Delta) = \sum_{n < 0} \hat{f}_t(n/\Delta)$$

since the terms with $n > 0$ are all zero. On the other hand, with $n < 0$ we get
$$\hat{f}_t(-|n|/\Delta) = e^{-2\pi(\nu+1)|n|/\Delta}p_N(te^{-2\pi|n|/\Delta}) \leqslant e^{-2\pi(\nu+1)|n|/\Delta}p_{\max}$$

where $p_{\max}$ denotes the maximum of $p_N(t)$ (an upper bound for $p_N(t)$ can be obtained from (2.19)).

Therefore, choosing
$$\Delta = \frac{2\pi}{\text{Digits} \log 10 + N \log 2 - \log t}$$

and setting $\nu = 0$ we have
$$\hat{f}_t(-|n|/\Delta) < (10^{-\text{Digits}}2^{-N}t)^{|n|}p_{\max}$$

Summing over $n = -1, -2, -3, \ldots$ we get an overall bound of
$$10^{-\text{Digits}}p_{\max}/(1 - 10^{-\text{Digits}}) \approx 10^{-\text{Digits}}p_{\max}.$$

We could choose ν to be larger, i.e. shift our line integral (2.19) to the right, and thus achieve more rapid decay of $\hat{f}_t(u)$ as $u \to -\infty$. However, this leads to precision issues. As ν increases, the integrand in (2.19) increases in size, yet $p_N(t)$ remains constant for given N and t. Therefore cancellation must occur when we evaluate the Riemann sum and higher precision is needed to capture this cancellation. We leave it as an excercise to determine the amount of precision needed for a given value of ν.

Another application appears in [RS] where Poisson summation is used to compute, on a logarithmic scale, the probability that $\pi(x)$, the number of primes up to x, exceeds $\text{Li}(x) = \int_2^\infty dt/\log(t)$. The answer turns out to be .00000026 ...

Later in this paper, we apply this method to computing certain complicated integrals that arise in the theory of general L-functions.

3 Analytic aspects of *L*-function computations

3.1 Riemann-Siegel formula

The Riemann Siegel formula expresses the Riemann ζ function as a main sum involving a truncated Dirichlet series and correction terms. The formula is often presented with $\Re s = 1/2$, but can be given for s off the critical line. See [OS] for a nice presentation of the formula for $1/2 \leqslant \Re s \leqslant 2$ and references. Here we stick to $\Re s = 1/2$.

Let

$$Z(t) = e^{i\theta(t)}\zeta(1/2+it)$$
$$e^{i\theta(t)} = \left(\frac{\Gamma(1/4+it/2)}{\Gamma(1/4-it/2)}\right)^{1/2}\pi^{-it/2}. \tag{3.1}$$

The rotation factor $e^{i\theta(t)}$ is chosen so that $Z(t)$ is real.

For $t > 2\pi$, let $a = (t/(2\pi))^{1/2}$, $N = \lfloor a \rfloor$, $\rho = \{a\} = a - \lfloor a \rfloor$ the fractional part of a. Then

$$Z(t) = 2\sum_{n=1}^{N} n^{-1/2}\cos(t\log(n)-\theta(t)) + R(t)$$

where

$$R(t) = \frac{(-1)^{N+1}}{a^{1/2}}\sum_{r=0}^{m}\frac{C_r(\rho)}{a^r} + R_m(t)$$

with

$$C_0(\rho) = \psi(\rho) := \cos(2\pi(\rho^2-\rho-1/16))/\cos(2\pi\rho)$$
$$C_1(\rho) = -\frac{1}{96\pi^2}\psi^{(3)}(\rho)$$
$$C_2(\rho) = \frac{1}{18432\pi^4}\psi^{(6)}(\rho) + \frac{1}{64\pi^2}\psi^{(2)}(\rho).$$

In general [E], $C_j(\rho)$ can be expressed as a linear combination of the derivatives of ψ. We also have

$$R_m(t) = O(t^{-(2m+3)/4}).$$

Gabcke [G] showed that

$$|R_1(t)| \leqslant .053t^{-5/4}, \quad t \geqslant 200.$$

The bulk of computational time in evaluating $\zeta(s)$ using the Riemann-Siegel formula is spent on the main sum $\sum_{n=1}^{N} n^{-1/2}\cos(t\log(n)-\theta(t))$. Odlyzko and Schönhage [OS] [O] developed an algorithm to compute the main sum for $T \leqslant t \leqslant T + T^{1/2}$ in $O(t^\epsilon)$ operations providing that a precomputation involving $O(T^{1/2+\epsilon})$ operations and bits of storage are carried out beforehand. This algorithm lies behind Odlyzko's monumental ζ computations [O] [O2]. An earlier implementation proceeded by using the fast Fourier transform to compute the main sum and its derivatives at equally spaced grid points to then compute the main sum in between using Taylor series. This was then improved [O, 4.4] to using just the values of the main sum at equally spaced points and an interpolation formula from the theory of band-limited functions.

Riemann used the saddle point method to obtain C_j, for $j \leqslant 5$. The reason that a nice formula works using a sharp cutoff, truncating the sum over n at N, is that all the Dirichlet coefficients are equal to one. Riemann starts with an expression for $\zeta(s)$ which involves the geometric series identity $1/(1-x) = \sum x^n$, the Taylor coefficients on the right being the Dirichlet coefficients of $\zeta(s)$. For general L-functions smoothing works better.

3.2 Smoothed approximate functional equations

Let

$$L(s) = \sum_{n=1}^{\infty} \frac{b(n)}{n^s}$$

be a Dirichlet series that converges absolutely in a half plane, $\mathfrak{Re}(s) > \sigma_1$, and hence uniformly convergent in any half plane $\mathfrak{Re}(s) \geqslant \sigma_2 > \sigma_1$ by comparison with the series for $L(\sigma_2)$.

Let

$$\Lambda(s) = Q^s \left(\prod_{j=1}^{a} \Gamma(\kappa_j s + \lambda_j) \right) L(s), \tag{3.2}$$

with $Q, \kappa_j \in \mathbb{R}^+$, $\mathfrak{Re}\lambda_j \geqslant 0$, and assume that:

1. $\Lambda(s)$ has a meromorphic continuation to all of $\mathbb{C}$ with simple poles at $s_1, \ldots, s_\ell$ and corresponding residues $r_1, \ldots, r_\ell$.

2. (functional equation) $\Lambda(s) = \omega \overline{\Lambda(1 - \bar{s})}$ for some $\omega \in \mathbb{C}$, $\omega \neq 0$.

3. For any $\alpha \leqslant \beta$, $L(\sigma + it) = O(\exp t^A)$ for some $A > 0$, as $|t| \to \infty$, $\alpha \leqslant \sigma \leqslant \beta$, with A and the constant in the 'Oh' notation depending on α and β.

Remarks . a) The 3rd condition, $L(\sigma + it) = O(\exp t^A)$, is very mild. Using the fact that $L(s)$ is bounded in $\Re es \geqslant \sigma_2 > \sigma_1$, the functional equation and the estimate (3.8), and the Phragmén-Lindelöf Theorem [Rud] we can show that in any vertical strip $\alpha \leqslant \sigma \leqslant \beta$,

$$L(s) = O(t^b), \quad \text{for some } b > 0$$

where both b and the constant in the 'Oh' notation depend on α and β.

b) If $b(n), \lambda_j \in \mathbb{R}$, then the second assumption reads $\Lambda(s) = \omega\Lambda(1 - s)$.

c) In all known examples the κ_j's can be taken to equal $1/2$. It is useful to know the Legendre duplication formula

$$\Gamma(s) = (2\pi)^{-1/2}2^{s-1/2}\Gamma(s/2)\Gamma((s+1)/2). \tag{3.3}$$

However, it is sometimes more convenient to work with (3.2), and we avoid specializing prematurely to $\kappa_j = 1/2$.

d) The assumption that $L(s)$ has at most simple poles is not crucial and is only made to simplify the presentation.

e) From the point of view of computing $\Lambda(s)$ given the Dirichlet coefficients and functional equation, we do not need to assume an Euler product for $L(s)$. Without an Euler product, however, it is unlikely that $L(s)$ will satisfy a Riemann Hypothesis.

To obtain a smoothed approximate functional equation with desirable properties we introduce an auxiliary function. Let $g : \mathbb{C} \to \mathbb{C}$ be an entire function that, for fixed s, satisfies

$$\left|\Lambda(z + s)g(z + s)z^{-1}\right| \to 0$$

as $|\Im mz| \to \infty$, in vertical strips, $-\alpha \leqslant \Re ez \leqslant \alpha$. The smoothed approximate functional equation has the following form.

Theorem 1. For $s \notin \{s_1, \ldots, s_\ell\}$, and $L(s)$, $g(s)$ as above,

$$\Lambda(s)g(s) = \sum_{k=1}^{\ell} \frac{r_k g(s_k)}{s - s_k} + Q^s \sum_{n=1}^{\infty} \frac{b(n)}{n^s} f_1(s, n)$$

$$+\omega Q^{1-s} \sum_{n=1}^{\infty} \frac{\overline{b}(n)}{n^{1-s}} f_2(1 - s, n) \tag{3.4}$$

where

$$f_1(s,n) = \frac{1}{2\pi i} \int_{\nu-i\infty}^{\nu+i\infty} \prod_{j=1}^{a} \Gamma(\kappa_j(z+s) + \lambda_j) z^{-1} g(s+z)(Q/n)^z dz$$

$$f_2(1-s,n) = \frac{1}{2\pi i} \int_{\nu-i\infty}^{\nu+i\infty} \prod_{j=1}^{a} \Gamma(\kappa_j(z+1-s) + \overline{\lambda_j}) z^{-1} g(s-z)(Q/n)^z dz$$

$$(3.5)$$

with $\nu > \max\{0, -\Re(\lambda_1/\kappa_1 + s), \ldots, -\Re(\lambda_a/\kappa_a + s)\}$.

Proof. Let C be the rectangle with verticies $(-\alpha, -iT)$, $(\alpha, -iT)$, (α, iT), $(-\alpha, iT)$, let $s \in \mathbb{C} - \{s_1, \ldots, s_\ell\}$, and consider

$$\frac{1}{2\pi i} \int_C \Lambda(z+s) g(z+s) z^{-1} dz. \qquad (3.6)$$

(integrated counter-clockwise). α and T are chosen big enough so that all the poles of the integrand are contained within the rectangle. We will also require, soon, that $\alpha > \sigma_1 - \Re s$. On the one hand (3.6) equals

$$\Lambda(s)g(s) + \sum_{k=1}^{\ell} \frac{r_k g(s_k)}{s_k - s} \qquad (3.7)$$

since the poles of the integrand are included in the set $\{0, s_1 - s, \ldots, s_\ell - s\}$, and are all simple. Typically, the set of poles will coincide with this set. However, if $\Lambda(s)g(s) = 0$, then $z = 0$ is no longer a pole of the integrand. But then $\Lambda(s)g(s)$ contributes nothing to (3.7) and the equality remains valid. And if $g(s_k) = 0$, then there is no pole at $z = s_k - s$ but also no contribution from $r_k g(s_k)/(s_k - s)$.

On the other hand, we may break the integral over C into four integrals:

$$\int_C = \int_{\alpha-iT}^{\alpha+iT} + \int_{\alpha+iT}^{-\alpha+iT} + \int_{-\alpha+iT}^{-\alpha-iT} + \int_{-\alpha-iT}^{\alpha-iT}$$

$$= \int_{C_1} + \int_{C_2} + \int_{C_3} + \int_{C_4}.$$

The integral over C_1, assuming that α is big enough to write $L(s+z)$ in terms of its Dirichlet series i.e. $\alpha > \sigma_1 - \Re s$, is

$$Q^s \sum_{n=1}^{\infty} \frac{b(n)}{n^s} \frac{1}{2\pi i} \int_{\alpha-iT}^{\alpha+iT} \prod_{j=1}^{a} \Gamma(\kappa_j(z+s) + \lambda_j) z^{-1} g(s+z)(Q/n)^z dz.$$

We are justified in rearranging summation and integration since the series for $L(z + s)$ converges uniformly on C_1. Further, by the functional equation, the integral over C_3 equals

$$\frac{\omega}{2\pi i} \int_{-\alpha+iT}^{-\alpha-iT} \overline{\Lambda(1 - \overline{z + s})} g(z + s) z^{-1} dz$$

$$= \omega Q^{1-s} \sum_{n=1}^{\infty} \frac{\overline{b}(n)}{n^{1-s}} \frac{1}{2\pi i} \int_{-\alpha+iT}^{-\alpha-iT} \prod_{j=1}^{a} \Gamma(\kappa_j(1 - s - z) + \overline{\lambda_j}) z^{-1} g(s + z)(Q/n)^{-z} dz$$

$$= \omega Q^{1-s} \sum_{n=1}^{\infty} \frac{\overline{b}(n)}{n^{1-s}} \frac{1}{2\pi i} \int_{\alpha-iT}^{\alpha+iT} \prod_{j=1}^{a} \Gamma(\kappa_j(1 - s + z) + \overline{\lambda_j}) z^{-1} g(s - z)(Q/n)^{z} dz.$$

Letting $T \to \infty$, the integrals over C_2 and C_4 tend to zero by our assumption on the rate of growth of $g(s)$, and we obtain (3.4). The integrals in (3.5) are, by Cauchy's Theorem, independent of the choice of ν, so long as $\nu > \max\{0, -\mathfrak{Re}(\lambda_1/\kappa_1 + s), \ldots, -\mathfrak{Re}(\lambda_a/\kappa_a + s)\}$.

$$\square$$

3.3   Choice of $g(z)$

Formulae of the form (3.4) are well known [L] [Fr]. Usually, one finds it in the literature with $g(s) = 1$. For example, for the Riemann zeta function this leads to Riemann's formula [R, pg 179] [Ti, pg 22]

$$\pi^{-s/2}\Gamma(s/2)\zeta(s) = -\frac{1}{s} - \frac{1}{1-s} + \pi^{-s/2} \sum_{n=1}^{\infty} \frac{1}{n^s} \Gamma(s/2, \pi n^2)$$

$$+ \pi^{(s-1)/2} \sum_{n=1}^{\infty} \frac{1}{n^{1-s}} \Gamma((1 - s)/2, \pi n^2)$$

where $\Gamma(s, w)$ is the incomplete gamma function (see Section 3.4).

However, the choice $g(s) = 1$ is not well suited for computing $\Lambda(s)$ as $|\mathfrak{Im}(s)|$ grows. By Stirling's formula [Ol, pg 294]

$$|\Gamma(s)| \sim (2\pi)^{1/2} |s|^{\sigma-1/2} e^{-|t|\pi/2}, \quad s = \sigma + it \tag{3.8}$$

as $|t| \to \infty$, and so decreases very quickly as $|t|$ increases. Hence, with $g(s) = 1$, the l.h.s. of (3.4) is extremely small for large $|t|$ and fixed σ. On the other hand, we can show that the terms on the r.h.s., though decreasing as $n \to \infty$, start off relatively large compared to the l.h.s.. Hence a tremendous amount of cancellation must occur on the r.h.s. and and we would need an unreasonable amount of precision. This problem is analogous to what happens if we try to sum $\exp(-x) = \sum(-x)^n/n!$ in a naive way. If x is positive and large, the l.h.s. is exponentially small, yet the terms on the r.h.s. are large before they become small and high precision is needed to capture the ensuing cancellation.

One way to control this cancellation is to choose $g(s)$ equal to δ^{-s} with $|\delta| = 1$ and chosen to cancel out most of the exponentially small size of the Γ factors. This idea appears in the work of Lavrik [L], and was also suggested by Lagarias and Odlyzko [LO] who did not implement it since it led to complications regarding the computation of (3.5). This method was successfully applied in the author's PhD thesis [Ru] to compute Dirichlet L-functions and L-functions associated to cusp forms and is used extensively in the author's L-function package [Ru3] More recently, this approach was used in the computation of Akiyama and Tanigawa [AT] to compute several elliptic curve L-functions.

In fact when there are multiple Γ factors it is better to choose a different δ for each Γ and multiply these together. For a given s let

$$t_j = \mathfrak{Im}(\kappa_j s + \lambda_j)$$

$$\theta_j = \begin{cases} \pi/2, & \text{if } |t_j| \leqslant 2c/(a\pi) \\ c/(a|t_j|), & \text{if } |t_j| > 2c/(a\pi) \end{cases}$$

$$\delta_j = \exp(i \operatorname{sgn}(t_j)(\pi/2 - \theta_j)). \tag{3.9}$$

Here $c > 0$ is a free parameter. Larger c means faster convergence of the sums in (3.4), but also more cancellation and loss of precision.

Next, we set

$$g(z) := \prod_{j=1}^{a} \delta_j^{-\kappa_j z - \mathfrak{Im}\lambda_j} = \beta \delta^{-z}. \tag{3.10}$$

Because δ_j depends on s, the constants δ and β depends on s. We can either use a fresh δ for each new s value, or else modify the above choice of t_j so as to use the same t_j for other nearby s's. The latter is prefered if we wish to carry out precomputations that can be recycled as we vary s. For simplicity, here we assume that a fresh δ is chosen as above for each new s.

The choice of g controls the exponentially small size of the Γ factors. Notice that the constant factor $\beta = \prod_{j=1}^{a} \delta_j^{-\mathfrak{Im}\lambda_j}$ in (3.10) appears in every term in (3.4), and hence can be dropped from $g(z)$ without any effect on cancellation or the needed precision. However, to analyze the size of the the l.h.s. of (3.4) and the terms on the r.h.s. this factor is helpful and we leave it in for now, but with the understanding that it can be omitted.

To see the effect of the function $g(z)$ on the l.h.s of (3.4) we have, by (3.8) and (3.9)

$$|\Lambda(s_0)g(s_0)| \quad \sim \quad * \cdot |L(s_0)| \prod_{|t_j| \leqslant 2c/\pi} \exp\left(-|t_j|\,\pi/2\right) \prod_{|t_j| > 2c/\pi} \exp\left(-c/a\right)$$

$$\geqslant \quad * \cdot |L(s_0)| \exp(-c)$$

where

$$* = Q^{\sigma_0} (2\pi)^{a/2} \prod_{j=1}^{a} |\kappa_j s_0 + \lambda_j|^{\kappa_j \sigma_0 + \Re\lambda_j - 1/2}.$$

We have thus managed to control the exponentially small size of $\Lambda(s)$ up to a factor of $\exp(-c)$ which we can regulate via the choice of c. We can also show that this choice of $g(z)$ leads to well balanced terms on the r.h.s. of (3.4).

3.4 Approximate functional equation in the case of one Γ-factor

We first treat the case $a = 1$ separately because it is the simplest, the greatest number of tools have been developed to handle this case, and many popular L-functions have $a = 1$.

Here we are assuming that

$$\Lambda(s) = Q^s \Gamma(\gamma s + \lambda) L(s).$$

According to (3.10) we should set

$$g(s) = \delta^{-s}$$

(we omit the factor β as described following (3.10)) with

$$\delta = \delta_1{}^{\gamma}$$

and

$$t_1 = \Im(\gamma s + \lambda)$$

$$\theta_1 = \begin{cases} \pi/2, & \text{if } |t_1| \leqslant 2c/\pi \\ c/|t_1|, & \text{if } |t_1| > 2c/\pi \end{cases}$$

$$\delta_1 = \exp(i\,\operatorname{sgn}(t_1)(\pi/2 - \theta_1)).$$

In that case, the function $f_1(s, n)$ that appears in Theorem 1 equals

$$f_1(s, n) = \frac{\delta^{-s}}{2\pi i} \int_{\nu-i\infty}^{\nu+i\infty} \Gamma(\gamma(z+s)+\lambda)z^{-1} \left(Q/(n\delta)\right)^z dz$$

$$= \frac{\delta^{-s}}{2\pi i} \int_{\gamma\nu-i\infty}^{\gamma\nu+i\infty} \Gamma(u+\gamma s+\lambda)u^{-1} \left(Q/(n\delta)\right)^{u/\gamma} du.$$

Now

$$\Gamma(v+u)u^{-1} = \int_0^\infty \Gamma(v,t)t^{u-1}dt, \quad \mathfrak{Re}\, u > 0, \quad \mathfrak{Re}(v+u) > 0 \qquad (3.11)$$

where

$$\Gamma(z, w) = \int_w^\infty e^{-x}x^{z-1}dx \quad |\arg w| < \pi$$

$$= w^z \int_1^\infty e^{-wx}x^{z-1}dx, \quad \mathfrak{Re}(w) > 0.$$

$\Gamma(z, w)$ is known as the incomplete gamma function. By Mellin inversion

$$f_1(s, n) = \delta^{-s}\Gamma\left(\gamma s + \lambda, (n\delta/Q)^{1/\gamma}\right).$$

Similarly

$$f_2(1-s, n) = \delta^{-s}\Gamma\left(\gamma(1-s)+\overline{\lambda}, (n/(\delta Q))^{1/\gamma}\right).$$

We may thus express, when $a = 1$ and $g(s) = \delta^{-s}$, (3.4) as

$$Q^s\Gamma(\gamma s + \lambda)L(s)\delta^{-s} = \sum_{k=1}^{\ell} \frac{r_k\delta^{-s_k}}{s-s_k}$$

$$+ (\delta/Q)^{\lambda/\gamma} \sum_{n=1}^\infty b(n)n^{\lambda/\gamma}G\left(\gamma s + \lambda, (n\delta/Q)^{1/\gamma}\right)$$

$$+ \frac{\omega}{\delta}(Q\delta)^{-\overline{\lambda}/\gamma} \sum_{n=1}^\infty \overline{b}(n)n^{\overline{\lambda}/\gamma}G\left(\gamma(1-s)+\overline{\lambda}, (n/(\delta Q))^{1/\gamma}\right)$$

$$\qquad (3.12)$$

where

$$G(z, w) = w^{-z}\Gamma(z, w) = \int_1^\infty e^{-wx}x^{z-1}dx, \quad \mathfrak{Re}(w) > 0. \qquad (3.13)$$

Note, from (3.10) with $a = 1$, we have $\mathfrak{Re}\,\delta^{1/\gamma} > 0$, so both $(n\delta/Q)^{1/\gamma}$ and $(n/(\delta Q))^{1/\gamma}$ have positive $\mathfrak{Re}$ part.

3.4.1 Examples

1) Riemann zeta function[2], $\zeta(s)$: the necessary background can be found in [Ti]. Formula (3.12), for $\zeta(s)$, is

$$\pi^{-s/2}\Gamma(s/2)\zeta(s)\delta^{-s} = -\frac{1}{s} - \frac{\delta^{-1}}{1-s} + \sum_{n=1}^{\infty} G\left(s/2, \pi n^2\delta^2\right)$$

$$+\delta^{-1}\sum_{n=1}^{\infty} G\left((1-s)/2, \pi n^2/\delta^2\right) \qquad (3.14)$$

2) Dirichlet L-functions[3], $L(s,\chi)$: (see [D, chapter 9]). When χ is primitive and even, $\chi(-1) = 1$, we get

$$\left(\frac{q}{\pi}\right)^{s/2}\Gamma(s/2)L(s,\chi)\delta^{-s} = \sum_{n=1}^{\infty} \chi(n)G\left(s/2, \pi n^2\delta^2/q\right)$$

$$+\frac{\tau(\chi)}{\delta q^{1/2}}\sum_{n=1}^{\infty} \overline{\chi}(n)G\left((1-s)/2, \pi n^2/(\delta^2 q)\right)$$

and when χ is primitive and odd, $\chi(-1) = -1$, we get

$$\left(\frac{q}{\pi}\right)^{s/2}\Gamma(s/2+1/2)L(s,\chi)\delta^{-s} = \delta\left(\frac{\pi}{q}\right)^{1/2}\sum_{n=1}^{\infty} \chi(n)nG\left(s/2+1/2, \pi n^2\delta^2/q\right)$$

$$+\frac{\tau(\chi)\pi^{1/2}}{iq\delta^2}\sum_{n=1}^{\infty} \overline{\chi}(n)nG\left((1-s)/2+1/2, \pi n^2/(\delta^2 q)\right)$$

Here, $\tau(\chi)$ is the Gauss sum

$$\tau(\chi) = \sum_{m=1}^{q} \chi(m)e^{2\pi im/q}.$$

3) Cusp form L-functions[4]: (see [Og]). Let $f(z)$ be a cusp form of weight k for $SL_2(\mathbb{Z})$, k a positive even integer:

   1. $f(z)$ is entire on $\mathbb{H}$, the upper half plane.

---

[2]Editors' comment: See also the lectures of D.R. Heath-Brown, page 1, Section 3, for more on the Riemann zeta function.

[3]Editors' comment: See also the lectures of D.R. Heath-Brown, page 1, Section 12, for more on these L-functions.

[4]Editors' comment: See also the lectures of J.B. Conrey (page 225, Section 3) and P. Michel (page 357, Section 2.2.2) for more on these L-functions.

2. $f(\sigma z) = (cz + d)^k f(z)$, $\sigma = \begin{pmatrix} a & b \\ c & d \end{pmatrix} \in SL_2(\mathbb{Z})$, $z \in \mathbb{H}$.

3. $\lim_{t \to \infty} f(it) = 0$.

Assume further that f is a Hecke eigenform, i.e. an eigenfunction of the Hecke operators. We may expand f in a Fourier series

$$f(z) = \sum_{n=1}^{\infty} a_n e^{2\pi i n z}, \quad \Im(z) > 0$$

and associate to $f(z)$ the Dirichlet series

$$L_f(s) := \sum_{1}^{\infty} \frac{a_n}{n^{(k-1)/2}} n^{-s}.$$

We normalize f so that $a_1 = 1$. This series converges absolutely when $\Re(s) > 1$ because, as proven by Deligne [Del],

$$|a_n| \leqslant \sigma_0(n) n^{(k-1)/2},$$

where $\sigma_0(n) := \sum_{d|n} 1 = O(n^\epsilon)$ for any $\epsilon > 0$.

$L_f(s)$ admits an analytic continuation to all of $\mathbb{C}$ and satisfies the functional equation

$$\Lambda_f(s) := (2\pi)^{-s} \Gamma(s + (k-1)/2) L_f(s) = (-1)^{k/2} \Lambda_f(1 - s).$$

With our normalization, $a_1 = 1$, the a_n's are real since they are eigenvalues of self adjoint operators, the Hecke operators with respect to the Petersson inner product (see [Og, III-12]). Furthermore, the required rate of growth on $L_f(s)$, condition 3 on page 443, follows from the modularity of f.

Hence, in this example, formula (3.12) is

$$(2\pi)^{-s} \Gamma(s + (k-1)/2) L_f(s) \delta^{-s} = (\delta 2\pi)^{(k-1)/2} \sum_{n=1}^{\infty} a_n G\left(s + (k-1)/2, 2\pi n \delta\right)$$

$$+ \frac{(-1)^{k/2}}{\delta} \left(\frac{2\pi}{\delta}\right)^{(k-1)/2} \sum_{n=1}^{\infty} a_n G\left(1 - s + (k-1)/2, 2\pi n/\delta\right)$$

4) Twists of cusp forms[5]: $L_f(s, \chi)$, χ primitive, $f(z)$ as in the previous example. $L_f(s, \chi)$ is given by the Dirichlet series

---

[5] Editors' comment: See also the lectures of J.B. Conrey, page 225, Section 3.3.

$$L_f(s,\chi) = \sum_1^\infty \frac{a_n \chi(n)}{n^{(k-1)/2}} n^{-s}.$$

$L_f(s,\chi)$ extends to an entire function and satisfies the functional equation

$$\begin{aligned}
\Lambda_f(s,\chi) &:= \left(\frac{q}{2\pi}\right)^s \Gamma(s + (k-1)/2) L_f(s,\chi) \\
&= (-1)^{k/2} \chi(-1) \frac{\tau(\chi)}{\tau(\overline{\chi})} \Lambda_f(1-s, \overline{\chi}).
\end{aligned}$$

In this example, formula (3.12) is

$$\begin{aligned}
\left(\frac{q}{2\pi}\right)^s &\Gamma(s + (k-1)/2) L_f(s,\chi)\delta^{-s} = \\
&\left(\frac{2\pi\delta}{q}\right)^{(k-1)/2} \sum_{n=1}^\infty a_n \chi(n) G\left(s + (k-1)/2, 2\pi n\delta/q\right) \\
&+ \frac{(-1)^{k/2}}{\delta} \chi(-1) \frac{\tau(\chi)}{\tau(\overline{\chi})} \left(\frac{2\pi}{q\delta}\right)^{(k-1)/2} \\
&\qquad\qquad \times \sum_{n=1}^\infty a_n \overline{\chi}(n) G\left(1 - s + (k-1)/2, 2\pi n/(\delta q)\right).
\end{aligned}$$

5) Elliptic curve L-functions[6]: (see [Kn, especially chapters X,XII]). Let E be an elliptic curve over $\mathbb{Q}$, which we write in global minimal Weierstrass form

$$y^2 + c_1 xy + c_3 y = x^3 + c_2 x^2 + c_4 x + c_6$$

where the c_j's are integers and the discriminant Δ is minimal.

To the elliptic curve E we may associate an Euler product

$$L_E(s) := \prod_{p|\Delta} (1 - a_p p^{-1/2-s})^{-1} \prod_{p\nmid\Delta} (1 - a_p p^{-1/2-s} + p^{-2s})^{-1} \qquad (3.15)$$

where, for $p \nmid \Delta$, $a_p = p + 1 - \#E_p(\mathbb{Z}_p)$, with $\#E_p(\mathbb{Z}_p)$ being the number of points (x,y) in $\mathbb{Z}_p \times \mathbb{Z}_p$ on the curve E considered modulo p, together with the point at infinity. When $p|\Delta$, a_p is either 1, -1, or 0. If $p \nmid \Delta$, a theorem of Hasse states that $|a_p| < 2p^{1/2}$. Hence, (3.15) converges when $\mathfrak{Re}(s) > 1$, and for these values of s we may expand $L_E(s)$ in an absolutely convergent Dirichlet series

$$L_E(s) = \sum_1^\infty \frac{a_n}{n^{1/2}} n^{-s}. \qquad (3.16)$$

---

[6]Editors' comment: See also the lectures of P. Michel, page 357, Section 4.

The Hasse-Weil conjecture asserts that $L_E(s)$ extends to an entire function and has the functional equation

$$\Lambda_E(s) := \left(\frac{N^{1/2}}{2\pi}\right)^s \Gamma(s+1/2) L_E(s) = -\varepsilon \Lambda_E(1-s).$$

where N is the conductor of E, and ε, which depends on E, is either ± 1. The Hasse-Weil conjecture and also the required rate of growth on $L_E(s)$ follows from the Shimura-Taniyama-Weil conjecture, which has been proven by Wiles and Taylor [TW] [Wi] for elliptic curves with square free conductor and has been extended, by Breuil, Conrad, Diamond and Taylor to all elliptic curves over $\mathbb{Q}$ [BCDT].

Hence we have

$$\left(\frac{N^{1/2}}{2\pi}\right)^s \Gamma(s+1/2) L_E(s) \delta^{-s} = \left(\frac{2\pi\delta}{N^{1/2}}\right)^{1/2} \sum_{n=1}^{\infty} a_n G\left(s+1/2, 2\pi n\delta/N^{1/2}\right)$$

$$-\frac{\varepsilon}{\delta}\left(\frac{2\pi}{N^{1/2}\delta}\right)^{1/2} \sum_{n=1}^{\infty} a_n G\left(1-s+1/2, 2\pi n/(\delta N^{1/2})\right).$$

6) Twists of elliptic curve L-functions: $L_E(s,\chi)$, χ a primitive character of conductor q, $(q,N)=1$. Here $L_E(s,\chi)$ is given by the Dirichlet series

$$L_E(s,\chi) = \sum_{1}^{\infty} \frac{a_n}{n^{1/2}} \chi(n) n^{-s}.$$

The Weil conjecture asserts, here, that $L_E(s)$ extends to an entire function and satisfies

$$\Lambda_E(s,\chi) := \left(\frac{qN^{1/2}}{2\pi}\right)^s \Gamma(s+1/2) L_E(s,\chi) = -\varepsilon\chi(-N)\frac{\tau(\chi)}{\tau(\overline{\chi})}\Lambda_E(1-s,\overline{\chi}).$$

Here N and ε are the same as for E. In this example the conjectured formula is

$$\left(\frac{qN^{1/2}}{2\pi}\right)^s \Gamma(s+1/2) L_E(s) \delta^{-s}$$

$$= \left(\frac{2\pi\delta}{qN^{1/2}}\right)^{1/2} \sum_{n=1}^{\infty} a_n \chi(n) G\left(s+1/2, 2\pi n\delta/(qN^{1/2})\right)$$

$$-\frac{\varepsilon}{\delta}\chi(-N)\frac{\tau(\chi)}{\tau(\overline{\chi})}\left(\frac{2\pi}{qN^{1/2}\delta}\right)^{1/2} \sum_{n=1}^{\infty} a_n \overline{\chi}(n) G\left(1-s+1/2, 2\pi n/(\delta qN^{1/2})\right).$$

We have reduced in the case $a = 1$ the computation of $\Lambda(s)$ to one of evaluating two sums of incomplete gamma functions. The $\Gamma(\gamma s + \lambda)\delta^{-s}$ factor on the left of (3.12) and elsewhere is easily evaluated using several terms of Stirling's asymptotic formula and also the recurrence $\Gamma(z+1) = z\Gamma(z)$ applied a few times. The second step is needed for small z. Some care needs to be taken to absorb the $e^{-\pi|\Im(\gamma s+\lambda)|/2}$ factor of $\Gamma(\gamma s + \lambda)$ into the $e^{\pi|\Im(\gamma s+\lambda)|/2}$ factor of δ^{-s}. Otherwise our effort to control the size of $\Gamma(\gamma s + \lambda)$ will have been in vain, and lack of precision will wreak havoc.

To see how many terms in (3.12) are needed we can use the rough bound

$$|G(z,w)| < e^{-\Re(w)} \int_0^\infty e^{-(\Re(w)-\Re(z)+1)t}dt = \frac{e^{-\Re(w)}}{\Re(w) - \Re(z) + 1},$$

valid for $\Re(w) > \Re(z) - 1 > 0$. We have put $t = x - 1$ in (3.13) and have used $t + 1 \leqslant e^t$. Also, for $\Re(w) > 0$ and $\Re(z) \leqslant 1$,

$$|G(z,w)| < \frac{e^{-\Re(w)}}{\Re(w)}.$$

These inequalities tells us that the terms in (3.12) decrease exponentially fast once n is sufficiently large.

For example, in equation (3.14) for $\zeta(s)$ we get exponential drop off roughly when

$$\Re \pi n^2 \delta^2 >> 1.$$

But

$$\Re \pi n^2 \delta^2 = \pi n^2 \Re \delta^2 \sim 2\pi n^2 c/t$$

so the number of terms needed is roughly

$$>> (t/c)^{1/2}.$$

3.4.2   Computing $\Gamma(z, w)$

Recall the definitions

$$\Gamma(z, w) = \int_w^\infty e^{-t} t^{z-1} dt, \quad |\arg w| < \pi$$
$$G(z, w) = w^{-z} \Gamma(z, w).$$

Let

$$\gamma(z,w) := \Gamma(z) - \Gamma(z,w) = \int_0^w e^{-x} x^{z-1} dx, \quad \Re e\, z > 0, \quad |\arg w| < \pi$$

be the complimentary incomplete gamma function, and set

$$g(z,w) = w^{-z} \gamma(z,w) = \int_0^1 e^{-wt} t^{z-1} dt \tag{3.17}$$

so that $G(z,w) + g(z,w) = w^{-z}\Gamma(z)$. The function $g(z,w)/\Gamma(z)$ is entire in z and w.

The incomplete Γ function undergoes a transition when $|w|$ is close to $|z|$. This will be described using Temme's uniform asymptotics for $\Gamma(z,w)$. The transition explains the difficulty in computing $\Gamma(z,w)$ without resorting to several different expressions or using uniform asymptotics.

A combination of series, asymptotics, and continued fractions are useful when $|z|$ is somewhat bigger than or smaller than $|w|$. When the two parameters are close in size to one another, we can employ Temme's more involved uniform asymptotics. We can also apply the Poisson summation method described in Section 2, or an expansion due to Nielsen. Below we look at a few useful approaches.

Integrating by parts we get

$$g(z,w) = e^{-w} \sum_{j=0}^{\infty} \frac{w^j}{(z)_{j+1}}$$

where

$$(z)_j = \begin{cases} z(z+1)\ldots(z+j-1) & \text{if } j > 0; \\ 1 & \text{if } j = 0. \end{cases}$$

(The case $j = 0$ occurs below in an expression for $G(z,w)$). While this series converges for $z \neq 0, -1, -2, \ldots$ and all w, it is well suited, say if $\Re e\, z > 0$ and $|w| < \alpha|z|$ with $0 < \alpha < 1$. Otherwise, not only does the series take too long to converge, but precision issues arise.

The following continued fraction converges for $\Re z > 0$

$$g(z,w) = \cfrac{e^{-w}}{z - \cfrac{zw}{z+1+ \cfrac{w}{z+2- \cfrac{(z+1)w}{z+3+ \cfrac{2w}{z+4- \cfrac{(z+2)w}{z+5+\cdots}}}}}}$$

The paper of Akiyama and Tanigawa [AT] contains an analysis of the truncation error for this continued fraction, as well as the continued fraction in (3.18) below, and show that the above is most useful when $|w| < |z|$, with poorer performance as $|w|$ approaches $|z|$.

Another series, useful when $|w| << 1$, is

$$g(z,w) = \sum_{j=0}^{\infty} \frac{(-1)^j}{j!} \frac{w^j}{z+j}.$$

This is obtained from (3.17) by expanding e^{-wt} in a Taylor series and integrating termwise. As $|w|$ grows, cancellation and precision become an issue in the same way it does for the sum $e^{-w} = \sum(-w)^j/j!$.

Next, integrate $G(z,w)$ by parts to obtain the asymptotic series

$$G(z,w) = \frac{e^{-w}}{w} \sum_{j=0}^{M-1} \frac{(1-z)_j}{(-w)^j} + \epsilon_M(z,w)$$

with

$$\epsilon_M(z,w) = \frac{(1-z)_M}{(-w)^M} G(z-M,w).$$

This asymptotic expansion works well if $|w| > \beta|z|$ with $\beta > 1$ and $|z|$ large. In that region the following continued fraction also works well

$$G(z,w) = \cfrac{e^{-w}}{w+ \cfrac{1-z}{1+ \cfrac{1}{w+ \cfrac{2-z}{1+ \cfrac{2}{w+ \cfrac{3-z}{1+ \cfrac{3}{w+\cdots}}}}}}} \qquad (3.18)$$

Temme's uniform asymptotics for $\Gamma(z, w)$ provide a powerful tool for computing the function in its transition zone and elsewhere. Following the notation in [T], let

$$
\begin{aligned}
Q(z, w) &= \Gamma(z, w)/\Gamma(z) \\
\lambda &= w/z \\
\eta^2/2 &= \lambda - 1 - \log \lambda
\end{aligned}
$$

where the sign of η is chosen to be positive for $\lambda > 1$. Then

$$
Q(z, w) = \frac{1}{2}\mathrm{erfc}(\eta(z/2)^{1/2}) + R_z(\eta)
$$

where

$$
\mathrm{erfc} = \frac{2}{\pi^{1/2}} \int_z^\infty e^{-t^2}\, dt,
$$

and R_z is given by the asymptotic series, as $z \to \infty$,

$$
R_z(\eta) = \frac{e^{-z\eta^2/2}}{(2\pi z)^{1/2}} \sum_{n=0}^\infty \frac{c_n(\eta)}{z^n}. \tag{3.19}
$$

Here

$$
\begin{aligned}
c_0(\eta) &= \frac{1}{\lambda - 1} - \frac{1}{\eta} \\
c_1(\eta) &= \frac{1}{\eta^3} - \frac{1}{(\lambda-1)^3} - \frac{1}{(\lambda-1)^2} - \frac{1}{12(\lambda-1)} \\
\eta c_n(\eta) &= \frac{d}{d\eta}c_{n-1}(\eta) + \frac{\eta}{\lambda-1}\gamma_n, \quad n \geqslant 1
\end{aligned}
$$

with

$$
\Gamma^*(z) = \sum_0^\infty \frac{(-1)^n \gamma_n}{z^n}
$$

being the asymptotic expansion of

$$
\Gamma^*(z) = (z/(2\pi))^{1/2}(e/z)^z \Gamma(z).
$$

The first few terms are $\gamma_0 = 1$, $\gamma_1 = -1/12$, $\gamma_2 = 1/288$, $\gamma_3 = 139/51840$. The singularities at $\eta = 0$, i.e. $\lambda = 1$, $z = w$, are removable. Unfortunately, explicit estimates for the remainder in truncating (3.19) when the parameters are complex have not been worked out, but in practice the expansion seems to work very well.

To handle the intermediate region $|z| \approx |w|$ we could also use the following expansion of Nielsen to step through the troublesome region

$$\gamma(z, w+d) = \gamma(z,w) + w^{z-1}e^{-w} \sum_{j=0}^{\infty} \frac{(1-z)_j}{(-w)^j}(1 - e^{-d}e_j(d)), \quad |d| < |w| \quad (3.20)$$

where

$$e_j(d) = \sum_{m=0}^{j} \frac{d^m}{m!}.$$

A proof can be found in [EMOT]. This expansion is very well suited, for example, for L-functions associated to modular forms, since in that case we increment w in equal steps from term to term in (3.12) and precomputations can be arranged to recycle data. Numerically, this expansion is unstable if $|d|$ is big. This can be overcome by taking many smaller steps, but this then makes Nielsen's expansion an inefficient choice for $\zeta(s)$ or Dirichlet L-functions.

In computing (3.20) some care needs to be taken to avoid numerical pitfalls. One pitfall is that, as j grows, $e^{-d}e_j(d) \to 1$. So once $\left|1 - e^{-d}e_j(d)\right| < 10^{-\text{Digits}}$, the error in computation of $1 - e^{-d}e_j(d)$ is bigger than its value, and this gets magnified when we multiply by $(1-z)_j/(-w)^j$. So in computing $((1-z)_j/(-w)^j)(1 - e^{-d}e_j(d))$ one must avoid the temptation to view this as a product of $(1-z)_j/(-w)^j$ and $1 - e^{-d}e_j(d)$. Instead, we let

$$a_j(z,w,d) = \frac{(1-z)_j}{(-w)^j}(1 - e^{-d}e_j(d)).$$

Now, $1 - e^{-d}e_j(d) = e^{-d}(e^d - e_j(d))$, and we get

$$a_{j+1}(z,w,d) = a_j(z,w,d)\frac{z - (j+1)}{w}\left(\sum_{j+2}^{\infty} d^m/m!\right) \bigg/ \left(\sum_{j+1}^{\infty} d^m/m!\right)$$

$$= a_j(z,w,d)\frac{z - (j+1)}{w}\left(1 - 1/\beta_j(d)\right), \quad j = 1,2,3,\ldots$$

where

$$\beta_j(d) = \sum_{m=0}^{\infty} d^m/(j+2)_m.$$

Furthermore

$$\beta_j(d) - 1 \sim d/(j+2), \quad \text{as } |d|/j \to 0.$$

Hence, for $|w| \approx |z|$, we *approximately* have (as $|d|/j \to 0$)

$$\left|\frac{z - (j+1)}{w}(1 - 1/\beta_j(d))\right| \leqslant \left(1 + \frac{j+1}{|w|}\right)\frac{|d|}{j+2} \leqslant \frac{|d|}{j+2} + \frac{|d|}{|w|}.$$

Thus, because $|d/w| < 1$, we have, for j big enough, that the above is < 1, and so the sum in (3.20) converges geometrically fast, and hence only a handful of terms are required.

One might be tempted to compute the $\beta_j(d)$'s using the recursion

$$\beta_{j+1}(d) = (\beta_j(d) - 1)(j + 2)/d$$

but this leads to numerical instability. The $\beta_j(d)$'s are all equal to $1 + O_d(1/(j+2))$ and are thus all roughly of comparable size. Hence, a small error due to roundoff in $\beta_j(d)$ is turned into a much larger error in $\beta_{j+1}(d)$, $(j+2)/|d|$ times larger, and this quickly destroys the numerics.

There seems to be some potential in an asymptotic expression due to Ramanujan [B, pg 193, entry 6]

$$G(z, w) \sim w^{-z}\Gamma(z)/2 + e^{-w} \sum_{k=0}^{M} p_k(w - z + 1)/w^{k+1}, \quad \text{as } |z| \to \infty,$$

for $|w - z|$ relatively small, where $p_k(v)$ is a polynomial in v of degree $2k + 1$, though this potential has not been investigated substantially.

We list the first few $p_k(v)$'s here:

$$p_0(v) = -v + 2/3$$

$$p_1(v) = -\frac{v^3}{3} + \frac{v^2}{3} - \frac{4}{135}$$

$$p_2(v) = -\frac{v^5}{15} + \frac{v^3}{9} - \frac{2v^2}{135} - \frac{4v}{135} + \frac{8}{2835}$$

$$p_3(v) = -\frac{v^7}{105} - \frac{v^6}{45} + \frac{v^5}{45} + \frac{7v^4}{135} - \frac{8v^3}{405} - \frac{16v^2}{567} + \frac{16v}{2835} + \frac{16}{8505}$$

$$p_4(v) = -\frac{v^9}{945} - \frac{2v^8}{315} - \frac{2v^7}{315} + \frac{8v^6}{405} + \frac{11v^5}{405} - \frac{62v^4}{2835} - \frac{32v^3}{1215} + \frac{16v^2}{1701}$$
$$+ \frac{16v}{2835} - \frac{8992}{12629925}$$

$$p_5(v) = -\frac{v^{11}}{10395} - \frac{v^{10}}{945} - \frac{2v^9}{567} - \frac{2v^8}{2835} + \frac{43v^7}{2835} + \frac{41v^6}{2835} - \frac{968v^5}{42525} - \frac{68v^4}{2835}$$
$$+ \frac{368v^3}{25515} + \frac{138064v^2}{12629925} - \frac{35968v}{12629925} - \frac{334144}{492567075}$$

It is worth noting that when many evaluations of $\Lambda(s)$ are required, we can reduce through precomputations the bulk of the work to that of computing a main sum. This comes from the identity

$$G(z, w) = w^{-z}\Gamma(z) - g(z, w).$$

The above discussion indicates that, in (3.12), we should use $g(z, w)$ and this identity to compute $G(z, w)$ roughly when $|w|$ is smaller than $|z|$. For example, with $\zeta(1/2 + it)$, the region $|w| < |z|$ corresponds in (3.14) to $|\pi n^2 \delta^2| < |1/4 + it/2|$ and $|\pi n^2/\delta^2| < |1/4 - it/2|$. Because $|\delta| = 1$ this leads to a main sum consisting of approximately $|t/(2\pi)|^{1/2}$ terms, the same as in the Riemann-Siegel formula.

3.5 The approximate functional equation when there is more than one Γ-factor, and $\kappa_j = 1/2$

In this case, the function $f_1(s, n)$ that appears in Theorem 1 is

$$f_1(s, n) = \frac{\delta^{-s}}{2\pi i} \int_{\nu-i\infty}^{\nu+i\infty} \prod_{j=1}^{a} \Gamma((z+s)/2 + \lambda_j) z^{-1} \left(Q/(\delta n)\right)^z dz. \qquad (3.21)$$

This is a special case of the Meijer G function and we develop some of its properties.

Let $M(\phi(t); z)$ denote the Mellin transform of ϕ

$$M(\phi(t); z) = \int_0^\infty \phi(t) t^{z-1}.$$

We will express $\prod_{j=1}^{a} \Gamma((z+s)/2 + \lambda_j) z^{-1}$ as a Mellin transform analogous to (3.11).

Letting $\phi_1 * \phi_2$ denote the convolution of two functions

$$(\phi_1 * \phi_2)(v) = \int_0^\infty \phi_1(v/t)\phi_2(t)\frac{dt}{t}$$

we have (under certain conditions on ϕ_1, ϕ_2)

$$M(\phi_1 * \phi_2; z) = M(\phi_1; z) \cdot M(\phi_2; z).$$

Thus

$$\prod_{j=1}^{a} M(\phi_j; z) = \int_0^\infty (\phi_1 * \cdots * \phi_a)(t) t^{z-1} dt, \qquad (3.22)$$

with

$$(\phi_1 * \cdots * \phi_a)(v) = \int_0^\infty \cdots \int_0^\infty \phi_1(v/t_1)\phi_2(t_1/t_2)\dots$$
$$\dots \phi_{a-1}(t_{a-2}/t_{a-1})\phi_a(t_{a-1})\frac{dt_1}{t_1}\dots\frac{dt_{a-1}}{t_{a-1}}.$$

Now

$$\prod_{j=1}^{a} \Gamma((z+s)/2+\lambda_j)z^{-1} = \left(\prod_{j=1}^{a-1} \Gamma((z+s)/2+\lambda_j)\right)\left(\Gamma((z+s)/2+\lambda_a)z^{-1}\right).$$

But

$$\Gamma((z+s)/2+\lambda) = M\left(2e^{-t^2}t^{2\lambda+s};z\right),$$

and (3.11) gives

$$\Gamma((z+s)/2+\lambda)z^{-1} = M\left(\Gamma(s/2+\lambda,t^2);z\right).$$

So letting

$$\phi_j(t) = \begin{cases} 2e^{-t^2}t^{2\lambda_j+s} & j=1,\ldots a-1; \\ \Gamma(s/2+\lambda_a,t^2) & j=a, \end{cases}$$

and applying Mellin inversion, we find that (3.21) equals

$$f_1(s,n) = \delta^{-s}(\phi_1 * \cdots * \phi_a)(n\delta/Q), \tag{3.23}$$

where

$$(\phi_1 * \cdots * \phi_a)(v) = v^{2\lambda_1+s}\int_0^\infty \cdots \int_0^\infty 2^{a-1}\prod_{j=1}^{a-1}t_j^{2(\lambda_{j+1}-\lambda_j)}e^{-\left(\frac{v^2}{t_1^2}+\frac{t_1^2}{t_2^2}+\cdots+\frac{t_{a-2}^2}{t_{a-1}^2}\right)}$$
$$\times \left(\int_1^\infty e^{-t_{a-1}^2 x}x^{s/2+\lambda_a-1}dx\right)\frac{dt_1}{t_1}\cdots\frac{dt_{a-1}}{t_{a-1}}.$$

Substituting $u_j = \frac{(v^2 x)^{j/a}}{v^2}t_j^2$ and rearranging order of integration this becomes

$$v^{2\mu+s}\int_1^\infty E_\lambda\left(xv^2\right)x^{s/2+\mu-1}dx,$$

where

$$\mu = \frac{1}{a}\sum_{l=1}^{a}\lambda_j, \tag{3.24}$$

$$E_\lambda(w) = \int_0^\infty \cdots \int_0^\infty \prod_{j=1}^{a-1}u_j^{\lambda_{j+1}-\lambda_j}e^{-w^{1/a}\left(\frac{1}{u_1}+\frac{u_1}{u_2}+\cdots+\frac{u_{a-2}}{u_{a-1}}+u_{a-1}\right)}\frac{du_1}{u_1}\cdots\frac{du_{a-1}}{u_{a-1}}. \tag{3.25}$$

So, returning to (3.23), we find that

$$f_1(s,n) = (n\delta/Q)^{2\mu}(n/Q)^s\int_1^\infty E_\lambda\left(x(n\delta/Q)^2\right)x^{s/2+\mu-1}dx.$$

Note that because (3.21) is symmetric in the λ_j's, so is E_λ.

Similarly

$$f_2(1-s,n) = \delta^{-1} \left(n/(\delta Q)\right)^{2\overline{\mu}} (n/Q)^{1-s} \int_1^\infty E_{\overline{\lambda}} \left(x \left(n/(\delta Q)\right)^2\right) x^{(1-s)/2+\overline{\mu}-1} dx.$$

Hence,

$$Q^s \prod_{j=1}^a \Gamma(s/2 + \lambda_j) L(s) \delta^{-s} = \sum_{k=1}^\ell \frac{r_k \delta^{-s_k}}{s - s_k}$$

$$+ (\delta/Q)^{2\mu} \sum_{n=1}^\infty b(n) n^{2\mu} G_\lambda \left(s/2 + \mu, (n\delta/Q)^2\right)$$

$$+ \frac{\omega}{\delta} (\delta Q)^{-2\overline{\mu}} \sum_{n=1}^\infty \overline{b}(n) n^{2\overline{\mu}} G_{\overline{\lambda}} \left((1-s)/2 + \overline{\mu}, (n/(\delta Q))^2\right)$$

$$(3.26)$$

with

$$G_\lambda(z,w) = \int_1^\infty E_\lambda(xw) x^{z-1} dx$$

(μ and E_λ are given by (3.24), (3.25)).

3.5.1   Examples

When $a = 2$

$$E_\lambda(xw) = \int_0^\infty t^{\lambda_2 - \lambda_1} e^{-(wx)^{1/2}(1/t+t)} \frac{dt}{t}$$

$$= 2K_{\lambda_2 - \lambda_1} \left(2(wx)^{1/2}\right) = 2K_{\lambda_1 - \lambda_2} \left(2(wx)^{1/2}\right), \qquad (3.27)$$

K being the K-Bessel function, so that G_λ is an incomplete integral of the K-Bessel function.

Note further that if $\lambda_1 = \lambda/2$, $\lambda_2 = (\lambda+1)/2$ then (3.27) is

$$2K_{1/2} \left(2(wx)^{1/2}\right) = \left(\pi^{1/2}/(wx)^{1/4}\right) e^{-2(wx)^{1/2}}$$

(see [EMOT]), so $G_{(\lambda/2,(\lambda+1)/2)}(z,w) = 2(2\pi)^{1/2}(4w)^{-z}\Gamma(2z-1/2, 2w^{1/2})$, i.e. the incomplete gamma function. This is what we expect since, using (3.3), we can write the gamma factor $\Gamma((s+\lambda)/2)\Gamma((s+\lambda+1)/2)$ in terms of $\Gamma(s+\lambda)$, for which the $a = 1$ expansion, (3.12), applies.

Maass cusp form L-functions[7]: (background material can be found in [Bu]). Let f be a Maass cusp form with eigenvalue $\lambda = 1/4 - v^2$, i.e. $\Delta f = \lambda f$, where $\Delta = -y^2(\partial/\partial x^2 + \partial/\partial y^2)$, and Fourier expansion

---

[7]Editors' comment: See also the lectures of J.B. Conrey, page 225, Section 3.4.

$$f(z) = \sum_{n \neq 0} a_n y^{1/2} K_v(2\pi |n| y) e^{2\pi i n x},$$

with $a_{-n} = a_n$ for all n, or $a_{-n} = -a_n$ for all n. Let

$$L_f(s) = \sum_{n=1}^{\infty} \frac{a_n}{n^s}, \quad \Re s > 1$$

(absolute convergence in this half plane can be proven via the Rankin-Selberg method), and let $\varepsilon = 0$ or 1 according to whether $a_{-n} = a_n$ or $a_{-n} = -a_n$. We have that

$$\Lambda_f(s) := \pi^{-s} \Gamma((s + \varepsilon + v)/2) \Gamma((s + \varepsilon - v)/2) L_f(s)$$

extends to an entire function and satisfies

$$\Lambda_f(s) = (-1)^\varepsilon \Lambda_f(1 - s).$$

Hence, formula (3.26), for $L_f(s)$, is

$$\pi^{-s} \Gamma((s + \varepsilon + v)/2) \Gamma((s + \varepsilon - v)/2) L_f(s) \delta^{-s} =$$

$$(\delta \pi)^\varepsilon \sum_{n=1}^{\infty} a_n n^\varepsilon G_{((\varepsilon+v)/2,(\varepsilon-v)/2)} \left(s/2 + \varepsilon/2, (n\delta\pi)^2 \right)$$

$$+ \frac{(-1)^\varepsilon}{\delta} (\pi/\delta)^\varepsilon \sum_{n=1}^{\infty} a_n n^\varepsilon G_{((\varepsilon+\overline{v})/2,(\varepsilon-\overline{v})/2)} \left((1-s)/2 + \varepsilon/2, (n\pi/\delta)^2 \right)$$

where, by (3.27),

$$G_{((\varepsilon+v)/2,(\varepsilon-v)/2)} \left(s/2 + \varepsilon/2, (n\delta\pi)^2 \right) = 4 \int_1^\infty K_v(2n\delta\pi t) t^{s+\varepsilon-1} dt$$

$$G_{((\varepsilon+\overline{v})/2,(\varepsilon-\overline{v})/2)} \left((1-s)/2 + \varepsilon/2, (n\pi/\delta)^2 \right) = 4 \int_1^\infty K_{\overline{v}}(2n\pi t/\delta) t^{-s+\varepsilon} dt.$$

Next, let

$$\Gamma_\lambda(z, w) = w^z G_\lambda(z, w) = \int_w^\infty E_\lambda(t) t^{z-1} dt,$$

$$\Gamma_\lambda(z) = \int_0^\infty E_\lambda(t) t^{z-1} dt, \qquad (3.28)$$

$$\gamma_\lambda(z, w) = \int_0^w E_\lambda(t) t^{z-1} dt,$$

with E_λ given by (3.25).

Lemma 1.

$$\Gamma_{\boldsymbol{\lambda}}(z) = \prod_{j=1}^{a} \Gamma(z - \mu + \lambda_j)$$

where $\mu = \frac{1}{a}\sum_{j=1}^{a}\lambda_j$.

Proof. Let $\psi_j(t) = e^{-t}t^{\lambda_j}$, $j = 1, \ldots, a$, and consider

$(\psi_1 * \cdots * \psi_a)(v)$

$$= v^{\lambda_1}\int_0^{\infty}\cdots\int_0^{\infty}\prod_{j=1}^{a-1}t_j^{\lambda_{j+1}-\lambda_j}e^{-\left(\frac{v}{t_1}+\frac{t_1}{t_2}+\cdots+\frac{t_{a-2}}{t_{a-1}}+t_{a-1}\right)}\frac{dt_1}{t_1}\cdots\frac{dt_{a-1}}{t_{a-1}}$$

$$= v^{\mu}\int_0^{\infty}\cdots\int_0^{\infty}\prod_{j=1}^{a-1}x_j^{\lambda_{j+1}-\lambda_j}e^{-v^{1/a}\left(\frac{1}{x_1}+\frac{x_1}{x_2}+\cdots+\frac{x_{a-2}}{x_{a-1}}+x_{a-1}\right)}\frac{dx_1}{x_1}\cdots\frac{dx_{a-1}}{x_{a-1}}.$$

(we have put $t_j = v^{1-j/a}x_j$). Thus, from (3.25)

$$E_{\boldsymbol{\lambda}}(v) = v^{-\mu}(\psi_1 * \cdots * \psi_a)(v),$$

and hence (3.28) equals

$$\int_0^{\infty}(\psi_1 * \cdots * \psi_a)(t)t^{z-\mu-1}dt$$

which, by (3.22) is $\prod_{j=1}^{a}\Gamma(z - \mu + \lambda_j)$.

□

Inverting, we get

$$E_{\boldsymbol{\lambda}}(t) = \frac{1}{2\pi i}\int_{\nu-i\infty}^{\nu+i\infty}\Gamma_{\boldsymbol{\lambda}}(z)\,t^{-z}dz$$

with ν to the right of the poles of $\Gamma_{\boldsymbol{\lambda}}(z)$. Shifting the line integral to the left, we can express $E_{\boldsymbol{\lambda}}(t)$ as a sum of residues, and hence obtain through termwise integration a series expansion for $\gamma_{\boldsymbol{\lambda}}(z, w)$. An algorithm for doing so is detailed in [Do], though with different notation. Such an expansion is useful for $|w| << 1$. That paper also describes how to obtain an asymptotic expansion for $E_{\boldsymbol{\lambda}}(t)$ and hence, by termwise integration, for $\Gamma_{\boldsymbol{\lambda}}(z, w)$, useful for $|w|$ large in comparison to $|z|$. The paper has, implicitly, $g(z) = 1$ and does not control for cancellation. Consequently, it does not provide a means to compute L-functions away from the real axis other than increasing precision.

If one wishes to use the methods of this paper to control for cancellation, then one will have w varying over a wide range of values for which the series expansion in [Do] is not adequate. We thus need an alternative method to compute $G_\lambda(z, w)$ especially in the transition zone $|z| \approx |w|$. It would be useful to have Temme's uniform asymptotics generalized to handle $G_\lambda(z, w)$. Alternatively, we can apply the naive but powerful Riemann sum technique described in section 2.

3.6 The functions $f_1(s, n)$, $f_2(1 - s, n)$ as Riemann sums

Substituting $z = v + iu$ into (3.5) we have

$$f_1(s, n) = \frac{1}{2\pi} \int_{-\infty}^{\infty} \prod_{j=1}^{a} \Gamma(\kappa_j(s + v + iu) + \lambda_j) \frac{g(s + v + iu)}{v + iu} (Q/n)^{v+iu} du.$$

Let

$$h(u) = \frac{1}{2\pi} \prod_{j=1}^{a} \Gamma(\kappa_j(s + v + iu) + \lambda_j) \frac{g(s + v + iu)}{v + iu} (Q/n)^{v+iu}.$$

With the choice of $g(z)$ as in (3.10), an analysis similar to that following (2.21) shows that $\hat{h}(y)$ decays exponentially fast as $y \to -\infty$, and doubly exponentially fast as $y \to \infty$. Hence, we can successfully evaluate $f_1(s, n)$, and similarly $f_2(1 - s, n)$ as simple Riemann sums, with step size inversely proportional to the number of digits of precision required.

The Riemann sum approach gives us tremendous flexibility. We are no longer bound in our choice of $g(z)$ to functions for which (3.5) has nice series or asymptotic expansions. For example, we can, with $A > 0$, set

$$g(z) = \exp(A(z - s)^2) \prod_{j=1}^{a} \delta_j^{-\kappa_j z}.$$

The extra factor $\exp(A(z - s)^2)$ is chosen so as to cut down on the domain of integration. Recall that in $f_1(s, n)$ and $f_2(1-s, n)$, g appears as $g(s \pm (v+iu))$, hence $\exp(A(z - s)^2)$ decays in the integral like $\exp(-Au^2)$. Ideally, we would like to have A large. However, this would cause the Fourier transform $\hat{h}(y)$ to decay too slowly. The Fourier transform of a product is a convolution of Fourier transforms, and the Fourier transform of $\exp(A(v + iu)^2)$ equals

$$(\pi/A)^{1/2} \exp(\pi y(2Av - \pi y)/A).$$

A large value of A leads to to a small $1/A$ and this results in poor performance of $\hat{h}(y)$. We also need to specify v, for the line of integration. Larger v means more rapid decay of $\hat{h}(y)$ but more cancellation in the Riemann sum and hence loss of precision.

Another advantage to the Riemann sum approach is that we can rearrange sums, putting the Riemann sum on the outside and the sum over n on the inside. Both sums are finite since we truncate them once the tails are within the desired precision. This then expresses, to within an error that we can control by our choice of stepsize and truncation, $\Lambda(s)$ as a sum of finite Dirichlet series evaluated at equally spaced points and hence gives a sort of interpolation formula for $\Lambda(s)$. Details related to this approach will appear in a future paper.

3.7   Looking for zeros

To look for zeros of an L-function, we can rotate it so that it is real on the critical line, for example working with $Z(t)$, see (3.1), rather than $\zeta(1/2 + it)$.

We can then advance in small steps, say one quarter the average gap size between consecutive zeros, looking for sign changes of this real valued function, zooming in each time a sign change occurs. Along the way, we need to determine if any zeros have been missed, and, if so, go back and look for them, using more refined step sizes. We can also use more sophisticated interpolation techniques to make the search for zeros more efficient [O]. If this search fails to turn up the missing zeros, then presumably a bug has crept into one's code, or else one should look for zeros of the L-function nearby but off the critical line in violation of the Riemann hypothesis.

To check for missing zeros, we could use the argument principle and numerically integrate the logarithmic derivative of the L-function along a rectangle, rounding to the closest integer. However, this is inefficient and difficult to make numerically rigorous.

It is better to use a test devised by Alan Turing [Tu] for $\zeta(s)$ but which seems to work well in general. Let $N(T)$ denote the number of zeros of $\zeta(s)$ in the critical strip above the real axis and up to height T:

$$N(T) = |\{\rho = \beta + i\gamma | \zeta(\rho) = 0, 0 \leqslant \beta \leqslant 1, 0 < \gamma \leqslant T\}|.$$

A theorem of von Mangoldt states[8] that

$$N(T) = \frac{T}{2\pi} \log(T/(2\pi)) - \frac{T}{2\pi} + \frac{7}{8} + S(T) + O(T^{-1}) \qquad (3.29)$$

with

$$S(T) = O(\log T).$$

However, a stronger inequality due to Littlewood and with explicit constants due to Turing [Tu] [Le] is given by

$$\left| \int_{t_1}^{t_2} S(t)dt \right| \leqslant 2.3 + .128 \log(t_2/\pi) \qquad (3.30)$$

for all $t_2 > t_1 > 168\pi$, i.e. $S(T)$ is 0 on average. Therefore, if we miss one sign change (at least two zeros), we'll quickly detect the fact. To illustrate this, Table 2 contains a list of the imaginary parts of the zeros of $\zeta(s)$ found naively by searching for sign changes of $Z(t)$ taking step sizes equal to two. We notice that near the ninth zero on our list a missing pair is detected, and similarly near the twenty fifth zero. A more refined search reveals the pairs of zeros with imaginary parts equal to 48.0051508812, 49.7738324777, and 94.6513440405, 95.8706342282 respectively.

It would be useful to have a general form of the explicit inequality (3.30) worked out for any L-function. The papers of Rumely [Rum] and Tollis [To] generalize this inequality to Dirichlet L-functions and Dedekind zeta functions respectively.

The main term, analogous to (3.29), for a general L-function is easy to derive. Let $L(s)$ be an L-function with functional equation as described in (3.2). Let $N_L(T)$ denote the number of zeros of $L(s)$ lying within the rectangle $|\Im m s| \leqslant T$, $0 < \Re e s < 1$. Notice here we are considering zeros lying both above and below the real axis since the zeros of $L(s)$ will not be located symmetrically about the real axis if its Dirichlet coefficients $b(n)$ are non-real.

Assume for simplicity that $L(s)$ is entire. The arguement principle and the functional equation for $L(s)$ suggests a main term for $N_L(T)$ equal to

$$N_L(T) \sim \frac{2T}{\pi} \log(Q) + \frac{1}{\pi} \sum_{j=1}^{a} \Im m \left(\log \left(\frac{\Gamma((1/2 + iT)\kappa_j + \lambda_j)}{\Gamma((1/2 - iT)\kappa_j + \lambda_j)} \right) \right).$$

---

[8] Editors' comment: For the proof see the lectures of D.R. Heath-Brown, starting on page 1, Theorem 10.

If we assume further that the λ_j's are all real, then the above is, by Stirling's formula, asymptotically equal to

$$N_L(T) \sim \frac{2T}{\pi} \log(Q) + \sum_{j=1}^{a} \left(\frac{2T\kappa_j}{\pi} \log(T\kappa_j/e) + (\kappa_j/2 + \lambda_j - 1/2) \right).$$

A slight modification of the above is needed if $L(s)$ has poles, as in the case of $\zeta(s)$. See Davenport [D, chapters 15,16] where rigorous proofs are presented for $\zeta(s)$ and Dirichlet L-functions (the original proof is due to von Mangoldt).

4   Experiments involving L-functions

Here we describe some of the experiments that reflect the random matrix theory philosophy, namely that the zeros and values of L-functions behave like the zeros and values of characteristic polynomials from the classical compact groups [KS2]. Consequently, we are interested in questions concerning the distribution of zeros, horizontal and vertical, and the value distribution of L-functions.

4.1   Horizontal distribution of the zeros

Riemann himself computed the first few zeros of $\zeta(s)$, and detailed numerical studies were initiated almost as soon as computers were invented. See Edwards [E] for a historical survey of these computations. To date, the most impressive computations for $\zeta(s)$ have been those of Odlyzko [O] [O2] and Wedeniwski [W]. The latter adapted code of van de Lune, te Riele, and Winter [LRW] for grid computing over the internet. Several thousand computers have been used to verify that the first $8.5 \cdot 10^{11}$ nontrivial zeros of $\zeta(s)$ fall on the critical line. Odlyzko's computations have been more concerned with examining the distribution of the spacings between neighbouring zeros, although the Riemann Hypothesis has also been checked for the intervals examined. In [O], Odlyzko computed 175 million consecutive zeros of $\zeta(s)$ lying near the 10^{20}th zero, and more recently, billions of zeros in a higher region [O2]. The Riemann-Siegel formula has been at the heart of these computations. Odlyzko also uses FFT and interpolation algorithms to allow for many evaluations of $\zeta(s)$ at almost the same cost of a single evaluation.

j	t_j	$\tilde{N}((t_j + t_{j-1})/2) - j + 1$
1	14.1347251417	-0.11752
2	21.0220396388	-0.04445
3	25.0108575801	-0.03216
4	30.4248761259	0.01102
5	32.9350615877	-0.01000
6	37.5861781588	-0.05699
7	40.9187190121	0.07354
8	43.3270732809	-0.07314
9	52.9703214777	0.81717
10	56.4462476971	2.01126
11	59.3470440026	2.12394
12	60.8317785246	1.90550
13	65.1125440481	1.95229
14	67.0798105295	2.11039
15	69.5464017112	1.94654
16	72.0671576745	1.90075
17	75.7046906991	2.09822
18	77.1448400689	2.10097
19	79.3373750202	1.82662
20	82.9103808541	1.99205
21	84.7354929805	2.09800
22	87.4252746131	2.03363
23	88.8091112076	1.88592
24	92.4918992706	1.95640
25	98.8311942182	3.10677
26	101.3178510057	4.03517
27	103.7255380405	4.11799

Table 2: Checking for missing zeros. The second column lists the imaginary parts of the zeros of $\zeta(s)$ found by looking for sign changes of $Z(t)$, advancing in step sizes equal to two. The third column compares the number of zeros found to the main term of $N(T)$, namely to $\tilde{N}(T) := (T/(2\pi)) \log(T/(2\pi e)) + 7/8$, evaluated at the midpoint between consecutive zeros, with t_0 taken to be 0. This detects a pair of missing zeros near the ninth and twenty fifth zeros on our list.

Dirichlet L-functions were not computed on machines until 1961 when Davies and Haselgrove [DH] looked at several $L(s, \chi)$ with conductor $\leqslant$ 163. Rumely [Rum], using summation by parts, computed the first several thousand zeros for many Dirichlet L-functions with small moduli. He both verified RH and looked at statistics of neighbouring zeros.

Yoshida [Y] [Y2] has also used summation by parts, though in a different manner, to compute the first few zeros of certain higher degree, with two or more Γ-factors in the functional equation, L-functions.

Lagarias and Odlyzko [LO] have computed the low lying zeros of several Artin L-functions using expansions involving the incomplete gamma function. They noted that one could compute higher up in the critical strip by introducing the parameter δ, as explained in section 3.3, but did not implement it since it led to difficulties concerning the computation of $G(z, w)$ with both z and w complex.

Other computations of L-functions include those of Berry and Keating [BK] and Paris [P] ($\zeta(s)$), Tollis [To] (Dedekind zeta functions), Keiper [Ke] and Spira [Sp] (Ramanujan τ L-function), Fermigier [F] and Akiyama-Tanigawa [AT] (elliptic curve L-functions), Strombergsson [St] and Farmer-Kranec-Lemurell [FKL] (Maass waveform L-functions), and Dokchister [Do] (general L-functions near the critical line).

The author has verified the Riemann hypothesis for various L-functions. These computations use the methods described in section 3 and are not rigorous in the sense that no attempt is made to obtain explicit bounds for truncation errors on some of the asymptotic expansions and continued fractions used, and no interval arithmetic to bound round off errors is carried out. Tables of the zeros mentioned may be obtained from the author's homepage [Ru4]. These include the first tens of millions zeros of all $L(s, \chi)$ with the conductor of χ less than 20, the first 300000 zeros of $L_\tau(s)$, the Ramanujan τ L-function, the first 100000 zeros of the L-functions associated to elliptic curves of conductors $11, 14, 15, 17, 19$, the first 1000 zeros for elliptic curves of conductors less than 1000, the first 100 zeros of elliptic curves with conductor less than 8000, and hundreds/millions of zeros of many other L-functions.

In all these computations, no violations of the Riemann hypothesis have been found.

4.2 Vertical distribution: correlations and spacings distributions

The random matrix philosophy predicts that various statistics of the zeros of L-functions will mimic the same statistics for the eigenvalues of matrices in the classical compact groups.

Montgomery [Mo] achieved the first result connecting zeros of $\zeta(s)$ with eigenvalues of unitary matrices. Write a typical non-trivial zero of ζ as

$$1/2 + i\gamma.$$

Assume the Riemann Hypothesis, so that the γ's are real. Because the zeros of $\zeta(s)$ come in conjugate pairs, we can restrict our attention to those lying above the real axis and order them

$$0 < \gamma_1 \leqslant \gamma_2 \leqslant \gamma_3 \cdots$$

We can then ask how the spacings between consecutive zeros, $\gamma_{i+1} - \gamma_i$, are distributed, but first, we need to 'unfold' the zeros to compensate for the fact that the zeros on average become closer as one goes higher in the critical strip. We set

$$\tilde{\gamma}_i = \gamma_i \frac{\log(\gamma_i/(2\pi e))}{2\pi} \tag{4.1}$$

and investigate questions involving the $\tilde{\gamma}$'s. This normalization is chosen so that the mean spacing between consecutive $\tilde{\gamma}$'s equals one. Summing the consecutive differences, we get a telescoping sum

$$\sum_{\gamma_i \leqslant T} (\tilde{\gamma}_{i+1} - \tilde{\gamma}_i) = \tilde{\gamma}(T) + O(1) = \gamma(T) \frac{\log(\gamma(T)/(2\pi e))}{2\pi} + O(1)$$

where $\gamma(T)$ is the largest γ less than or equal to T. By (3.29), the r.h.s above equals

$$N(\gamma(T)) + O(\log(\gamma(T))) = N(T) + O(\log(T)),$$

hence $\tilde{\gamma}_{i+1} - \tilde{\gamma}_i$ has mean spacing equal to one.

From a theoretical point of view, studying the consecutive spacings distribution is difficult since this assumes the ability to sort the zeros. The tool that is used for studying spacings questions about the zeros, namely the explicit formula, involves a sum over all zeros of $\zeta(s)$, and it is easier to consider the pair correlation, a statistic incorporating differences between all pairs of zeros.[9] Montgomery conjectured that for $0 \leqslant \alpha < \beta$ and $M \to \infty$,

$$M^{-1}|\{1 \leqslant i < j \leqslant M : \tilde{\gamma}_j - \tilde{\gamma}_i \in [\alpha, \beta)\}|$$
$$\sim \int_\alpha^\beta \left(1 - \left(\frac{\sin \pi t}{\pi t}\right)^2\right) dt. \qquad (4.2)$$

Notice that M^{-1}, and not, say, $\binom{M}{2}^{-1}$, is the correct normalization. For any j there, are just a handful of i's with $\tilde{\gamma}_j - \tilde{\gamma}_i \in [\alpha, \beta)$.

Montgomery was able to prove that

$$M^{-1} \sum_{1 \leqslant i < j \leqslant M} f(\tilde{\gamma}_j - \tilde{\gamma}_i) \to \int_0^\infty f(t) \left(1 - \left(\frac{\sin \pi t}{\pi t}\right)^2\right) dt. \qquad (4.3)$$

as $M \to \infty$, for test functions f satisfying the stringent restriction that $\hat{f}$ be supported in $(-1, 1)$.

An equivalent way to state the conjecture as $M \to \infty$, and one which Odlyzko uses in his numerical experiments, is to let

$$\delta_i = (\gamma_{i+1} - \gamma_i)\frac{\log(\gamma_i/(2\pi))}{2\pi}. \qquad (4.4)$$

and replace the condition $\tilde{\gamma}_j - \tilde{\gamma}_i \in [\alpha, \beta)$ with the condition $\delta_i + \delta_{i+1} + \cdots + \delta_{i+k} \in [\alpha, \beta)$ for $1 \leqslant i \leqslant M, k \geqslant 0$. The main difference is the absence of the $1/e$ in the logarithm. This is done so as to maintain a mean spacing tightly asymptotic to one. Set

$$C(T) = \sum_{\gamma_i \leqslant T} (\gamma_{i+1} - \gamma_i),$$

and sum by parts

$$\sum_{\gamma_i \leqslant T} \delta_i = C(T)\frac{\log(T/(2\pi))}{2\pi} - \frac{1}{2\pi}\int_{\gamma_1}^T C(t)\frac{dt}{t}.$$

Now, $C(t)$ telescopes, and von Mangoldt's formula (3.29) implies that $C(t) = t + O(1)$, so that the r.h.s above equals $N(T) + O(\log(T))$, and δ_i is on average equal to one. In carrying out numerical experiments with zeros one can either

[9]Editors' comment: The corresponding statistic for random matrix eigenvalues, often called the 2-point correlation function, is described in the lectures by Y.V. Fyodorov, page 31, Section 3.

use the normalization given in (4.1) or (4.4). For the theoretical purpose of examining leadings asymptotics of, say, the pair correlation, the factors appearing in these normalizations in the logarithm, $1/(2\pi e)$ or $1/(2\pi)$, are not important as they only affect lower order terms. However, for the purpose of comparing numerical data to theoretical predictions it is crucial to include them.

On a visit by Montgomery to the the Institute for Advanced Study, Freeman Dyson pointed out that large unitary matrices have the same pair correlation. Let

$$e^{i\theta_1}, e^{i\theta_2}, \ldots, e^{i\theta_N}$$

be the eigenvalues of a matrix in $U(N)$, sorted so that

$$0 \leqslant \theta_1 \leqslant \theta_2 \leqslant \ldots \leqslant \theta_N < 2\pi.$$

Normalize the eigenangles

$$\tilde{\theta}_i = \theta_i N/(2\pi) \tag{4.5}$$

so that $\tilde{\theta}_{i+1} - \tilde{\theta}_i$ equals one on average. Then, a classic result in random matrix theory [M] asserts that

$$N^{-1}|\{1 \leqslant i < j \leqslant N, \tilde{\theta}_j - \tilde{\theta}_i \in [\alpha, \beta)\}|$$

equals, when averaged according to Haar measure over $U(N)$ and letting $N \to \infty$,

$$\int_\alpha^\beta \left(1 - \left(\frac{\sin \pi t}{\pi t}\right)^2\right) dt.$$

Odlyzko [O] [O2] has carried out numerics to verify Montgomery's conjecture (4.2). His most extensive data to date involves billions of zeros near the 10^{23}rd zero of $\zeta(s)$. With kind permission we reproduce [O4] Odlyzko's pair correlation picture in figure 1.

This picture compares the l.h.s. of (4.2) for many bins $[a, b)$ of size $b - a = .01$ to the curve

$$1 - \left(\frac{\sin \pi t}{\pi t}\right)^2.$$

Odlyzko's histogram fits the theoretical prediction beautifully. Bogomolny and Keating [K] [BoK], using conjectures of Hardy and Littlewood, have explained the role played by secondary terms in the pair correlation of the zeros of $\zeta(s)$ and these terms are related to $\zeta(s)$ on the one line. A nice description of these results are contained in [BK2]. Recently, Conrey and Snaith [CS] obtained the

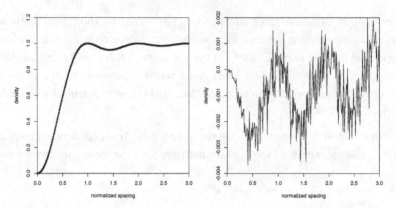

Figure 1: The first graph depicts Odlyzko's pair correlation picture for 2×10^8 zeros of $\zeta(s)$ near the 10^{23}rd zero. The second graph shows the difference between the histogram in the first graph and $1 - ((\sin \pi t)/(\pi t))^2$. In the interval displayed, the two agree to within about .002.

main and lower terms of the pair correlation using a conjecture for the full asymptotics of the average value of a ratio of four zeta functions rather than the Hardy-Littlewood conjectures.

Montgomery's pair correlation theorem (4.3) has been generalized by Rudnick and Sarnak [RudS] to any primitive L-function, i.e. one which does not factor as a product of other L-functions, as well as to higher correlations which are defined in a way similar to the pair correlation. Again, there are severe restrictions on the Fourier transform of the allowable test functions, and further, for L-functions of degree greater than three, Rudnick and Sarnak assume a weak form of the the the Ramanujan conjectures. Bogomolny and Keating provide a heuristic derivation of the higher correlations of the zeros of $\zeta(s)$ using the Hardy-Littlewood conjectures [BoK2].

The author has tested the pair correlation conjecture for a number of L-functions. Figure 2 depicts the same experiment as in Odlyzko's figure, but for various Dirichlet L-functions and L-functions associated to cusp forms. Altogether there are eighteen graphs.

The first twelve graphs depict the pair correlation for all primitive Dirichlet L-functions, $L(s, \chi)$ for conductors $q = 3, 4, 5, 7, 8, 9, 11, 12, 13, 15, 16, 17$. Each graph shows the average pair correlation for each q, i.e. the pair correlation was computed individually for each $L(s, \chi)$, and then averaged over $\chi \bmod q$.

In the case of $q = 3, 4$ there is only one primitive L-function for either q, and approximately five million zeros were used for each ($4,772,120$ and $5,003,411$ zeros respectively to be precise). In the case of $q = 5, 7, 8, 9, 11, 12, 13, 15, 16, 17$ there are $3, 5, 2, 4, 9, 1, 11, 3, 4, 15$ primitive L-functions respectively. For $q = 5, 7, 8, 9, 11, 12$ either $2,000,000$ zeros or $1,000,000$ zeros were computed for each $L(s, \chi)$, depending on whether χ was real or complex. In the case of $q = 16, 17$ half as many zeros were computed.

The last six graphs are for L-functions associated to cusp forms. The first of these six shows the pair correlation of the first $284,410$ zeros of the Ramanujan τ L-function, corresponding to the cusp form of level one and weight twelve. The next five depict the pair correlation of the first $100,000$ zeros of the L-functions associated to the elliptic curves of conductors $11, 14, 15, 17, 19$. These last six graphs use larger bins since data in these cases is more limited.

The quality of the fit is comparable to what one finds with zeros of $\zeta(s)$ up to the same height. See, for example, figures 1 and 3 in [O3]. It would be possible to extend the $L(s, \chi)$ computations and obtain data near the 10^{20}th or higher zero, at least for reasonably sized q. Using the methods of section 3 the time required to compute $L(1/2 + it, \chi)$ is $O(|qt|^{1/2})$, compared to $O(|t|^{1/2})$ for $\zeta(1/2 + it)$. Adapting the Odlyzko-Schönhage algorithm would allow for many evaluations of these L-functions at essentially the cost of a single evaluation. While such a computation might be manageable for Dirichlet L-functions, it is hopeless for cusp form L-functions where the time and also the number of Dirichlet coefficients required is $O(|N^{1/2}t|)$, i.e. linear in t. Here N is the conductor of the L-function. Using present algorithms and hardware, it might be possible to extend these cusp form computations to $t = 10^8$ or 10^9.

Slight care is needed to normalize these zeros correctly as the formula for the number of zeros of $L(s)$ depends on the degree of the L-function and on its conductor. For Dirichlet L-functions $L(s, \chi)$, $\chi \bmod q$, we should normalize its zeros $1/2 + i\gamma$ as follows:

$$\tilde{\gamma} = \gamma \frac{\log(|\gamma|q/(2\pi e))}{2\pi}$$

For a cusp form L-function of conductor N, we should take the following normalization:

$$\tilde{\gamma} = \gamma \frac{\log(|\gamma|N^{1/2}/(2\pi e))}{\pi}$$

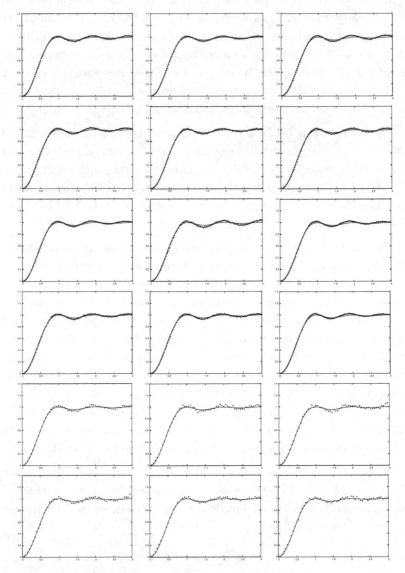

Figure 2: Pair correlation for zeros of all primitive $L(s, \chi)$, $3 \leqslant q \leqslant 17$, the
Ramanujan τ L-function, and five elliptic curve L-functions

From a graphical point of view, it is hard to display information concerning higher order correlations. Instead one can look at a statistic that involves knowing [KS] all the n-level correlations for zeros of characteristic polynomials, namely the nearest neighbour spacings distribution.

In Figure 3 we display Odlyzko's picture for the distribution of the normalized spacings δ_j for 2×10^8 zeros of $\zeta(s)$ near the 10^{23}rd zero. This is computed by breaking up the x-axis into small bins and counting how many δ_j's fall into each bin, and then comparing this against the nearest neighbour spacings distribution of the normalized eigenangles of matrices in $\mathrm{U}(N)$, as $N \to \infty$, again averaged according to Haar measure on $\mathrm{U}(N)$. The density function for this distribution is given [M] as

$$\frac{d^2}{dt^2} \prod_n (1 - \lambda_n(t))$$

where $\lambda_n(t)$ are the eigenfunctions of the integral operator

$$\lambda(t)f(x) = \int_{-1}^{1} \frac{\sin(\pi t(x-y))}{\pi(x-y)} f(y)dy, \qquad (4.6)$$

sorted according to $1 \geqslant \lambda_0(t) \geqslant \lambda_1(t) \geqslant \ldots \geqslant 0$. See [O3] for a description of how the density function can be computed.

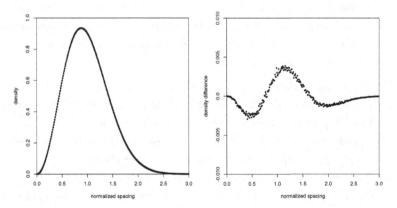

Figure 3: The first graph shows Odlyzko's nearest neighbour spacings distribution for 2×10^8 zeros of $\zeta(s)$ near the 10^{23}rd zero. The second graph shows the difference he computed between the histogram and the predicted density function. Recently, Bogomolny, Bohigas, Leboeuf and Monastra [BBLM] have explained the role of secondary terms in shaping the difference displayed.

In Figure 4 we display the nearest neighbour spacings distribution for the sets of zeros described above, namely millions of zeros of primitive $L(s, \chi)$, with conductors $3 \leqslant q \leqslant 17$, and hundreds of thousands of zeros of six cusp form L-functions. We also depict the nearest neighbour spacings for the first $500,000$ zeros of each of the 16 primitive $L(s, \chi)$ with $\chi \bmod 19$ complex, and $1,000,000$ zeros for the one primitive real $\chi \bmod 19$.

Eight graphs are displayed. The first is for the $4,772,120$ zeros of $L(s, \chi)$, $\chi \bmod 3$. The second one depicts the average spacings distribution for all 76 primitive $L(s, \chi)$, $\chi \bmod q$ with $3 \leqslant q \leqslant 19$, i.e. the spacings distribution was computed individually for each of these L-functions and then averaged. The next six graphs show the spacings distribution for the Ramanujan τ L-function, and the L-functions associated to the elliptic curves of conductors $11, 14, 15, 17, 19$. Again, the fit is comparable to the fit one gets with the same number of zeros of $\zeta(s)$.

4.3   Density of zeros

Rather than look at statistics of a single L-function, we can form statistics involving a collection of L-functions. This has the advantage of allowing us to study the behaviour of our collection near the critical point where specific information about the collection may be revealed. This idea was formulated by Katz and Sarnak [KS] [KS2] who studied function field zeta functions and conjectured that the various classical compact groups should be relevant to questions about L-functions.

While the eigenvalues of matrices in all the classical compact groups share, on average, the same limiting correlations and spacings distributions, their characteristic polynomials do exhibit distinct behaviour near the point $z = 1$. Using the idea that the unit circle for characteristic polynomials in the classical compact groups correponds to the critical line, with the point $z = 1$ on the unit circle corresponding to the critical point, Katz and Sarnak were led to formulate conjectures regarding the density of zeros near the critical point for various collections of L-functions. This is detailed in section 4.3.1 below.

The fact that different families of L-functions exhibit distinct behaviour near the critical point is illustrated in figure 5. This plot depicts the imaginary parts of the zeros of many $L(s, \chi)$ with χ a generic non-real primitive Dirichlet character for the modulus q, with $5 \leqslant q \leqslant 10000$. Other than the fact that, at

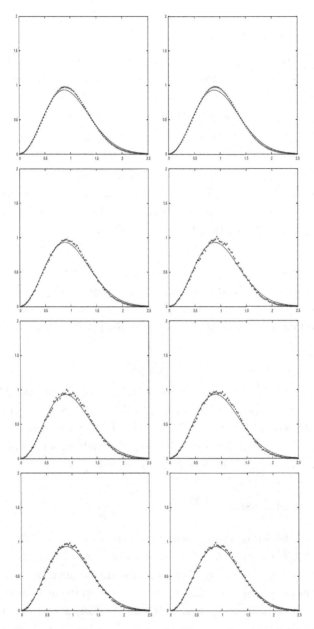

Figure 4: Nearest neighbour spacings distribution for several Dirichlet and cusp form L-functions. The first is for $L(s,\chi)$, $q = 3$. The second is the average nearest neighbour spacing for all primitive $L(s,\chi)$, $3 \leqslant q \leqslant 19$. The last six are for the Ramanujan τ L-function, and five L-functions associated to elliptic curves.

a fixed height, the zeros become more dense proportionally to $\log q$, the zeros appear to be uniformly dense.

This contrasts sharply with the plot in figure 6 which depicts the zeros of $L(s, \chi_d)$ where χ_d is a real primitive character (the Kronecker symbol), and d ranges over fundamental discriminants with $-20000 < d < 20000$. Here we see the density of zeros fluctuating as one moves away from the real axis.

Other features can be seen in the plot. First, from the white band near the x-axis we notice that the lowest zero for each $L(s, \chi_d)$ tends to stay away from the critical point. We can also see the effect of secondary terms on this repulsion. The lowest zero for $d > 0$ tends to be higher than the lowest zero for $d < 0$. This turns out to be related to the fact that the Γ-factor in the functional equation for $L(s, \chi_d)$ is $\Gamma(s/2)$ if $d > 0$, but is $\Gamma((s+1)/2)$ when $d < 0$.

We can also see slightly darker regions appearing in horizontal strips. The first one occurs roughly at height 7., half the height of the first zero of $\zeta(s)$. These horizontal strips are due to secondary terms in the density of zeros for this collection of L-functions which include [Ru3] [CS] a term that is proportional to

$$\mathfrak{Re} \frac{\zeta'(1 + 2it)}{\zeta(1 + 2it)}.$$

This is large when $\zeta(1 + 2it)$ is small. Surprisingly, $\zeta(1 + iy)$ and $\zeta(1/2 + iy)$ track each other very closely, see figure 7, and the minima of $|\zeta(1 + iy)|$ appear close to the zeros of $\zeta(1/2 + iy)$. This is similar to a phenomenon that occurs when we look at secondary terms in the pair correlation of the zero of $\zeta(s)$ which also involves $\zeta(s)$ on the one line [BoK] [BK2].

4.3.1 n-level density

The n-level density is used to measure the average density of the zeros of a family of L-functions or matrices. It is arranged to be sensitive to the low lying zeros in the family, i.e. those near the critical point if we are dealing with L-functions, and those near the point $z = 1$ on the unit circle if we are dealing with characteristic polynomials from the classical compact groups.

Let A be an $N \times N$ matrix in on of the classical compact groups. Write the eigenvalues of A as $\lambda_j = e^{i\theta_j}$ with

$$0 \leqslant \theta_1 \leqslant \ldots \leqslant \theta_N < 2\pi.$$

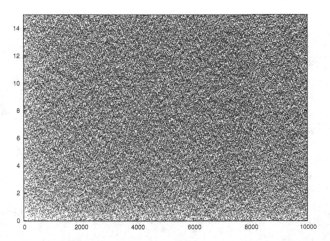

Figure 5: Zeros of $L(s,\chi)$ with χ a generic non-real primitive Dirichlet character for the modulus q, with $5 \leqslant q \leqslant 10000$. The horizontal axis is q and, for each $L(s,\chi)$, the imaginary parts of its zeros up to height 15 are listed.

Let

$$H^{(n)}(A,f) = \sum_{\substack{1 \leqslant j_1, \ldots, j_n \leqslant N \\ \text{distinct}}} f\left(\theta_{j_1} N/(2\pi), \ldots, \theta_{j_n} N/(2\pi)\right)$$

with $f : \mathbb{R}^n \to \mathbb{R}$, bounded, Borel measurable, and compactly supported. Because of the normalization by $N/(2\pi)$, and the assumption that f has compact support, $H^{(n)}(A,f)$ only depends on the small θ_j's.

Katz and Sarnak [KS] proved the following family dependent result:

$$\lim_{N \to \infty} \int_{G(N)} H^{(n)}(A,f)dA = \int_0^\infty \cdots \int_0^\infty W_G^{(n)}(x)f(x)dx \qquad (4.7)$$

for the following families:

G	$W_G^{(n)}$
$U(N), U_\kappa(N)$	$\det\left(K_0(x_j, x_k)\right)_{\substack{1 \leqslant j \leqslant n \\ 1 \leqslant k \leqslant n}}$
$USp(N)$	$\det\left(K_{-1}(x_j, x_k)\right)_{\substack{1 \leqslant j \leqslant n \\ 1 \leqslant k \leqslant n}}$
$SO(2N)$	$\det\left(K_1(x_j, x_k)\right)_{\substack{1 \leqslant j \leqslant n \\ 1 \leqslant k \leqslant n}}$
$SO(2N+1)$	$\det\left(K_{-1}(x_j, x_k)\right)_{\substack{1 \leqslant j \leqslant n \\ 1 \leqslant k \leqslant n}} + \sum_{\nu=1}^n \delta(x_\nu)\det\left(K_{-1}(x_j, x_k)\right)_{\substack{1 \leqslant j \neq \nu \leqslant n \\ 1 \leqslant k \neq \nu \leqslant n}}$

with

$$K_\varepsilon(x,y) = \frac{\sin(\pi(x-y))}{\pi(x-y)} + \varepsilon\frac{\sin(\pi(x+y))}{\pi(x+y)}.$$

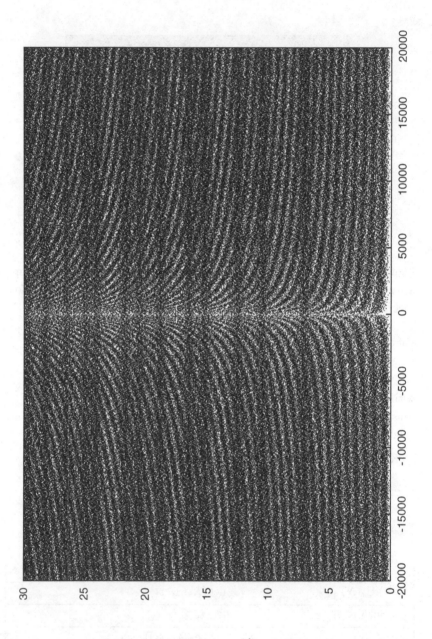

Figure 6: Zeros of $L(s, \chi_d)$ with $\chi_d(n) = \left(\frac{d}{n}\right)$, the Kronecker symbol. We restrict d to fundamental discriminants $-20000 < d < 20000$. The horizontal axis is d and, for each $L(s, \chi_d)$, the imaginary parts of its zeros up to height 30 are listed.

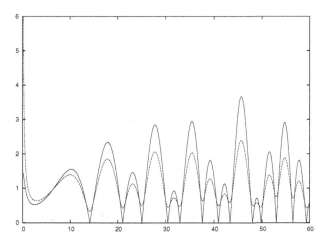

Figure 7: A graph illustrating that, at least initially, the minima of $|\zeta(1+iy)|$ occur very close the zeros of $|\zeta(1/2+iy)|$. The dashed line is the graph of the former while the solid line is the graph of the the latter.

Here

$$\mathrm{U}_\kappa(N) = \{A \in \mathrm{U}(N) : \det(A)^\kappa = 1\}.$$

The delta functions in the SO($2N+1$) case are accounted for by the eigenvalue at 1. Removing this zero from (4.7) yields the same $W_G^{(n)}$ as for USp.

Let

$$D(X) = \{d \text{ a fundamental discriminant} : |d| \leqslant X\}$$

and let $\chi_d(n) = \left(\frac{d}{n}\right)$ be Kronecker's symbol. Write the non-trivial zeros of $L(s, \chi_d)$ as

$$1/2 + i\gamma_j^{(d)}, \qquad j = \pm 1, \pm 2, \ldots$$

sorted by increasing imaginary part, and

$$\gamma_{-j}^{(d)} = -\gamma_j^{(d)}.$$

The author proved [Ru2] that

$$\lim_{X \to \infty} \frac{1}{|D(X)|} \sum_{d \in D(X)} \sum_{\substack{j_i \geqslant 1 \\ \text{distinct}}} f\left(l_d\gamma_{j_1}^{(d)}, l_d\gamma_{j_2}^{(d)}, \ldots, l_d\gamma_{j_n}^{(d)}\right)$$

$$= \int_0^\infty \cdots \int_0^\infty f(x) W_{\mathrm{USp}}^{(n)}(x) dx, \qquad (4.8)$$

where

$$l_d = \frac{\log(|d|/\pi)}{2\pi}.$$

Here, f is assumed to be smooth, and rapidly decreasing with $\hat{f}(u_1, \dots, u_n)$ supported in $\sum_{i=1}^n |u_i| < 1$. This generalized the $n = 1$ case that had been achieved earlier [OzS, KS2]. Assuming the Riemann Hypothesis for all $L(s, \chi_d)$, the $n = 1$ case has been extended to $\hat{f}$ supported in $(-2, 2)$ [OzS2] [KS3]. Chris Hughes has an alternate derivation of (4.8) appearing Theorem 3 of his lectures, which start on page 337.

This result confirms the connection between zeros of $L(s, \chi_d)$ and eigenvalues of unitary symplectic matrices and explains the repulsion away from the critical point and the fluctuations seen in figure 6, at least near the real axis, because, when $n = 1$, the density of zeros is described by the function $W_{\text{USp}}^{(1)}(x)$ which equals

$$1 - \frac{\sin(2\pi x)}{2\pi x}.$$

At height x, we therefore also expect, as we average over larger and larger $|d|$, for the fluctuations to diminish proportional to $1/x$. However, if we allow x to grow with d then the fluctuations actually persist due to secondary fluctuating terms that can be large if x is allowed to grow with d [Ru3] [CS].

The above suggests that the distribution of the lowest zero, i.e. the one with smallest imaginary part, in this family of L-functions ought to be modeled by the distribution of the smallest eigenangle of characteristic polynomials in USp(N), with $N \to \infty$. Similary we expect that the distribution, say, of the second lowest zero ought to fit the distribution of the second smallest eigenangle.

The probability densities describing the distribution of the smallest and second smallest eigenangles, normalized by $N/(2\pi)$, for characteristic polynomials in USp(N), with N even and tending to ∞ are given [KS] respectively by

$$\nu_1(\text{USp})(t) = -\frac{d}{dt} E_{-,0}(t)$$

and

$$\nu_2(\text{USp})(t) = -\frac{d}{dt}(E_{-,0}(t) + E_{-,1}(t)),$$

where

$$E_{-,0}(t) = \prod_{j=0}^{\infty}(1 - \lambda_{2j+1}(2t))$$

$$E_{-,1}(t) = \sum_{k=1}^{\infty}\lambda_{2k+1}(2t)\prod_{\substack{j=0 \\ j \neq k}}^{\infty}(1 - \lambda_{2j+1}(2t)).$$

Here, the $\lambda_j(t)$'s are the eigenvalues of the integral equation in (4.6).

This also suggests that the means of the the first and second lowest zeros are given by

$$\lim_{X \to \infty}\frac{1}{|D(X)|}\sum_{d \in D(X)}\gamma_1^{(d)}l_d = \int_0^{\infty}t\nu_1(\text{USp})(t)dt = .78\ldots$$

$$\lim_{X \to \infty}\frac{1}{|D(X)|}\sum_{d \in D(X)}\gamma_2^{(d)}l_d = \int_0^{\infty}t\nu_2(\text{USp})(t)dt = 1.76\ldots$$

However, the convergence to the predicted means is logarithmically slow due to secondary terms of size $O(1/\log(X))$. Consequently, when comparing against the random matrix theory predictions, one gets a better fit by making sure the lowest zero has the correct mean. This can be achieved by rescaling the data, further multiplying, for a set D of fundamental discriminants, $\gamma_1^{(d)}l_d$ by

$$.78\left(\frac{1}{|D|}\sum_{d \in D}\gamma_1^{(d)}l_d\right)^{-1} \tag{4.9}$$

and $\gamma_2^{(d)}l_d$ by

$$1.76\left(\frac{1}{|D|}\sum_{d \in D}\gamma_2^{(d)}l_d\right)^{-1} \tag{4.10}$$

In figures 8 and 9, we use the normalization described above. For our data set, the denominator in (4.9) equals .83, and, in (4.10) equals 1.84.

In figure 8 we depict the 1-level density of the zeros of $L(s, \chi_d)$ for 7243 prime $|d|$ lying in the interval $(10^{12}, 10^{12} + 200000)$. These zeros were computed in 1996 as part of the authors PhD thesis [Ru]. Here we divide the x-axis into small bins, count how many normalized zeros of $L(s, \chi_d)$ lie in each bin, divide that count by the number of d, namely 7243, and compare that to the graph of $1 - \sin(2\pi x)/(2\pi x)$.

In figure 9 we depict the distribution of the lowest and second lowest normalized zero for the set of zeros just described. These are compared against ν_1

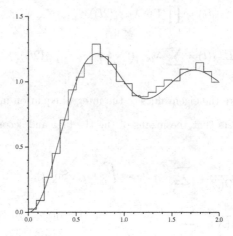

Figure 8: Density of zeros of $L(s, \chi_d)$ for 7243 prime values of $|d|$ lying in the interval $(10^{12}, 10^{12} + 200000)$. Compared against the random matrix theory prediction, $1 - \sin(2\pi x)/(2\pi x)$.

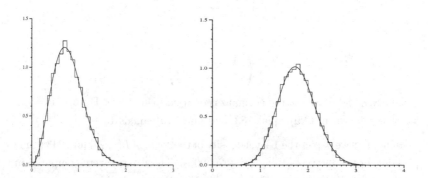

Figure 9: Distribution of the lowest and second lowest zero of $L(s, \chi_d)$ for 7243 prime values of $|d|$ lying in the interval $(10^{12}, 10^{12} + 200000)$. Compared against the random matrix theory predictions.

and ν_2 which were computed using the same program, obtained from Andrew Odlyzko, that was used in [O3].

In figure 10 we depict the 1-level density and distribution of the lowest zeros for quadratic twists of the Ramanujan τ L-function, $L_\tau(s, \chi_d)$, $d > 0$. For this family of L-functions, one can prove [Ru2] a result similar to (4.8) but with W_{USp} replaced with $W_{\text{SO(even)}}$, and the support of $\hat{f}$ reduced to $\sum_{i=1}^{n} |u_i| < 1/2$. The 1-level density is therefore given by $1 + \sin(2\pi x)/(2\pi x)$ and the probability density for the distribution of the smallest eigenangle, normalized by $2N/(2\pi)$, for matrices in $SO(2N)$, with $N \to \infty$, is given [KS] by

$$\nu_1(\text{SO(even)})(t) = -\frac{d}{dt} \prod_{j=0}^{\infty} (1 - \lambda_{2j}(2t)),$$

whose mean is .32. The figure uses 11464 prime values of $|d|$ lying in $(350000, 650000)$, and the zeros were normalized by $2l_d$, and then rescaled so as to have mean .32 rather than .29. The choice of using $2l_d$ for normalizing the zeros is the correct one up to leading term, but is slightly adhoc and by now a better understanding of a tighter normalization up to lower terms has emerged [CFKRS] [Ru3].

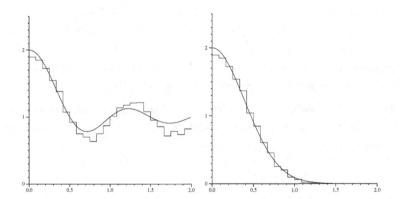

Figure 10: One-level density and distribution of the lowest zero of even quadratic twists of the Ramanujan τ L-function, $L_\tau(s, \chi_d)$, for 11464 prime values of $d > 0$ lying in the interval $(350000, 650000)$.

4.4   Value distribution of *L*-functions

Keating and Snaith initiated the use of random matrix theory to study the
value distribution of *L*-functions with their important paper [KeS] where they
consider moments of characteristic polynomials of unitary matrices and conjecture the leading-order asymptotics for the moments of $\zeta(s)$ on the critical line.
This was followed by a second paper [KeS2] along with a paper by Conrey and
Farmer [CF] which provide conjectures for the leading-order asymptotics of
moments of various families of *L*-functions by examining analogous questions
for characteristic polynomials of the various classical compact groups.

Keating and Snaith's technically impressive work also represents a philosophical breakthrough. Until their paper appeared, one would compare, say,
statistics involving zeros of $\zeta(s)$ to similar statistics for eigenvalues of $N \times N$
unitary matrices, with $N \to \infty$. However, their work compares the average
value of $\zeta(1/2 + it)$ to the average value of $N \times N$ unitary characteristic polynomials evaluated on the unit circle, with $N \sim \log(t/(2\pi))$. This choice of N
is motivated by comparing local spacings of zeros, for example (4.4) v.s. (4.5).
A slightly different approach to this choice of N proceeds by comparing functional equations of *L*-functions to functional equations of characteristic polynomials [CFKRS].

At first sight, it seems strange to compare the Riemann zeta function which
has infinitely many zeros to characteristic polynomials of finite size matrices.
However, this suggests that a given height t, the Riemann zeta function can
be modeled locally by just a small number of zeros, as well as by more global
information that incorporates the role played by primes. Recently, Gonek,
Hughes, and Keating have developed such a model [GHK].

Below we describe three specific examples where random matrix theory has
led to important advances in our understanding of the value distribution of
L-functions. These concern the families:

1. $\zeta(1/2 + it)$, where we average over t.

2. $L(1/2, \chi_d)$, where we average over fundamental discriminants d.

3. $L_E(1/2, \chi_d)$, quadratic twists of the *L*-function associated to an elliptic
   curve E over $\mathbb{Q}$, where we average over fundamental discriminants d.

These three are examples of unitary, unitary unitary symplectic, and even orthogonal families respectively [KS2] [CFKRS]. Note that in the last example, we normalize the Dirichlet coefficient of the L-function as in (3.16) so that the functional equation of $L_E(1/2, \chi_d)$ brings s into $1 - s$ with the critical point being $s = 1/2$.

We first illustrate that these three examples exhibit distinct behaviour by contrasting their value distributions. The first-order asymptotics for the moments of $|\zeta(1/2 + it)|$ are conjectured by Keating and Snaith [KeS] to be given by

$$\frac{1}{T} \int_0^T |\zeta(1/2 + it)|^r dt \sim a_{r/2} \prod_{j=1}^N \frac{\Gamma(j)\Gamma(j+r)}{\Gamma(j+r/2)^2}, \quad \Re r > -1, \qquad (4.11)$$

with $N \sim \log(T)$ and $a_{r/2}$ defined by (2.13).

For quadratic Dirichlet L-functions Keating and Snaith [KeS2] conjecture that

$$\frac{1}{|D(X)|} \sum_{d \in D(X)} L(1/2, \chi_d)^r \qquad (4.12)$$

$$\sim b_r 2^{2Nr} \prod_{j=1}^N \frac{\Gamma(N+j+1)\Gamma(j+\frac{1}{2}+r)}{\Gamma(N+j+1+r)\Gamma(j+\frac{1}{2})}, \quad \Re r > -3/2,$$

with $N \sim \log(X)/2$, where the sum runs over fundamental discriminants $|d| \leqslant X$, and, as suggested by Conrey and Farmer [CF],

$$b_r = \prod_p \frac{(1 - \frac{1}{p})^{\frac{r(r+1)}{2}}}{1 + \frac{1}{p}} \left(\frac{(1 - \frac{1}{\sqrt{p}})^{-r} + (1 + \frac{1}{\sqrt{p}})^{-r}}{2} + \frac{1}{p} \right).$$

Next, let q be the conductor of the elliptic curve E. Averaging over fundamental discriminants and restricting to discriminants for which $L_E(s, \chi_d)$ has an even functional equation, the conjecture asserts [CF] [KeS2] that

$$\frac{1}{|D(X)|} \sum_{\substack{d \in D(X) \\ (d,q)=1 \\ \text{even funct eqn}}} L_E(1/2, \chi_d)^r \qquad (4.13)$$

$$\sim c_r 2^{2Nr} \prod_{j=1}^N \frac{\Gamma(N+j-1)\Gamma(j-\frac{1}{2}+r)}{\Gamma(N+j-1+r)\Gamma(j-\frac{1}{2})}, \quad \Re r > -1/2,$$

with $N \sim \log(X)$ and

$$c_r = \prod_p \left(1 - \frac{1}{p}\right)^{k(k-1)/2} R_{r,p}$$

where, for $p \nmid q$,

$$R_{k,p} = \left(1 + \frac{1}{p}\right)^{-1} \left(\frac{1}{p} + \frac{1}{2}\left(\left(1 - \frac{a_p}{p} + \frac{1}{p}\right)^{-k} + \left(1 + \frac{a_p}{p} + \frac{1}{p}\right)^{-k}\right)\right).$$

In the above equation, a_p stands for the pth coefficient of the Dirichlet series of L_E.

In the case of the Riemann zeta function we take absolute values, $|\zeta(1/2 + it)|$, otherwise the moments would be zero. In the other two cases, the L-values are conjectured to be non-negative real numbers, hence we directly take their moments.

We should also observe that, while statistics such as the pair correlation or density of zeros involving zeros of L-functions have arithmetic information appearing in the secondary terms, moments already reveal such behaviour at the level of the main term. This reflects the global nature of the moment statistic as compared to the local nature of statistics of zeros that have been discussed.

Using the above conjectured asymptotics we can naively plot value distributions. Figure 11 compares numerical value distributions for data in these three examples against the counterpart densities from random matrix theory. Notice that these three graphs behave distinctly near the origin. The solid curves are computed by taking inverse Mellin transforms, as in (2.19), of the right hand sides of equations (4.11), (4.12), and (4.13), but without the arithmetic factors a_k, b_k, c_k. Shifting the inverse Mellin transform line integral to the left, the location of the first pole in each integrand dictates the behaviour of the corresponding density functions near the origin. The locations of these three poles are at $r = -1, -3/2$, and $-1/2$ respectively. Taking t to be the horizontal axis, near the origin the first density is proportional to a constant, the second to $t^{1/2}$, and the third to $t^{-1/2}$. In forming these graphs one takes N as described above, so that the proportionality constants do depend on N. As N grows, these graphs tend to get flatter.

The first graph is reproduced from [KeS]. In the second and third graphs displayed, a slight cheat was used to get a better fit. The histograms were rescaled linearly along both axis until the histogram matched up nicely with the solid curves. We must ignore the arithmetic factors when taking inverse Mellin transforms since these factors are known [CGo] to be functions of order two and cause the inverse Mellin transforms to diverge. To properly plot the correct value distributions we would need to use more than just the leading-order asymptotics. Presently, our knowledge of the moments of various families of L-functions extends beyond the first-order asymptotics, but only for positive integer values of r (even integer in the case of $|\zeta(1/2 + it)|^r$), however, one would need to apply full asymptotics for complex values of r. The paper by Conrey, Farmer, Keating, the author, and Snaith [CFKRS] conjectures the full asymptotics, for example, of the three moment problems above, but for integer r, with corresponding theorems in random matrix theory given in [CFKRS2]. The paper of Conrey, Farmer and Zirnbauer goes even beyond this stating conjectures for the full asymptotics of moments of ratios of L-functions, and, using methods from supersymmetry, proving corresponding theorems in random matrix theory [CFZ1, CFZ2]. Another paper, by Conrey, Forrester, and Snaith, uses orthogonal polynomials to obtain alternative proofs of the random matrix theory theorems for ratios [CFS].

4.4.1 Moments of $|\zeta(1/2 + it)|$

Next we describe the full moment conjecture from [CFKRS] for $|\zeta(1/2 + it)|$. In that paper, the conjecture is derived heuristically by looking at products of zetas shifted slightly away from the critical line and then setting the shifts equal to zero.

The formula is written in terms of contour integrals and involves the Vandermonde determinant:

$$\Delta(z_1, \ldots, z_m) = \prod_{1 \le i < j \le m} (z_j - z_i).$$

Suppose $g(t) = f(t/T)$ with $f : \mathbb{R}^+ \to \mathbb{R}$ non-negative, bounded, and integrable. The conjecture of [CFKRS] states that, as $T \to \infty$,

$$\int_0^\infty |\zeta(1/2 + it)|^{2k} g(t)\, dt \sim \int_0^\infty P_k \left(\log(t/(2\pi)) \right) g(t)\, dt,$$

where P_k is the polynomial of degree k^2 given by the $2k$-fold residue

$$P_k(x) = \frac{(-1)^k}{k!^2} \frac{1}{(2\pi i)^{2k}} \oint \cdots \oint \frac{G(z_1, \ldots, z_{2k}) \Delta^2(z_1, \ldots, z_{2k})}{\displaystyle\prod_{j=1}^{2k} z_j^{2k}}$$

$$\times e^{\frac{x}{2} \sum_{j=1}^k z_j - z_{k+j}}\, dz_1 \ldots dz_{2k},$$

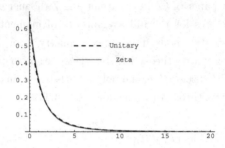

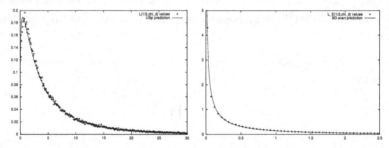

Figure 11: Value distribution of L-functions compared to the random matrix theory counterparts. The first picture, depicts the value distribution of $|\zeta(1/2+it)|$, with t near 10^6, the second of $L(1/2, \chi_d)$ with $800000 < |d| < 10^6$, and the third of $L_{E_{11}}(1/2, \chi_d)$, with $-85000000 < d < 0$, $d = 2, 6, 7, 8, 10 \bmod 11$.

where one integrates over small circles about $z_i = 0$, with

$$G(z_1, \ldots, z_{2k}) = A_k(z_1, \ldots, z_{2k}) \prod_{i=1}^{k} \prod_{j=1}^{k} \zeta(1 + z_i - z_{k+j}),$$

and A_k is the Euler product

$$A_k(z) = \prod_p \prod_{i=1}^{k} \prod_{j=1}^{k} \left(1 - \frac{1}{p^{1+z_i-z_{k+j}}} \right) \int_0^1 \prod_{j=1}^{k} \left(1 - \frac{e^{2\pi i\theta}}{p^{\frac{1}{2}+z_j}} \right)^{-1}$$

$$\times \left(1 - \frac{e^{-2\pi i\theta}}{p^{\frac{1}{2}-z_{k+j}}} \right)^{-1} d\theta$$

$$= \prod_p \sum_{m=1}^{k} \prod_{i \neq m} \frac{\prod_{j=1}^{k} \left(1 - \frac{1}{p^{1+z_j-z_{k+i}}} \right)}{1 - p^{z_{k+i}-z_{k+m}}}.$$

When $k = 1$ or 2, this conjecture agrees with theorems for the full asymptotics as worked out by Ingham [I] and Heath-Brown respectively [H]. In the first case $A_1(z) = 1$ and in the second case $A_2(z) = \zeta(2 + z_1 + z_2 - z_3 - z_4)^{-1}$, and one can write down the coefficients of the polynomials $P_k(x)$ in terms of known constants. When $k = 3$ the product over primes becomes rather complicated. However, one can numerically evaluate [CFKRS3] the coefficients of $P_3(x)$ and the polynomial is given by:

$$\begin{aligned}
P_3(x) = {} & 0.000005708527034652788398376841445252313\,x^9 \\
& + 0.000405021330884114403312153320259849\,x^8 \\
& + 0.011072455215246998350410400826667\,x^7 \\
& + 0.148400730801502726808514015187774\,x^6 \\
& + 1.045925177905488343938532379805\,x^5 \\
& + 3.984385094823534724747964073429\,x^4 \\
& + 8.607319145781206756148347636297\,x^3 \\
& + 10.274330830703446134183009522\,x^2 \\
& + 6.593913020649975810465713392\,x \\
& + 0.916515507637893059017854\,3.
\end{aligned}$$

In the $k = 3$ case the moments of $|\zeta(1/2 + it)|$ have not been proven, and it makes sense to test the moment conjecture numerically. Table 3, reproduced from [CFKRS], depicts

$$\int_C^D |\zeta(1/2 + it)|^6 dt \tag{4.14}$$

as compared to

$$\int_C^D P_3(\log(t/2\pi))dt, \tag{4.15}$$

along with their ratio, for various blocks $[C, D]$ of length 50000, as well as a larger block of length 2,350,000.

4.4.2 Moments of $L(1/2, \chi_d)$

Another conjecture listed in [CFKRS] concerns the full asymptotics for the moments of $L(1/2, \chi_d)$. We quote the conjecture here:

Suppose $g(t) = f(t/T)$ with $f : \mathbb{R}^+ \to \mathbb{R}$ non-negative, bounded, and integrable. Let $X_d(s) = |d|^{\frac{1}{2}-s} X(s, a)$ where $a = 0$ if $d > 0$ and $a = 1$ if $d < 0$, and

$$X(s, a) = \pi^{s-\frac{1}{2}} \Gamma\left(\frac{1 + a - s}{2}\right) / \Gamma\left(\frac{s + a}{2}\right).$$

That is, $X_d(s)$ is the factor in the functional equation $L(s, \chi_d) = X_d(s)L(1 - s, \chi_d)$. Summing over negative fundamental discriminants d we have, as $T \to \infty$,

$$\sum_{d<0} L(1/2, \chi_d)^k g(|d|) \sim \sum_{d<0} Q_k(\log|d|)g(|d|)$$

where Q_k is the polynomial of degree $k(k + 1)/2$ given by the k-fold residue

$$Q_k(x) = \frac{(-1)^{k(k-1)/2}2^k}{k!}\frac{1}{(2\pi i)^k}\oint \cdots \oint \frac{G(z_1,\ldots,z_k)\Delta(z_1^2,\ldots,z_k^2)^2}{\displaystyle\prod_{j=1}^k z_j^{2k-1}}$$

$$\times e^{\frac{x}{2}\sum_{j=1}^k z_j}\, dz_1 \ldots dz_k,$$

where

$$G(z_1,\ldots,z_k) = B_k(z_1,\ldots,z_k)\prod_{j=1}^k X(1/2 + z_j, 1)^{-\frac{1}{2}}\prod_{1 \le i \le j \le k}\zeta(1 + z_i + z_j),$$

and B_k is the Euler product, absolutely convergent for $|\mathfrak{Re}\, z_j| < \frac{1}{2}$, defined by

$$B_k(z_1,\ldots,z_k) = \prod_p \prod_{1 \le i \le j \le k}\left(1 - \frac{1}{p^{1+z_i+z_j}}\right)$$

$$\times \left(\frac{1}{2}\left(\prod_{j=1}^k\left(1 - \frac{1}{p^{\frac{1}{2}+z_j}}\right)^{-1} + \prod_{j=1}^k\left(1 + \frac{1}{p^{\frac{1}{2}+z_j}}\right)^{-1}\right) + \frac{1}{p}\right)\left(1 + \frac{1}{p}\right)^{-1}.$$

$[C, D]$	conjecture (4.15)	reality (4.14)	ratio
[0,50000]	7236872972.7	7231005642.3	.999189
[50000,100000]	15696470555.3	15723919113.6	1.001749
[100000,150000]	21568672884.1	21536840937.9	.998524
[150000,200000]	26381397608.2	26246250354.1	.994877
[200000,250000]	30556177136.5	30692229217.8	1.004453
[250000,300000]	34290291841.0	34414329738.9	1.003617
[300000,350000]	37695829854.3	37683495193.0	.999673
[350000,400000]	40843941365.7	40566252008.5	.993201
[400000,450000]	43783216365.2	43907511751.1	1.002839
[450000,500000]	46548617846.7	46531247056.9	.999627
[500000,550000]	49166313161.9	49136264678.2	.999389
[550000,600000]	51656498739.2	51744796875.0	1.001709
[600000,650000]	54035153255.1	53962410634.2	.998654
[650000,700000]	56315178564.8	56541799179.3	1.004024
[700000,750000]	58507171421.6	58365383245.2	.997577
[750000,800000]	60619962488.2	60870809317.1	1.004138
[800000,850000]	62661003164.6	62765220708.6	1.001663
[850000,900000]	64636649728.0	64227164326.1	.993665
[900000,950000]	66552376294.2	65994874052.2	.991623
[950000,1000000]	68412937271.4	68961125079.8	1.008013
[1000000,1050000]	70222493232.7	70233393177.0	1.000155
[1050000,1100000]	71984709805.4	72919426905.7	1.012985
[1100000,1150000]	73702836332.4	72567024812.4	.984589
[1150000,1200000]	75379769148.4	76267763314.7	1.011780
[1200000,1250000]	77018102997.5	76750297112.6	.996523
[1250000,1300000]	78620173202.6	78315210623.9	.996121
[1300000,1350000]	80188090542.5	80320710380.9	1.001654
[1350000,1400000]	81723770322.2	80767881132.6	.988303
[1400000,1450000]	83228956776.3	83782957374.3	1.006656
[0,2350000]	3317437762612.4	3317496016044.9	1.000017

Table 3: Sixth moment of ζ versus the conjecture. The 'reality' column, i.e. integrals involving ζ, were computed using Mathematica.

We can also sum over $d > 0$ but then need to replace $X(1/2 + z_j, 1)$ with $X(1/2 + z_j, 0)$.

This conjecture agrees with theorems in the case of $k = 1, 2, 3$ [J] [S] (only the leading term has been checked in the case of $k = 3$, but in principle the lower terms could be verified).

Figure 12, reproduced from [CFKRS], depicts, for $k = 1, \ldots, 8$ and $X = 10000, 20000, \ldots, 10^7$,

$$\sum_{0 < d \leqslant X} L(1/2, \chi_d)^k$$

divided by

$$\sum_{0 < d \leqslant X} Q_k(\log d).$$

4.4.3 Vanishing of $L_E(1/2, \chi_d)$

In [CKRS], Conrey, Keating, the author, and Snaith apply the moment conjecture (4.13) to the problem of predicting asymptotically the number of vanishings of $L_E(1/2, \chi_d)$. Using the fact that these L-values are discretized, for example via the Birch and Swinnerton-Dyer conjecture or the theorem of Kohnen-Zagier [KZ], and by studying, up to leading term and for small values, the density function predicted by (4.13) they conjectured that

$$\sum_{\substack{d \in D(X) \\ (d,q)=1 \\ \text{even funct eqn} \\ L_E(1/2, \chi_d)=0}} 1 \sim \alpha_E X^{3/4} \log(X)^{\beta_E}.$$

The power on the logarithm depends on the underlying curve E because, in the Birch and Swinnerton-Dyer conjecture, the Tamagawa factors can contribute powers of 2 depending on the prime factors of d and on E and this affects the discretization. The constant α_E depends on $c_{-1/2}$ and the real period of E, but also on some extra subtle arithmetic information that seems to be related to Delaunay's heuristics for Tate-Shafarevich groups [De] and is not yet fully understood. Numerical evidence in favour of this conjecture is presented in [CKRS2]. One can skirt these delicate issues, the power on the logarithm and the constant α_E, as follows.

Let $p \nmid q$ be prime. Sort the d's for which $L_E(1/2, \chi_d) = 0$ by residue classes

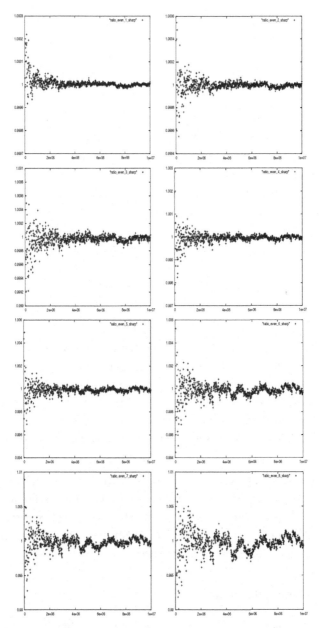

Figure 12: Horizontal axis in each graph is X. These graphs depict the first eight moments, sharp cutoff, of $L(1/2, \chi_d)$, $0 < d \leqslant X$ divided by the conjectured value, sampled at $X = 10000, 20000, \ldots, 10^7$. We see the graphs fluctuating above and below one. Notice that the vertical scale varies from graph to graph

mod p, according to whether $\chi_d(p) = 1$ or -1, and consider the ratio

$$
R_p(X) = \frac{\sum\limits_{\substack{d \in D(X) \\ (d,q)=1 \\ \text{even funct eqn} \\ L_E(1/2,\chi_d)=0 \\ \chi_d(p)=1}} 1}{\sum\limits_{\substack{d \in D(X) \\ (d,q)=1 \\ \text{even funct eqn} \\ L_E(1/2,\chi_d)=0 \\ \chi_d(p)=-1}} 1}.
$$

One can formulate [CKRS2] [CFKRS] conjectures for the moments in these two subfamilies and the moments agree except for a factor that depends on p. By considering this ratio, the powers of X, of $\log X$, and the constant α_E should all cancel out, except for a single factor that depends on p. This leads to a conjecture [CKRS] for $R_p(X)$:

$$
R_p = \lim_{X \to \infty} R_p(X) = \sqrt{\frac{p+1-a_p}{p+1+a_p}},
$$

where a_p denotes the pth coefficient of the Dirichlet series for L_E. The square root in this conjecture is a consequence of the moments having a pole at $r = -1/2$.

We end this paper with a plot that substantiates this conjecture. Figure 13 compares, for one hundred elliptic curves E, the predicted value of R_p to the actual value $R_p(X)$, with $X = 10^8$ and the set of d's restricted to certain residue classes depending on E as described in [CKRS2]. The L-values were computed in this special case by exploiting their connection to the coefficients of certain weight three halves modular forms and using a table of Rodriguez-Villegas and Tornaria [RT].

The horizontal axis is p. For each p, and each of the one hundred elliptic curves E we plot $R_p(X) - R_p$. We see the values fluctuating about zero, most of the time agreeing to within about two percent. The convergence in X is predicted from secondary terms to be logarithmically slow and one gets a better fit by including more terms [CKRS2].

4.4.4   Acknowledgements

The author wishes to thank Andrew Odlyzko for providing him with figures 1 and 3 and the solid curves used in figures 2 and 4. Nina Snaith supplied the first graph in figure 11. Atul Pokharel assisted in the preperation of figure 13. He thanks Fernando Rodriguez-Villegas and Gonzalo Tornaria for giving him

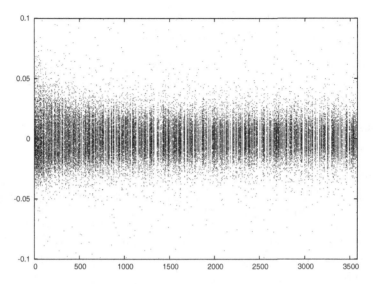

Figure 13: A plot for one hundred elliptic curves of $R_p(X) - R_p$ for $2 \leqslant p < 3500$, $X = 10^8$.

a list of ternary quadratic forms that he used in computing the L-values for section 4.4.3. He is grateful for the kind use of Andrew Granville's and William Stein's computer clusters on which some of the computations described were run. Brian Conrey, David Farmer and Ralph Furmaniak provided feedback on the manuscript. He also wishes to thank the organizers of the Random Matrix Approaches in Number Theory program for inviting him to the Newton Institute to participate.

References

[AT] S. Akiyama and Y. Tanigawa, *Calculation of values of L-functions associated to elliptic curves*, Math. Comp. **68** (1999) no. 227, 1201–1231

[B] B. Berndt, *Ramanujan's notebooks. Part II*, Springer-Verlag, New York, 1989.

[BK] M. Berry and J. Keating, *A new asymptotic representation for $\zeta(\frac{1}{2} + it)$ and quantum spectral determinants*, Proc. Roy. Soc. London Ser. A **437** (1992) no. 1899, 151–173.

[BK2] M. Berry and J. Keating, *The Riemann zeros and eigenvalue asymptotics*, Siam Review **41** (1999) no. 2, 236–266.

[BoK] E. Bogomolny and J. Keating, *Gutzwiller's trace formula and spectral statistics: beyond the diagonal approximation*, Physical Review Letters **77** (1996) no.8, 1472–1475.

[BoK2] E. Bogomolny and J. Keating, *Random matrix theory and the Riemann zeros II: n-point correlation*, Nonlinearity **9** (1996) 911–935.

[BBLM] Bogomolny E., Bohigas O., Leboeuf P., Monastra A.G., *On the spacing distribution of the Riemann zeros: corrections to the asymptotic result*, J. Phys. A **39** (34) (2006) 10743–54.

[Bu] D. Bump, *Automorphic forms and representations*, Cambridge Studies in Advanced Mathematics, vol. 55, Cambridge University Press, Cambridge, 1997.

[BCDT] C. Breuil, B. Conrad, F. Diamond, and R. Taylor, J. Amer. Math. Soc. **14** (2001) no. 4, 843–939.

[C] H. Cohen, *High precision computation of Hardy-Littlewood constants*, draft. Available at www.math.u-bordeaux.fr/~cohen.

[CF] B. Conrey and D. Farmer, *Mean values of L-functions and symmetry*, Internat. Math. Res. Notices **17** (2000) 883–908.

[CFKRS] B. Conrey, D. Farmer, J. Keating, M. Rubinstein, and N. Snaith, *Integral moments of L-functions*, Proc. Lon. Math. Soc. **91** (2005) 1, 33–104.

[CFKRS2] B. Conrey, D. Farmer, J. Keating, M. Rubinstein, and N. Snaith, *Autocorrelation of random matrix polynomials*, Commun. Math. Phys **237** (2003) 3, 365–395.

[CFKRS3] B. Conrey, D. Farmer, J. Keating, M. Rubinstein, and N. Snaith, *Lower order terms in the moments of L-functions*, J. Num. Theory **128** (2008) 6, 1516-1554.

[CFZ1] B. Conrey, D. Farmer, and M. Zirnbauer *Howe Pairs, supersymmetry, and ratios of random characteristic polynomials for the unitary groups U(N)*, preprint arXiv:math-ph/0511024.

[CFZ2] B. Conrey, D. Farmer, and M. Zirnbauer *Autocorrelation of ratios of L-functions*, preprint arXiv:0711.0718.

[CFS] B. Conrey, P. Forrester, and N. Snaith *Averages of ratios of characteristic polynomials for the compact classical groups*, Int. Math. Res. Not. **7** (2005) 397–431.

[CG] B. Conrey and A. Ghosh, *Mean values of the Riemann zeta-function*, Mathematika **31** (1984) 159–161.

[CG2] B. Conrey and A. Ghosh, *A conjecture for the sixth power moment of the Riemann zeta-function*, Int. Math. Res. Not. **15** (1998) 775–780.

[CGo] B. Conrey and S. Gonek, *High moments of the Riemann zeta-function*, Duke Math. Jour. (2001) **107** 577–604.

[CKRS] B. Conrey, J. Keating, M. Rubinstein, and N. Snaith, *On the frequency of vanishing of quadratic twists of modular L-functions*, in *Number Theory for the Millennium I: Proceedings of the Millennial Conference on Number Theory*, editor, M.A. Bennett et al., pages 301–315. A K Peters, Ltd, Natick, 2002.

[CKRS2] B. Conrey, J. Keating, M. Rubinstein, and N. Snaith, *Random Matrix Theory and the Fourier Coefficients of Half-Integral Weight Forms*, Experimental Mathematics **15** (2006) 1, 67–82.

[CS] B. Conrey and N.Snaith, *Applications of the L-functions ratios conjectures*, Proc. Lond. Math. Soc. **94** (2007) 3, 594–646.

[D] H. Davenport, Multiplicative Number Theory, GTM 74 Springer-Verlag, New York, NY (2000).

[DH] D. Davies and C. Haselgrove, *The evaluation of Dirichlet L-functions*, Proc. Roy. Soc. Ser. A **264** (1961), 122–132.

[De] C. Delaunay, *Heuristics on Tate-Shafarevitch groups of elliptic curves defined over* $\mathbb{Q}$, Experiment. Math., **10** (2001), 2 , 191–196.

[Del] P. Deligne, *La conjecture de Weil. I*, Inst. Hautes Études Sci. Publ. Math. (1974), no. 43, 273–307.

[Do] T. Dokchister, *Computing Special Values of Motivic L-Function*, Experimental Mathematics **13** (2004) no. 2, 137–149.

[E] H. Edwards, *Riemann's Zeta Function*, Academic Press (1974)

[EMOT] A. Erdélyi, W. Magnus, F. Oberhettinger, and F. Tricomi, *Higher transcendental functions. Vol. II*, Robert E. Krieger Publishing Co. Inc., Melbourne, Fla., 1981, Based on notes left by Harry Bateman, Reprint of the 1953 original.

[FKL] D. Farmer, W. Kranec, and S. Lemurell, *Maass forms on* $\Gamma_0(11)$, draft.

[F] S. Fermigier, *Zéros des fonctions L de courbes elliptiques*, Experiment. Math. **1** (1992), no. 2, 167–173.

[Fr] E. Friedman, *Hecke's integral formula*, Séminaire de Théorie des Nombres, 1987–1988, Exp. No. 5, 23, Univ. Bordeaux I.

[G] W. Gabcke, *Neue Herleitung und explicite Restabschatzung der Riemann-SiegelFormel*, Ph.D. Dissertation, Gottingen (1979).

[GHK] S. Gonek, C. Hughes, J, Keating, *A hybrid Euler-Hadamard product for the Riemann Zeta Function*, Duke Math. Jour. **136** (2007) 3, 507–549.

[H] R. Heath-Brown, *The fourth power moment of the Riemann zeta-function*, Proc. London Math. Soc. (3) **38** (1979) 385 – 422.

[I] A. E. Ingham, *Mean-value theorems in the theory of the Riemann zeta-function*, Proceedings of the London Mathematical Society (92) **27** (1926) 273–300.

[J] M. Jutila, *On the mean value of* $L(\frac{1}{2}, \chi)$ *for real characters*, Analysis **1** (1981) 149–161.

[KS] N. Katz and P. Sarnak, *Random matrices, Frobenius eigenvalues, and monodromy*, Amer. Math. Soc., Providence, RI (1999).

[KS2] N. Katz and P. Sarnak, *Zeroes of zeta functions and symmetry*, Bull. Amer. Math. Soc. (N.S.) **36** (1999), no. 1, 1–26.

[KS3] N. Katz and P. Sarnak, *Zeros of zeta functions, their spacings and their spectral nature*, 1997 preprint of KS2.

[K] J. Keating, *Periodic orbits, spectral statistics, and the Riemann zeros*, in Supersymmetry and Trace Formulae: Chaos and Disorder, J. Keating, D. Khmelnitskii, and I. Lerner, eds., Plenum, New York, 1998, 1–15.

[KeS]   J. Keating and N. Snaith, *Random matrix theory and* $\zeta(\frac{1}{2} + it)$, Comm. Math. Phys. **214** (2000) 57–89.

[KeS2] J. P. Keating and N. C. Snaith, *Random matrix theory and L-functions at* $s = \frac{1}{2}$, Comm. Math. Phys. **214** (2000) 91–110.

[Ke]   J. Keiper, *On the zeros of the Ramanujan* τ*-Dirichlet series in the critical strip*, Math. Comp. **65** (1996), no. 216, 1613–1619.

[KZ]   W. Kohnen and D. Zagier, *Values of L-series of modular forms at the center of the critical strip*, Invent. Math., **64** (1981), 175–198.

[Kn]   A. Knapp, *Elliptic curves*, Mathematical Notes, vol. 40, Princeton University Press, Princeton, NJ, 1992.

[LO]   J. Lagarias and A. Odlyzko, *On computing Artin L-functions in the critical strip*, Math. Comp. **33** (1979), no. 147, 1081–1095.

[L]   A. Lavrik, *Approximate functional equation for Dirichlet Functions*, Izv. Akad. Nauk SSSR **32** (1968), 134–185.

[Le]   R. Lehman, *On the distribution of the zeros of the Riemann zeta function*, Proc. London Math. Soc. (3) **20** (1970), 303-320. MR 41:3414

[LRW] J. van de Lune, H. te Riele, and D. Winter, *On the zeros of the Riemann zeta function in the critical strip. IV*, Math. Comp. **46** (1986), no. 174, 667–681.

[M]   M. Mehta, *Random Matrices*, 2nd edition, Academic Press, 1991.

[Mo]   H. Montgomery, *The pair correlation of zeros of the zeta function*, Analytic number theory (Proc. Sympos. Pure Math., Vol. XXIV, St. Louis Univ., St. Louis, Mo., 1972) (Providence, R.I.), Amer. Math. Soc., 1973, 181–193.

[O]   A. Odlyzko, *The* 10^{20}*-th zero of the Riemann zeta function and 175 million of its neighbors*, unpublished. www.dtc.umn.edu/~odlyzko

[O2]   A. Odlyzko, *The* 10^{22}*-nd zero of the Riemann zeta function*, Dynamical, Spectral, and Arithmetic Zeta Functions, M. van Frankenhuysen and M. L. Lapidus, eds., Amer. Math. Soc., Contemporary Math. series, **290**, 2001, 139–144.

[O3] A. Odlyzko, *On the distribution of the spacings between zeros of the zeta function*, Math. Comp., **48** (1987), 273–308.

[O4] A. Odlyzko, private communication.

[OS] A. Odlyzko and A. Schönhage, *Fast algorithms for multiple evaluations of the Riemann zeta function* Trans. Am. Math. Soc., **309** (1988), 797–809.

[Og] A. Ogg, *Modular forms and Dirichlet series*, W. A. Benjamin, Inc., New York-Amsterdam, 1969.

[Ol] F. Olver, *Asymptotics and special functions*, Academic Press [A subsidiary of Harcourt Brace Jovanovich, Publishers], New York-London, 1974, Computer Science and Applied Mathematics.

[OzS] A. Özlük and C. Snyder, *Small Zeroes of Quadratic L-Functions*, Bull. Aust. Math. Soc. **47** (1993), 307–319.

[OzS2] A. Özlük and C. Snyder, *On the distribution of the nontrivial zeros of quadratic L-functions close to the real axis*, Acta Arith. **91** (1999), no. 3, 209–228.

[P] R. Paris, *An asymptotic representation for the Riemann zeta function on the critical line*, Proc. Roy. Soc. London Ser. A **446** (1994), no. 1928, 565–587.

[R] Bernhard Riemann, *Gesammelte mathematische Werke, wissenschaftlicher Nachlass und Nachträge*, Springer-Verlag, Berlin, 1990, Based on the edition by Heinrich Weber and Richard Dedekind, Edited and with a preface by Raghavan Narasimhan.

[RT] F. Rodriguez-Villegas and G. Tornaria, private communication.

[Ru] M. Rubinstein, *Evidence for a spectral interpretation of the zeros of L-functions*. Princeton Ph.D. Dissertation, 1998.

[Ru2] M. Rubinstein, *Low lying zeros of L-functions and random matrix theory*. Duke Mathematical Journal **109** (2001), no. 1, 147–181.

[Ru3] M. Rubinstein, *Lower terms in the density of zeros of quadratic Dirichlet L-functions*, preprint.

[Ru4] M. Rubinstein, *The L-function class library and command line interface*,
www.math.uwaterloo.ca/~mrubinst/L_function/L.html.

[RS] M. Rubinstein and P. Sarnak, *Chebyshev's Bias*, Experimental Mathematics **3** (1994), no. 3, 173–197.

[Rud] W. Rudin, *Real and complex analysis*, third ed., McGraw-Hill Book
Co., New York, 1987.

[RudS] Z. Rudnick and P. Sarnak, *Zeros of principal L-functions and random
matrix theory*, Duke Mathematical Journal (2) **81** (1996), 269–322.

[Rum] R. Rumely, *Numerical computations concerning the ERH*, Math. Comp.
61 (1993), 415–440.

[S] K. Soundararajan, *Non-vanishing of quadratic Dirichlet L-functions at
$s = \frac{1}{2}$* , Ann. of Math. (2) **152** (2000) 447–488.

[Sp] R. Spira, *Calculation of the Ramanujan τ-Dirichlet series*, Math. Comp.
27 (1973), 379–385.

[St] A. Strombergsson, *On the zeros of L-functions associated to Maass wave-
forms*, IMRN (1999), No. 15.

[TW] R. Taylor and A. Wiles, *Ring-theoretic properties of certain Hecke algebras*, Ann. of Math. (2) **141** (1995), no. 3, 553–572.

[T] N. Temme, *The asymptotic expansions of the incomplete gamma functions*,
SIAM J. Math. Anal. **10** (1979), 757–766.

[Ti] E. Titchmarsh, *The theory of the Riemann zeta-function*, second ed., The
Clarendon Press Oxford University Press, New York, 1986, Edited and
with a preface by D. R. Heath-Brown.

[To] E. Tollis, *Zeros of Dedekind zeta functions in the critical strip*, Math.
Comp. **66** (1997), no. 219, 1295–1321.

[Tu] A. Turing, *Some calculations of the Riemann zeta function*, Proc. London
Math. Soc. (3) **3** (1953), 99–117.

[W] S. Wedeniwski, *Verification of the Riemann Hypothesis*,
www.zetagrid.net.

[Wi]   A. Wiles, *Modular elliptic curves and Fermat's last theorem*, Ann. of Math. (2) **141** (1995), no. 3, 443–551.

[Y]   H. Yoshida, *On calculations of zeros of L-functions related with Ramanujan's discriminant function on the critical line*, J. Ramanujan Math. Soc. **3** (1988), no. 1, 87–95.

[Y2]   H. Yoshida, *On calculations of zeros of various L-functions*, J. Math. Kyoto Univ. **35** (1995), no. 4, 663–696.

Pure Mathematics,
University of Waterloo
200 University Ave. W.,
Waterloo N2L 3G1,
Ontario, Canada

Front row	Second row	Third row	Back row
PF Scudo	N Søndergaard	NC Snaith	P Kurlberg
X Li	Y Chen	MV Berry	B Winn
D Farmer	S Gonek	MM Kotecha	L Addamiano
Y Sakellaridis	EL Basor	V Perez-Abreu	DA Goldston
P-O Dehaye	E Bogomolny	PJ Forrester	BD Sutton
A Nikeghbali	I Mostafa	MB Milinovich	CD Sinclair
B Conrey	P Dreher	AK Swallow	P Howard
T Oliker	R Heath-Brown	JC Orum	J Bolte
S Rahav	YV Fyodorov	HA Ledoan	D Vandeth
M Harmer	TW Hilberdink	F Rilke	R Schubert
O Bohigas	S Kristensen	IG Williams	SC O'Keefe
A Monti	K Hare	D Kelmer	CT Wheeler
HC Rosu	G Ergun	F Brumley	E Daems
PG Michel	A Majumdar	L Rosenzweig	ML Smith
M Sheinman	RR Khan	P Zhao	TD Browning
	LB Pierce	A Fujii	AV Kontorovich
	H Kadiri	DJ Whitehouse	M Young
	JG Leo	J Dioses	I Wigman
	N Ng	T Claeys	J Harrison
	T Schwaibold	F Götze	C Hughes
	F Mezzadri	M Degli Esposti	BE Odgers
		M Sieber	
		T Grava	
		A Gambini	
		AG Monastra	
		A Gamburd	
		M Dieng	
		I Smolyarenko	
		M Feigin	

Table 4: Participants of the school "Recent perspectives in Random Matrix Theory and Number Theory" at the Isaac Newton Institute for Mathematical Sciences, March 28 to April 8, 2004. The participants are listed from left to right in each row.

Index

Airy function, 59, 60, 63, 75, 333
Airy kernel, 63, 65
analytic conductor, 364
analytic continuation, 382
  of L-functions, *226*, 363
  of the Riemann zeta function, 11
analytic properties
  of L-functions, 361
Andréief's identity, 123, 125, 330
approximate functional equation, 190–192, 211, 219, 220, 448, 460
  for $\zeta^2(s)$, 191
  for $\zeta^k(s)$, 191, 219
  smoothed, 443, 444
arithmetic group, 172
arithmetic progressions, 6
arithmetic-geometric mean inequality, 203, 206, 208
Artin conductor, 382
Artin conjecture, 382, 383, 385
Artin L-function, 360, 361, 373, 381, 383, 385, 406, 410, 412
  computation, 470
  Euler product, 382
automorphic forms, 372
  L-function, 368, 373, 385, 410

Baker-Campbell-Hausdorf formula, 315
Balmer formula, 165, 169
Banach algebra, 318, 326
Barnes' G-function, 262, 321, 327
Bell number, 339
Bernoulli number, *428*
Bernoulli polynomial, *427*
  special values, *429*
Bernoulli's formula, 430
Bessel function, 333
Bessel kernel, 75
billiard
  Bunimovich's stadium, 165–167
  circle, 165
  quantum, 166
  Sinai, 166, 167
Birch and Swinnerton-Dyer conjecture, 271, 367, 397, 401, 496
Bohigas-Giannoni-Schmit Conjecture, 168
Bohr-Sommerfeld quantization, 169
Brownian motion, 39, 40
Bunimovich's stadium billiard, 165–167

Cauchy transform of orthogonal polynomials, 72, 74
Cauchy's Theorem, 425, 446
Cauchy-Schwarz inequality, 210
central limit theorem, 256, 261, 265, 271, 334
Chadwick's neutron chamber, 151
chaotic motion, 164
character
  principal, 230
characteristic polynomial, 31, 32, 65–67, 69, 71, 219, 259, 263, 271, 272, 439, 468, 477, 478, 480, 484, 488
  moments, 195
Chebotareff density theorem, 379
Chebyshev function, 81
Chebyshev polynomial, 119, 122
chiral ensemble, 71
chiral GUE, 75
Christoffel-Darboux, 295
Christoffel-Darboux formula, 52, 54, 66
circle method, 221
circular billiard, 165
circular ensembles
  eigenvalue probability density function, 290
Circular Unitary Ensemble, 156, 252, 309
class function, 115, 252, 260
class number formula, 367
classical compact groups, 252, 263, 337, *342*, 468, 471, 478, 480, 488
  Haar measure, 342
Classical Mean Value Theorem, 210, 212
cluster function, *44*, 50, 62, 74
coefficient correlation sum, 217
compactified
  elliptic curve, 388
complex multiplication, 390
compound nucleus hypothesis, 151
computing zeros of an L-function, 466
conductor, 364
  analytic, 364
  Artin, 382
  log-conductor, 227, 240
  of an L-function, 475
  of cusp form L-functions, 475
  of Dirichlet characters, 228, 453, 470, 474, 478
  of elliptic curve, 234, 400, 453, 470, 475, 478, 489
convexity bound, 367

convexity estimate, 194
convolution, 460, 465
convolution L-function, 239
correlation function, 337
    n-point, 42, 44, 45, 48, 50, 73, 156, 287
    of characteristic polynomials, 69
correlations
    of the zeros of L-functions, 474
counting function
    moments of smooth counting function,
        338
    of eigenvalues, 337
    of the Riemann zeros, $N(T)$, *13*, 14, 82,
        228, 348, 466
    of zeros of L-functions, 337, 467
Cramér explicit formula, 85
Cramér's model, 4, 281
critical line, *11*, 28, 66, 104, 107, 158, 160,
    173, 185, 186, 189, 191, 201, 204,
    206, 207, 209, 261, 263, 268, 299,
    415, 442, 466, 478, 488, 491
critical point, 268, 478, 480, 484, 489
critical strip, *11*, 17, 28, 82, 104, 195, 205,
    207, 228, 365–367, 466, 471
cross section, 151, 152
CUE, 156, 252
cumulants, 344
cusp form, 232, 235, 237, 267, 371, 462
    primitive, 371
cusp form L-function, 352, 447, 474, 475, 478,
    479
    computation, 450
cyclotomic character, 404
cyclotomic fields, 378

Dedekind zeta function, 160, 384, 467
    computation, 470
degree of an L-function, 227, *362*, 475
Deligne's equidistribution theorem, 422
densities
    n-level, 337
density conjecture for low-lying zeros of L-
        functions, 240
density of zeros, 478
    of an L-function, 227
determinant
    Toeplitz, 309
difference equation
    for Bernoulli polynomials, *428*
Dirichlet character, *24*, 229, 481
    orthogonality relations, 230
    real quadratic, 230, 267
Dirichlet L-function, 24, *27*, 228, 229, 266,
    434, 467, 474–476, 478, 479

computation, 450, 470
    functional equation, 229
    real characters, 351, 480, 488, 489
Dirichlet polynomial, 185, *187*, 205, 220
    mean value, 201, 202, 221
    moments, 187
Dirichlet series, 202, *361*, 443
    elliptic curve L-function, 452
    for $\zeta(s)^{-1}$, 82, 205
    for L-functions, 226
    for L-functions associated to Hecke forms,
        451
    for the Riemann zeta function, 80
    for $\zeta^k(s)$, 219
    twists of cusp form L-functions, 451
    twists of elliptic curve L-functions, 453
discrete mean value, 202, 205
discriminant, 387
disordered metallic particle, 157, 173
distribution function
    k-point, 280, *285*, 287
divisor function, 191, 219
divisor sum, 221
dual isogeny, 390
Dyson kernel, 60, 62, 65, 73

eigenvalues
    bulk, 283, 286
    density for GOE matrices, 283
    in intervals, 140
    joint probability density function, 41, 42
    jth lowest, 142
    nearest neighbour spacing, 298
    probability density function, 287, 288
        circular ensembles, 290
        symplectic-invariant ensemble, 290
    statistical properties, 282
Eisenstein series, 233, 237
elliptic curve, 267, 273, 386, 387, 389–392,
    400, 413, 420, 452, 489
    compactified, 388
    complex multiplication, 400
    conductor, 234, 400, 453, 470, 475, 478,
        489
    family, 248
    function field, 389, 415
    Galois representation, 401
    group law, 388
    L-function, 234, 248, 267, 360, 386, 398,
        470, 475, 476, 478, 479, 498
        computation, 452, 470
        vanishing at the central point, 271
    over finite fields, 393, 396
    over the rationals, 396

rank, 248, 396, 397, 401
  reduction, 399
  Tate module, 404
Essential Simplicity Conjecture, 94
Estermann phenomenon, 198
Euler product, *362*, 444, 493, 494
  for Artin *L*-functions, 382
  for elliptic curve *L*-functions, 452
  for *L*-functions, *226*
  for the Riemann zeta function, 7, 80, 228
  local factor, *362*
Euler totient function, $\varphi(q)$, 25, 195, 229
Euler's identity, 432
Euler-Maclaurin summation, 425, 427, 430, 432, 437
exceptional Lie groups, 272
explicit formula, 23, 216, 241
  Cramér, 85
  Montgomery, 87
  Riemann-von Mangoldt, 83
  Weil, 366
exterior square *L*-function, 239

family of *L*-functions, 225, 252, 263, 267, 268, 480, 484, 487, 488
fast Fourier transform, 425, 443, 468
Feynman path integral, 169
finite fields
  elliptic curves, 393, 396
Fisher-Hartwig conjecture, 326–328, 332, 334
Fisher-Hartwig symbol, 321, 324, 326, 331, 333
form factor, 89, 93, 171, 173
Fourier series
  for the Bernoulli polynomials, *428*
Fourier transform, 438, 465
Fredholm determinant, 50, 287, 295–299, 309, 310
Fredholm integral equation, 288
frobenius, 361, 377–380, 382, 393, 395, 404, 406, 408, 417, 418, 420, 423
function field, 413
  elliptic curve, 389, 415
  zeta function, 478
functional equation, 363, 382, 488, 489
  Bernoulli polynomials, *428*
  Dirichlet *L*-function, 229
  elliptic curve *L*-function, 453
  L-function, *226*, 443
  L-functions associated to Hecke forms, 233, 451
  L-functions of elliptic curves over finite fields, 398
  Riemann zeta function, 11, 208, 227

theta-series, 369
  twists of cusp form *L*-functions, 452
  twists of elliptic curve *L*-functions, 268, 453
fundamental discriminant, 230, 480, 482, 483, 485, 488, 494
Fundamental Theorem of Arithmetic, *1*

Galois group, 360, *373*, 377–379, 381, 401, 403, 413
  absolute, 379, 401, 405
  arithmetic, 417
  geometric, 417
Galois representation, 360, 361, 376, 409, 410, 420
  elliptic curve, 401
gamma distribution, Gamma[s, σ], *283*
gap probability, *284*, 289, 291, 296, 299
  bulk, 296, 297
  hard edge, 296
gaps
  between consecutive primes, 7, 85, 98
  between zeros of the Riemann zeta function, 90, 91, 471
  small gaps between zeros, 94
Gaudin's Lemma, 46, 115, 126, 128–134, 140
Gauss sum, 28, 229
Gaussian ensemble, 31, 39, 73
  joint distribution of eigenvalues, 153
Gaussian moments of $S(T)$, 104
Gaussian Orthogonal Ensemble (GOE), 153, 283
Gaussian Symplectic Ensemble (GSE), 153
Gaussian Unitary Ensemble (GUE), 31, 54, 75, 153, 332
Gaussian white noise, 39, 41
Generalized Lindelöf Hypothesis, 205, 210, 367
Generalized Prime Number Theorem, 366
Generalized Riemann Hypothesis, 218, 366
generating function, 285
  for the Bernoulli polynomials, *428*
geometric frobenius, 417
Geronimo-Case-Borodin-Okounkov identity, 319
GOE, 153, 283
Goldbach's Conjecture, 6
Gram's identity, 115, 138, 140
group law
  elliptic curve, 388
GSE, 153
GUE, 31, 60, 75, 153

Haar measure, 34, 115, 119, 124
  on classical compact groups, 342, 380

$SO(2N)$, 120
$SO(2N+1)$, 122
$U(N)$, 118, 252, 329, 439, 473, 477
$USp(2N)$, 122
Hadamard product, 12, 365
Hamilton equations, 293
Hankel matrix, 318, 334
Hankel operator, 312, 317, 318, 323
Hannay-Ozorio sum-rule, 172
Hardy space, 310
Hardy-Littlewood conjecture, 93, 174, 435, 473
Hasse's bound, 393
Hasse-Weil conjecture, 453
Hecke congruence group, 235, 370
Hecke eigenform, 232, 451
Hecke L-functions, 370
Hecke operator, 371, 372, 451
Heisenberg matrix mechanics, 169
helium atom, 169, 173
Hermite polynomial, 31, 32, 51–54, 58–60, 65, 73
    integral representations, 53
    orthogonality relations, 51
    Plancherel-Rotach asymptotics, 31, 53, 54, 57, 63
    recurrence relations, 51
Hermitian matrix, 34, 35
Hilbert space, 310, 311, 314
Hilbert-Polya conjecture, 174, 175
Hilbert-Schmidt norm, 314
Hilbert-Schmidt operator, 314, 318, 324
hole probability, 44–46
holomorphic cusp forms, 352
hydrogen atom, 169, 173

ideal, 324, 374
    prime, 375
incomplete gamma function, 434, 446, 449, 462
    computation, 455
integrable systems, 291
integral operator, 141, 288
    kernel, 288
invariant tori, 164
isogeny, 389
    dual, 390

Jensen's Formula, 202
joint probability density function, 41, 42, 48, 50, 66, 72, 284
jth lowest eigenvalue, 142

K-Bessel function, 462

KAM Theorem, 164
Katz-Sarnak philosophy, 114, 252, 263, 268, 269, 338, 478
Keating-Snaith conjecture, 262, 435
kernel, 49, 50, 60
    GUE, 61
        asymptotic form, 62
kernel function, 113
Kloosterman sum, 205, 234, 248
Kronecker's symbol, 351, 480, 482, 483

L-function, 161, 175, *226*, 268, 468
    analytic continuation, *226*, 363
    analytic properties, 361
    Artin, 360, 361, 373, 381, 383, 385, 406, 410, 412
    computation, 442, 447, 470
    computing zeros, 466
    conductor, 475
    convolution, 239
    cusp form, 352, 447, 450, 474, 475, 478, 479
    degree, 227, *362*, 475
    density of zeros, 227
    Dirichlet, 268, 434, 450, 458, 467, 474–476, 478, 479
    Dirichlet series, 226
    elliptic curve, 234, 248, 267, 268, 360, 452, 470, 475, 476, 478, 479, 489, 498
    Euler product, *226*
    exterior square, 239
    families, 225, 263, 267, 268, 480, 484, 487, 488
        G_2 symmetry, 274
    functional equation, *226*, 443
    log-conductor, 227
    Maass form, 238, 462
    meromorphic continuation, 443
    modular form, 368, 458
    of elliptic curves, 386, 398
    of elliptic curves over finite fields, 398
        functional equation, 398
    of elliptic curves over the rationals, 399
    primitive, *227*, 475
    real Dirichlet L-function, 351
    Selberg class, *225*, 229
    symmetric square, 239
    twists of cusp forms, 451
    twists of elliptic curve L-functions, 267, 453, 488
    vanishing at the central point, 271
    zero statistics, 471
Laguerre ensemble, 333

Lamb shift, 161
$\Lambda(n)$, *8*
Landau formula, 84
Langlands Philosophy, 10
Laplace-Beltrami operator, 172
lattice, 391
Legendre duplication formula, 444
Legendre symbol, 267
level density, 170
Levinson's method, 206, 207
Levinson-Montgomery Theorem, 207
Lie group, 115
Lindelöf Hypothesis
    Generalised, 367
linear statistic, 329–333, 343
Littlewood's Lemma, 203, 205, 208
log-conductor, 240
    of an L-function, 227
logarithmic derivative
    of the Riemann zeta function, 80

Maass form, 237, 238, 243, 372, 462, 470
MacDonald constant term identities, 274
Maslov index, 170
matrix
    orthogonal, 116
    symplectic, 117
    unitary, 115
mean density, 43
mean eigenvalue density, 60
mean value, 204, 215, 217
    Classical Mean Value Theorem, 210, 212
    Dirichlet polynomial, 201, 202, 210, 221
    discrete, 202, 205
    mollified, 205
    Montgomery-Vaughan Mean Value Theorem, 210, 211, 217, 220
mean value theorem, 201, 202, 222
mean value theorem for Dirichlet polynomials, 188
mean value theorem of Montgomery and Vaughan, *88*
Mellin inversion, 449, 461
Mellin transform, *23*, 263, 440, 460
meromorphic continuation, 363
    of an L-function, 443
    of the Riemann zeta function, 431
Mersenne primes, 6
microwave cavity, 168
Möbius function, *82*, 204, 229, 437
Möbius inversion, 425, 435, 437
mock-Gaussian, 338
modular form, 10, 232, 370
modular group, 232

mollified mean values, 205
mollify, 205, 208
moment generating function, 261
    for the logarithm of characteristic poyno-mials, 259
moments, 491, 498
    for primes in short intervals, 100
    of characteristic polynomials, 66, 69–71, 195, 439
    $O(N)$, 264
    $U(N)$, 261
    $USp(2N)$, 264
    of Dirichlet L-functions, 494
    of Dirichlet polynomials, 187
    of L-functions, 251, 269
    of smooth counting function, 338
    of the logarithm of characteristic poly-nomials
    $U(N)$, 261
    of the logarithm of the Riemann zeta function, 257
    of the Riemann zeta function, 185, 201, 218, 435, 488, 489, 491
    of traces of matrices, 346
monodromy group, 418
Montgomery's conjecture, 92, 256, 473
Montgomery's explicit formula, 87
Montgomery's Theorem, 87, *89*
Montgomery-Odlyzko law, 161, 174, 299
Montgomery-Vaughan Mean Value Theorem, 210, 211, 217, 220
moveable singularities, 292

n-correlation, 135, 136
    for $U(N)$, 133
    $USp(2N)$, 134
n-level densities, 337, 480
    large matrix limit, 130
    $SO(2N)$, 131
    $SO(2N + 1)$, 132
    $U(N)$, 130
    $USp(2N)$, 132
n-point correlation function, *42*, 44, 45, 48, 50, 73, 156
nearest neighbour spacing distribution, 152, 166, 167, 477
    for $U(N)$, 135
    of eigenvalues, 298, 477
    of Riemann zeros, 477
    of zeros of L-functions, 478
newforms, 352
Nielsen's expansion, 458
$N(T)$, *13*, 14, 82, 228, 348, 466
Nuclear Data Ensemble, 163

nuclear resonances, 152, 161, 163, 166
number variance, *44*, 46, 62

Odlyzko-Schönhage algorithm, 475
$O(N)$, 263
one-level density, 225, 247, 343, 485, 487
one-point correlation function, 70
operator norm, 325
orthogonal (even) symmetry, 489
orthogonal group, *342*
orthogonal matrix, 116
orthogonal polynomial, 31, 32, 46, 49–51, 65,
       66, 69, 70, 491
   monic, 287
orthogonal projection, 314
orthogonal symmetry, 225, 268, 352
   even, 116
   odd, 116
orthogonality relations
   of Dirichlet characters, 230

Painlevé III' equation, 297, 302, 303
Painlevé VI equation, 303
Painlevé V equation, 293, 302, 303
Painlevé differential equations, 292
Painlevé equation, 156, *293*
Painlevé theory, 300
Painlevé transcendent, 291, 297
pair correlation, 159, 175
   for eigenvalues of $U(N)$, 132
   of Riemann zeros, 96, 132, 201, 472, 473,
       480
   of zeros of L-functions, 474–476
Pair Correlation Conjecture, 94
Pauli principle, 162
periodic orbit theory, 169–171, 173–175
Perron's formula, 193
Petersson formula, 234
Petersson inner product, 451
Pfaffian, 289
$\varphi(q)$, 25, 195, 229
Phragmén-Lindelöf Theorem, 444
point process, 280, *281*, 284
Poisson distribution, *281*
Poisson process, 281, 286
Poisson summation, 425, 455
   numerical integration, 438
Poisson Summation Formula, *9*
Polya-Vinogradov inequality, 230
prime, *1*
   average spacing, 7
   gaps between consecutive primes, 7, 85,
       98
   ideal, 375

Mersenne, 6
   ramified, 231, 377
   spacing, 280
   split, 231
prime ideal, 375
Prime Number Theorem, *2*, 81, 82, 281, 351
   Generalised, 366
   proof, 19
prime twins, 5
primes as periodic orbits, 174
primes in short intervals
   second moment, 100
primitive character, 453
primitive cusp forms, 371
primitive Dirichlet character, 229
primitive L-function, *227*
primitive root, 229
principal character, *25*, *230*
probability density function, 32, 38
   of values of the characteristic polyno-
       mial, 440
   on Hermitian matrices, 36
   Orstein-Uhlenbeck, 40
   unitary-invariant, 37, 38
projective plane, 387

quadratic Dirichlet L-function, 489
quantum chaos, 147, 173
quantum electrodynamics, 149

Ramanujan bound, 226
Ramanujan's tau-function, $\tau(n)$, 232
   computation, 470
   L-function, 470, 475, 476, 478
   quadratic twist L-function, 487
Ramanujan/Petersson bound, 364
ramified, 377
random matrix
   circular ensembles
       eigenvalue probability density function,
           290
   complex Hermitian, 31, 153, 286
   ensemble, 279
   independent entries, 32, 37, 38
   invariant ensembles, 32
   orthogonally-invariant ensemble, 288
   real quaternion, 153, 286
   real symmetric, 31, 153, 279, 282, 283,
       288
   symplectic-invariant ensemble
       eigenvalue probability density function,
           290
   unitary-invariant ensemble, 287
Wishart, 75

Random Matrix Conjecture, 168
random polynomial, 283
random variable, 329, 331, 332
 distribution function, 329
rank
 of an elliptic curve, 248, 396, 397, 401
Rankin-Selberg method, 463
ratios
 of characteristic polynomials, 71–73, 76
real Dirichlet L-function, 351
reduction of an elliptic curve, 399
resolvent kernel, 300
resolvent matrix, 71
resolvent operator, 295
resonance peaks, 151
resonances, 151
Riccati equation, 293
Riemann Hypothesis, *11*, 82, 174, 204, 205,
 207, 210, 228, 256, 268, 348, 444,
 466, 471, 484
 Generalised, 366
 computational verification, 470
 for L-functions of elliptic curves over finite fields, 398
Riemann sum, 438, 440, 465
Riemann zeta function, *7*, 65, 66, 80, 157,
 159, 173, 174, 201, 227, 256, 299,
 348, 368, 458, 488
 analytic continuation, 11
 approximate formula, 186
 average value on the critical line, 186
 computation, 450
 Dirichlet series, 80
 Euler product, 7, 80
 functional equation, 11
 logarithmic derivative, 80
 moments, 185
 trivial zeros, 11
 zero counting function, *13*, 14, 348
Riemann-Hilbert, 63, 74–76, 157
Riemann-Siegel formula, 433, 434, 442, *442*,
 460, 468
Riemann-von Mangoldt formula, 14, 82
Riemannian metric, 32–34
ring of integers, 374
root number, 364
Rydberg atom, 173

saddle-point method, 54, 57, 58
Sato-Tate law, 270
Sato/Tate conjecture, 408
Sato/Tate measure, 380
Schrödinger equation, 161, 169, 171
Selberg class, *225*, 229

Selberg integral, 260, 264, 439
Selberg trace formula, 172
semiclassical limit, 172
set partition, *339*
shell-model, 161, 162
sieve of Eratosthenes, 437
simple zeros, 209, 210
Simple Zeros Conjecture, 94
Sinai's billiard, 166, 167
sine kernel, 288, 295, 299
singular symbol, 331
Small Gaps Conjecture, 94
smooth counting function, 343, 349
$SO(2N)$, 116, 225, 265, 268, *342*
$SO(2N+1)$, 116, 225, *342*
$SO(N)$, 116
spacing distribution, 279, 280, 283, 286, 296
 empirical determination, 282
 in the bulk, 301, 303
spectral fluctuations, 162
spectral rigidity, 163, 167
spectrum
 bulk, 65, 73
 edge, 63, 65
 hard edge, 296, *296*
 soft edge, *296*
 unfold, 282
Speiser's Theorem, 207
$S(T)$, *13*, 83, 97, 103, 105, 228, 348, 467
 Gaussian moments, 104
stationary phase, 54
steepest descent, 54
Stieltjes integral, 426
Stirling number, 339
Stirling's formula, 60, 62, 446, 468
Strong Pair Correlation Conjecture, 93
Strong Szegő Limit Theorem, 319–322, 330–
 332
subconvexity, 368
summation by parts, *426*
supersymmetric non-linear σ-model, 173
supersymmetry, 491
symbol, 313
 Fisher-Hartwig, 321, 324, 326, 331, 333
 singular, 309, 331
 smooth, 309
symmetric square L-function, 239
symplectic group, 117
symplectic matrix, 117
symplectic symmetry, 225, 268, 351, 488

Tamagawa factor, 496
Tate module, 402, 405
Tate-Shafarevich group, 496

Tchebychev polynomial, 119
theta-function, $\theta(v)$, 10
theta-series, 368
Toeplitz determinant, 309, 311, 329–331
Toeplitz matrix, 309, 310, 313, 322, 327, 332, 334
Toeplitz operator, 311–314, 324
torsion subgroup, 396
torus quantization, 173
trace class, 314–317, 323, 325
trace norm, 314, 315, 323, 325
transposing lemma, 118, 121–123
trivial zeros
    of the Riemann zeta function, 11, 228
Twin Prime Conjecture, 217
twin primes, 5, 101
twists of cusp form L-functions
    computation, 451
twists of elliptic curve L-functions
    computation, 453
two-body random ensemble, 162
two-norm, 311
two-point correlation function, 43, 50, 171, 173, 253
    of eigenvalues from the CUE ($U(N)$), 252
    of the Riemann zeros, 256

$U(N)$, 115, 156, 225, 252, *342*
unfold, *283*
unilateral sequence, 310
unitary group, 252, *342*
unitary symmetry, 225, 348, 488
unitary symplectic group, *342*
unitry matrix, 115
$USp(2N)$, 117, 225, 263, 268, *342*

value distribution, 489, 490, 492
    of characteristic polynomials
        $O(N)$, 264
        $SO(2N)$, 266
        $U(N)$, 263, 264
        $USp(2N)$, 264, 265
    of Dirichlet L-functions, 269
    of elliptic curve L-functions, 269, 271
    of L-functions, 488
    of the logarithm of characteristic polynomials
        $U(N)$, 260, 261
    of the logarithm of the Riemann zeta function, 256, 257
    of the Riemann zeta function, 256, 264
Vandermonde determinant, 48, 67, 69, 117, 329

Virasoro constraints, 300
von Mangoldt, 366
von Mangoldt function, $\Lambda(n)$, *8*, 81, 195

Weierstrass $\mathcal{P}$-function, 293
Weierstrass form, 452
weight, 364
Weil explicit formula, 366
Weil pairing, 405
Weyl equidistribution criterion, 381
Weyl integration formula, 131, 252, 254, 260, 264, 274, 439
    $SO(2N)$, 116
    $SO(2N+1)$, 117
    $U(N)$, 115
    $USp(2N)$, 117
white noise
    Gaussian, 39, 41
Wiener-Hopf operator, 332
Wigner semi-circle law, 60, 61, 156, 162, *283*
Wigner surmise, 153, 284, 286, 290, 303

zero-density estimate, 205, 206
zero-free region, 367
zeros
    near the critical point, 478
    of L-functions
        computation, 466
    of the Riemann zeta function, 207
    simple, 209, 210
    spacing between zeros of the Riemann zeta function, 471
$\zeta(s)^{-1}$, 82, 205
$\zeta(s)$, 80, 227

Printed in the United States
By Bookmasters